T0192592

STOCHASTIC PHYSICS AND CLIMATE MODELLING

This is the first book to promote the use of stochastic, or random, processes to understand, model and predict our climate system.

One of the most important applications of this technique is in the representation in comprehensive climate models of processes that, although crucial, are too small or fast to be explicitly modelled. The book shows how stochastic methods can lead to improvements in climate simulation and prediction, compared with more conventional bulk-formula parameterisation procedures.

Beginning with expositions of the relevant mathematical theory, the book moves on to describe numerous practical applications. It covers the complete range of time scales of climate variability, from seasonal to decadal, centennial and millennial.

With contributions from leading experts in climate physics, this book is invaluable to anyone working on climate models, including graduate students and researchers in the atmospheric and oceanic sciences, numerical weather forecasting, climate prediction, climate modelling and climate change.

TIM PALMER is a Royal Society Research Professor at the University of Oxford and a Senior Scientist at the European Centre for Medium-Range Weather Forecasts (ECMWF). He has won the Royal Society Esso Energy Award, the Royal Meteorological Society Adrian Gill Prize and the American Meteorological Society Jule Charney Award. He is a fellow of the Royal Society, the Royal Meteorological Society, the American Meteorological Society and Academia Europaea. He is a lead author of the Intergovermental Panel on Climate Change (IPCC), co-chair of the Scientific Steering Group of the UN World Meteorological Organisation's Climate Variability and Predictability (CLIVAR) project and coordinator of two European Union climate prediction projects (PROVOST and DEMETER). He has had numerous appearances on radio and TV, in relation to weather, climate and chaos theory, and has co-edited another book with Cambridge University Press – *Predictability of Weather and Climate* – in 2006.

PAUL WILLIAMS is a Royal Society Research Fellow at the National Centre for Atmospheric Science (NCAS) and the Department of Meteorology, University of Reading. He has won the Royal Astronomical Society Blackwell Prize (2004) and the Royal Meteorological Society Rupert Ford Travel Award (2005) and has received a prestigious Crucible Fellowship from the National Endowment for Science, Technology and the Arts (2007). He was the lead author of a climate change report commissioned and published by the European Parliament (2004). He is a Fellow of the Royal Meteorological Society, the Institute of Physics, and the Royal Astronomical Society. His research findings have been reported widely in the media, including feature articles in *New Scientist* and the *Financial Times*, and a panel discussion on BBC Radio 4.

STOCHASTIC PHYSICS AND CLIMATE MODELLING

STOCHASTIC PHYSICS AND CLIMATE MODELLING

TIM PALMER
European Centre for Medium-Range Weather Forecasts, UK and the University of Oxford, UK

PAUL WILLIAMS
National Centre for Atmospheric Science and the University of Reading, UK

Originating from a Theme Issue published in *Philosophical Transactions of the Royal Society A: Mathematical, Physical & Engineering Sciences*

CAMBRIDGE
UNIVERSITY PRESS

University Printing House, Cambridge CB2 8BS, United Kingdom

One Liberty Plaza, 20th Floor, New York, NY 10006, USA

477 Williamstown Road, Port Melbourne, VIC 3207, Australia

4843/24, 2nd Floor, Ansari Road, Daryaganj, Delhi - 110002, India

79 Anson Road, #06-04/06, Singapore 079906

Cambridge University Press is part of the University of Cambridge.

It furthers the University's mission by disseminating knowledge in the pursuit of education, learning and research at the highest international levels of excellence.

www.cambridge.org
Information on this title: www.cambridge.org/9781108446990

First published 2010
First paperback edition 2017

A catalogue record for this publication is available from the British Library

Library of Congress Cataloging in Publication data
Stochastic physics and climate modelling / [edited by] Paul Williams, Tim Palmer.
p. cm.
Includes index.
ISBN 978-0-521-76105-5 (hardback)
1. Meteorology – Statistical methods. 2. Meteorology – Mathematical models.
3. Climatology – Statistical methods. 4. Climatology – Mathematical models.
5. Weather forecasting. I. Palmer, Tim (Tim N.) II. Williams, Paul, 1952– III. Title.
QC874.5.S76 2009
551.601′51923 – dc22 2009040945

ISBN 978-0-521-76105-5 Hardback
ISBN 978-1-108-44699-0 Paperback

Contents

Contributors

JULIE ALEXANDER, School of Earth and Ocean Sciences, University of Victoria, PO Box 3055 STN CSC, Victoria BC, V8W 3P6, Canada

THOMAS ALLEN, Met Office, FitzRoy Road, Exeter, EX1 3PB, UK

MICHAEL A. W. BALL, Department of Meteorology, University of Reading, PO Box 243, Earley Gate, Reading, RG6 6BB, UK

JUDITH BERNER, ECMWF, Shinfield Park, Reading, RG2 9AX, UK

PETER C. CHU, Department of Oceanography, Code OC/CU, Naval Postgraduate School, 833 Dyer Road, Monterey, CA 93943, USA

TIMOTHY DELSOLE, Center for Ocean-Land-Atmosphere Sciences, 4041 Powder Mill Rd, Suite 302, Calverton, MD 20705, USA

FRANCESCA DI GIUSEPPE, ARPA-SIM, viale Silvani, 6, 40122 Bologna, Italy

HENK A. DIJKSTRA, Institute for Marine and Atmospheric Research, Department of Physics and Astronomy, Utrecht University, Princetonplein 5, 3584 CC Utrecht, The Netherlands

FRANCISCO J. DOBLAS-REYES, ECMWF, Shinfield Park, Reading, RG2 9AX, UK

JINQIAO DUAN, Laboratory for Stochastics and Dynamics, Illinois Institute of Technology, Department of Applied Mathematics, Chicago, IL 60616, USA

BRIAN D. EWALD, Department of Mathematics, Florida State University, Room 205C, 1017 Academic Way, Tallahassee, FL 32306–4510, USA

LEELA M. FRANKCOMBE, Institute for Marine and Atmospheric Research, Department of Physics and Astronomy, Utrecht University, Princetonplein 5, 3584 CC Utrecht, The Netherlands

CHRISTIAN FRANZKE, British Antarctic Survey, Natural Environment Research Council, High Cross, Madingley Road, Cambridge, CB3 0ET, UK

MICHAEL GHIL, Geosciences Department & Laboratoire de Météorologie Dynamique (CNRS and IPSL), Ecole Normale Supérieure, F-75231 Paris Cedex 05, France

KATRINA HALES, University of California, Los Angeles, Department of Atmospheric and Oceanic Sciences, Institute of Geophysics and Planetary Physics, 405 Hilgard Ave, Los Angeles, CA 90095-1565, USA

CHRISTOPHER E. HOLLOWAY, Department of Meteorology, University of Reading, PO Box 243, Earley Gate, Reading, RG6 6BB, UK

JOCHEM I. JONGMA, Faculty of Earth and Life Sciences, Vrije Universiteit Amsterdam, De Boelelaan 1085, NL-1081 Amsterdam, The Netherlands

BOUALEM KHOUIDER, Department of Mathematics and Statistics, University of Victoria, PO BOX 3045 STN CSC, Victoria, BC, V8W 3P4, Canada

RICHARD KLEEMAN, Courant Institute of Mathematical Sciences, New York University, 251 Mercer Street, New York, NY 10012, USA

DMITRI KONDRASHOV, Department of Atmospheric and Oceanic Sciences, Institute of Geophysics and Planetary Physics, University of California at Los Angeles, Los Angeles, CA 90095, USA

SERGEY KRAVTSOV, Department of Mathematical Sciences, Atmospheric Science Group, University of Wisconsin-Milwaukee, PO Box 413, Milwaukee, WI 53201, USA

JOHNNY W.-B. LIN, North Park University, Physics Department-Box 30, 3225 W. Foster Ave, Chicago, IL 60625, USA

ANDREW J. MAJDA, Department of Mathematics and Climate, Atmosphere, Ocean Science (CAOS), Courant Institute of Mathematical Sciences, New York University, 251 Mercer Street, New York, NY 10012, USA

ADAM H. MONAHAN, School of Earth and Ocean Sciences, University of Victoria, PO Box 3055 STN CSC, Victoria BC, V8W 3P6, Canada

BALASUBRAMANYA T. NADIGA, MS-B296, Los Alamos National Laboratory, Los Alamos, NM 87544, USA

J. DAVID NEELIN, University of California, Los Angeles, Department of Atmospheric and Oceanic Sciences, Institute of Geophysics and Planetary Physics, 405 Hilgard Ave, Los Angeles, CA 90095-1565, USA

TIM N. PALMER, ECMWF, Shinfield Park, Reading, RG2 9AX, UK

CÉCILE PENLAND, NOAA/ESRL/Physical Sciences Division, 325 Broadway, Boulder, CO 80305, USA

OLE PETERS, University of California, Los Angeles, Department of Atmospheric and Oceanic Sciences, Institute of Geophysics and Planetary Physics, 405 Hilgard Ave, Los Angeles, CA 90095-1565, USA

ROBERT S. PLANT, Department of Meteorology, University of Reading, PO Box 243, Earley Gate, Reading, RG6 6BB, UK

MATTHIAS PRANGE, MARUM (Center for Marine Environmental Sciences) and Faculty of Geosciences, Universität Bremen, Fachbereich 5, Klagenfurter Str., D-28334 Bremen, Germany

MICHAEL SCHULZ, MARUM (Center for Marine Environmental Sciences) and Faculty of Geosciences, Universität Bremen, Fachbereich 5, Klagenfurter Str., D-28334 Bremen, Germany

GLENN J. SHUTTS, Met Office, FitzRoy Road, Exeter, EX1 3PB, UK

CHRISTOPHER J. SMITH, Met Office, FitzRoy Road, Exeter, EX1 3PB, UK

ADRIAN M. TOMPKINS, Abdus Salam International Centre for Theoretical Physics, Strada Costiera 11, 34014 Trieste, Italy

GEOFFREY VALLIS, AOS Program, Princeton University, Princeton, NJ 08544, USA

ANNA S. VON DER HEYDT, Institute for Marine and Atmospheric Research, Department of Physics and Astronomy, Utrecht University, Princetonplein 5, 3584 CC Utrecht, The Netherlands

ANDREW J. WEAVER, School of Earth and Ocean Sciences, University of Victoria, PO Box 3055 STN CSC, Victoria BC, V8W 3P6, Canada

ANTJE WEISHEIMER, ECMWF, Shinfield Park, Reading, RG2 9AX, UK

DANIEL S. WILKS, Department of Earth & Atmospheric Sciences, Cornell University, Ithaca, NY 14853, USA

Preface

Eleven chapters of this book were originally published as an issue of the *Philosophical Transactions of the Royal Society A: Mathematical, Physical & Engineering Sciences* (Volume 366; Issue 1875). Several chapters have been materially changed and updated. Seven new chapters have been added, which were commissioned specially for this book. We are grateful for the assistance of our Senior Commissioning Editors at CUP, Dr Matt Lloyd and Dr Susan Francis; our Production Editor at CUP, Anna-Marie Lovett; and our copy-editor, Zoë Lewin.

The dynamical evolution equations for weather and climate are formally deterministic. As such, one might expect that solutions of these dynamical evolution equations are uniquely determined by the imposed initial condition. A key purpose of this book is to suggest otherwise.

Before expanding on this seemingly paradoxical claim, let us first outline the reason why the theme of this book is of enormous practical importance. As discussed below, we could legitimately call it a trillion-dollar topic.

While weather forecasting has a long and perhaps chequered history, the present era, whereby predictions are made from numerical solutions of the underlying dynamic and thermodynamic equations, can be traced back to the pioneering work of L. F. Richardson in the early years of the twentieth century. As is well known, the notion that detailed weather forecasts could be made arbitrarily far into the future was dealt a practical blow through the discovery that weather was chaotic, i.e. that weather forecasts are sensitive to small errors in their initial conditions. To some people, the fact that the weather is chaotic seemed to imply that it is hopeless to try to forecast it. However, a fundamental property of any chaotic system is that the degree to which it is predictable is itself a function of the initial state; forecasts from some initial states can be very predictable, even though the system as a whole is chaotic.

To exploit this property of weather as a chaotic dynamical system, methods based on ensemble forecasting have been developed to try to predict when the weather is predictable and when it is unpredictable. The method is conceptually simple: an ensemble is a collection of forecasts made from almost, but not quite, identical initial conditions. The spread among members of the ensemble gives an estimate of flow-dependent predictability.

In recent years, the ensemble method has become a backbone of numerical weather prediction and is used not only by weather forecasters but also by commercial traders whose activities depend on weather. For example, weather is a dominant driver of many commodities traded in liberalised markets (electricity, gas, coal, oil, crops). Having an estimate of flow-dependent uncertainty in forecasts of weather is critical to the success of such trading, and ensemble weather forecasting is the tool used by the traders to determine this.

Developing practical tools for estimating the uncertainty of a forecast requires a detailed knowledge of the sources of forecast uncertainty. The simple chaotic paradigm discussed above suggests that the only relevant uncertainty lies in the weather observations that determine the initial state of the forecast, e.g. that the measuring instruments are never perfectly accurate or never sufficiently dense in space to determine every small fluctuation in the initial atmospheric state. However, the problem is not nearly as simple as this. Another key source of uncertainty in any weather forecast is the numerical model used to make the predictions.

So let us return to the beginning of this preface. The dynamic and thermodynamic equations are given as deterministic partial differential equations, but are solved by discretisation onto some sort of grid (or spectral or other equivalent representation). Since there are inevitably scales of motion and indeed key processes that are not resolved by this discretisation, methods must be found to represent approximately the subgrid features of the flow. For example, if a global numerical weather prediction problem has a typical grid spacing of 50 km, then all individual cloud systems will be unresolved. For this reason, the numerical equations are 'closed' by adding empirically based subgrid parameterisation formulae to represent the effects of the unresolved scales. Hence, for example, convective clouds (e.g. associated with thunderstorms) are represented by convective subgrid parameterisation formulae. Other subgrid parameterisation formulae represent the effects of flow over and around small-scale topography, boundary-layer turbulence and the absorption and emission of radiation in various relevant parts of the electromagnetic spectrum by radiatively active constituents in our atmosphere.

The formulation of these parameterisation formulae is motivated by notions in statistical mechanics. So, just as the momentum transfer by the bulk effects of molecular motions is represented by a diffusive formula, so a similar type of formula might represent the bulk effects of cumulus clouds on vertical temperature, humidity and momentum transfer on the grid scale. However, there is a problem with such an approach. Within a typical 50 km square grid box, there often exist sufficiently few individual cumulus clouds for the parameterised bulk formula to be an accurate estimate of the subgrid effects.

How can we represent this source of error in ensemble forecasts? This is where the concept of stochastic modelling of the subgrid scales is relevant. By representing model uncertainty through stochastic equations (or more generally by stochastic–dynamic models) the resulting ensemble forecasts can sample the effects of both initial observation uncertainties and forecast model uncertainties. The resulting ensemble weather forecasts are more reliable (in a precise statistical sense) than those associated with only a sampling of initial observation error, and this has made the whole process of predicting uncertainty more valuable to the real-world customers of weather forecasts.

But this is only half the story! Although weather forecasting has a long history, it is only in recent years that the world has become aware of the threat of climate change. Many regard this as the most serious threat facing humanity – a threat literally to our civilisation. Others, while perhaps acknowledging that the world has warmed in recent years and that some of this could be due to human activities, believe that the climate-change problem is not as important as other problems facing society. To some extent, extreme views about climate change, the cataclysmic and the dismissive, arise because there remains considerable uncertainty in the magnitude of future global warming, e.g. as reflected in the Intergovernmental Panel on Climate Change (IPCC) assessment reports. Certainly the IPCC assessment reports show that among the range of model predictions, there is a quantifiable risk of dangerous climate change in the coming century; most sensible observers deduce from this that the world needs to take action, first to reduce emissions of greenhouse gases and second to start preparing to adapt to inevitable climate change.

Climate-change predictions will play a key role in both mitigation and adaptation policies in years to come. For mitigation, policy makers need more precise predictions about how much more likely dangerous climate change will occur, as a function of anticipated atmospheric greenhouse-gas concentrations. For adaptation, predictions are needed to guide decisions on infrastructure investment. For example, how will patterns of precipitation change; what parts of the world need to be prepared for water shortages and what parts of the world need to be prepared for more frequent and devastating flooding?

Reducing uncertainty in climate prediction, both global and regional, requires improvements in the models used to predict climate. These models are similar in many respects to the types of weather forecast model discussed above, but differ in two key respects. First, because climate models have to be run over century time scales, rather than days, they must include processes like dynamic sea ice and biogeochemistry, processes that are not especially relevant for weather prediction. This makes the climate models intrinsically more complex than weather prediction models. Owing to this additional complexity and the need to simulate climate on longer time scales than numerical weather prediction models, climate models typically have much coarser grid resolution than weather prediction models: hundreds of kilometres rather than tens of kilometres.

On the other hand, as with weather prediction, neglecting the small-scale motions causes problems. For climate models, it causes the models to drift compared with reality, even for variables that, in principle, are well resolved in terms of the model's grid spacing. The problem of systematic error is an endemic problem in climate modelling. One of the primary goals of any climate-modelling centre is to eliminate, or at least minimise, this systematic drift. To give one example, many climate models have difficulty simulating the atmospheric phenomenon known as persistent anticyclonic blocking. However, such persistent anticyclonic blocks are the primary cause of drought in many locations; a persistent block causes rain-bearing weather systems to be diverted away from the region of interest. Hence, in order to know whether such a region is likely to be more prone to drought under climate change, it is necessary to know whether the frequency of occurrence of persistent blocking anticyclones will increase in that region as a result of increases in greenhouse-gas concentrations. However, if the models have difficulty simulating the blocking phenomenon in the first place, due to systematic drift, they are not well placed to answer this key question.

Clearly a potential solution to the problem of model drift is to reduce the grid spacing, e.g. to that of contemporary numerical weather prediction models. However, to do this would require computing resources beyond the means of most climate institutes. For example, to run century-long integrations with a 10 km grid would require sustained multi-petaflop computing capability.

This raises a fundamental theoretical question. How can we expect uncertainty in our predictions of climate change to reduce as the grid spacing reduces? If we look to our knowledge of the mathematical properties of the Navier–Stokes equations for guidance, we are left with a potential dilemma: a simple scaling argument based on the Kolmogorov turbulence suggests that any systematic truncation error, no matter how small scale it may be, can infect the large-scale systematic error of the model in finite time. Whether the Navier–Stokes equations really have this property is the topic of one of the unsolved million-dollar Clay Mathematics Millennium Prize problems.

This analysis suggests that, effectively, solutions of the dynamic and thermodynamic equations may have some irreducible uncertainty. In this case, it makes sense to try to treat at least the small-scale components of the flow by computationally simple stochastic processes, rather than by the conventional deterministic bulk formula.

This should not be seen as a council of despair, but as a way forward for a problem, climate prediction, that is arguably the most challenging of problems in computational science. For example, let us return to the problem of simulating persistent blocking anticyclones. One way of thinking of the persistent blocking anticyclone is as a preferred regime in the state space of our climate. However, it is secondary to the normal westerly flow that could be viewed as defining the dominant flow regime. Hence think of a double-well potential, the deeper of which represents normal westerly flow, the shallower representing blocking anticyclonic flow. With a highly resolved model, it should be possible not only to represent this potential well but also the right transition frequency between regimes. With a lower resolution model, perhaps the potential well structure is resolved, but the model is sufficiently damped and inactive that the state resides too frequently in the dominant, deeper, westerly flow regime. As a result, this low-resolution model will exhibit a westerly systematic bias, and be poor at simulating spells of persistent anticyclonic weather. However, if this is the case, then injecting stochastic noise into the near-grid scale may be sufficient to lead to a significant improvement in simulating the correct regime statistics.

Hence, as well as exploring the benefits of high resolution (and this work must certainly be done), climate modellers should additionally explore the benefits of improving the representation of near and subgrid flow in lower-resolution models by stochastic processes. In practice, it is quite probable that these pursuits are not mutually exclusive: as explicit resolution approaches that associated with individual convective cloud systems, the unresolved sub-cloud dynamics will then be represented stochastically.

In his study of the economics of climate change, Lord Stern has shown that the climate problem is, globally, a trillion-dollar problem. Reliable global and regional climate predictions with accurate error bars are an essential element in trying to combat the threat of climate change. This is the reason why, at the beginning of this preface, we suggested that the theme of this book is itself a trillion-dollar theme!

We believe we are at the beginning of a new era in weather and climate modelling – an era that recognises that although the equations of motion are formally deterministic, the best predictions, whether of weather on time scales of days, or climate on time scales of a century or more, may be based on models that are at least partially stochastic.

Tim Palmer and Paul Williams
Reading, UK

1

Mechanisms of climate variability from years to decades

GEOFFREY K. VALLIS

This chapter discusses and reviews of some of the mechanisms that may be responsible for climate variability on yearly to decadal time scales. The discussion is organised around a set of mechanisms that primarily involve the atmosphere, the ocean, or the coupling between the two. We choose an example of each, try to explain what the underlying mechanism is, and set it in the context of climate variability as a whole. All of the mechanisms are in principle deterministic, although we may not always care about the details of the process that give rise to the variability and in that case a stochastic description may be the most economical and insightful.

One person's signal is another person's noise.

1.1 Preamble

This is an essay on the mechanisms of natural climate variability on time scales of years to decades. It is meant to serve both as an introductory chapter to the articles appearing later in this book that delve into the mechanisms and modelling in greater depth, and as a stand-alone article for those requiring an overview, or at least a perspective, of the subject. In this preamble I'll discuss rather generally the nature of stochastic and deterministic processes and their role in weather and climate, and in the following sections I will focus more explicitly on climate variability on time scales from years to decades, emphasising processes that primarily involve the atmosphere and/or ocean.

Variability of climate – indeed variability of many systems – is often partitioned into two categories, *stochastic* and *deterministic*, each associated with rather different mechanisms. According to one dictionary, stochastic means 'randomly determined, having a random probability distribution or pattern that may be analysed statistically, but may not be predicted precisely'. Another dictionary defines stochastic as 'involving a random variable' or 'involving chance or probability'. Deterministic, on the other hand, is usually taken to refer to a phenomenon whose outcome is causally determined, at least in principle, by preceding events in conjunction with the laws of nature; thus, a deterministic sequence is one that may be predicted to a specified degree of accuracy, using appropriate equations of motion, if the initial conditions are given. Now, discounting quantum effects, the laws of nature are wholly deterministic – they may be cast as equations of motion that predict the evolution of objects given their state at some instant. Thus, *nearly all phenomena in climate dynamics are deterministic.* This statement, although true, is, however, perhaps not the whole truth, for whereas a system may be deterministic in principle, in practice we may not be able to predict it for at least two reasons:

(i) We are unable to compute the details of the evolution of part or all of the system because the system is chaotic. No matter how well we know the initial conditions, if not perfectly, then the future outcome is unpredictable and may be best described statistically.
(ii) Our *knowledge* of the system is imperfect and so we represent possible outcomes by probabilities, as if the system were stochastic. The probabilities then reflect our uncertain knowledge of the system, rather than an inherent indeterminism.

The weather and the climate, respectively, provide illustrations of these two points. As is well known, the Earth's atmosphere is a chaotic system and even if we could know the initial conditions extremely accurately (and had a very good numerical weather prediction model) the details of the future weather would still be unpredictable after a couple of weeks. The climate (as usually defined as some kind of average of the weather, or the statistics of the weather) is more predictable in this sense: for example, if we knew how much carbon dioxide we were to put in the atmosphere then the degree of global warming should be predictable, and the fact that it is not reflects our ignorance of climate dynamics. Roe & Baker (2007) argue that even a small amount of ignorance may lead inevitably to large uncertainties in climate projections. But even if this hypothesis is granted, the ensuing probability distribution for a climate projection is still of a somewhat different nature than that of weather forecasts: the climate probability distribution primarily reflects our ignorance of how the laws of physics and chemistry apply to the Earth's climate, whereas the weather probability distribution reflects the amplification of small fluctuations by chaos.

Nevertheless, there are similarities in the two cases, in the sense that both reflect an ignorance of some aspect of the system, an ignorance that is amplified either by the chaos of the system in the case of weather, or the feedbacks within the system in the case of climate; both then lend themselves to probabilistic approaches. But there is another reason – perhaps the main one in our context – for studying stochastic processes; it is that we *don't care about the details of a particular process.* We care only about its statistical properties, and sometimes only its variance. One example of this lies in the small scales of turbulence – by and large we don't care about the path of a small eddy near the viscous scale, but we do care about the statistical properties of eddies in cascading energy and/or enstrophy to small scales where they may be dissipated. There is a similar aspect to climate variability on the decadal time scale, in that we don't care about the weather that is taking place on time scales of weeks. Now, weather can be explicitly modelled far better than it can be parameterised by a stochastic process, but the details of the weather are generally irrelevant to decadal-scale climate variability – only its statistics likely matter. We may therefore choose to model weather as some kind of stochastic process, running the risk that it might be improperly modelled, in the hopes of isolating the mechanisms that might give rise to climate variability on longer time scales. Similarly, for those whose interest is variability on time scales of millions of years then even centennial variability might best be treated as noise.

In the rest of this chapter I focus on climate variability on time scales of years to decades, with the exclusion of the El Niño phenomenon for that deserves an article unto itself. The discussion is organised around mechanisms; thus, following a brief look at some observations, I summarise the general classes of mechanisms that might give rise to such variability. Each of the subsequent sections is then devoted to one type of mechanism, illustrating it with one or two examples.

1.2 Observations and classes of mechanisms

1.2.1 A few observations

Climate variability exists on time scales of seasons to millennia, but in this article the emphasis will be on time scales of years to centuries, or the decadal time scale. A rough indication of such variability is shown in Fig. 1.1, where the globally averaged surface temperature is plotted for the period 1850–2007. In addition to the evident general warming trend one seems to see variability on the decadal scale, which also seems evident if one restricts attention to a particular region of the globe, as in the central England temperatures shown in Fig. 1.2, which is perhaps the longest continuous instrumental record in climate.

However, we need to be rather careful that we are not deceived by a casual visual inspection of such time series into thinking that there is more decadal variability than might be expected by chance. To illustrate this, a Monte Carlo simulation is performed by taking the time series of successive winter temperatures

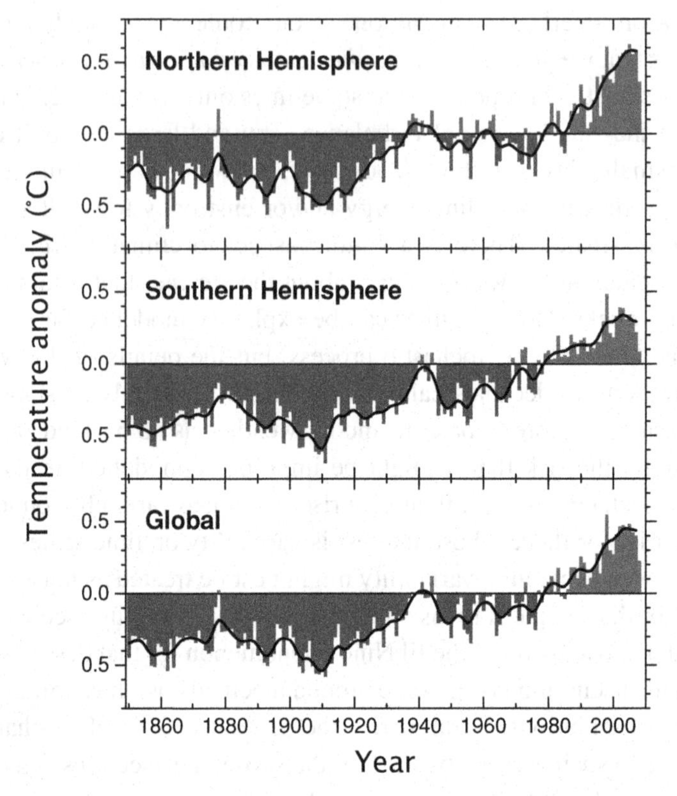

Figure 1.1 The global average surface temperature (anomaly from 1961–1990 average) from the HadCRUT3 data, from Brohan *et al.* (2006). In addition to the general warming trend some decadal variability seems apparent, presumably due to natural variability in the system.

from the central England time series and shuffling the temperatures randomly; in the resulting time series any mechanistic decadal variability has manifestly been removed. After detrending to remove secular changes we plot a realisation of a shuffled time series alongside the original time series, and we see that the two series look remarkably similar (upper right panel of Fig. 1.2). The power spectra of the original series is fairly white for periods from about 1 year to 200 years, with some apparent peaks at about 10 years and 100 years, but the power spectra of the shuffled time series are not qualitatively different from the original (lower right panel of Fig. 1.2). In this figure we show the power spectra of the original time series (thick solid line), a sampling of the spectra computed from shuffled time series (thin solid lines), the mean of the these (thick dashed line) and the mean plus or minus one standard deviation (thin dashed lines). Some of the shuffled time series have as much or more decadal scale variability as the original one, although the original time series does just stand out from the noise at long time periods – although

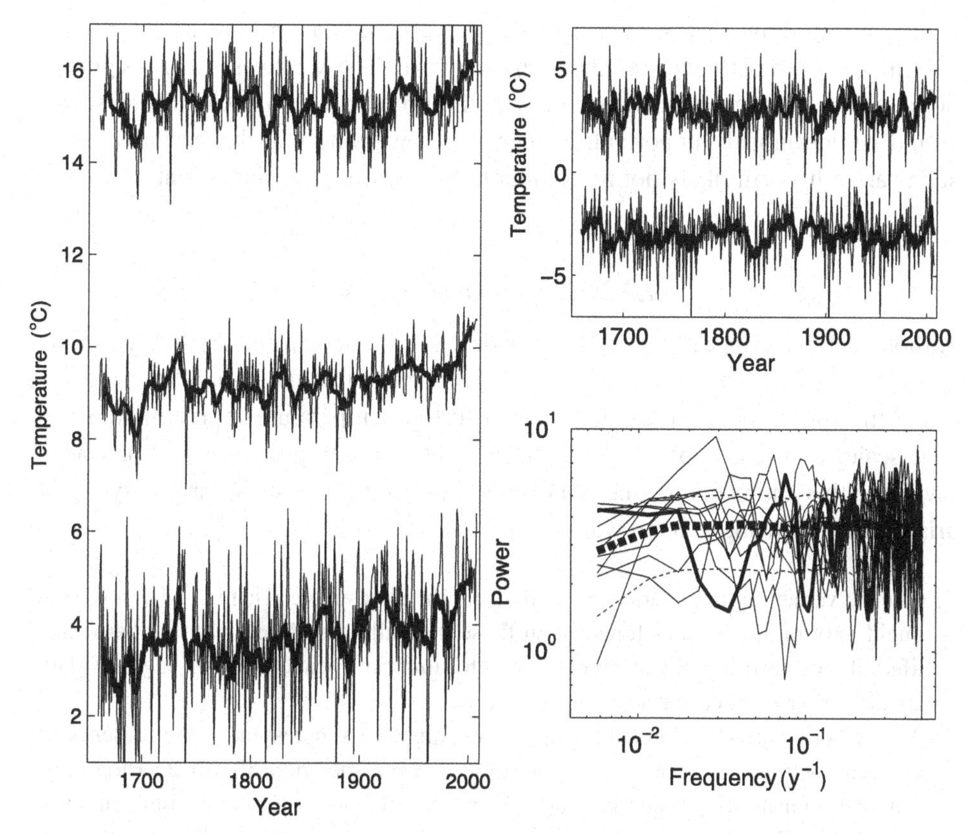

Figure 1.2 Left: central England temperature from 1650 to 2007, using the HadCET data (Parker *et al.* 1992). The three sets of curves show, from the top, summer (JJA), annual average, and winter (DJF) temperatures. The thin curve shows the average over the season or year, and the thicker curve shows the ten-year running mean. Right top: detrended winter temperature anomalies; the thin line is the seasonal temperature, and the thicker line is the temperature after a six-year running mean. The lower curve is the same for a sample shuffled time series, in which the ordering of the years is random. The real and shuffled series are offset from the origin by plus and minus 3 °C respectively. Right bottom: Power spectra of real and shuffled central England temperature. The thick solid line is the power spectrum of the actual winter (DJF) temperatures, the thick dashed line is the mean power spectrum of 1000 shuffled series, with plus and minus one standard deviation marked by thin dashed lines to either side. The multiple thin lines are the power spectra of 10 of the shuffled time series.

the sceptic may certainly argue that decadal variability has not been demonstrated from this time series alone. Other, more complete, observational analyses have detected decadal-scale variability, especially when account is taken of the spatial patterns in the data (e.g. Mann & Park 1994; Tourre *et al.* 1999). Similarly, Biondi *et al.* (2001) conclude the climate in and around the Pacific region has undergone real decadal-scale variability over the past few centuries. Overall, it is a defensible conclusion to draw that decadal and longer variability is present in the climate system, but by most measures the signal is weak. The weakness of the signal is not a reason for neglecting it, since any skill at prediction on decadal time scales would be enormously important. Rather, it is a warning that the mechanisms of such variability will likely not reveal themselves easily to the investigator.

1.2.2 General mechanisms

Although climate variability is difficult to define without being either overly general or overly prescriptive, in this article we will regard it as the variability of large-scale atmospheric or oceanic fields (such as surface temperature or precipitation) on time scales of a season or longer. Restricting attention to processes that primarily involve either the ocean or the atmosphere, the source of such variability could arise in the following general ways.

1. Climate variability might arise primarily from the atmosphere. That is, the atmosphere might vary on time scales longer than those normally associated with the baroclinic lifecycle, or have long-lived regimes of behaviour, independent of varying boundary conditions such as sea-surface temperature.
2. Atmospheric variability on short time scales might be suppressed by the presence of an ocean with a large heat capacity, leading to a red spectrum of climate variability. This mechanism, as essentially proposed by Hasselmann (1976) and Frankignoul & Hasselmann (1977), has become a de facto null hypothesis for climate variability.
3. Climate variability might arise via coupled modes, that is via non-trivial interactions between the ocean and atmosphere. The EI Niño/Southern Oscillation (ENSO) cycle is one example, perhaps even the only uncontroversial example.
4. Climate variability might have a primarily oceanic origin. Ocean variability might affect the atmosphere, and so the climate, without the need for coupled modes of the kind envisioned in item 3.
5. Secular changes in climate can be caused by changes in forcings external to the ocean–atmosphere system. This includes changes in atmospheric composition (such as carbon dioxide concentration), incoming solar radiation, volcanism, and changes in land surface and distribution.

In the next few sections I will discuss these mechanisms, excluding the last item which is well documented elsewhere. The El Niño phenomenon is not discussed for similar reasons. I don't aim to provide a comprehensive review, but nor is my aim to be provocative for its own sake. Rather, the goal is to provide a perspective, to illustrate some of the mechanisms with results from coupled ocean–atmosphere models, and to see how deterministic or stochastic ideas might fit in with them.

1.3 Atmospheric variability

In the extra-tropical atmosphere the primary mechanism of variability on large scales is baroclinic instability, the basic lifecycle of which, from genesis to maturation to decay, is about 10 days (e.g. Simmons & Hoskins 1978). The baroclinic time scale stems from the growth rate of baroclinic instability, and the simplest measure of this is the Eady growth rate,

$$\sigma \equiv \frac{0.3\Lambda H}{L_d} = \frac{0.3U}{L_d} \tag{1.3.1}$$

where Λ is the shear, H a vertical scale, U a horizontal velocity and L_d is the deformation radius. For values of $H = 10$ km, $U = 10$ m s^{-1} and $L_d = 1000$ km we obtain $\sigma \approx 1/4\,\mathrm{d}^{-1}$. (If $\beta \neq 0$ the height scale of the instability may be changed – it is no longer necessarily the height of the troposphere – but in practice a similar time scale emerges. Here β is the rate of change of Coriolis parameter with latitude.) The advective time scale of a baroclinic disturbance can similarly be expected to be about L_d/U, or a few days, and the total lifecycle, although not exactly an advective time scale, might be expected to be a multiple of it. (In the ocean the baroclinic lifecycle is longer, primarily because the oceanic U is two orders of magnitude smaller – 10 cm s^{-1} as opposed to 10 m – and so even though the oceanic L_d is one order of magnitude smaller – 100 km as opposed to 1000 km – the oceanic eddy time scales are roughly ten times longer that the atmospheric ones.)

Of course baroclinic eddies are nonlinear, so that time scales considerably longer than the advective scales can in principle be produced. It is known, for example, that baroclinic waves interact with the stationary wave pattern, produced by flow over topography and over large-scale heat anomalies, such as cold continental land masses, to produce slowly varying planetary waves as well storm tracks (e.g. Chang *et al.* 2002) to produce intraseasonal variability. The zonal index will also vary on intraseasonal time scales by way of an interaction between the baroclinic eddies and the zonally averaged flow – this type of interaction is often invoked to explain the variability associated with the North Atlantic Oscillation and with so-called annular modes (Feldstein & Lee 1998; Hartmann & Lo 1998; Vallis *et al.* 2004; Vallis & Gerber 2008). There is evidence that such interactions can involve feedbacks that give rise to time scales longer than those normally associated with the baroclinic lifecycle (Robinson 2000; Gerber & Vallis 2007), although as currently understood they do not give rise to any predictable time scales longer than a few weeks, or months at most. Nevertheless, we can also expect some interannual variability essentially as a residual of the intraseasonal variability (e.g. Feldstein 2000), but such interannual variability will be weak and unpredictable.

However, for the atmosphere to produce variability on time scales significantly longer than a few weeks – for example with some peak in the power spectrum at interannual time scales – would likely require there to be some kind of regime behaviour, in which the gross atmospheric behaviour changes on time scales independent of those associated with baroclinic instability or stationary waves. Such behaviour is certainly not impossible, for, to give one example, atmospheric blocks appear to have a time scale not closely associated with baroclinic waves. Even more strikingly, zonal jets, once formed, can have an extremely long time scale (Panetta 1993; Vallis & Maltrud 1993). However, it seems unlikely that the atmosphere alone could give rise to significant, predictable, natural interannual variability, for two reasons:

(i) No mechanism is apparent that could produce such variability, except as a residual of intraseasonal variability or, perhaps, mechanisms associated with the slow variability of persistent jets or the quasi-biennial oscillation, whose climate relevance is unclear.
(ii) Suppose that the atmosphere *were* able to produce regime-like behaviour when steadily forced. The difference between any two realistic regimes would still likely be much smaller than the seasonal cycle, and it would seem likely that a seasonal cycle would disrupt any regime behaviour that persisted beyond a few months.

The first argument is rather weak, being an example of what has been called 'an argument from personal incredulity'. The second argument is a little stronger, for it does propose a mechanism that would prevent long time scales from emerging. Most integrations with atmospheric general circulation models (GCMs) do not produce significant variability on interannual time scales (an issue we revisit in later sections), but a notable exception was described by James & James (1992). They performed fairly long (decades and centuries) integrations with a dry primitive equation atmospheric model with very idealised forcing (a Newtonian relaxation), and found a red spectrum of various atmospheric fields, with power increasing as the time period increases from 10 days to 10 years, as illustrated in Fig. 1.3. James and James call this 'ultra-low-frequency variability'.

The structure of the variability, as represented by the first empirical orthogonal function (EOF) of the zonally averaged zonal wind, represents equivalent barotropic (i.e. no tilting in the vertical) fluctuations in the strength, and to a lesser degree the position, of the subtropical jet. The time variations of the first principal component can be modelled fairly well by a first-order Markov, AR(1), process (discussed more in the next section), but the shoulder of the spectrum occurs at a time scale of about 1 year, which is considerably longer than what might be expected to occur as a result of frictional spin-down effects; these would typically produce reddening on time scales of a few weeks or less. In the James–James simulations, the power at very low frequencies appears to come from a transition between a two-jet state, with a subtropical jet distinct from an eddy-driven midlatitude jet, and a single or merged jet state. It seems either state is a near-equilibrium state, and

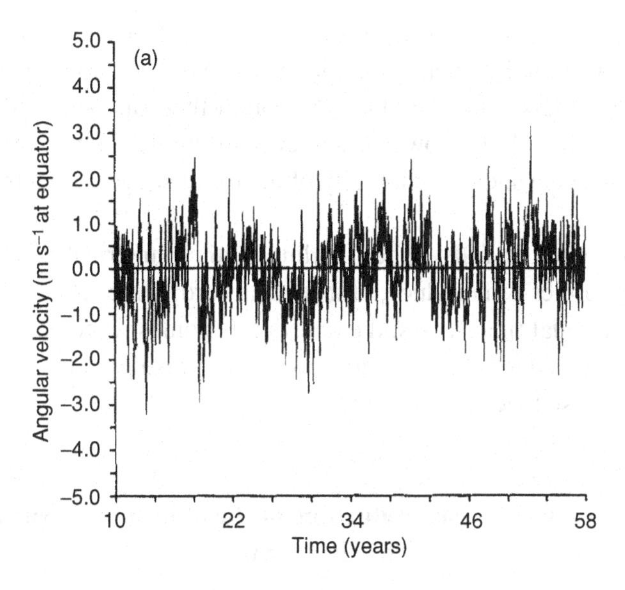

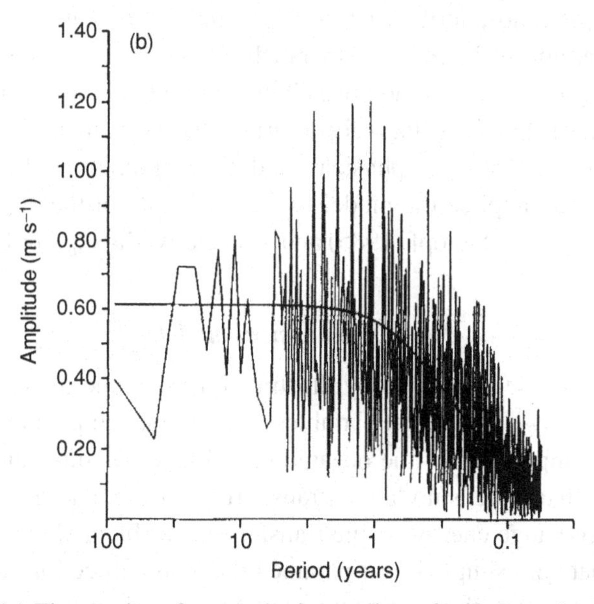

Figure 1.3 (*a*) Time series of atmospheric angular velocity in an integration of a low resolution (T31) primitive equation model with idealised, time-independent forcing. Note the variability at decadal time scales. (*b*) The power spectra of the first principal component of the zonal mean zonal wind, along with a first-order Markov ('red noise') process (thick smooth line). From James & James (1992).

that transitions between the two equilibria can occur after rather long intervals in one state. This would fall under the rubric of 'regime' behaviour discussed above. Somewhat surprisingly, James & James reported that this persisted even with a seasonal cycle, although the power is somewhat diminished compared to the run with no seasonal cycle and the ambiguity of the result suggests that the work bears repeating.

If we accept, without fully understanding it, the numerical evidence that the atmospheric dynamics can produce, of its own accord, some variability on interannual and even decadal time scales, the question becomes whether such variability is important compared to other mechanisms that can also produce such variability, and we discuss these mechanisms next.

1.4 The null hypothesis: reddening of the atmospheric variability by the ocean

The most unequivocal mechanism for producing climate variability, if not predictability, on time scales longer than those of atmospheric weather comes by way of the reddening of atmospheric variability by its interaction with the oceanic mixed layer. Climate variability then arises rather in the manner of Brownian motion, as the integrated response to a quasi-random excitation provided by atmospheric weather. The mechanism was first quantitatively described by Hasselmann (1976), although without reference to a specific physical model, and Frankignoul & Hasselmann (1977) subsequently applied the model to the variability of the upper ocean. Our treatment of this follows Schopf (1985) and, especially, Barsugli & Battisti (1998).

1.4.1 The physical model

The mechanism can be economically illustrated using a one-dimensional climate model with one or two dependent variables, namely the temperature of the atmosphere and the temperature of the oceanic mixed layer, as illustrated in Fig. 1.4. We will assume that there is no lateral transport of energy, and that the ocean and atmosphere interchange energy by the transfer of sensible and latent heat and by radiation and that very simple linear parameterisations suffice for these.

The physical parameterisations of the model are as follows:

absorption of solar energy at surface: $S(1-\alpha)$ (1.4.1)

sensible, latent and radiative flux from surface to atmosphere: $A_s + B_s T_s$ (1.4.2)

downwards infrared radiation from atmosphere to surface: $A_d + B_d T_a$ (1.4.3)

upwards infrared radiation from atmosphere to space: $A_u + B_u T_a$ (1.4.4)

upwards infrared radiation from surface escaping to space: $C T_s$. (1.4.5)

Table 1.1 *Typical values of radiative parameters for the energy balance model of Fig. 1.4. The derived quantities are $B_2 = B_s - C$, $B_3 = B_d + B_u$ and $B_4 = B_3 - B_2 B_d / B_s = B_u + C B_d / B_s$. The units of the B parameters and C are $Wm^{-2}\,K^{-1}$ and the units of the C parameters are $J\,m^{-2}\,K^{-1}$; C_s corresponds to a 60 m deep slab.*

Parameter	B_d	B_u	B_s	C	B_2	B_3	B_4	C_a	C_s
Value	11.3	2.83	10.4	0.54	9.86	14.13	3.42	1.2×10^7	3.6×10^8

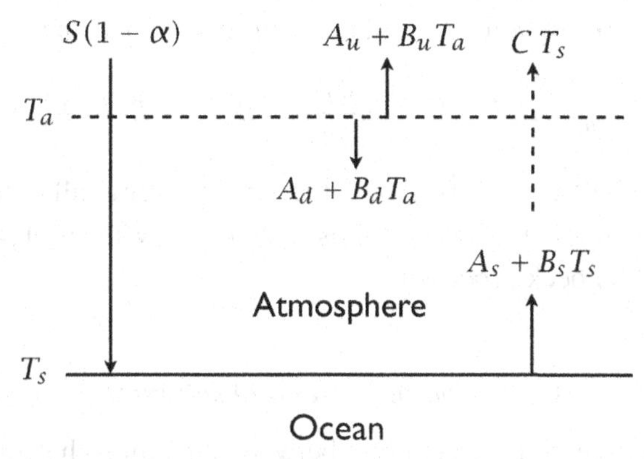

Figure 1.4 A schematic of a simple atmosphere–ocean energy balance model.

These are obviously caricatures of the real processes, but they will suffice to illustrate the point. The parameters A, B and C are constants, as are S and α, with typical values given in Table 1.1; the precise values do not affect the results or conclusions. The sensible and latent energy flux from the surface to the atmosphere might reasonably be taken to have the form $A_s + B_s(T_s - T_a)$, so depending on the difference between the atmospheric and surface temperatures. Here, we have absorbed the $B_s T_a$ factor in the term $B_d T_a$, giving the forms above.

The evolution equations corresponding to the above parameterisations are

$$C_s \frac{\partial T_s}{\partial t} = S(1-\alpha) + (A_d + B_d T_a) - (A_s + B_s T_s), \tag{1.4.6a}$$

$$C_a \frac{\partial T_a}{\partial t} = A_s + (B_s - C)T_s - (A_d + B_d T_a) - (A_u + B_u T_a), \tag{1.4.6b}$$

where C_s and C_a are effective heat capacities of the atmosphere and surface (e.g. the mixed layer of the ocean), respectively. Typically, $C_s \gg C_a$, and indeed if C_s represents an oceanic mixed layer of depth 60 m, then $C_s \approx 30 C_a$. It is this difference in heat capacity that can lead to the seeming generation of climate variability on long time scales.

The above equations may be written as

$$C_s \frac{\partial T_s}{\partial t} = A_1 + B_d T_a - B_s T_s, \tag{1.4.7a}$$

$$C_a \frac{\partial T_a}{\partial t} = A_2 + B_2 T_s - B_3 T_a, \tag{1.4.7b}$$

where $A_1 = S(1-\alpha) + A_d - A_s$, $A_2 = A_s - A_d - A_u$, $B_2 = B_s - C$ and $B_3 = B_d + B_u$. We can write (1.4.7) as anomaly equations by writing $T_s = \overline{T}_s + T_s'$ and $T_a = \overline{T}_a + T_a'$, where $\overline{T}_s$ and $\overline{T}_a$ are the steady solutions of (1.4.7), and if we also introduce a random forcing, $\sigma \dot{W}$, on the right-hand side of (1.4.7b), representing the weather in the atmosphere, the equations of motion become

$$C_s \frac{\partial T_s'}{\partial t} = B_d T_a' - B_s T_s', \qquad C_a \frac{\partial T_a'}{\partial t} = B_2 T_s' - B_3 T_a' + \sigma \dot{W}. \tag{1.4.8a,b}$$

We will take $\dot{W}$ to be 'white noise', with the same power at all time scales, so that the Fourier transform of $\dot{W}$ is unity. ($\dot{W}$ is related to a Wiener process and (1.4.8) to an Ornstein–Uhlenbeck process.)

1.4.2 Limiting cases and solutions

First consider the case with no coupling between the atmosphere and ocean. With common configurations of atmospheric GCMs in mind, we might implement this in two possible ways, the simplest being just to hold the sea-surface temperature (SST) at its mean value given $\overline{T}_a$, whence $T_s' = 0$ in (1.4.8), so that the equation for T_a is just

$$C_a \frac{\partial T_a'}{\partial t} = -B_3 T_a' + \sigma \dot{W}. \tag{1.4.9}$$

Another way of supposing there are no dynamical interactions between atmosphere and ocean is to suppose that the heat capacity of the ocean is zero, so that the ocean is slaved to the atmosphere; the surface temperature is then given by setting the left-hand side of (1.4.8a) to zero and $B_d T_a' = B_s T_s'$. This model might be best considered as representing an atmosphere overlying a land surface that has a small heat capacity. The equation for T_a is then

$$C_a \frac{\partial T_a'}{\partial t} = -B_4 T_a' + \sigma \dot{W}, \tag{1.4.10}$$

where $B_4 = B_3 - B_2 B_d / B_s$. This is of the same form as (1.4.9), but evidently the damping is reduced. Physically, if the atmosphere is warm then the ocean is warmed by the atmosphere, so reducing the damping on the atmosphere. (Note that $B_4 = B_3 - B_2 B_d / B_s = B_u + C B_d / B_s$, which is positive, so the system is always damped.)

Quantitatively, using the values given in Table 1.1, $B_3/B_4 \approx 4$, so that the reduction in damping is potentially significant, at least on long time scales when the SST can respond, but a more quantitative calculation, using for example a GCM, would be needed to see if the effect is truly important.

Note that (1.4.9) and (1.4.10) are closely related to first-order autoregressive, or AR(1), models. A simple forward-in-time finite differencing of (1.4.9) gives

$$T_{i+1} = \left(1 - \frac{B_3 \Delta t}{C_a}\right) T_i + \sigma W_i \tag{1.4.11}$$

where Δt is the time step and W_i a white-noise process with zero mean. Equation (1.4.11) is an example of – indeed, it virtually defines – an AR(1) process.

The power spectrum of (1.4.9) may be calculated by Fourier transforming, whence

$$\mathrm{i}\, C_a \omega \widetilde{T}_a = -B_3 \widetilde{T}_a + \sigma, \tag{1.4.12}$$

where $\widetilde{T}$ is the complex Fourier amplitude of the temperature. The power spectral density may be defined by $P_a(\omega) = |\widetilde{T}_a|^2$, giving

$$P_a^U = \frac{\sigma^2}{B_3^2 + C_a^2 \omega^2}, \tag{1.4.13}$$

where we add a superscript U ('uncoupled') to differentiate it from subsequent atmospheric spectra. Equation (1.4.13) describes a standard 'red-noise' power spectrum. At longer times the spectrum of the response is white, with higher frequencies being damped, the shoulder occurring approximately at the frequency $\omega = B_3/C_a \approx 1/10^6\ \mathrm{s}^{-1} \approx 1/10\ \mathrm{d}^{-1}$. Of course, this calculation is very approximate, but it indicates that on time scales longer than a week or so – longer than the spin-down time scale C_a/B_3 – the atmospheric spectrum can be expected to be white.

The power spectrum resulting from (1.4.10) is of the same form as (1.4.13), being

$$P_a^{C0} = \frac{\sigma^2}{B_4^2 + C_a^2 \omega^2}, \tag{1.4.14}$$

differing from (1.4.13) only in the coefficient in the denominator, with the superscript $C0$ denoting 'coupled, zero heat capacity ocean'. The coupling has two related effects. (i) Because $B_4 < B_3$, the damping is reduced on long time scales, and the atmospheric variance is stronger in the coupled case. At low frequencies the ratio of the two variances is just $B_3^2/B_4^2 \approx 16$ using the values given in Table 1.1. (ii) The shoulder of the spectrum occurs at a frequency $\omega = B_4/C_a \approx 1/40\ \mathrm{d}^{-1}$, which is lower than that of (1.4.13).

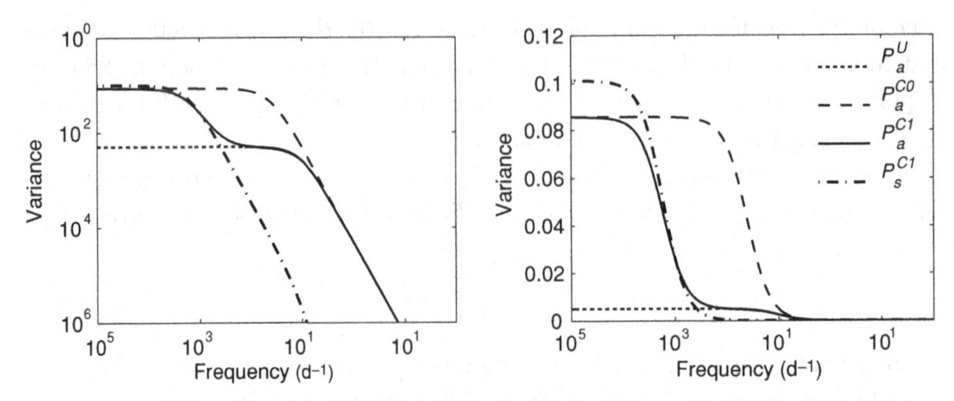

Figure 1.5 Power spectra of the model ocean and atmosphere obtained from (1.4.8) in various limits. Both panels show the same spectra, on the left in a log-log plot and on the right in a log-linear plot. P_a^U corresponds to an atmosphere-only model with $T_a' = 0$, and a spectrum given by (1.4.13). P_a^{C0} corresponds to a model with a zero heat capacity surface and a spectrum given by (1.4.14). P_a^{C1} corresponds to a model with a finite heat capacity surface (e.g. an oceanic mixed layer) and a spectrum given by (1.4.16). P_s^{C1} is the oceanic spectrum that accompanies P_a^{C1}, and is given by (1.4.17).

1.4.3 Finite-capacity oceanic mixed layer

A more realistic situation is that in which the ocean has a finite heat capacity that is much larger than that of the atmosphere; that is, with $C_a \ll C_s$. The Fourier transformed equations of motion are then

$$\mathrm{i}\omega C_s \widetilde{T}_s = B_d \widetilde{T}_a - B_s \widetilde{T}_s, \qquad \mathrm{i}\omega C_a \widetilde{T}_a = B_2 \widetilde{T}_s - B_3 \widetilde{T}_a + \sigma. \qquad (1.4.15\text{a,b})$$

The power spectra of the atmosphere and ocean are then found to be

$$P_a^{C1} = \frac{\sigma^2 \mid \omega_s \mid^2}{\mid \omega_a \omega_s - B_2 B_d \mid^2}, \qquad (1.4.16)$$

and

$$P_s^{C1} = \frac{\sigma^2 \mid B_d \mid^2}{\mid \omega_a \omega_s - B_2 B_d \mid^2} = \frac{\mid B_d \mid^2}{\mid \omega_s \mid^2} P_a^{C1}, \qquad (1.4.17)$$

where $\omega_s \equiv \mathrm{i}\omega C_s + B_s$ and $\omega_a \equiv \mathrm{i}\omega C_a + B_3$, and these quantities are plotted in Fig. 1.5. The atmospheric spectrum P_a^{C1} (solid black line) lies between the uncoupled spectrum, P_a^U, and the spectrum of the atmosphere coupled to a zero heat capacity ocean, P_a^{C0}, to which it is equal in the low-frequency limit. Physically, on time scales long enough that the ocean can change its SST, the response of the ocean to the atmosphere reduces damping on the atmosphere (because the ocean warms up if the atmosphere is warm), so increasing the variance of the atmosphere. In the zero-frequency limit, the atmospheric spectrum then displays considerably more power than the power spectra of the process with SST fixed, by the ratio

$$\frac{P_a^{C0}}{P_a^U} = \frac{P_a^{C1}}{P_a^U} = \frac{|\, B_3 \,|^2}{|\, B_4 \,|^2} \approx 17. \tag{1.4.18}$$

As regards the oceanic spectra, we may note that for time scales much longer than C_a/B_3 (about 10 days) the atmosphere is in a quasi-equilibrium state and the left-hand side of (1.4.8b) may be neglected. The atmospheric spectrum is white and the oceanic spectrum is given by

$$P_s^{C2} = \frac{\sigma^2 B_d^2}{B_4^2 B_s^2 + B_3^2 C_s^2 \omega^2}. \tag{1.4.19}$$

This is a red spectrum, with a shoulder at the frequency ω_{sh} given by

$$\omega_{sh} = \frac{B_4 B_s}{B_3 C_s} \approx 7 \times 10^{-9}\,\mathrm{s}^{-1} \approx 1/1600\,\mathrm{d}^{-1}. \tag{1.4.20}$$

On time scales longer than C_a/B_3 the above spectrum (not plotted) is almost indistinguishable from that given by (1.4.17). It is evidently the fact that the heat capacity of the oceanic mixed layer is so much greater than that of the atmosphere that gives the long time scale here. Even given the simplicity of the model and the uncertainty or possible inappropriateness of the parameters, the result suggests that there will be variability in the SST on annual time scales that is forced purely by the 'noise' of the weather in the atmosphere.

1.4.4 Forcing in the oceanic component

A variant of the above calculations is to assume that the natural variability arises in the *oceanic* component of the system. If so then it is instructive to put a forcing on the right-hand side of (1.4.8a). For simplicity we suppose this to be a white noise as before (and because it is a linear equation with additive noise each frequency component is in any case independent) and that there is no noise in the atmosphere. The equations of motion then become,

$$C_s \frac{\partial T_s'}{\partial t} = B_d T_a' - B_s T_s' + \sigma \dot{W}, \qquad C_a \frac{\partial T_a'}{\partial t} = B_2 T_s' - B_3 T_a', \tag{1.4.21a,b}$$

plainly having the same form as those previously analysed. There are two particularly relevant limits:

(i) An 'uncoupled case', with atmosphere of fixed temperature and so $T_a{}' = 0$. Such a case roughly corresponds to an ocean model forced by fluxes from a specified, unchanging atmosphere.

(ii) An atmosphere with zero heat capacity, and so in thermal equilibrium with the ocean beneath with $B_2 T_s' = B_3 T_a'$.

In these two cases we obtain, respectively,

$$C_s \frac{\partial T_s'}{\partial t} = -B_s T_s' + \sigma \dot{W}, \qquad C_s \frac{\partial T_s'}{\partial t} = -B_5 T_s' + \sigma \dot{W}, \qquad (1.4.22a,b)$$

where $B_5 = B_s - B_2 B_d / B_3 = B_4 B_2 / B_3$. The parameter B_5 is, like B_4, always positive, but it is smaller than B_s so that the system is damped less than in the fixed atmosphere case. Physically, if the ocean is anomalously warm because of the forcing (or more generally because of its internal dynamics) then the atmosphere warms up, and the damping of the ocean is subsequently less than it otherwise would be.

The corresponding power spectra (with superscript U again denoting 'uncoupled') are

$$P_s^U = \frac{\sigma^2}{B_s^2 + C_s^2 \omega^2}, \qquad P_s^{C2} = \frac{\sigma^2}{B_5^2 + C_s^2 \omega^2}. \qquad (1.4.23)$$

These are both red-noise spectra, with more power in the case with the coupled atmosphere – on long time scales the ratio of their respective variances is given by

$$\frac{P_s^{C2}}{P_s^U} = \frac{B_s^2}{B_5^2} \approx 20. \qquad (1.4.24)$$

This is potentially significant, but its importance in reality can probably be best assessed by performing experiments with coupled ocean–atmosphere models.

1.4.5 Observations and predictability

If the ultimate cause of variability in the coupled system described above is weather noise, and not the longer term dynamics inherent in the ocean, then there is little predictability in the system on time scales longer than the predictability inherent in weather, even though variables may appear to vary on long time scales; it is simply that the high frequencies have been filtered out leaving only the slow variations. Another way to see this is that one need only note that 'white noise' is essentially equivalent to a sequence of random numbers with each value independent of all the others; we plainly cannot predict future values by smoothing the sequence, which is more-or-less what reddening the spectrum by damping is equivalent to. Of course, if a stochastic variable is far from its mean, its persistence away from the mean will appear to increase if its value is smoothed, as the period between zero crossings increases with redness.

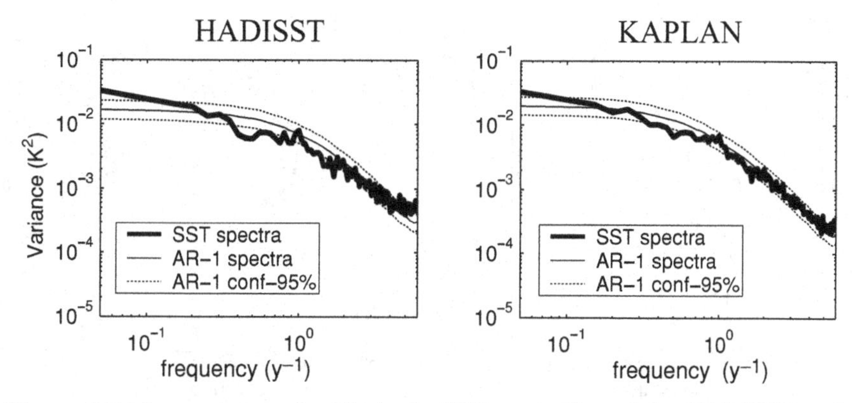

Figure 1.6 Mean spectra of midlatitude SST anomalies of the HADISST and Kaplan SST data sets (thick lines), along with the best-fit spectra from an AR(1) process (thin solid central line) with 95% confidence levels (thin dotted outer lines). Adapted from Dommenget & Latif (2002).

It is the nature of a null hypothesis that the mechanisms upon which it relies certainly do exist – otherwise it would not be a null hypothesis. But this is not to say that these mechanisms are the only ones that operate, or that the predictions of the hypothesis are borne out by reality. In the case at hand we may ask, do observations indicate that the integration of the atmospheric variability by the oceanic mixed layer, so producing a red spectrum in the oceanic mixed layer, is 'all there is'? The HADISST set (Folland *et al.* 1999) and the Kaplan set (Kaplan *et al.* 1998) both extend over a hundred years, and Fig. 1.6 shows their mean spectra as computed by Dommenget & Latif (2002). Neither of the spectra conform very well to AR(1) spectra with 95% confidence limits. The deviations do not occur through a single spectral peak indicating some periodic oscillation, but the general shape of the spectrum is different, having a shallower slope than is predicted at seasonal to interannual time scales, but at the same time the spectrum fails to flatten into a white spectrum at long time scales; rather, it continues to redden at decadal time scales, suggesting perhaps that there are dynamical processes that can directly produce variance on these long periods. We discuss what these might be in the next few sections.

1.5 Coupled modes of interaction

Let us now look at the evidence for dynamically coupled modes of interaction between the ocean and atmosphere, but omitting discussion of the single unambiguous example in the climate system, namely El Niño and the Southern Oscillation (ENSO). Our reason for such a seemingly egregious omission is that the ENSO

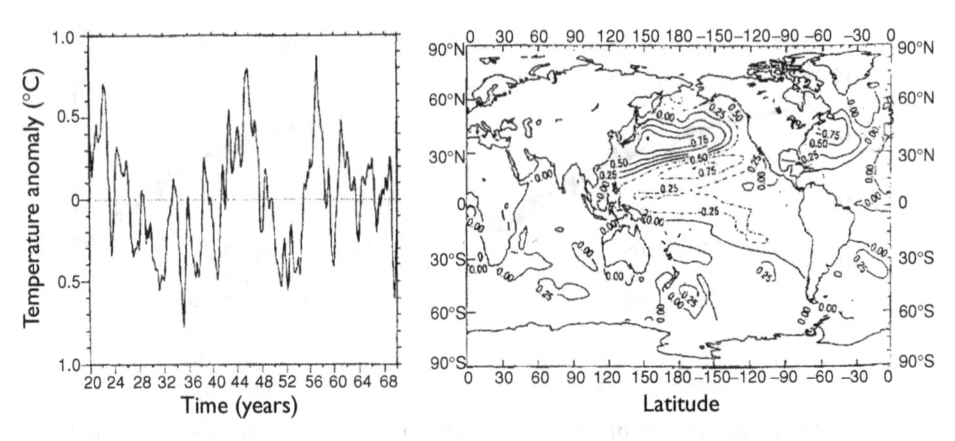

Figure 1.7 Left: time series of the SST in a coupled ocean–atmosphere model. The SST is averaged over the region from 150° E to 180° E and 25° N to 35° N, and smoothed with a nine-month running mean. Right: the regression of the SST on the time series shown in the left panel. From Latif & Barnett (1994).

phenomenon is well documented, reasonably well understood, and discussed at great length elsewhere.

A striking example of apparent mid- and high-latitude ocean–atmosphere coupling was described by Latif & Barnett (1996). Using a then state-of-the-art coupled ocean–atmosphere model, in conjunction with observations, they posited a cycle of unstable air–sea interactions involving the North Pacific subtropical gyre and the Aleutian low-pressure system. This cycle appeared to be responsible for about a third of the variability of the climate system and implies some genuine predictability within the system. The mechanism they suggested was roughly as follows. Suppose that, for whatever reason, the subtropical gyre is anomalously strong. The western boundary current (the Kuroshio in the Pacific) transports more warm water polewards and this produces a positive SST anomaly in the North Pacific. This heats the atmosphere leading to a weakened Aleutian low, and this in turn leads to air–sea fluxes that reinforce the initial anomaly. At the same time, according to Latif and Barnett, the storm track is weakened and the anomalous Ekman heat transport also contributes to maintaining the positive SST anomaly. Their model produced examples of SST variability varying on time and space scales as illustrated in Fig. 1.7.

The cycle above is mechanistic and deterministic and, in principle, predictable, at least on the time scale of a single cycle – that is, on the decadal time scale. The memory of the system resides in the ocean, and it is the ocean dynamics that determines the time scales on which anomalies come and go, as in Fig. 1.7.

Other coupled mechanisms

Other possible mechanisms exist for the coupling between ocean and atmosphere that might give rise to climate variability, and one such was identified by Cessi (2000). In a numerical study, she noted the presence of a strong thermal front at the boundary between the subtropical and subpolar gyre in the ocean, associated with the weaker oceanic meridional heat transfer at the gyre boundary. This creates, on average, a sharper atmospheric gradient and a stronger meridional atmospheric heat transport by midlatitude baroclinic eddies, compensating in part for the weaker ocean heat transport, resulting in a well-defined atmospheric storm track. The atmospheric surface winds are, of course, responsible for the location of the inter-gyre boundary, for at zeroth order this is determined by the latitude at which the wind-stress curl changes sign. Changes in the position of the gyre boundary lead to changes in the storm track, and this may feed back onto the location of the gyre boundary and so on, and Cessi found that, because of the delayed adjustment of the gyres to the wind stress, the whole system oscillated with a period of about 18 years.

1.5.1 Air–sea interaction: evidence and models

Although the coupled mechanisms described above are quite striking they have not been robustly reproduced by other coupled atmosphere–ocean models. Such a coupled mechanism requires, broadly speaking, the following sequence of events:

1. The generation of some large-scale pattern of sea-surface temperature anomalies.
2. The SST anomaly pattern must imprint itself on the atmosphere.
3. The atmospheric dynamics must then act in such a way as to reinforce or maintain the SST anomaly pattern.

Although all of these steps seem possible, there are difficulties with each, and so the likelihood of an entire chain of events occurring becomes somewhat delicate, as we discuss in the next few subsections.

1.5.2 Generation of SST anomalies

The first issue is the generation of SST anomaly patterns. Although there are small-scale SST patterns associated with oceanic mesoscale eddies, observations show that, at least on time scales shorter than a year, most of the variance in the SST is forced by the atmosphere, and not vice versa. One of the first demonstrations of this was that of Davis (1976), using 28-year records of SST and sea-level pressure (SLP) anomalies in the North Pacific. Davis tested three hypotheses: (i) that SST anomalies can be predicted, in part, from previous SST anomalies; (ii) that SLP anomalies can be determined, in part, from contemporaneous SST observations; and (iii) that (i) and (ii) may be combined to allow a prediction of SLP from present and past SST observations. Davis found that, although (i) and (ii) are true to some degree, indicating that SLP and SST are related, future SLP anomalies *cannot* be predicted from present SST data. He concluded that, at least as regards SLP and SST, the atmosphere drives the ocean and that the 'back-forcing' of the ocean on the atmosphere is very weak. That is, in a nutshell, that the observed SST anomalies on seasonal time scales are the result of, not the cause of, SLP anomalies. Such a study is broadly consistent with a physical model of the SST as being governed by the null hypothesis of Section 1.4 and equations qualitatively similar to (1.4.8), as well with the model of Frankignoul & Hasselmann (1977) that SST anomalies represent the integral response to short-time-scale atmospheric forcing. The higher heat capacity of the oceanic mixed layer, compared to that of the atmosphere, gives SST anomalies a longer time scale than the atmospheric anomalies that cause them. Later studies have also supported this picture – for example Cayan (1992) essentially showed that SST anomalies are positive when there is a heat flux into the ocean, implying an atmospheric driving.

This picture does not, however, preclude the ocean driving the atmosphere on still longer time scales, and it should be said (because it is often forgotten) that the atmosphere must and does respond to the ocean on very long time scales, including the infinitely long, statistical equilibrium state. This is because the ocean transports heat polewards, especially in the Northern Hemisphere where in the extra-tropics it transfers about one third of that of the atmosphere (Trenberth & Caron 2001). Without such a transfer the pole–equator temperature difference would almost certainly be larger than it in fact is, and the atmosphere and climate would ipso facto be different. Indeed Winton (2003) found using a GCM that an Earth-like planet without an ocean became largely ice-covered: a 'snowball Earth'. In the unlikely event that the atmosphere compensated entirely for the ocean's lack of transport and kept the pole–equator temperature gradient the same, then again the atmosphere would have responded to the ocean. And since the ocean's transport is made apparent to the atmosphere primarily through the sea-surface temperature, clearly the sea-surface temperature does influence the atmosphere.

1.5.3 Effect of SST anomalies on the atmosphere

The question therefore becomes, by what mechanisms and on what time scale do *variations* in the SST affect the atmosphere? The problem is, at least in part, one of signal to noise. On the infinite time scale the ocean does significantly affect the behaviour of the atmosphere; if, for example, we could engineer a large-scale SST anomaly covering the North Pacific for several years we could certainly expect that it would influence the atmosphere, including the sea-level pressure, because on a time scale shorter than the lifetime of the SST anomaly, the atmosphere would move into a new quasi-equilibrium state consistent with the new thermal boundary conditions. However, on time scales of a year or less the response to an SST anomaly seems likely to be smaller than the natural variability of the atmosphere (as we discuss further below), regardless of whether or not the atmosphere is responsible for producing SST anomalies. We conclude that SST anomalies likely have a weak but pervasive influence on atmospheric dynamics.

The mechanisms and degree of influence of SST anomalies on the atmosphere are matters on which there is a large amount of literature but less consensus, at least in midlatitudes. [For a discussion of mechanisms, experiments with general circulation models and observations and reviews, see among others Webster (1981); Lau & Nath (1994); Kushnir & Held (1996); Lunkeit & von Detten (1997); Czaja & Frankignoul (2002); Deser *et al.* (2004); these papers constitute but a tiny sampling of the literature, and Frankignoul (1985) and Kushnir *et al.* (2002) provide more comprehensive reviews.] Most GCMs show a rather *small* response to the imposition of SST anomalies, perhaps somewhat stronger in the summer when the mean flow is small. For example, using a fairly coarse-resolution atmospheric GCM Lau & Nath (1994) and Kushnir & Held (1996) found that their GCM exhibited quite weak sensitivity to midlatitude SST anomalies, with results that are generally consistent with linear quasi-geostrophic theory (e.g. Hoskins & Karoly 1981; Held 1983; Vallis 2006) with a shallow heating anomaly. In this case the heating anomaly at the surface will be balanced locally by zonal horizontal advection, producing a baroclinic response in the vicinity of and slightly downstream of the SST anomaly, with a warm ocean leading to a warm atmosphere and the response decreasing with height. In the far field, quasi-geostrophic theory suggests that the flow should be dominated by the homogeneous solution, and equivalent barotropic wave trains propagating around the globe may be generated.

If a sea-surface temperature anomaly is able to generate a *deep* heating source in the atmosphere then the response may be larger. In this case theory suggests that the heating source may be balanced by meridional advection, with a positive heating anomaly producing equatorwards advection, and a warm low downstream of the heating region. The far-field response is the same as that for shallow heating, with equivalent barotropic wave trains. Although the fields produced by GCMs seem consistent with this general picture the response is not strong for realistic values and scales of SST anomalies. The equivalent barotropic downstream response is typically not robustly found in GCMs, and generally the response is more typical of that of a shallow anomaly. It may be that the GCM's response to SST anomalies is unrealistically weak because of inadequacies in its boundary-layer formulation (perhaps confining the response too much to the near-surface region) but such a remark is speculative.

Even though the response of GCMs to SST anomalies may be weak, some GCMs (Rodwell *et al.* 1999; Robertson *et al.* 2000) do seem to show that the North Atlantic Oscillation (NAO) *is* influenced by sea-surface temperature when sea-surface temperature is imposed, and there is also some observational support for this (e.g. Czaja & Frankignoul 2002). Such results should probably not be regarded as contradicting or being wholly inconsistent with the GCM studies because the response to SST anomalies remains rather weak, but the pervasive nature of the anomalies does bring about systematic atmospheric response over the course of a season. Nonlinear effects likely play a role in the response also, as seems to have been found by Lunkeit & von Detten (1997).

Such a response does suggest a degree of predictability to the NAO, presuming of course that SST can be predicted. However, it is this last phrase that indicates a potential weakness in the argument: as discussed above, on the seasonal time scale SSTs seem likely to be a response to the atmosphere, and not a driver of the atmosphere. Thus, the SSTs themselves are likely not predictable over the course of a season, and so any skill in an atmospheric forecast dependent on such prediction is lost; Bretherton & Battisti (2000) discuss this interpretation further.

1.5.4 Effect of the atmosphere in maintaining SST anomalies

On the seasonal to yearly time-scale SST anomalies are, as mentioned, the consequence of atmospheric anomalies. For example, the North Atlantic Oscillation, whose dynamics are primarily tropospheric, will produce anomalous heat and momentum fluxes, both of which may generate SST anomalies, on seasonal time scales. Such anomalies may be generated both by anomalous winds – and so anomalous Ekman transports and mixed layers – and by anomalous sensible and latent heat fluxes. However (given the evident weakness on the seasonal time scale of the atmospheric response to SST anomalies) the subsequent dynamical feedback of the atmosphere to the ocean, maintaining and reinforcing the initial SST anomaly, seems unlikely except possibly in one regard: the reduced local thermal damping provided by the atmosphere if the ocean is anomalously warm. This effect is captured by the null hypothesis of Section 1.4: in Section 1.4.4 we noted that having an atmosphere, even with zero heat capacity, coupled to an ocean provides a reduced damping compared to a ocean underneath a constant-temperature atmosphere, and, as in (1.4.24), more variance at low frequencies. Even if this effect is not large in the context of maintaining seasonal SST anomalies, it may be an important effect in producing and maintaining decadal and longer climate fluctuations, as we discuss in the next section.

1.6 Climate variability with an oceanic origin

In this section we discuss the degree to which climate variability on decadal to centennial time scales might have a predominantly oceanic origin, possibly modified by the atmosphere. Most of the results come from coupled GCMs and so that is where our focus lies.

1.6.1 Some results from coupled GCMs

Coupled ocean–atmosphere models, when integrated for long periods of time after having reached equilibrium, often do display oscillations on time scales of decades to centuries. An early example showing this in a full GCM is the integration described by Delworth *et al.* (1993), in which a fully coupled, then state-of-the-art, ocean–atmosphere model was found to have oscillations in its meridional overturning circulation (MOC) with a period of about 50 years. In these simulations, the mean overturning circulation had an amplitude of about 18 Sv and the variability was about ± 1 Sv. These oscillations led to variations in SST of about 0.5 °C, mainly in the Atlantic subpolar gyre, and similar differences in surface air temperature. The surface air temperature anomalies are similarly concentrated at high latitudes, although they spread over a larger longitudinal range on either side of the North Atlantic. The patterns of both SST variations and the surface air temperature variations seemed qualitatively consistent with observations, or at least what observations there are of the difference in temperature between warm decades (e.g. the 1950s) and cold decades (e.g. the 1970s). Making a connection with observations is especially difficult given that in the real world natural variability is conflated with anthropogenic change (global warming), and even the general cooling trend from the 1940s to the 1970s may have had an anthropogenic origin in the emission of aerosols; thus, any agreement or disagreement between model and observations regarding decadal variability must be regarded with caution.

Other, but not all, coupled climate models have displayed a qualitatively similar variability. For example, using the Hadley Centre coupled ocean–atmosphere Climate Model (HadCM3) Shaffrey & Sutton (2006) also found decadal climate variability associated with, and possibly caused by, fluctuations in the MOC of the Atlantic. Shaffrey and Sutton noted that the meridional heat transport in the atmosphere was, on decadal time scales and in mid- and high-latitudes, significantly anti-correlated with the fluctuations in transport in the ocean, suggesting that the atmosphere may be compensating for the long-term changes in the ocean heat transport, a phenomenon also found by Magnusdottir & Saravanan (1999) and an issue we revisit in the next subsection.

1.6.2 An idealised coupled model

The difficulty of understanding decadal-scale oscillations in the ocean–atmosphere system is essentially as follows. Fully coupled GCMs, which one might expect to provide a reasonably comprehensive and perhaps realistic climate simulation, are very complex and are difficult to understand in themselves. If anything, they have become experimental tools, and rather cumbersome experimental tools at that, rather than theoretical tools that provide some conceptual understanding. The main other line of investigation has been to use quite simple models, including ocean-only models, linear analyses and (with somewhat more complexity) ocean models coupled to energy-balance atmospheric models. However, these are all potentially unrealistic in their response. This duality of approaches is not ideal, and ideally one would like there to be a continuous spectrum of models from the very simple to the very complex; indeed, the notion of a hierarchy of models has become something of a cliché, often talked about but rarely implemented. However, ideas and sayings often become clichés because they are *good* ideas and sayings; a bad idea is usually ultimately forgotten, at least in science.

In an attempt to fill in part of the spectrum, that part adjacent to but simpler than comprehensive GCMs, Farneti & Vallis (2009) constructed a coupled ocean–atmospheric model with two general simplifications.

(i) The geometry was much simplified, having a single, two-hemisphere sector ocean with a periodic channel near the southern edge, crudely representing the Antarctic Circumpolar Channel.
(ii) The 'physics' of the atmospheric model was much simplified, having no interactive clouds and a simple, semi-grey radiation scheme. The vertical resolution (seven levels) is also less than is typically used in comprehensive GCMs.

Both the ocean and atmospheric models solve the three-dimensional primitive equations, and include salinity and moisture, with the only external forcing being the incoming solar radiation at the top of the atmosphere. The advantages over a comprehensive GCM lie in both the computational economy (so allowing multiple numerical experiments in clean settings to be performed) and in the relative ease of interpretation of the results. Nevertheless, such a model remains very complex and difficult to understand.

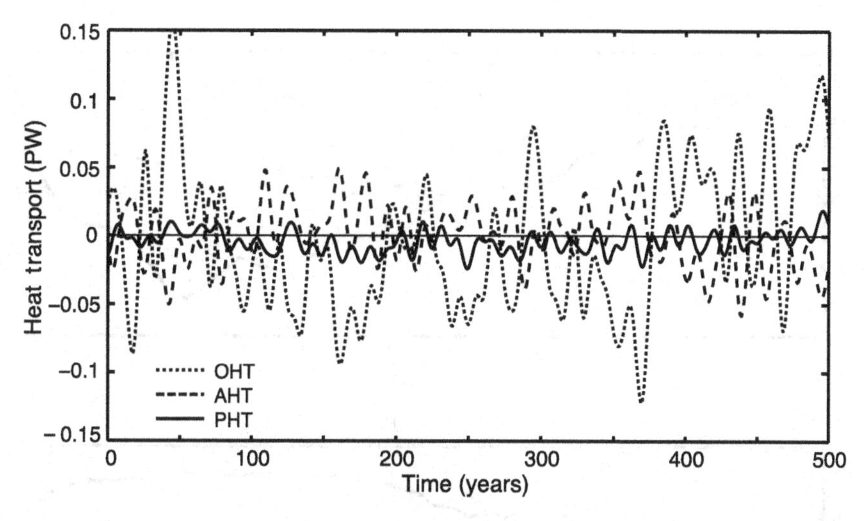

Figure 1.8 The meridional energy transport of the Farneti–Vallis model, averaged from 20° N to 70° N, and low-pass filtered to show only decadal variability; OHT (dotted line) is ocean heat transport, AHT (dashed line) is atmospheric heat transport (sensible plus latent) and PHT (solid) is the planetary (atmospheric plus oceanic) heat transport. The variation in PHT is much less than that of either atmosphere or ocean. (Because of the spatial and temporal averaging, PHT is not exactly the sum of AHT and OHT.)

Consistent with results from a number of other more comprehensive GCMs, Farneti and Vallis found decadal-to-multidecadal oscillations involving the overturning circulation; some illustrative results are given in Figs. 1.8 and 1.9. They find that, first, decadal oscillations are primarily oceanic, in that the time scales, memory and predictability arise from the ocean dynamics; and second that this oceanic variability is indeed imprinted on the atmosphere. Two lines of evidence support the oceanic origin of the variability:

(i) The meridional heat flux of the ocean–atmosphere system is positively correlated with that of the ocean, and negatively correlated with that of the atmosphere, on decadal time scales. This does suggest that the ocean is the driver and the pacemaker for the variability, and the atmosphere is compensating. On annual and shorter time scales the total heat flux is correlated with that of the atmosphere, reflecting the variability ultimately caused by baroclinic eddies (including NAO-type phenomena) in the atmosphere. Of course, these conclusions should be tempered with the usual caveats that correlation does not imply causation. Nevertheless, the interpretation that the ocean is the pacemaker of variability on time scales longer than a year or two, but the atmosphere itself is primarily responsible for interannual and shorter-term variability, is the most straightforward one to draw in the absence of other evidence.

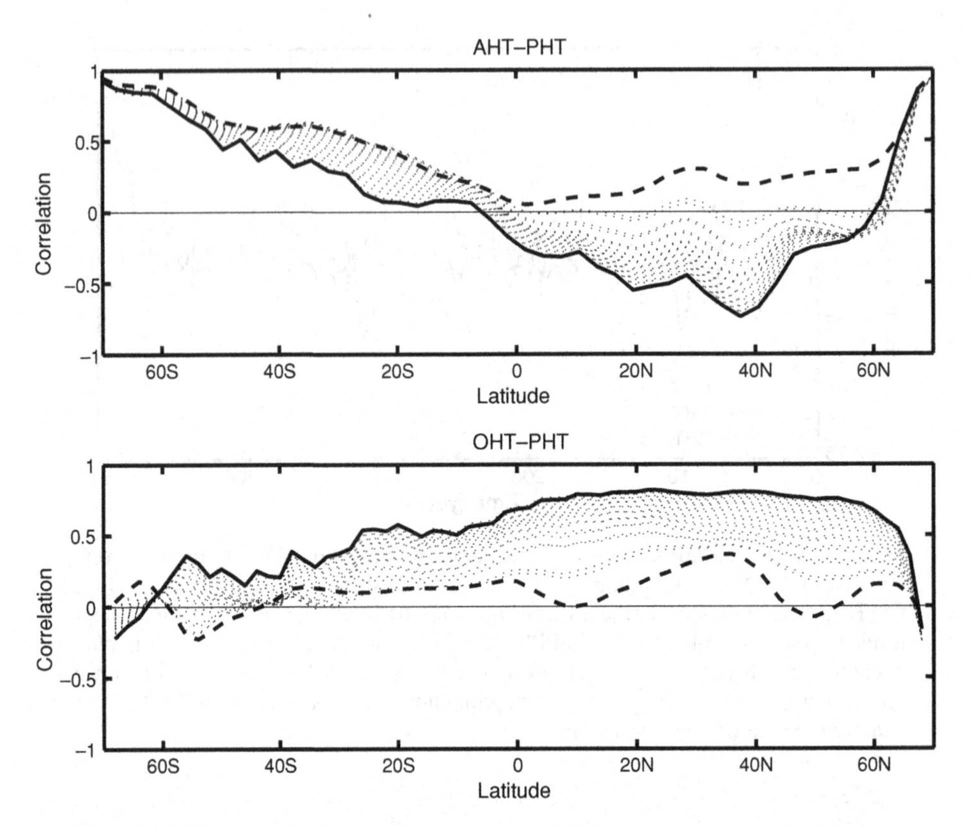

Figure 1.9 The correlation between atmospheric heat transport and planetary (atmospheric plus oceanic) heat transport (top panel) and the correlation between oceanic heat transport and planetary heat transport (bottom panel), calculated using a multi-century integration of the idealized coupled model of Farneti and Vallis. The solid line in each panel shows the correlation after the time series has been filtered to allow time scales of 20 years and longer, and the dashed lines the correlation at successively shorter time scales. The end member of the dashed lines allows time scales of one year and longer.

(ii) Atmosphere-only models do not display significant variability at time scales longer than the interannual (Fig. 1.10). Thus, any process similar to that found in James & James (1992) either does not operate in this model or is insignificant compared to the variability arising from the ocean.

However, the atmosphere does play a role in the oscillations, in two regards.

(i) A catalytic aspect, which seems the most important, is that an active atmosphere reduces the damping felt by the ocean, as described in Section 1.4.4. That is, an anomalously warm ocean leads to an anomalously warm atmosphere, which damps the ocean less than would a fixed atmosphere and allows it to oscillate. The effect is not required in all parameter regimes; thus, with a high diapycnal diffusivity the ocean spontaneously

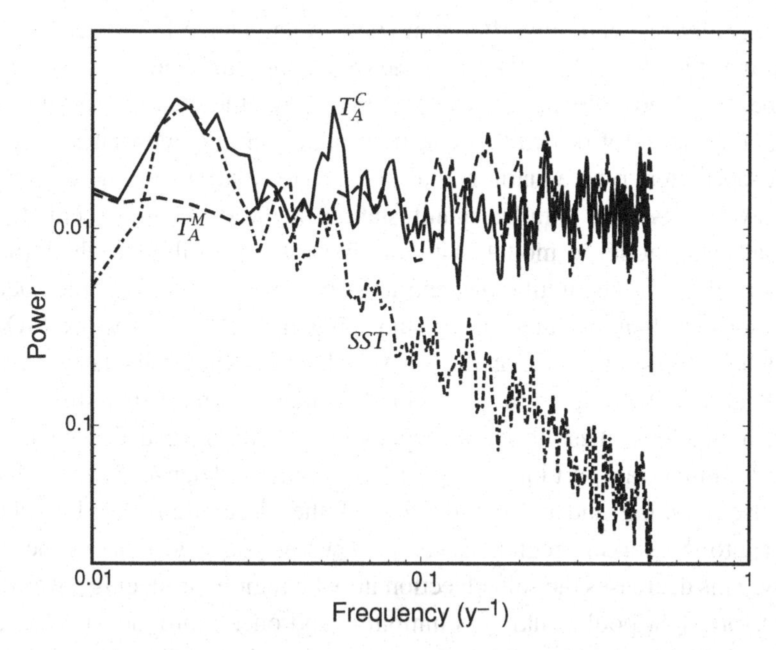

Figure 1.10 Power spectra of atmospheric temperature at 450 mb in a fully coupled integration (solid, T_A^C) and an integration in which an atmosphere is coupled to a mixed-layer ocean (dashed, T_A^M), using the Farneti–Vallis model. The dot-dashed curve shows the spectrum of the SST in the subpolar region in the coupled model. The atmosphere–mixed-layer (T_A^M) simulation shows much less variability on decadal time scales than does the atmosphere–ocean coupled model (T_A^C).

oscillates even with a constant atmosphere, whereas at very low diapycnal diffusivity the oscillations are smaller, even with an interactive atmosphere.

(ii) An active aspect is that stochastic forcing from the atmosphere seems, in part, to enable the oceanic oscillations. Thus, given an ocean-only simulation that evolves into a steady state, the addition of daily fluxes taken from a coupled-simulation (but with the long-time variability removed so that the fluxes do represent only 'weather noise') gives rise again to decadal-scale variability in the ocean's MOC. This is suggestive of a damped oscillator that may be excited by noise, much as in the box model of Griffies & Tziperman (1995). However, the oceanic variability is relatively weak compared to that in a fully coupled model.

1.6.3 Robustness and mechanisms

The mechanisms of decadal oceanic oscillations remain somewhat obscure, except in a broad sense. Ultimately, oscillations seem to arise because of a delay in the response of the MOC and the associated transport of heat to the meridional temperature gradient (Huck *et al.* 2001). If one posits that the heat transport of the

MOC will fully respond only after some years to a change in temperature gradient, and that the change in MOC will then transport sufficient heat meridionally to compensate, and potentially overcompensate, for the changes in temperature gradient, then it is not difficult to construct, albeit on a somewhat ad hoc basis, 'delayed-oscillator' type equations that produce oscillations on the decadal time scale. Nevertheless, the detailed manifestation of the oscillation seems to differ somewhat from model to model, with varying roles for salinity, which plays an important role in some simulations and less so in others. For example, Delworth *et al.* described their mechanism by supposing first there is a weak MOC; the reduced heat transport associated with this leads, some years later, to a cold North Atlantic extending from the surface to about 1 km. This appears to generate an anomalous quasi-horizontal cyclonic circulation, and this results in an enhanced salinity transport into the region. The enhanced density that results leads to increased sinking and a strengthening of the circulation that, by transporting heat into the sinking region, develops a warm pool and reduces the density anomaly. This decreases the salt advection into the region, producing a warm, fresh and therefore light pool. Sinking is inhibited, and once more the MOC weakens. Selten *et al.* (1999) also described a mechanism in which salinity anomalies were important, using a simple dynamical model. In their model the forcing by the atmosphere, in particular the NAO pattern, was responsible for producing an initial SST anomaly, following which ocean dynamics was responsible for producing a decadal oscillation. Their mechanism seems to involve shallower dynamics than those in the Geophysical Fluid Dynamics Laboratory (GFDL) model but, like the GFDL result, the subsequent response of the atmosphere to SST anomalies does not seem essential in maintaining the oscillation and the mechanism may be thought of as essentially oceanic. The role of salinity in these oscillations suggests a connection to the so-called 'great salinity anomalies' (GSAs; Dickson *et al.* 1988; Walsh & Chapman 1990), which, through their effect on the meridional overturning circulation, may affect the path of the Gulf Stream and so have an influence on climate on decadal time scales (Zhang & Vallis 2006), but the mechanisms of the genesis of observed GSA events remains obscure.

In the more idealised integrations of Farneti & Vallis (2009) the dynamical mechanism is also primarily oceanic, except that by having an interactive atmosphere the reduced damping allows the ocean to oscillate more readily than it might if the forcing on the ocean were held fixed. Indeed, Huck & Vallis (2001) suggest that the oscillations might be the manifestation of a linear instability of the three-dimensional overturning circulation, and depending on the boundary conditions applied to the ocean model, they find that the structure of the linearly unstable modes resembles that of the oscillation found in the forward integration of a fully nonlinear model.

The presence of oscillations does seem fairly robust to model formulation and parameters, having appeared across a variety of models from the comprehensive to the very idealised (Greatbatch & Zhang 1995; Winton 1996; Huck *et al.* 2001), but the period and details vary considerably from model to model. The presence of realistic bottom topography also seems to have a damping effect (Winton 1997), and as noted previously the oscillations are also sensitive to the value of the diapycnal diffusivity and to a thermally interactive atmosphere. As a consequence, no clearly accepted single explanation has emerged that is akin to baroclinic instability as the cause of weather in the atmosphere. Can we expect one to do so? I am optimistic; we are now rather at an analogous stage in development as occurred in the 1950s with regard to the understanding of rotating annuli, mapping out regimes of flow. Increasing computer power will make the mapping easier, although the theoretical challenge of understanding such a complex system remains daunting.

1.7 Discussion

Natural climate variability remains a rather poorly understood phenomenon, compared for example with our understanding of weather variability for which the underlying mechanism – namely baroclinic instability and its nonlinear consequences – is relatively well understood.

In what follows I'll set out some conclusions about the matter, trying to be clear about what is generally accepted and what is not. First of all, over the past few decades of research it has become clear that on time scales of a year or less the atmosphere is the dominant driver of variability. The dominant extratropical modes of variability – the North Atlantic Oscillation and related phenomena such as the variability in the zonal index and the 'annular modes' (both in the Northern Hemisphere and Southern Hemisphere) – are uncontroversially primarily atmospheric in origin. They are related to the variability of the major storm tracks and to the stationary wave pattern, and although the mechanisms of such variability are still not fully understood [see Wallace (2000); Feldstein (2003) and Vallis & Gerber (2008) and references therein for discussions], and may be influenced both by the ocean and the stratosphere, the troposphere itself is almost certainly the primary origin of the variability. There are still nevertheless a number of weaknesses in our understanding: notably in the precise mechanisms that seem to give rise to variability on time scales rather longer than that of a baroclinic lifecycle and the possible influences of the stratosphere and the ocean; as well as the mechanisms that give rise to an oscillation (or at least a variability) that is in some ways more prominent in the North Atlantic than the North Pacific.

It seems likely that it is this atmospheric variability that gives rise to variability in the sea-surface temperature on time scales of up to a year, as was discussed in

Section 1.5.2; if so, it is likely that there is very little predictability in this process. What to the meteorologist is a signal, namely the atmospheric variability on daily to monthly time scales, becomes to the oceanographer a source of noise. The oceanic mixed layer integrates this noise, generating sea-surface temperature anomalies that might be quite long lived and of large scale. This integration is perhaps more usefully described as suppression of fast-time variability, so reddening the spectrum of the near-surface temperature but providing no genuinely new predictable source of information on longer time scales aside from that due to persistence. That there is no additional information coming in on long times is apparent when one realises that white noise is essentially the same as a sequence of random numbers, each uncorrelated with its neighbour; subsequently applying a low-pass filter to give red noise plainly cannot induce new predictability, apart from that associated with the suppression of noise.

Whether significant variability on time scales of much longer than a year can arise purely from an atmosphere, whether with fixed SSTs, an oceanic mixed layer or still more idealised forcing seems, on balance, unlikely. In order for this to happen the atmospheric dynamics would have to produce regimes of flow that persisted longer than any natural time scales in the system. Of course, such behaviour *is* known to happen in some dynamical systems, and it does seem to happen in the simulations of James & James (1992), so the situation is not wholly settled. Jet merger and separation is one process that can demonstrably produce regimes with long time scales (Panetta 1993; Vallis & Maltrud 1993), especially when dynamics lie on the border between one and two eastward jets, but if this were the case it would seem that the seasonal cycle would still prevent multiannual time scales from emerging. Evidently, further investigation is called for.

The ocean certainly has dynamical time scales of decades, centuries and more, and ocean models seem to be able to oscillate, fairly robustly, on decadal and longer time scales. Whether the ocean is responsible, in whole or in part, for climate variability on these time scales depends on two things: (i) the existence of variability in the sea-surface temperature on these time scales; and (ii) this variability affecting the atmosphere. Regarding the latter, results from most GCM experiments seem to show a weak but persistent effect; the subsequent feedback from the atmosphere to the ocean seems weak, effective mainly in reducing the damping felt by the ocean. (In this sense the oscillations may be thought as weakly coupled, with the time scales and memory residing in the ocean.) Still, the results are not wholly uniform from GCM to GCM and it may still be possible that GCMs underestimate the effects of SST anomalies if their boundary-layer schemes are inadequate. In any case, it is on the decadal and longer time scale that the oceanic effects are more likely to be felt, because now the oceanic signal may be felt over the atmospheric noise. On these time scales the main robust atmospheric response may simply be a warming at latitudes of anomalously warm ocean, so affecting the storm tracks and producing weak but potentially predictable behaviour in such indices as the NAO or Southern Annular Mode.

Finally, let me make a comment on the relation of natural variability on the decadal–centennial time scale to anthropogenic global warming. Two lines of evidence argue *against* natural variability being the cause of the global warming seen in the last century. The first is that the variability produced by most models of the climate system is much smaller than the warming observed. Over a given decade or conceivably over a few decades the natural variability may match or even outweigh the trend, and so we should not be surprised if in the future sometime a cool decade occurs. But over the course of half a century or longer, the natural variability is and will remain (we assume) too small to overcome the current and predicted global warming. The second is that there is no evidence that there has been a flux of heat from the ocean to the atmosphere (Levitus *et al.* 2001; Barnett *et al.* 2001); if anything, there has been a flux of heat *into* the ocean, implying that the ocean is warming *because* of a radiative imbalance and not that the ocean is producing the warming by fluxing heat into the atmosphere.

Acknowledgements

I am grateful to Riccardo Farneti for many discussions on this topic, and to the National Science Foundation (NSF) and the National Oceanic and Atmospheric Administration (NOAA) for funding.

References

Barnett, T., Pierce, D. W. & Schnur, R. 2001 Detection of anthropogenic climate change in the world's oceans. *Science*, **292**, 270–273.

Barsugli, J. J. & Battisti, D. S. 1998 The basic effects of atmosphere–ocean thermal coupling on midlatitude variability. *J. Atmos. Sci.*, **55**, 477–493.

Biondi, F., Gershunov, A. & Cayan, D. R. 2001 North Pacific decadal climate variability since 1661. *J. Climate*, **14**, 5–10.

Bretherton, C. S. & Battisti, D. S. 2000 An interpretation of the results from atmospheric general circulation models forced by the time history of the observed sea surface temperature distribution. *Geophys. Res. Lett.*, **27**, 767–770.

Brohan, P., Kennedy, J. J., Harris, I., Tett, S. F. B. & Jones, P. D. 2006 Uncertainty estimates in regional and global observed temperature changes: a new dataset from 1850. *J. Geophys. Res.*, **111**, D12106 (doi:10.1029/2005JD00654).

Cayan, D. R. 1992 Latent and sensible heat flux anomalies over the northern oceans: driving the sea surface temperature. *J. Phys. Oceanogr.*, **22**, 859–881.

Cessi, P. 2000 Thermal feedback on wind stress as a contributing cause of climate variability. *J. Climate*, **13**, 232–244.

Chang, E. K. M., Lee, S. & Swanson, K. L. 2002 Storm track dynamics. *J. Climate*, **15**, 2163–2183.

Czaja, A. & Frankignoul, C. 2002 Observed impact of Atlantic SST anomalies on the North Atlantic Oscillation. *J. Climate*, **15**, 606–623.

Davis, R. E. 1976 Predictability of sea surface temperature and sea level pressure anomalies over the North Pacific Ocean. *J. Phys. Oceanogr.*, **6**, 249–266.

Delworth, T., Manabe, S. & Stouffer, R. 1993 Interdecadal variations of the thermohaline circulation in a coupled ocean–atmosphere model. *J. Climate*, **6**, 1993–2011.

Deser, C., Magnusdottir, G., Saravanan, R. & Phillips, A. 2004 The effects of North Atlantic SST and sea ice anomalies on the winter circulation in CCM3. Part II: direct and indirect components of the response. *J. Climate*, **17**, 877–889.

Dickson, R. R., Meincke, J., Malmberg, S.-A. & Lee, A. J. 1988 The 'Great Salinity Anomaly' in the northern North Atlantic 1968–1982. *Prog. Oceanogr.*, **20**, 103–151.

Dommenget, D. & Latif, M. 2002 Analysis of observed and simulated SST spectra in the midlatitudes. *Clim. Dyn.*, **19**, 277–288.

Farneti, R. & Vallis, G. K. 2009 An intermediate complexity climate model based on GFOL's flexible modeling system. *Geosci. Mod. Development*, **2**, 73–88.

Feldstein, S. B. 2000 Is interannual zonal mean flow variability simply climate noise? *J. Climate*, **13**, 2356–2362.

Feldstein, S. B. 2003 The dynamics of NAO teleconnection pattern growth and decay. *Q. J. R. Meteorol. Soc.*, **129**, 901–924.

Feldstein, S. B. & Lee, S. 1998 Is the atmospheric zonal index driven by an eddy feedback? *J. Atmos. Sci.*, **55**, 3077–3086.

Folland, C. K., Parker, D. E., Colman, A. W. & Washington, R. 1999 Large-scale modes of ocean surface temperature since the late nineteenth century. In *Beyond El Niño*, ed. A. Navarra, pp. 73–102. Springer.

Frankignoul, C. 1985 Sea surface temperature anomalies, planetary waves, and air–sea feedback in the middle latitudes. *Rev. Geophys.*, **23**, 357–390.

Frankignoul, C. & Hasselmann, K. 1977 Stochastic climate models, part II. Application to sea-surface temperature anomalies and thermocline variability. *Tellus*, **29**, 289–305.

Gerber, E. P. & Vallis, G. K. 2007 Eddy–zonal flow interactions and the persistence of the zonal index. *J. Atmos. Sci.*, **64**, 3296–3311.

Greatbatch, R. J. & Zhang, S. 1995 An interdecadal oscillation in an idealized ocean basin forced by constant heat flux. *J. Climate*, **8**, 81–91.

Griffies, S. M. & Tziperman, E. 1995 A linear thermohaline oscillator driven by stochastic atmospheric forcing. *J. Climate*, **8**, 2440–2453.

Hartmann, D. L. & Lo, F. 1998 Wave-driven zonal flow vacillation in the Southern Hemisphere. *J. Atmos. Sci.*, **55**, 1303–1315.

Hasselmann, K. 1976 Stochastic climate models: Part I, Theory. *Tellus*, **28**, 473–485.

Held, I. M. 1983 Stationary and quasi-stationary eddies in the extratropical troposphere: theory. In *Large-scale Dynamical Processes in the Atmosphere*, ed. B. Hoskins & R. P. Pearce, pp. 127–168. Academic Press.

Hoskins, B. J. & Karoly, D. J. 1981 The steady linear response of a spherical atmosphere to thermal and orographic forcing. *J. Atmos. Sci.*, **38**, 1179–1196.

Huck, T. & Vallis, G. K. 2001 Linear stability analysis of the three-dimensional thermally driven ocean circulation: application to interdecadal oscillations. *Tellus*, **53A**, 526–545.

Huck, T., Vallis, G. K. & de Verdière, A. C. 2001 On the robustness of interdecadal modes of the thermohaline circulation. *J. Climate*, **14**, 940–963.

James, I. N. & James, P. M. 1992 Spatial structure of ultra-low frequency variability of the flow in a simple atmospheric circulation model. *Q. J. R. Meteorol. Soc.*, **118**, 1211–1233.

Kaplan, A., Kane, M. A., Kushnir, Y. *et al.* 1998 Analysis of global sea-surface temperature 1856–1991. *J. Geophys. Res.*, **103**, 18567–18589.

Kushnir, Y. & Held, I. 1996 Equilibrium atmospheric response to North Atlantic SST anomalies. *J. Climate*, **9**, 1208–1220.

Kushnir, Y., Robinson, W. A., Bladé, I. *et al.* 2002 Atmospheric GCM response to extratropical SST anomalies: synthesis and evaluation. *J. Climate*, **15**, 2233–2256.

Latif, M. & Barnett, T. 1994 Causes of decadal climate variability over the North Pacific and North America. *Science*, **266**, 634–637.

Latif, M. & Barnett, T. 1996 Decadal climate variability over the North Pacific and North America: dynamics and predictability. *J. Climate*, **10**, 219–239.

Lau, N.-C. & Nath, M. J. 1994 A modeling study of the relative roles of tropical and extratropical SST anomalies in the variability of the global atmosphere–ocean system. *J. Climate*, **7**, 1184–1207.

Levitus, S., Antonov, J. I., Wang, J. *et al.* 2001 Anthropogenic warming of Earth's climate system. *Science*, **292**, 267–270.

Lunkeit, F. & von Detten, Y. 1997 The linearity of the atmospheric response to North Atlantic sea surface temperature anomalies. *J. Climate*, **10**, 3003–3014.

Magnusdottir, G. & Saravanan, R. 1999 The response of atmospheric heat transport to zonally averaged SST trends. *Tellus A*, **51**, 815–832.

Mann, M. E. & Park, J. 1994 Global scale modes of surface temperature variability on interannual to century time scales. *J. Geophys. Res.*, **99**, 25819–25833.

Panetta, R. L. 1993 Zonal jets in wide baroclinically unstable regions: persistence and scale selection. *J. Atmos. Sci.*, **50**, 2073–2106.

Parker, D. E., Legg, T. P. & Folland, C. K. 1992 A new daily central England temperature series. *Int. J. Climate*, **12**, 317–342.

Robertson, A. W., Mechoso, C. R. & Kim, Y. J. 2000 The influence of Atlantic sea surface temperature anomalies on the North Atlantic Oscillation. *J. Climate*, **13**, 122–138.

Robinson, W. A. 2000 A baroclinic mechanism for eddy feedback on the zonal index. *J. Atmos. Sci.*, **57**, 415–422.

Rodwell, M. J., Rowell, D. P. & Folland, C. K. 1999 Oceanic forcing of the wintertime North Atlantic Oscillation and European climate. *Nature*, **398**, 320–323.

Roe, G. & Baker, M. 2007 Why is climate sensitivity so unpredictable? *Science*, **318**, 629–632.

Schopf, P. S. 1985 Modeling tropical sea-surface temperature: implications of various atmospheric responses. In *Coupled Ocean–Atmosphere Models*, ed. J. C. J. Nihoul; pp. 727–734. Elsevier.

Selten, F. M., Haarsma, R. J. & Opsteegh, J. D. 1999 On the mechanism of North Atlantic decadal variability. *J. Climate*, **12**, 1956–1973.

Shaffrey, L. & Sutton, R. 2006 Bjerknes compensation and the decadal variability of the energy transports in a coupled climate model. *J. Climate*, **19**, 1167–1181.

Simmons, A. & Hoskins, B. 1978 The life-cycles of some nonlinear baroclinic waves. *J. Atmos. Sci.*, **35**, 414–432.

Tourre, Y., Rajagopalan, B. & Kushnir, Y. 1999 Dominant patterns of climate variability in the Atlantic Ocean during the last 136 years. *J. Climate*, **12**, 2285–2299.

Trenberth, K. E. & Caron, J. M. 2001 Estimates of meridional atmosphere and ocean heat transports. *J. Climate*, **14**, 3433–3443.

Vallis, G. K. 2006 *Atmospheric and Oceanic Fluid Dynamics*. Cambridge University Press.
Vallis, G. K. & Gerber, E. P. 2008 Local and hemispheric dynamics of the North Atlantic Oscillation, annular patterns and the zonal index. *Dyn. Atmos. Oceans*, **44**, 184–212.
Vallis, G. K. & Maltrud, M. E. 1993 Generation of mean flows and jets on a beta plane and over topography. *J. Phys. Oceanogr.*, **23**, 1346–1362.
Vallis, G. K., Gerber, E. P., Kushner, P. J. & Cash, B. A. 2004 A mechanism and simple dynamical model of the North Atlantic Oscillation and annular modes. *J. Atmos. Sci.*, **61**, 264–280.
Wallace, J. M. 2000 North Atlantic Oscillation/Annular Mode: two paradigms – one phenomenon. *Q. J. R. Meteorol. Soc.*, **126**, 791–805.
Walsh, J. E. & Chapman, W. L. 1990 Arctic contribution to upper-ocean variability in the North Atlantic. *J. Climate*, **3**, 1462–1473.
Webster, P. J. 1981 Mechanisms determining the atmospheric response to sea surface temperature anomalies. *J. Atmos. Sci.*, **38**, 554–571.
Winton, M. 1996 The role of horizontal boundaries in parameter sensitivity and decadal scale variability of coarse-resolution ocean general circulation models. *J. Phys. Oceanogr.*, **26**, 289–304.
Winton, M. 1997 The damping effect of bottom topography on internal decadal-scale oscillations of the thermohaline circulation. *J. Phys. Oceanogr.*, **267**, 203–208.
Winton, M. 2003 On the climatic impact of ocean circulation. *J. Climate*, **16**, 2875–2889.
Zhang, R. & Vallis, G. K. 2006 Impact of great salinity anomalies on the low-frequency variability of the North Atlantic climate. *J. Climate*, **19**, 470–481.

2

Empirical model reduction and the modelling hierarchy in climate dynamics and the geosciences

SERGEY KRAVTSOV, DMITRI KONDRASHOV AND MICHAEL GHIL

Modern climate dynamics uses a two-fisted approach in attacking and solving the problems of atmospheric and oceanic flows. The two fists are: (i) observational analyses; and (ii) simulations of the geofluids, including the coupled atmosphere–ocean system, using a hierarchy of dynamical models. These models represent interactions between many processes that act on a broad range of spatial and time scales, from a few to tens of thousands of kilometers, and from diurnal to multidecadal, respectively. The evolution of virtual climates simulated by the most detailed and realistic models in the hierarchy is typically as difficult to interpret as that of the actual climate system, based on the available observations thereof. Highly simplified models of weather and climate, though, help gain a deeper understanding of a few isolated processes, as well as giving clues on how the interaction between these processes and the rest of the climate system may participate in shaping climate variability. Finally, models of intermediate complexity, which resolve well a subset of the climate system and parameterise the remainder of the processes or scales of motion, serve as a conduit between the models at the two ends of the hierarchy.

We present here a methodology for constructing intermediate models based almost entirely on the observed evolution of selected climate fields, without reference to dynamical equations that may govern this evolution; these models parameterise unresolved processes as multivariate stochastic forcing. This methodology may be applied with equal success to actual observational data sets, as well as to data sets resulting from a high-end model simulation. We illustrate this methodology by its applications to: (i) observed and simulated low-frequency variability of atmospheric flows in the Northern Hemisphere; (ii) observed evolution of tropical sea-surface temperatures; and (iii) observed air–sea interaction in the Southern Ocean. Similar results have been obtained for (iv) radial-diffusion model simulations of Earth's radiation belts, but are not included here because of space restrictions. In each case, the reduced stochastic model represents surprisingly well a variety of linear and nonlinear statistical properties of the resolved fields. Our methodology thus provides an efficient means of constructing reduced, numerically inexpensive climate models. These models can be thought of as stochastic–dynamic prototypes of more complex deterministic models, as in examples (i) and (iv), but work just as well in the situation when the actual governing equations are poorly known, as in (ii) and (iii). These models can serve as competitive prediction tools, as in (ii), or be included as stochastic parameterisations of certain processes within more complex climate models, as in (iii). Finally, the methodology can be applied, with some modifications, to geophysical problems outside climate dynamics, as illustrated by (iv).

2.1 Introduction

Comprehensive general circulation models (GCMs) are governed by a nonlinear set of partial differential equations that, given the climate's state at an initial time t_0, predict the climatic fields at future times $t_0 + k\Delta t$. These climatic fields – such as atmospheric winds, temperature and humidity, oceanic currents, temperatures and salinities, sea-ice areas and concentrations, among others – are discretised on a spatial grid spanning the entire volume of the Earth's fluid envelope. The governing equations so discretised are but approximations to the original partial differential equations, and are complemented by a large number of semi-empirical relations that connect locally the large-scale fields with subgrid-scale processes, like clouds and radiation. In the absence of the small-scale processes, these approximations could become progressively more accurate as the grid size Δx and time step Δt tend to zero in a judicious manner. Limited computer power poses restrictions on the minimal grid size and, therefore, on the maximal dimension D of the grand climate-state vector, $\mathbf{X} \equiv (X_1, X_2, \ldots, X_D)$; there are also restrictions on the appropriate treatment of the small-scale processes as the grid size is progressively reduced.

While D is finite, it is still very large. To facilitate analyses of climatic variability, it is reasonable to use data compression techniques, such as empirical orthogonal function (EOF) analysis, also known as principal component (PC) analysis; see Preisendorfer (1988). Suppose that we have archived a long simulation of a climate model, $\{\mathbf{X}^{(n)}\}$, $n = 1, 2, \ldots, N$, as an $N \times D$ array, whose d-th column represents N consecutive values of a climate variable, X_d: $(X_d^{(1)}, X_d^{(2)}, \ldots, X_d^{(N)})$. Assuming that the climate is stationary, we first subtract from each of our D variables their time-mean values, thus redefining the climate state in terms of anomalies $X_d^{(n)} \to X_d^{(n)} - N^{-1}\sum_{n=1}^{N} X_d^{(n)}$. The climate-state anomaly vector $\mathbf{X}$ will thus have zero time mean.

Let us now define a new time series, $\mathbf{x}_1 \equiv (x_1^{(1)}, x_1^{(2)}, \ldots, x_1^{(N)})$, as a weighted average of D original time series: $x_1^{(n)} = e_1^{(1)} X_1^{(n)} + e_2^{(1)} X_2^{(n)} + \cdots + e_D^{(1)} X_D^{(n)}$, where the weights $e_d^{(1)}$ are chosen to maximise the variance of the $\mathbf{x}_1$ time series, subject to the normalisation constraint $\mathbf{e}^{(1)} \cdot \mathbf{e}^{(1)} \equiv e_1^{(1)} e_1^{(1)} + e_2^{(1)} e_2^{(1)} + \cdots + e_D^{(1)} e_D^{(1)} = 1$, where $\mathbf{a} \cdot \mathbf{b}$ denotes the inner product of the vectors $\mathbf{a}$ and $\mathbf{b}$. The time series $\mathbf{x}_1$ so obtained is called the leading PC, while the set of weights $\mathbf{e}^{(1)}$ represents a spatial pattern referred to as the leading EOF of our multivariate climate field. One can compute the second PC–EOF pair, $\mathbf{x}_2$, $\mathbf{e}^{(2)}$, by maximising the variance of $\mathbf{x}_2$ subject to the normalisation constraint $\mathbf{e}^{(2)} \cdot \mathbf{e}^{(2)} = 1$ and the additional orthogonality constraint $\mathbf{e}^{(1)} \cdot \mathbf{e}^{(2)} = 0$. Higher-order PCs and EOFs, up to D, can be found in a similar way, with $\mathbf{e}^{(d)}$ computed to satisfy d orthonormality constraints. In other areas of fluid dynamics, in particular in turbulence theory, this statistical approach goes by the name of proper orthogonal decomposition (Tennekes & Lumley 1972). Thus, the multivariate climate field can be decomposed into the sum of orthonormal spatial patterns $\mathbf{e}^{(d)}$, or EOFs, whose corresponding time series $\mathbf{x}_d$, or PCs, turn out to be orthogonal, while their variances decrease monotonically with d.

The latter property has important consequences, since typically a limited number M of leading EOFs, $M \ll D$, accounts for a major fraction of climate variance. It thus appears reasonable to think of climate evolution in terms of variability associated with leading EOF modes. A few leading EOF modes are sometimes referred to as *teleconnection* patterns (Wallace & Gutzler 1981), since they reflect correlations between variables at spatial locations separated by distances much larger than the typical decorrelation radius of atmospheric fields. Our description of the EOF analysis above is intentionally oversimplified. Even if it were computationally feasible to apply this statistical method directly to the full climate-state vector of a comprehensive GCM, many technical challenges would have to be addressed: treatment of physically and thus dimensionally distinct variables, for example, wind and temperature; choice of optimal inner product in computing covariances while respecting quadratic invariants, like energy or enstrophy and several others. A more comprehensive coverage of these issues is beyond the scope of the present discussion.

Let us consider multivariate dynamical equations with quadratic nonlinearity as a prototype of the discretised equations describing climate evolution (Lorenz 1963; Ghil & Childress 1987, ch. 5):

$$\dot{\mathbf{X}} = \hat{\mathbf{F}} + \hat{\mathbf{L}}\mathbf{X} + \hat{\mathbf{N}}(\mathbf{X}, \mathbf{X}); \tag{2.1}$$

here $\mathbf{X}$ is a D-dimensional state vector and the dot denotes the time derivative. Denoting the matrix transpose by superscript T and substituting the EOF decomposition

$$\mathbf{X} = \mathbf{x} \cdot \mathbf{e}^{\mathrm{T}} \tag{2.2}$$

into (2.1), multiplying these equations on the right by $\mathbf{e}$ and using the orthonormality of the EOFs, we rewrite the governing equations in the EOF basis (Schubert 1985; Selten 1995, 1997; Kwasniok 1996, 2004; Branstator & Haupt 1998; Achatz & Branstator 1999; D'Andrea & Vautard 2001; Achatz & Opsteegh 2003a,b; Franzke *et al.* 2007):

$$\dot{\mathbf{x}} = \tilde{\mathbf{F}} + \tilde{\mathbf{L}}\mathbf{x} + \tilde{\mathbf{N}}(\mathbf{x}, \mathbf{x}) \tag{2.3a}$$

or, componentwise,

$$\dot{x}_i = \tilde{F}_i + \tilde{L}_{ij}x_j + \tilde{N}_{ijk}x_j x_k; \tag{2.3b}$$

the repeated indices, throughout the chapter, imply summation, and i, j, k vary from 1 to D.

Since a moderate number, M, of leading EOFs account for a major fraction of climate-state vector variance, it is natural to consider only the first M of (2.3b), while neglecting in them all the terms that involve higher-order PCs, x_i, $i > M$. This is the simplest approach to obtaining a reduced climate model and it is called the bare-truncation model. It turns out, however, that while the trailing EOF modes may not account for a large fraction of variance, their interaction with the resolved modes over time is important for the dynamics of the resolved modes, as the bare-truncation models typically experience systematic biases. These biases can be partially corrected for empirically by introducing ad hoc linear damping terms of the form $-r_{ij}\, x_j$ to parameterise the neglected interactions between the resolved and unresolved modes (Selten 1995; Achatz & Branstator 1999). Strounine *et al.* (2008) advance this idea one step further by combining it with data assimilation methods (Dee *et al.* 1985; Kondrashov *et al.* 2008) and estimating, in addition to the linear damping coefficients, the parameters of additive 'random-noise' error associated with the statistical linear fit; this random term is subsequently introduced into the reduced model as stochastic forcing. The parameterisation of interactions between the resolved and unresolved modes here involves a stochastic component and increased linear damping, as in Farrell & Ioannou (1993, 1995) but, unlike in these authors' work, retains the quadratic nonlinearity of the bare-truncation equations.

A systematic theoretical approach to model reduction is due to Majda *et al.* (1999, 2001, 2002, 2003, 2006; hereafter MTV). In the limit of significant time scale separation between the fastest resolved and slowest unresolved EOF modes, the MTV procedure derives the functional form of the reduced equations, based on standard projection methods from the theory of stochastic differential equations (Khasminsky 1963; Kurtz 1973; Gardiner 1985). The self-interaction of the fast, unresolved modes is modelled by a stochastic process, and the reduced equations include modified forcing, linear terms, quadratic and cubic nonlinearities, as well as additive and multiplicative noise terms; the latter terms consist of products of stochastic and resolved variables. The reduced-model coefficients are formally predicted by the MTV approach, provided the lag-covariance structure of all unresolved modes is given. Franzke *et al.* (2005) and Franzke & Majda (2006) applied this approach to the analysis of intermediate complexity models describing variability of midlatitude atmospheric flow, but managed to achieve only modest agreement between the statistical characteristics of the full and reduced models. The reason behind this partial success may lie in the nature of climatic EOF spectra: overall, higher-order modes do have shorter time scales, and somewhat smaller spatial scales, than the leading modes, but there is no pronounced time-scale separation between the resolved and unresolved modes, as required by the stochastic-process theory on which the MTV approach to model reduction is based. While Majda and colleagues have shown that the MTV formulation may still be approximately valid for some idealised systems without such a spectral gap (e.g. Majda *et al.* 2008), this is apparently not the case for the prototype models of atmospheric variability considered by Franzke *et al.* (2005) and Franzke & Majda (2006).

The problem of constructing a reduced model that describes key features of the climate system can be addressed in a data-driven, rather than model-driven approach, by using inverse stochastic modelling. This data-driven approach lacks the dynamical appeal inherent in model-based reduction methods; it offers, though, greater practical flexibility by allowing one to work directly with appropriate subsets of climatic fields. The inverse modelling approach does not require either the equations that govern the fields of interest, nor the laws that couple a given climate subsystem to the rest of the system. By the same token, empirical methodologies are in no way restricted by the necessity to explicitly separate between slow and fast dynamics within a climate subsystem of interest.

The simplest type of inverse stochastic model is the so-called linear inverse model (LIM; Penland 1989, 1996; Penland & Ghil 1993), which has the form

$$\mathrm{d}x_i = L_{ij}x_j\,\mathrm{d}t + \mathrm{d}\xi_i(t), \tag{2.4}$$

where $\mathrm{d}\xi_i(t)$, $i = 1, 2, \ldots, M$ is a vector-valued white-noise process characterised by the $M \times M$ noise covariance matrix $\mathbf{Q}$, and $\mathbf{L}$ is the $M \times M$ dynamics matrix, which is assumed to be constant and stable. The LIM procedure aims at finding $\mathbf{Q}$ and $\mathbf{L}$ given observations of a vector-valued time series $\mathbf{x}(t) \equiv [x_1(t), x_2(t), \ldots, x_M(t)]$ that represents, for example, M leading PCs of the field(s) of interest. The matrices $\mathbf{Q}$ and $\mathbf{L}$ in (2.4) satisfy a fluctuation–dissipation relation, which involves $\mathbf{C}(\tau)$, the lag-covariance matrix of the process $\mathbf{x}$ at lag τ (Gardiner 1985):

$$\mathbf{LC}(0) + \mathbf{C}(0)\mathbf{L}^{\mathrm{T}} + \mathbf{Q} = 0. \tag{2.5}$$

The Green's function for (2.4), $\mathbf{G}(\tau) = \exp(\mathbf{L}\tau)$, can be expressed in terms of $\mathbf{C}(\tau)$ as

$$\mathbf{G}(\tau) = \mathbf{C}(\tau)\mathbf{C}^{-1}(0), \tag{2.6}$$

while the optimal forecast of $\mathbf{x}_{\mathrm{f}}(\tau)$ given the initial state $\mathbf{x}(0)$ is

$$\mathbf{x}_{\mathrm{f}}(\tau) = \mathbf{G}(\tau)\mathbf{x}(0). \tag{2.7}$$

Equations (2.5, 2.6) are valid exactly for the stochastic differential equation (2.4). In linear inverse modelling, the true lag-covariance matrix $\mathbf{C}$ is replaced by the sample covariance matrix using the actual available data, while (2.6) can be exploited to estimate $\mathbf{L}$ using different lags τ and thus check whether the linear form of (2.4) is supported by the data (Penland & Ghil 1993; DelSole 2000). Note that the dynamics operator $\mathbf{L}$ in (2.4) is in general different from $\tilde{\mathbf{L}}$ in (2.3), since it represents not only the linear part of the bare-truncation operator, but also parameterises, in a linear fashion and along with the white-noise forcing term, the nonlinear interactions between the resolved modes, as well as linear and nonlinear effects associated with the unresolved modes. Linear inverse models have shown some success in predicting seasonal-to-interannual variability associated with the El Niño/Southern Oscillation (ENSO: Penland & Sardeshmukh 1995), variability of sea-surface temperatures in the tropical Atlantic (Penland & Matrosova 1998), and even the much more nonlinear and chaotic extra-tropical atmospheric variability in the Northern Hemisphere (Penland & Ghil 1993; Winkler *et al.* 2001).

In most geophysical situations, however, the assumptions of linear, stable dynamics and white-noise forcing used to construct LIMs are only valid to a certain degree of approximation. In particular, when nonlinearity is strong enough, the matrices $\mathbf{L}$ and $\mathbf{Q}$ obtained from data can exhibit substantial dependence on the time scales considered (Penland & Ghil 1993): in this case, estimates of the matrices $\mathbf{L}$ and $\mathbf{Q}$ via (2.5, 2.6) using lag-covariance information for different lags, τ, produce different results. Another problem has to do with serial correlations in the model's estimated stochastic forcing.

Let us consider N observations of a vector time series

$$\mathbf{x}^{(n)} = \left(x_1^{(n)},\ x_2^{(n)}, \ldots, x_M^{(n)}\right), \qquad n = 1, 2, \ldots, N, \tag{2.8}$$

sampled at time intervals of Δt. If we denote the time increments in x as

$$\Delta x_i^{(n)} \equiv x_i^{(n+1)} - x_i^{(n)}, \tag{2.9}$$

the discrete representation of (2.4) has the same symbolic form as the original stochastic differential equations, with Δ replacing d. The i-th row of the dynamics matrix $\mathbf{L}$ can be estimated by multiple linear regression (MLR; Wetherill 1986) of the response time series, Δx_i, using the vector-valued time series of $\mathbf{x}$ as predictors, while the residual time series, $r_i^{(n)}$, can be defined according to

$$r_i^{(n)}\Delta t \equiv \Delta x_i^{(n)} - L_{ij}x_j^{(n)}\Delta t. \tag{2.10}$$

The serial correlation problem arises if the lag-correlation function of $\mathbf{r}$ has long tails, thus contradicting the LIM assumption of white-noise forcing. The standard way of dealing with serial correlations is to use higher-order autoregressive models, referred to in the literature as autoregressive–moving average (ARMA) models (Wetherill 1986; Box *et al.* 1994). DelSole (1996, 2000) considered ARMA models in his study of stochastic parameterisations of quasi-geostrophic turbulence.

Recently, Kravtsov *et al.* (2005b) proposed an empirical model formulation that addresses both of the above weaknesses of LIMs by introducing nonlinear, multilevel extensions of (2.4). An important application of this methodology is to diagnose simulations of complex dynamical models by studying their stochastic–dynamic prototypes, derived empirically using the output from the full model (Kondrashov *et al.* 2006); hence this methodology has been called Empirical Model Reduction (EMR). The EMR approach has also been applied to observational data sets, including Northern Hemisphere geopotential heights (Kravtsov *et al.* 2005b), tropical sea-surface temperatures (SSTs) (Kondrashov *et al.* 2005) and a combined SST–sea-level wind data set over the Southern Ocean (Kravtsov *et al.* 2008).

The purpose of the present chapter is to overview the EMR methodology and applications, compare it with other available model reduction and data modelling approaches, and evaluate its role in the climate modeling hierarchy (Ghil & Robertson 2000; Ghil 2001; Held 2005; McWilliams 2007). We describe the general EMR formulation in Section 2.2, while referring interested readers to Appendices A and B for technical details. Section 2.3 discusses EMR models of simulated and observed atmospheric low-frequency variability in the Northern Hemisphere (Kravtsov *et al.* 2005b; Kondrashov *et al.* 2006), with emphasis on the application of this approach to the analysis of a detailed dynamical model. This section also provides comparisons with the results of MTV model reduction (Franzke & Majda 2006; Strounine *et al.* 2008).

The predictive capabilities of EMR models are illustrated in Section 2.4 using the example of tropical SST modelling. In Section 2.5, we outline yet another potential application of EMR models to stochastic parameterisation of a subset of processes within a more complex climate model; we deal in this case with air–sea interaction over the Southern Ocean (Kravtsov *et al.* 2008). Empirical model prediction applications are not restricted to climate dynamics, and an example of such an application to a space-physics problem (Shprits 2009, personal communication) is quite instructive, but beyond the space allotted to this chapter. Section 2.6 contains a summary of the paper's results and the authors' outlook on the problems of model reduction in the geosciences.

2.2 Empirical Model Reduction (EMR)

We consider N observations of the M-valued vector $\mathbf{x}$, as in (2.8), sampled at intervals Δt, and define the increments $\Delta\mathbf{x}$ as the differences between consecutive observations according to (2.9). In most of the following examples, the vector time series $\mathbf{x}$ will be represented by M leading PCs of the field(s) of interest. The first step of EMR uses multiple linear regression to find a set of coefficients N_{ijk}, L_{ij} and F_i $(i, j, k = 1, 2, \ldots, M)$ that minimise, for each i separately, the root-mean-square (rms) distance, χ_i, between the discrete time series of $\Delta x_i^{(n)}$, $n = 1, 2, \ldots, N$, and the test function $(N_{ijk}\, x_j\, x_k + L_{ij}\, x_j + F_i)\Delta t$:

$$\chi_i^2 \equiv \sum\nolimits_{n=1}^{N} \left(r_i^{(n)}\right)^2; \quad r_i^{(n)}\Delta t \equiv \Delta x_i^{(n)} - \left(N_{ijk}x_j^{(n)}x_k^{(n)} + L_{ij}x_j^{(n)} + F_i\right)\Delta t, \tag{2.11}$$

where $i = 1, 2, \ldots, M$, $n = 1, 2, \ldots, N$, and $r_i^{(n)}$ is the discrete N-valued time series of the i-th regression residual.

For large enough M, the distribution of the residuals will typically tend to Gaussian, with all non-Gaussian features of $\mathbf{x}$ accounted for by the nonlinear terms in (2.11). At the same time, inspection of the lag-covariance structure of $\mathbf{r}$ often identifies long-tailed autocorrelations, which indicates that modelling of the residual as white-noise forcing is not justified in such cases. In order to address this issue, we introduce an extended state vector $[\mathbf{x}; \mathbf{r}]$ of dimension $2M$, and fit a multiple linear regression model $\Delta r_i \to L_{ij}^{(1)}[\mathbf{x}; \mathbf{r}]_j \Delta t$, where the summation over j now runs from 1 to $2M$. This model defines the first-level residual $r_{1,i}^{(n)}$, with $i = 1, 2, \ldots, M$, and $n = 1, 2, \ldots, N$, as

$$r_{1,i}^{(n)}\Delta t \equiv \Delta r_i^{(n)} - L_{ij}^{(1)}[\mathbf{x}; \mathbf{r}]_j \Delta t. \tag{2.12}$$

The discrete vector time series, $\mathbf{r}_1$, is in turn tested for whiteness; if the test fails, additional model levels are introduced as necessary, to model the evolution of the previous-level residuals as a linear function of the extended state vector that involves the variables of all preceding levels, until the L-th level residuals $\mathbf{r}_L$ become white. In subsequent inverse modelling, this residual is substituted by the vector-valued, discrete-time white-noise process $\Delta\xi_i\,(t)$, $i = 1, 2, \ldots, M$, with lag-0 covariance matrix $\mathbf{Q}$ computed from the sample covariance of $\mathbf{r}_L$.

The EMR, discrete-time model with L levels in addition to the main level has thus the following form:

$$
\begin{aligned}
\Delta x_i^{(n)} &= \left(N_{ijk}x_j^{(n)}x_k^{(n)} + L_{ij}x_j^{(n)} + F_i\right)\Delta t + r_i^{(n)}\Delta t,\\
\Delta r_i^{(n)} &= L_{ij}^{(1)}[\mathbf{x};\mathbf{r}]_j\Delta t + r_{1,\,i}^{(n)}\Delta t,\\
\Delta r_{1,\,i}^{(n)} &= L_{ij}^{(2)}[\mathbf{x};\mathbf{r};\mathbf{r}_1]_j\Delta t + r_{2,\,i}^{(n)}\Delta t,\\
&\dots\\
\Delta r_{L-1,\,i}^{(n)} &= L_{ij}^{(L)}[\mathbf{x};\mathbf{r};\mathbf{r}_1;\mathbf{r}_2;\dots;\mathbf{r}_{L-1}]_j\Delta t + \Delta\xi_i(t).
\end{aligned}
\tag{2.13}
$$

In all applications discussed herein, we use a time step of $\Delta t = 1$, thus rescaling our time by the sampling interval. When linearised and rewritten as a single equation, the system (2.13) is formally equivalent to an ARMA model (Box *et al.* 1994; DelSole 1996, 2000), but the multilevel form offers greater algorithmic simplicity and is easier to interpret dynamically as well as statistically. In particular, the dependence of the 'hidden variables' $\mathbf{r}_l$ on the observable state vector $\mathbf{x}$ may account for two-way feedbacks between the resolved and unresolved variables. Berloff & McWilliams (2002) used similar multilevel strategy to account for stochastic ocean-eddy effects in statistical simulations of tracer trajectories produced by a model of the wind-driven ocean gyres.

The major technical difficulty in applying (2.13) to geophysical problems lies in the large number of regression coefficients that need to be estimated at the main level of the EMR model: the total number of coefficients for each of the M main-level equations is $M(M+1)/2 + M + 1$. This number makes the dimension of matrices used in the multiple linear regression inversion uncomfortably large even for moderate values of M; more importantly, it may result in overfitting due to lack of linear independence among the predictors. As a result, the estimated regression coefficients will not be statistically robust; that is, they may change substantially if a different data subsample is used to estimate them. This is called the collinearity or multicollinearity problem (Wetherill 1986; Press *et al.* 1994). The numerical procedures that choose an optimal subset of predictors to avoid multicollinearity and overfitting are referred to as *regularisation* techniques.

A key regularisation tool is cross-validation, in which one chooses randomly a subset of the vector time series (typically 80% of original data points), applies a given regression technique, and then uses the regression model to reconstruct the segments of the time series that were omitted in the model identification step. The performance of the regression technique may then be assessed according, for example, to the smallness of the differences between the regression-based prediction and the actual values of the time series. One can use cross-validation in a number of different ways when constructing an EMR model; see Appendices A and B.

Appendix A introduces the methods of principle component regression (PCR; Wetherill 1986) and partial least-squares (PLS; Abdi 2003) regression, which linearly transform original predictor variables into a much smaller set of optimal predictors. This transformation means that the actual number of regression coefficients determined is much less than the number of original predictor variables; thus the $M(M+1)/2 + M + 1$ regression coefficients in each of the main-level equations (2.13) are in fact linear combinations of a much smaller number of independent coefficients. Appendix B describes a fine-tuning procedure for model selection developed by Kravtsov *et al.* (2008), which combines the usage of PCR and PLS with iterative identification of trivial regression coefficients in the original predictor space. Removing the latter coefficients may help study the interactions that give rise to the nonlinear dynamical features simulated by the EMR models (see Section 2.3).

2.3 Application to extra-tropical atmospheric variability

We analyse the output of a global, three-level quasi-geostrophic (QG3) atmospheric model with topography (Marshall and Molteni 1993; D'Andrea & Vautard 2001; Kondrashov *et al.* 2004). The model equations describe the evolution of winds at each of three pressure-coordinate levels, representing the lower, middle and upper troposphere; the troposphere contains about 80% of atmospheric mass and has an average thickness of about 10 km. The equations that govern the QG3 model have the form

$$\dot{q} = -J(\psi, q) - D\psi + S, \qquad q = \Lambda\psi, \tag{2.14}$$

where ψ is the streamfunction, and q is the potential vorticity; q is related to ψ via the linear Laplace–Beltrami operator Λ, while the linear operator D parameterises frictional and radiative damping effects, and S is the constant forcing. The equations also include a quadratic advective nonlinearity written in terms of the Jacobian operator J.

Despite its simple form, the QG3 model has a fairly realistic climatology and complex variability, which also compares favourably with the observed atmospheric behaviour; it has been used therefore extensively in theoretical studies of extra-tropical flows. In addition to synoptic variability associated with baroclinic eddies and a time scale of a few days, the model is characterised on longer time scales by the existence of a few persistent and recurrent flow patterns, or regimes (Reinhold & Pierrehumbert 1982; Legras & Ghil 1985; Molteni 2002), as well as by intraseasonal oscillations (Kondrashov *et al.* 2004). Selten & Branstator (2004) identified signatures of nonlinearity in the model's phase-space mean tendencies (see also Franzke *et al.* 2007); they argued that the dimension of the subspace in which these nonlinear effects are apparent is as low as three.

Such a low-dimensionality of the model's low-frequency variability prompts the development of reduced models, which have considerably fewer degrees of freedom compared to the full QG3 model (Selten 1997; D'Andrea & Vautard 2001; D'Andrea 2002; Kravtsov *et al.* 2005b; Franzke & Majda 2006; Kondrashov *et al.* 2006; Strounine *et al.* 2008). The performance of these models should be judged according to their ability to represent accurately both linear and nonlinear aspects of the full model's behaviour, in particular intraseasonal oscillations and multiple regimes. The reduced models can be studied further to track down dynamical causes of each type of behaviour. The various reduced models differ in part by the way they parameterise the effect of higher-frequency synoptic transients on lower-frequency modes. This effect is referred to in the literature as the synoptic-eddy feedback (Robinson 1996, 2000; Lorenz & Hartmann 2001, 2003; Kravtsov *et al.* 2003, 2005a).

The original QG3 model is global and projected onto spherical harmonics with a total wavenumber not exceeding 21; this so-called T21 version has $D = 3 \times 483 = 1449$ scalar variables (Marshall & Molteni 1993; D'Andrea & Vautard 2001; D'Andrea 2002; Kondrashov *et al.* 2004, 2006; Kravtsov *et al.* 2005b; Strounine *et al.* 2008). A hemispheric version has also been investigated (Selten 1997; Selten & Branstator 2004; Franzke & Majda 2006; Franzke *et al.* 2007). The two model versions produce somewhat different Northern Hemisphere variability; both exhibit, however, similar teleconnection patterns and are also characterised by intraseasonal oscillations. The goal of model reduction methodologies is to construct a model that captures as well as possible the evolution of $M \propto O(10)$ leading EOFs of the full model, and thus reproduces the key features of the full model's variability.

Before even choosing the dynamics coupling a set of the leading EOFs, one may want to optimise the choice of the EOFs, which depends on the inner product used in defining the covariances. The standard choice in the above-mentioned studies of the QG3 model's global version was the use of an inner product consistent with the energy norm. A reduced model constructed in the subspace of the leading EOFs so obtained, though, does not conserve any quadratic invariants of the flow, unless additional approximations are made to project model fields into the subspace of truncated EOFs (Rinne & Karhila 1975; Schubert 1985; Selten 1995); the latter approximations, however, have a detrimental effect on the performance of such a reduced model. Franzke & Majda (2006) proposed to use the energy norm (Ehrendorfer 2000) instead of the streamfunction norm. This choice ensures that the projected equations will conserve the total energy at any truncation, in the absence of forcing and dissipation. Strounine *et al.* (2008) have also considered an inner product consistent with the potential-enstrophy norm; the equations written in this basis conserve the potential enstrophy at any truncation.

We now present and discuss results produced by the EMR and MTV methods in reducing the QG3 model. To make the comparison as fair as possible, both reduced models have 10 resolved components ($M = 10$) that represent the leading energy-norm EOFs of the global QG3 version. For the EMR, we use a three-level ($L = 2$) model (2.13), constructed based on a 30 000-day archive of the QG3 simulation documented by Kondrashov *et al.* (2004, 2006). In deriving this EMR model, PCR-and-PLS regularisation (see Appendix A) resulted in reducing the number of independent regression coefficients from $M[M(M+1)/2 + M + 1 + 2M + 3M] = 1160$ to $O(100)$.

To deal with the lack of scale separation in the QG3 model, Franzke & Majda (2006) had introduced a few free parameters in front of the various groups of terms predicted by the MTV theory and then applied a trial-and-error procedure to 'tune' these parameters in order to achieve better approximations of the statistics of the original, full model's behavior. Following Strounine *et al.* (2008), we replaced this tuning by sequential estimation of the parameters (Dee *et al.* 1985; Ghil 1997; Kondrashov *et al.* 2008).

Franzke and Majda's (2006) empirical fitting also addresses another practical issue in applying the MTV approach to climate problems. While the mathematical expressions for MTV model coefficients are predicted by the theory, these coefficients are given in terms of integrals that involve lagged autocovariances of unresolved modes, over all lags. Accurate numerical computation of these integrals requires very long and frequently sampled libraries of the full model's evolution: Franzke & Majda (2006) used in fact a 1 000 000-day model simulation sampled at half-day intervals. Strounine *et al.*'s (2008) sequential parameter estimation can use QG3 model simulations that are as short as 30 000 days, a number comparable with actual atmospheric data sets, and achieve better statistical fits than those of Franzke & Majda's (2006).

Figures 2.1 and 2.2 compare the QG3 and EMR models in terms of one-dimensional probability density functions (PDFs) and autocorrelation functions (ACFs), respectively, for each of the nine leading PCs. Analogous comparisons between the QG3 and MTV models are given in Figs. 2.3 and 2.4. The EMR model generates time series with PDFs that are almost indistinguishable from those of the QG3 model (Fig. 2.1), including a strongly skewed PDF for EOF-1 and slightly skewed PDF for EOF-4. The fit for the ACFs is not quite as good (Fig. 2.2), but still fairly tight for short lags, up to five days. At longer lags, the EMR model underestimates the time scale of the QG3 model's first PC, and exhibits smaller, slightly oscillatory deviations for the other leading PCs.

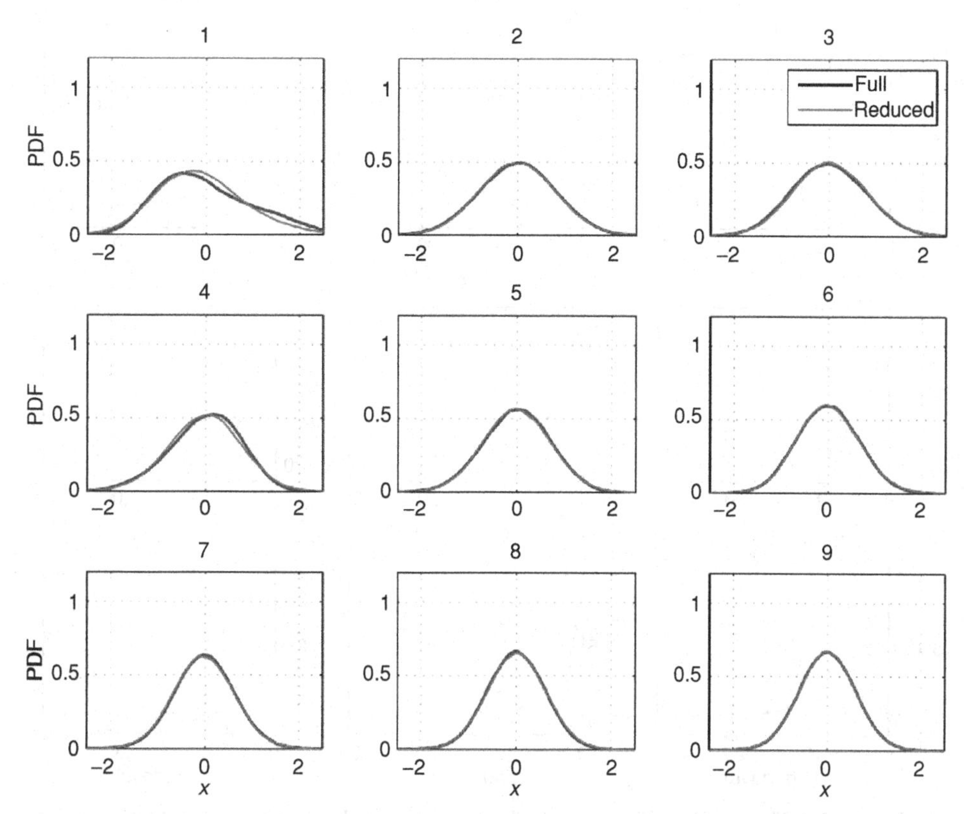

Figure 2.1 Individual probability density functions (PDFs) of the nine leading PCs for the EMR model constructed in the phase space of energy-norm EOFs based on the QG3 model simulation; EOF indices are given in the caption of each panel.

The results for the MTV model are uniformly worse relative to the EMR model for all PCs, with respect to both the PDF (Fig. 2.3) and ACF (Fig. 2.4) comparisons. The MTV model without the sequential parameter estimation of its semi-empirical coefficients (not shown) exhibits even more substantial biases in its PDFs, while its ACFs are fairly similar to those in Fig. 2.4. The quantitative correspondence between the ACFs of the QG3 and MTV models is similar to that reported by Franzke & Majda (2006) for their hemispheric version of the QG3 model (their Fig. 13). On the other hand, their results for PDFs (their Fig. 14) show greater deviations from those of the QG3 model than those in Fig. 2.3 here; the improvement reported herein is presumably due to our use of sequential parameter estimation, rather than trial-and-error tuning.

The performance of our EMR model in capturing nonlinear features of the QG3 model's behaviour is better illustrated by comparing multidimensional PDFs of the full and reduced model solutions. Figure 2.5 shows the PDFs of the data sets

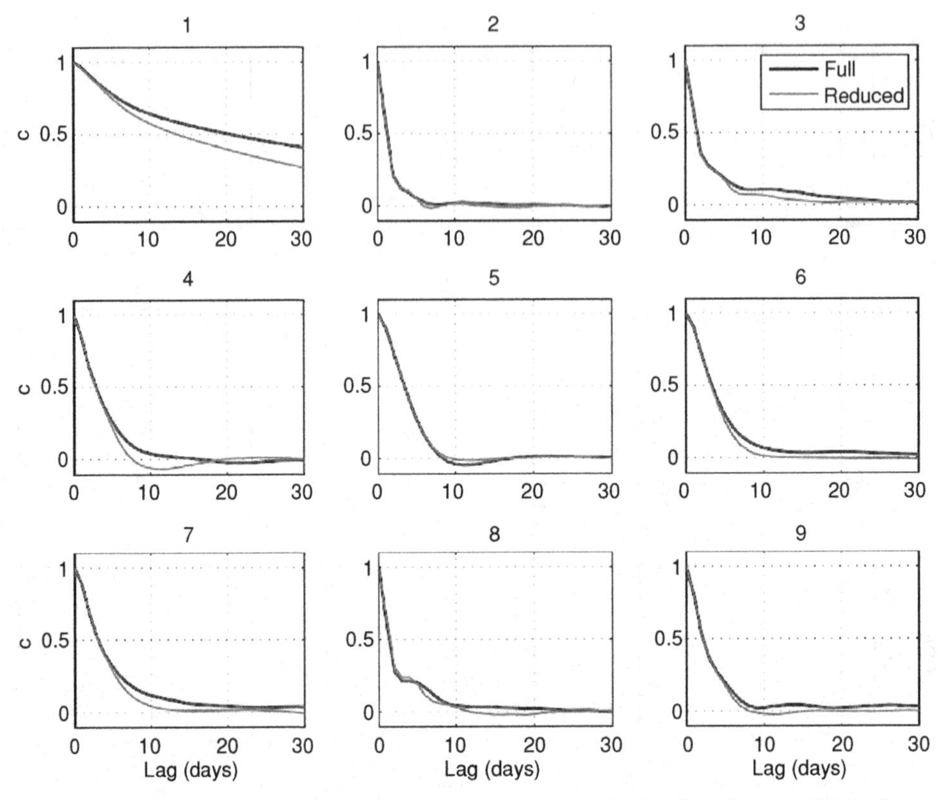

Figure 2.2 The same as Fig. 2.1, but for the autocorrelation functions (ACFs) of the the nine leading PCs for EMR model.

produced by the QG3 and the EMR model, constructed in the phase space of 15 leading streamfunction-norm EOFs (Kondrashov *et al.* 2004; Kravtsov *et al.* 2005b; Kondrashov *et al.* 2006). The clusters were found by mixture modelling of the PDFs (Smyth *et al.* 1999; Hannachi & O'Neill 2001) using an optimal mix of $k = 4$ Gaussian components in a phase subspace of the four leading EOFs. The locations, shapes and sizes of clusters, and hence the general shape of the PDF, are reproduced quite well by the EMR model in Fig. 2.5.

The composites over the data points that belong to each of the clusters in Fig. 2.5 represent, in physical space, the patterns of four planetary flow regimes (Legras & Ghil 1985; Ghil & Childress 1987, ch. 6; Mo & Ghil 1987; Cheng & Wallace 1993; Kimoto & Ghil 1993a,b; Hannachi 1997; Smyth *et al.* 1999; Hannachi & O'Neill 2001; Molteni 2002). In Fig. 2.5a, cluster AO^- (labelled 2 in the figure) occupies a distinctive region on the PDF ridge that stretches along PC-1. It corresponds to the low-index phase of the well-known Arctic Oscillation (AO), which may be related to a more regional North Atlantic Oscillation (NAO) (Deser 2000; Wallace 2000).

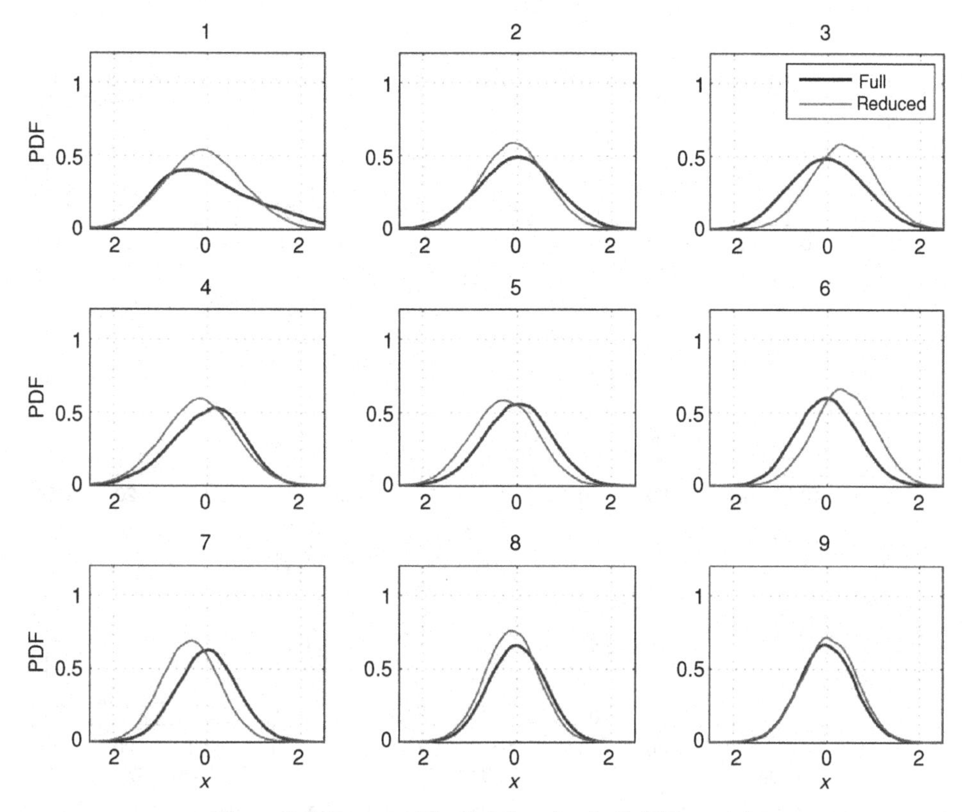

Figure 2.3 Same as Fig. 2.1, but for the MTV model.

The clusters AO$^+$, NAO$^-$ and NAO$^+$ are located around the global PDF maximum, with the centroid of AO$^+$ to the left and below, NAO$^+$ above, and NAO$^-$ slightly to the right of this maximum, respectively. These four regimes are not identical to but in fairly good agreement with the observational results of Cheng & Wallace (1993) and Smyth *et al.* (1999); see also Ghil & Robertson (2002) and Kondrashov *et al.* (2004, 2006).

The streamfunction anomalies associated with each regime centroid of the QG3 model are plotted in Fig. 2.6. The spatial correlations between these anomaly patterns and those obtained from the EMR model (not shown) all exceed 0.9. They are thus much higher than the correlations obtained by D'Andrea & Vautard (2001) and D'Andrea (2002), who used a reduced deterministic model obtained by a statistical–dynamical approach to reproduce the behaviour of the largest scales in the QG3 model. We have also computed Gaussian-mixture PDFs of the MTV model simulations (not shown), which however failed to reproduce the PDFs of the full QG3 model (see Fig. 16 of Strounine *et al.* 2008).

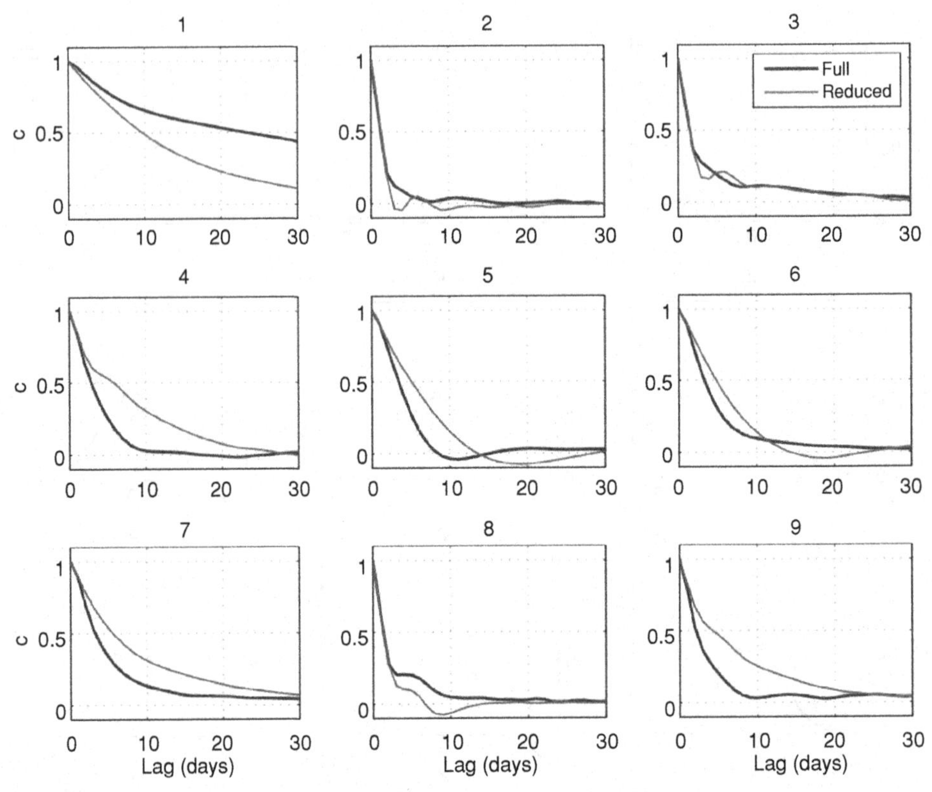

Figure 2.4 Same as Fig. 2.2, but for the MTV model.

The results above indicate a fair degree of success of the EMR models in capturing key characteristics of the QG3 model. Kondrashov *et al.* (2006) built upon this success and conducted a detailed study of the origin of the QG3 model's multiple regimes and low-frequency oscillations, and of possible connections between them. They showed how to use the EMR models' much greater flexibility in studying the dynamic and stochastic contributions to these episodic and oscillatory features present in the QG3 model, as well as in atmospheric observations of low-frequency variability (LFV); see also Ghil and Robertson (2002) and Ghil *et al.* (2003). Kondrashov *et al.* (2006) applied standard tools from numerical bifurcation theory for deterministic dynamical systems to the quadratically nonlinear deterministic operator of their optimal EMR model, and used a continuation method on the variance of the multilevel noise process. This somewhat ad hoc combination of continuation methods allowed them to move all the way from fixed points of the deterministic operator to the multiple regimes of the complete EMR model, with its optimal parameter values.

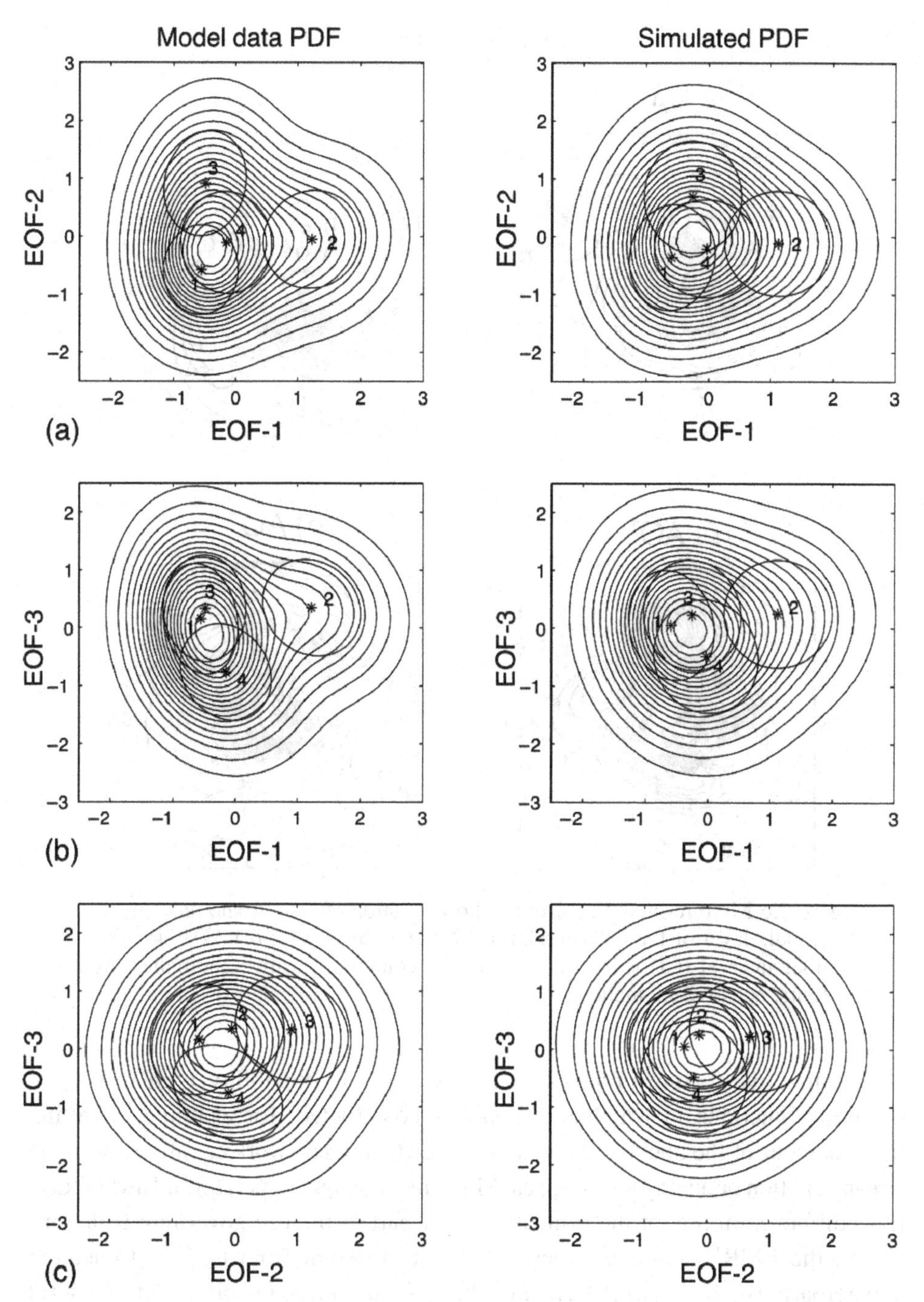

Figure 2.5 PDFs of the QG3 model (left panels) and EMR model (right panels), projected onto the planes of the three leading EOFs: (*a*) EOF-1–EOF-2; (*b*) EOF-1–EOF-3; and (*c*) EOF-2–EOF-3. The ellipses superimposed on the PDFs are obtained by Gaussian-mixture modelling. Cluster centroids are plotted as asterisks, while projections of cluster boundaries are shown as ellipses, the semi-axes of which equal one standard deviation of the cluster, in each direction. Shown are projections onto EOF planes: the cluster centroid indices correspond to AO^+, AO^-, NAO^+ and NAO^-, in this order (see text and Fig. 2.6). Reproduced from Kravtsov *et al.* (2005b), with the permission of the American Meteorological Society.

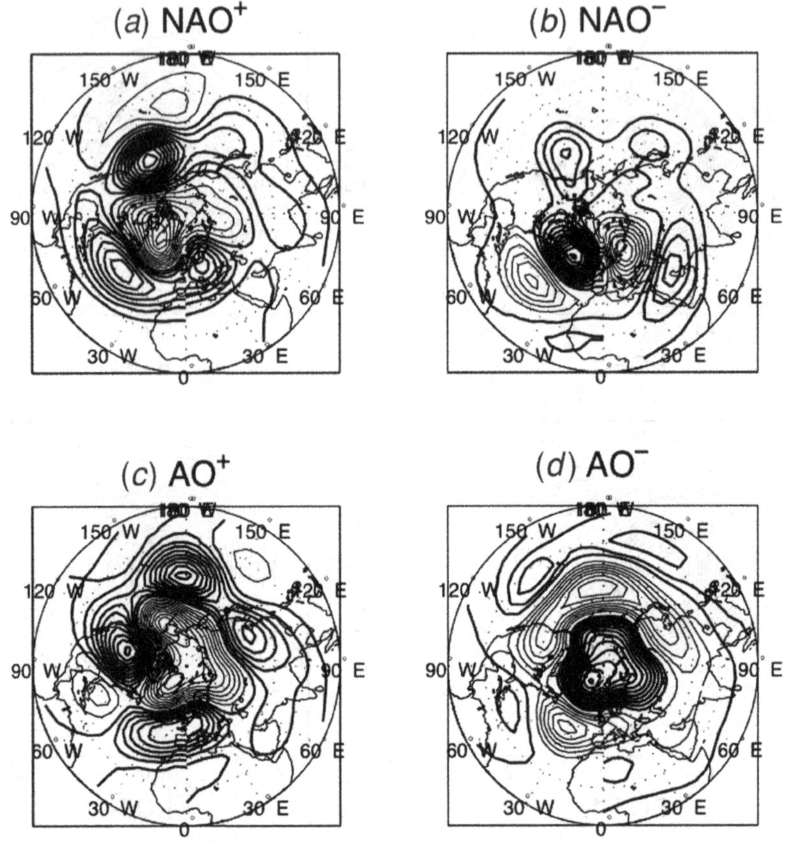

Figure 2.6 Mixture-model centroids, showing streamfunction anomaly maps at the middle level of the QG3 model: (a) NAO$^+$; (b) NAO$^-$; (c) AO$^+$; (d) AO$^-$. Contour interval (CI) is 10^6 m^2 s^{-1}; heavy contours: positive anomalies, light contours: negative anomalies.

An attractive feature of EMR methodology is its flexibility with respect to the basis functions in which the model is constructed; for example, in fitting the behaviour of the above QG3 model, the EMR models work equally well in the streamfunction or energy-norm bases. Much like the linear operator of LIM models (2.4) differs from that of the truncated linear part of the full governing equations (2.3b), the EMR's nonlinear operator is also different from its bare-truncation counterpart. Therefore, irrespective of the basis used and in spite of other useful features of the EMR models, a cautionary note is in order: the stability of the EMR model is not guaranteed, since the empirical fit does not conserve quadratic invariants of the QG3 model, but rather parameterises all the important interactions within a given functional form of the EMR model. In contrast, an MTV model constructed in a basis that conserves quadratic invariant(s) is stable, as long as this stability is not affected by the additional, semi-empirical coefficients. We will return to a point-by-point comparison of EMR with other mode reduction methodologies in Section 2.6.

Since the EMR procedure is entirely data based, it can be applied directly to the observational data sets, for which the full model does not have to be known. For example, Kravtsov *et al.* (2005b) studied low-frequency variability in the observed Northern Hemisphere's geopotential heights by constructing a nine-variable, two-level EMR model thereof; they were able to reproduce the multidimensional PDF and power spectra of the observed fields, with a quantitative success that is quite similar to the QG3 model fit discussed here. It turned out, however, that the EMR model's dynamical operator in this case does possess unstable directions, which led to occasional run-away model realisations with unphysically large values of the model variables, consistent with the lack of quadratic invariants mentioned above. These authors developed a simple strategy to avoid such rare situations altogether, by tracking the instantaneous norm of the EMR model's state vector. If the values of this norm exceeded a certain threshold, the time series being modelled were 'rewound' by a few time steps and restarted with a different realisation of the random forcing.

In summary, the EMR methodology provides a means to construct skillful nonlinear reduced models with stochastic forcing that need not be white in time. Empirical model reduction models are based solely on the data output from a full dynamical model, or directly on observational data. Regularisation techniques ensure that the number of independent regression coefficients to be estimated in an EMR model is much smaller than the number of variables in the full model or the number of coefficients in the absence of regularisation. While the stability of a nonlinear EMR model is not guaranteed a priori, simple engineering fixes are available to avoid unstable directions in the empirical model's dynamical operator. The EMR model based on QG3 output reproduces key features of the full model better than the models based on alternative methodologies, such as that of Selten (1995), D'Andrea (2002) or MTV, and can thus be used for a detailed dynamical and stochastic diagnosis of the full QG3 model (Kondrashov *et al.* 2006).

2.4 Modelling of tropical sea-surface temperatures (SSTs)

The EMR methodology can be used to simulate and predict phenomena whose dynamical modelling requires fairly complex and computationally expensive models. Among such phenomena is the ENSO, which dominates interannual climate signals centered in the tropical Pacific Ocean (Philander 1990), and has a substantial effect on the atmospheric circulation and air–sea interaction through many parts of the globe, via atmospheric or oceanic teleconnections (Alexander *et al.* 2002).

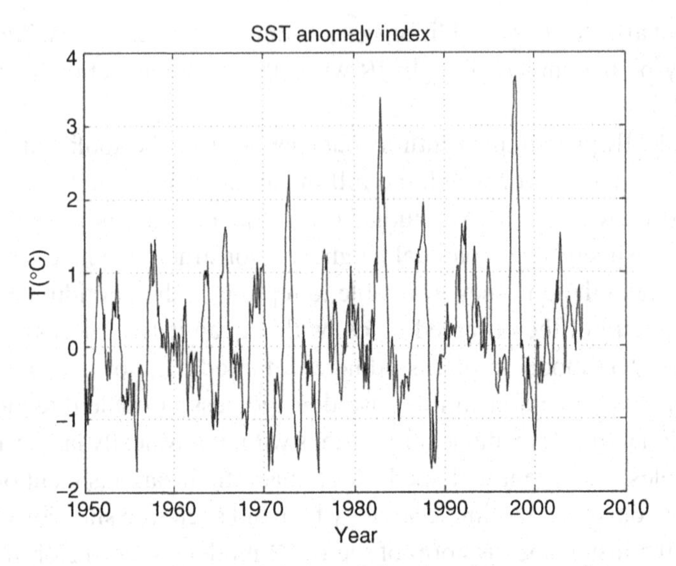

Figure 2.7 Time series of Niño-3 index, defined as the sea-surface temperature (Kaplan *et al.* 1998) averaged over the rectangular box (5° S–5° N, 150°–90° W); this box is also outlined in light solid black in Fig. 2.8a,b.

Figure 2.7 shows a widely used index of ENSO activity, the so-called Niño-3 index, computed as the average of SST over a rectangular box located in the tropical Pacific, with coordinates (5° S–5° N, 150 °–90° W). The pronounced positive-anomaly events in this time series are associated with warm, El Niño episodes in the eastern tropical Pacific, while the negative events point to La Niña conditions. A striking property of ENSO is that El Niño events are in general stronger than La Niñas, thus suggesting that the dynamics of ENSO involves nonlinear processes (Neelin *et al.* 1994, 1998; Ghil & Robertson 2000). At the same time, most detailed numerical models used for operational ENSO predictions significantly underestimate this nonlinearity (Hannachi *et al.* 2003), and the quality of their forecasts is still far from satisfactory (Barnston *et al.* 1994, 1999; Ghil & Jiang 1998; Landsea & Knaff 2000).

Kondrashov *et al.* (2005) constructed a 20-variable, single-level (1-L) and two-level (2-L) EMR models of ENSO based on a 645-month-long monthly time series of SST anomalies given on a $5° \times 5°$ grid over the 30° S–60° N latitude belt (Kaplan *et al.* 1998). Despite the seasonal cycle having been removed from the observed time series, the EMR models were extended to include seasonal dependence by making the main-level linear operator's coefficients 12-month periodic and estimating additional regression parameters in the usual way.

Figure 2.8 displays the cross-validated hindcast skill of the 1-L and 2-L EMR models; the skill was defined in terms of anomaly correlation between

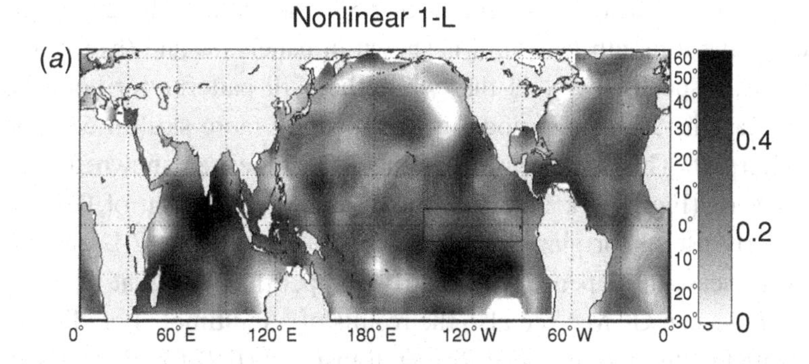

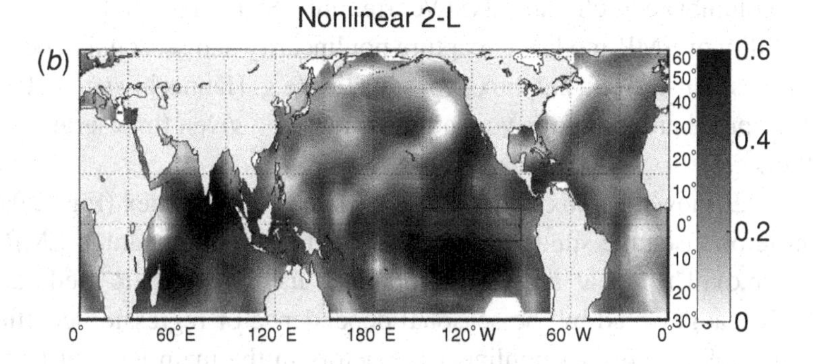

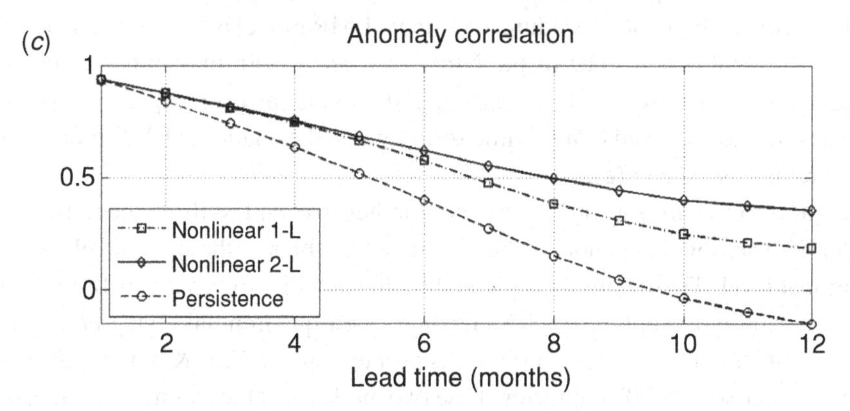

Figure 2.8 Comparison of predictive skill for one-level (1-L) and two-level (2-L) EMR models. Anomaly correlation map for nine-month-lead cross-validated hindcasts using (*a*) the 1-L model and (*b*) the 2-L model. (*c*) Niño-3 hindcast skill in terms of anomaly correlation for the 1-L model (dash-dotted line with squares) and the 2-L model (solid line with diamond symbols), with damped persistence forecast (dashed line with circles) being used as a reference skill. The Niño-3 SST anomaly is defined as the area average over the rectangular box shown in panels (*a*) and (*b*). Adapted from Kondrashov *et al.* (2005), with the permission of the American Meteorological Society.

the actual and forecasted time series of SST anomalies. The geographical distribution of the 9-month-lead skill is shown in panels (a) and (b) for the 1-L and 2-L models, respectively. Note that while the skill patterns for both models are fairly similar, the 2-L EMR model is significantly more skillful compared to its 1-L counterpart. This is the case for other lead times, as shown for the Niño-3 index forecast (Fig. 2.8c); note that the anomaly correlation of 0.5–0.6 is considered fairly useful for planning purposes. The skill of the 2-L quadratic EMR model with seasonal dependence is fairly competitive with that of fully coupled atmosphere–ocean GCMs; see also the results of a multimodel prediction scheme implemented by the International Research Institute (IRI) for climate and society: http://iri.columbia.edu/climate/ENSO/currentinfo/SST_table.html.

The ability of EMR models to capture nonlinearity and seasonal dependence of ENSO is best illustrated using box-plot statistics (e.g. Hannachi *et al.* 2003), which show the spread around the mean and skewness of a given time series, as well as its outliers.

Figure 2.9 shows the box plots of the observed Niño-3 index (Fig. 2.9a), along with those of ensemble simulations of the 2-L linear and quadratic EMR models (Figs. 2.9b,c). The linear model was obtained using full EMR methodology of Section 2.2, augmented by the seasonal dependence of regression coefficients at the main level, except that nonlinear predictors in the main level of (2.13) were neglected throughout the procedure. The linear EMR model is successful in describing the seasonal dependence of the Niño-3 variance, with maximal variability in boreal winter; compare panels (a) and (b). It does not, however, capture the positive skewness of the observed Niño-3 time series, while the quadratic EMR model does; compare panels (a) and (c).

Kondrashov *et al.*'s (2005) most comprehensive and skilful quadratic EMR model with seasonal dependence had 20 state variables at the main level, and one additional level. These authors showed that this model's forecast capabilities were due to its ability to capture ENSO's leading quasi-quadrennial (Jiang *et al.* 1995) and quasi-biennial oscillatory modes. As pointed out by Ghil & Jiang (1998) and Ghil & Robertson (2000), capturing these two modes is of the essence for successful ENSO prediction beyond six months. Kondrashov *et al.* (2005) demonstrated that these modes are present as inherent periodicities within the dynamical operator of their EMR model, by carrying out month-by-month linear stability analyses, as well as Floquet analysis of the model's seasonal cycle (Strong *et al.* 1995; Jin *et al.* 1996).

To summarise, EMR methodology allows one to come up with a skillful predictive model and then use it to study the dynamical causes of the observed variability in situations in which comprehensive dynamical models based on first principles are either unavailable or else very complex and difficult to interpret.

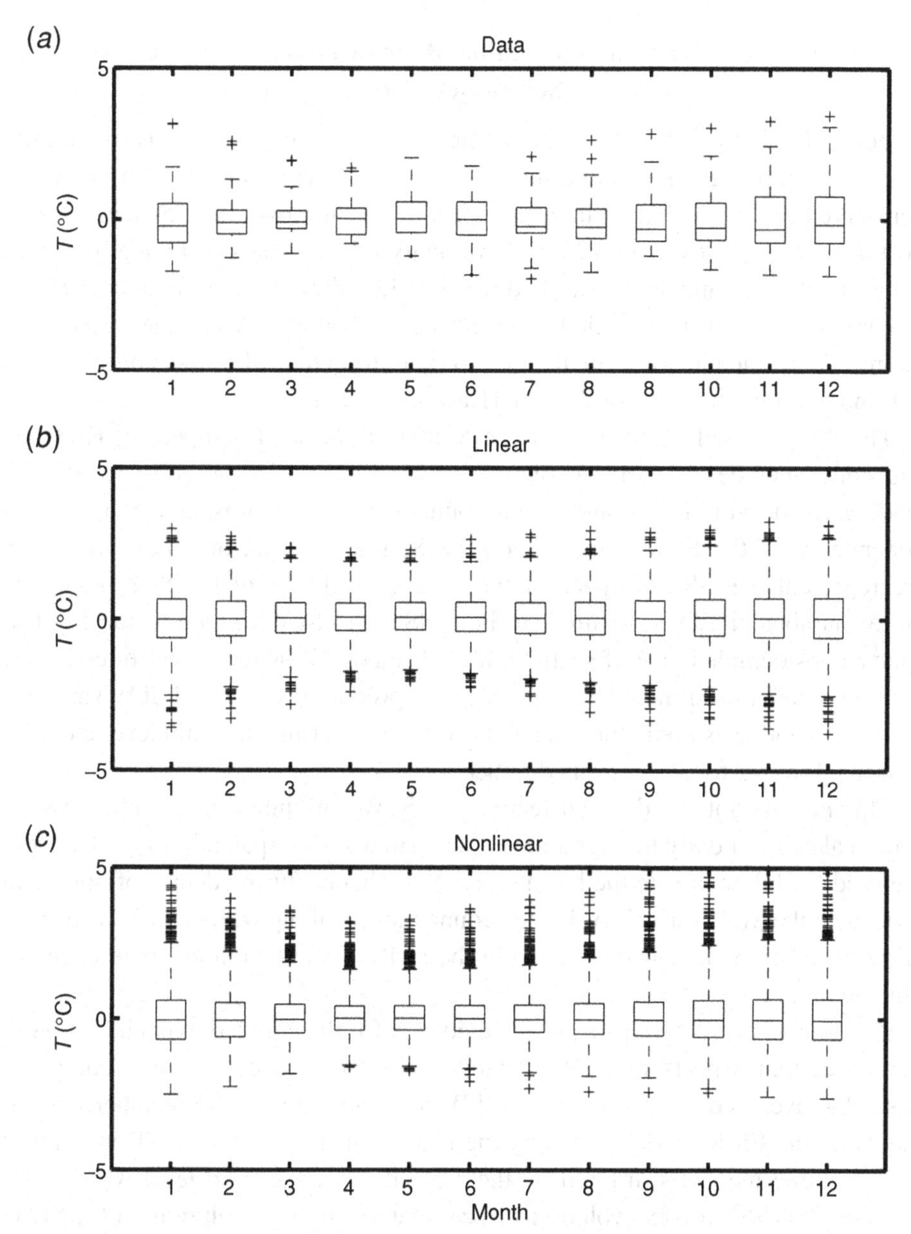

Figure 2.9 Seasonal dependence of EMR model fit to ENSO statistics, visualised as box-plot statistics for each month of the year, for (a) the observed Niño-3 index (1950–2003); (b) a 100-member ensemble of 645-month-long runs of the 2-L linear EMR model; and (c) same as in (b), but for the quadratic EMR model. The horizontal line within each box marks the median, while the height of the box represents the interquartile range (IQR), which is defined as the distance between the first and the third quartiles and is a robust measure of spread. The whiskers represent the most extreme data values within 1.5 × IQR, while points beyond the last of the whiskers represent outliers and are indicated by individual tick marks. Reproduced from Kondrashov *et al.* (2005), with the permission of the American Meteorological Society.

2.5 Stochastic parameterisation of air–sea interaction over the Southern Ocean

Another possible application of the EMR models we discuss here is to parameterise certain processes within a more complex dynamical model. The processes considered in this section pertain to air–sea interaction over the Southern Ocean. To this end, Kravtsov *et al.* (2008) have analysed five years of remotely sensed data sets of SSTs and sea-level wind (SLW: Liu 2002) over the Southern Ocean. The microwave sensors installed on recently launched NASA satellites provide an unprecedented quantity and quality of observations in this otherwise poorly known but important part of the world ocean (Kawanishi *et al.* 2003).

The EMR model of SLW over the Southern Ocean developed by Kravtsov and colleagues describes the evolution of winds in the phase subspace of 100 leading vector-wind EOFs, and also accounts for ocean–atmosphere coupling via dependence of the SLW equations on the SST anomalies; the latter anomalies are represented in the subspace of the 75 leading EOFs of the SST field. The larger number of EOFs required, in both SST and SLW, is due to the need to span a considerable range of spatial scales. The best EMR model has three levels, including the bilinear main level, in which the products of SST and SLW variables are also included as predictors, in addition to linear terms; the main level includes seasonal forcing, too, as described in Section 2.4.

This model captures detailed features of SLW variability on a wide range of time scales, from daily to interannual, and spatial scales spanning the range from hundreds of kilometres to the basin scale. Note that capturing details of non-local aspects of the SLW variability, i.e. teleconnections – along with the local aspects – is essential for the intended coupling of the EMR model to a more comprehensive climate model.

Kravtsov *et al.* (2008) also showed evidence for the coupled dynamics at work behind certain aspects of SLW variability. In order to do so, they computed ensemble-averaged evolution of the SLW anomalies for a 100-member ensemble using the EMR model forced by the history of the observed SST anomalies; the ensemble members differed by the realisation of the third-level white-noise process. This SST-driven evolution was compared with the evolution for the SLW-only stochastic model, in which SST anomalies were artificially set to zero. The authors then computed the standard deviation, in time, of the ensemble-averaged wind speed for both cases, at each grid point: the results of this computation for the SST-forced model are shown in Fig. 2.10.

The standard deviation in the SST-dependent case is much larger, at all grid points, than that in the SLW-only case (not shown). Its distinctive large-scale spatial pattern suggests that this SLW variability is forced by long-term, ocean-induced

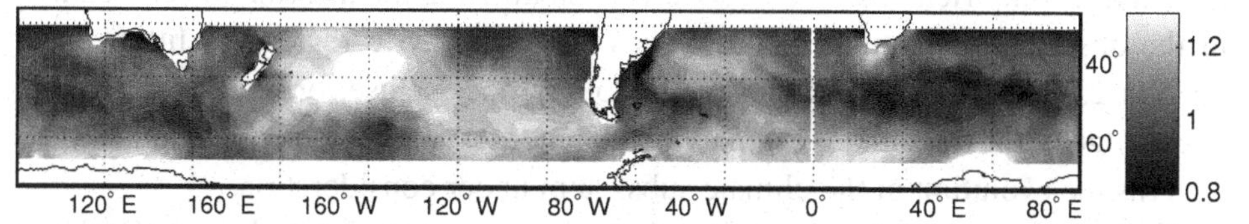

Figure 2.10 Sea-surface temperature effect on the sea-level wind as modelled by the EMR model of air–sea interaction over the Southern Ocean (65°–30° S). Shown is the standard deviation of the wind speed time series near sea level. These time series, at each spatial grid point, were computed as the ensemble average of 100 simulations of the EMR model constructed in the phase space of 100 leading EOFs of sea-level wind and forced by the observed history of SST anomalies.

SST anomalies. Bretherton & Battisti (2000) proposed alternative explanations to such findings in the atmospheric GCMs forced by North Atlantic SSTs. On the other hand, Goodman & Marshall (1999) formulated a theory of interannual-to-decadal coupled variability that is potentially applicable to the Southern Ocean. Their theory predicts the existence of coupled modes, given a certain spatial phase relationship between SST patterns and SST-induced SLW anomalies; this phase relationship gives rise to Ekman pumping anomalies that force and modify the oceanic circulation and the associated SST field. One possible use of the EMR model would be to check whether it supports such a phase relationship between SLP and SLW on interannual time scales.

Another very promising way to apply the EMR model of SLW is, however, to couple it to an oceanic general circulation model. This can be achieved by blending the SST-dependent SLW model with the atmospheric boundary layer model of Seager *et al.* (1995). The latter model needs the specification of boundary-layer winds to compute ocean–atmosphere heat fluxes. These winds can be supplied by the EMR model, and would then be used to compute the atmosphere–ocean momentum flux. The ocean model forced by heat, moisture and momentum fluxes will, in turn, predict the evolution of the SST field, which will affect the future SLW anomalies. Experiments with such a hybrid coupled GCM of the Southern Ocean regions are currently underway; they may provide valuable insights into the dynamics of climate variability there.

2.6 Concluding remarks

2.6.1 Summary

We developed a systematic, albeit empirical strategy for constructing parsimonious, dynamic–stochastic models that are able to capture key aspects of a given climate subsystem's evolution. Our approach, called Empirical Model Reduction (EMR) is indeed fully empirical, being based entirely on utilising the information embedded in the observed or simulated time series, via parametric regression fitting (see Section 2.2); in other words, our methodology doesn't use, or require at all, the knowledge of the true, and presumably much more complex, dynamical model that governs the evolution of the climate subsystem under consideration.

The EMR models are extensions of the linear inverse models (LIMs) of Penland and associates (Penland 1989, 1996; Penland & Ghil 1993) that may include quadratic – and higher-order polynomial, if necessary – combinations of predicted variables in the dynamical operator of the main model level; additional model levels are also included to capture the lagged autocorrelations possibly present in the main-level stochastic forcing. The number of model levels is chosen to ensure that the forcing at the last level can be well approximated by a vector-valued white-noise process. The stochastically forced simulations of an EMR model can then be exploited to analyse various aspects of the actual, observed evolution of the system or one that is generated by a high-end model thereof. The actual regression fitting of an EMR model's coefficients is carried out by regularisation methods, such as principal component regression (PCR; Press *et al.* 1994) and partial least-squares (PLS; Abdi 2003); these methods substantially reduce – by as much as two orders of magnitude, depending on the particular problem at hand – the number of independent coefficients to be estimated (see Appendices A and B).

Section 2.3 documents a remarkable success of EMR methodology in reducing a fairly sophisticated, nonlinear model of the extra-tropical atmosphere, with more than a thousand degrees of freedom (the QG3 model: Marshall & Molteni 1993) to a dynamic–stochastic model with ten main-level variables and $O(100)$ independent coefficients. The original QG3 model is well known for having a fairly realistic climatology and complex variability, which also compares favourably with the observed atmospheric behaviour. The coefficients here were estimated based on a 30 000-day-long simulation of the full QG3 model.

The EMR model accurately reproduces non-Gaussian features of the PDF of the full model, computed in the phase subspace of the ten resolved variables (see Figs. 2.1, 2.2, 2.5 and 2.6 here). It also reproduces surprisingly well the intraseasonal oscillations (Ghil *et al.* 2003) that characterise the full QG3 model's low-frequency variability (Kravtsov *et al.* 2005b; Kondrashov *et al.* 2006). Kondrashov *et al.* (2006) studied the deterministic operator of the EMR model, as well as its stochastic forcing, to analyse the dynamical causes behind the persistent and recurrent states associated with the non-Gaussian PDF, the intraseasonal oscillations, as well as the connections between the two (see also Ghil & Robertson 2002). Kondrashov and colleagues applied standard tools from numerical bifurcation theory for deterministic dynamical systems to the quadratically nonlinear deterministic operator of their optimal EMR model, and used a continuation method on the variance of the multilevel noise process. This somewhat ad hoc combination of continuation methods allowed them to move all the way from fixed points of the deterministic operator to the multiple regimes of the complete EMR model.

In situations when the full dynamical model is not available, or otherwise is very complex and difficult to implement efficiently or to interpret, the EMR methodology can be used to set up a competitive analysis and prediction scheme for the phenomena of interest. Section 2.4 illustrates the predictive capabilities of an EMR-based model using the example of tropical sea-surface temperature (SST) evolution (Kondrashov *et al.* 2006; see Fig. 2.7 here). Their 20-variable, three-level EMR model with quadratic main-level and seasonally-dependent coefficients in its linear part has a forecast skill in predicting El Niño–Southern Oscillation (ENSO) events that is comparable with that of state-of-the-art dynamical and statistical models (see Fig. 2.8). The EMR prediction scheme of these authors is currently a member of the multimodel forecast ensemble developed by the International Research Institute (IRI) for Climate and Society.

The successful EMR forecasts are rooted in the EMR model's ability to capture major oscillatory signals associated with ENSO behaviour; namely, the quasi-quadrennial and quasi-biennial oscillations (Jiang *et al.* 1995). These signals can be explained in terms of eigenmodes of the EMR model's dynamical operator (Kondrashov *et al.* 2006). In addition, the nonlinear EMR model successfully reproduces the observed asymmetry between larger positive (El Niño) and smaller negative (La Niña) anomalies of the ENSO cycle (see Fig. 2.9), as well as the seasonal dependence of ENSO predictability, including the well-known 'spring barrier' to interannual forecast skill.

The EMR models can also serve to parameterise stochastically a subset of processes within a more complex dynamical model. As an example, Section 2.5 outlines a strategy for coupling an EMR model of air–sea interaction over the Southern Ocean with a comprehensive ocean model. In this case, the EMR modelling bypasses dynamical consideration of a fairly complex chain of SST effects on the sea-level winds (SLWs) to construct a statistical model of SST-dependent SLW evolution (Kravtsov *et al.* 2008). The subsequent coupling uses SST fields produced by the ocean model to force the EMR and predict SLW distribution. This distribution affects in turn the atmospheric boundary layer and atmosphere–ocean heat, moisture and momentum fluxes, and, therefore, the oceanic variables.

The EMR model used here has 100 SLW variables and 75 SST variables, and features a bilinear SST–SLW main-level structure. This fairly large model dimension is necessary since coupling to the ocean model requires a wide range of spatial scales in the simulated SLW field. Note that, despite a fairly short five-year training interval and large number of predictors, the regularisation techniques described in Appendices A and B are successful in obtaining robust model coefficients: the model's dynamical operator is stable and simulates well both local and non-local statistical properties of the observed SLW. These results (see Fig. 2.10) demonstrate the applicability of the EMR methodology to problems of intermediate size.

Finally, the EMR methodology is in no way restricted to the problems of climate dynamics; EMR modelling has also been applied to describe the variability of the Earth's radiation belts (Shprits 2009, personal communication).

2.6.2 Discussion

The EMR models belong to a class of multivariate parametric stochastic models forced by additive, state-independent noise. Alternative statistical formulations used in climate dynamics involve empirical fitting to the Fokker–Planck equation (Gardiner 1985). Such fits result in non-parametric, univariate or bivariate models that include state-dependent or multiplicative noise (Sura 2003; Sura & Gille 2003; Sura *et al.* 2006; Sura & Newman 2008; Sura & Sardeshmukh 2008). The extreme truncation of these models to one or two dimensions only is due to the limited amount of data, which is typically insufficient to obtain reliable estimates of higher-dimensional analogues of drift and diffusion coefficients.

This approach is therefore applied either to scalar teleconnection indices (Feldstein 2000; Stephenson *et al.* 2000) or to the time series of a climatic field at a single location. In the latter case, in order to get reliable estimates of model parameters given relatively sparse observations, one may also concatenate data sets from multiple locations, which are situated far enough from each other so that their respective time series may be assumed to be uncorrelated (Sura 2003). The scalar stochastic differential equations so obtained describe local features of interactions between processes evolving on different time scales. They are successful in interpreting certain non-Gaussian aspects of SLW (Sura 2003; Monahan 2004, 2006a,b), as well as of SST (Sura *et al.* 2006; Sura & Newman 2008; Sura & Sardeshmukh 2008) variability. These stochastic models are not suitable, however, for modelling non-local aspects of climatic variability (compare the results of Deloncle *et al.* 2007, with those of Sura *et al.* 2005); the ability to capture non-local effects is a considerable advantage of the EMR-based models.

Sura *et al.* (2005) applied the above methodology to explain non-Gaussian features in the phase subspace of the observed atmospheric winds in terms of dynamics involving multiplicative noise. Berner (2005) and Branstator & Berner (2005) independently used similar ideas to analyse long output from an atmospheric GCM. Kravtsov & Branstator (2006, personal communication) performed EMR modelling of the same GCM simulation and produced surrogate time series, whose statistical properties were subsequently analysed by estimating the drift and diffusion matrices of planar subspaces of the EMR-simulated data. The drift and diffusion coefficient structure, as well as the mean phase-space tendencies computed for the EMR-based surrogate were very similar to those based on the GCM-generated data set.

These results demonstrate that the multiplicative-noise explanations of non-Gaussian atmospheric behaviour depend on how the climate 'signal' and 'noise' are defined; in the EMR approach, the signal and the noise are equivalent to the unresolved and resolved variables. For example, for the EMR model of Kravtsov and Branstator, with its nine resolved variables, the non-Gaussianity is clearly due to quadratic nonlinearity in the model's deterministic propagator. On the other hand, if only the two leading EMR variables are interpreted as signal, and the others as 'noise,' the same quadratic combinations of the first two variables with the rest become manifestations of multiplicative noise. The same comments apply to attribution of various features in the phase-subspace mean tendencies to either interaction between the resolved modes or that between the resolved and unresolved modes (Franzke *et al.* 2007; Majda *et al.* 2008).

Purely empirical models, such as the EMR models, are very flexible in their choice of predictor fields, including bases other than EOFs, such as those proposed by Kwasniok (1996, 2004) and DelSole (2001); see also Crommelin & Majda (2004). These models are in general not limited by various restrictions and constraints inherent to reduction methods that are based entirely or to a large extent on properties of the full governing equations. This flexibility of empiricism comes, however, at the expense of the lack of an easy, immediate interpretation of results in terms of formal model properties, such as that offered, at least at first sight, by the latter methods. The simplest approach of incorporating dynamical information into simplified models of geophysical flows is to linearise the full governing equations with respect to their long-term time-mean state and introduce a combination of linear damping and stochastic forcing terms to achieve time-variable climates that resemble those of the full nonlinear model (Branstator 1992, 1995; Farrell & Ioannou 1993, 1995; Zhang & Held 1999; DelSole 2004).

Another possibility is to follow an empirical–dynamical approach and write the governing equations in a truncated EOF basis, while parameterising the interaction between the resolved and unresolved modes empirically (Selten 1993, 1995; Achatz & Branstator 1999; D'Andrea & Vautard 2001; D'Andrea 2002; Achatz & Opsteegh 2003a,b). Finally, Majda *et al.* (1999, 2001, 2002, 2003, 2006; abbreviated as MTV) presented a model reduction methodology based on standard projection methods for stochastic differential equations (Khasminsky 1963; Kurtz 1973; Gardiner 1985), which is rigorously correct for systems with substantial time-scale separation between the resolved and unresolved modes; in some cases, these results carry over to the situations without such a spectral gap (Majda *et al.* 2002, 2003; Majda & Timofeyev 2004). However, when applied to prototype barotropic (Franzke *et al.* 2005) and baroclinic (Franzke and Majda 2006) atmospheric models, the results are less striking (see our Section 2.3 here and Strounine *et al.* 2008), presumably due to the fairly continuous atmospheric power spectra violating the main assumption of the MTV methodology.

In general, Strounine *et al.* (2008) found that reduced models with an equal number of resolved variables perform better when a larger amount of statistical information is used in model construction. Thus, the EMR models with quadratic nonlinearity and additive noise reproduce very well spectral properties of the full model's variability, such as autocorrelations and spectra, as well as the model's, and the extra-tropical atmosphere's, multiple flow regimes that induce non-Gaussian features in the model's PDF. The empirical–dynamical models capture some of the basic statistical properties of the full model's variability, such as the variance and integral correlation time scales of the leading PCs, as well as some of the regime-behaviour features; but they fail to reproduce the detailed structure of autocorrelations and distort the statistics of the regimes. The MTV-type models that use sequential estimation of additional parameters (see also Section 2.3 here) do capture the univariate statistics of the leading PCs to a degree comparable with that of empirical–dynamical models, but do much less well on the full model's nonlinear dynamics, in particular on its multivariate PDF.

The EMR methodology thus occupies by now an important position at the lower and intermediate rungs of the full climate modelling hierarchy (Schneider & Dickinson 1974; Ghil & Robertson 2000; Held 2005). Further improvements in reduced models of various types can only benefit from, and to, a more complete and systematic exploration of this hierarchy.

Acknowledgements

We are grateful to A. W. Robertson, I. Kamenkovich and Y. Shprits for sharing their invaluable insights at various stages of this research. Our work was supported by NASA grant NNG-06AG66G-1 and DOE grant DE-FG02-02ER63413 (all co-authors). MG was also supported by the US Department of Energy grant DE-FG02-07ER64439 from its Climate Change Prediction Program, and by the European Commission's No. 12975 (NEST) project 'Extreme Events: Causes and Consequences (E2-C2),' while SK's partial support was provided by 2006 and 2007 grants within the University of Wisconsin–Milwaukee Research Growth Initiative program.

Appendix A. PCR and PLS regression

The multicollinearity problem can be avoided by finding linear combinations of original predictors whose time series are uncorrelated, while each linear combination accounts for the maximum possible fraction of the total variance. A natural way to determine this modified set of predictors is to apply PC analysis to the original vector of predictors, and then use cross-validation for finding the optimal number of PCs to retain in the regression; this procedure is called principal component regression (PCR; Wetherill 1986). Note that for the LIM model constructed in the phase space of the data set's EOFs (Penland 1989, 1996), the predictor variables are already uncorrelated. On the other hand, the predictors in the main level of (2.13) are the original set of PCs augmented by their quadratic combinations. Therefore, applying PC analysis to this new multivariate data set generally produces a different set of predictors.

Principal component regression does a fairly good job in picking the smallest set of uncorrelated predictors that capture most of the variance. However, the choice of the PCR predictors does not involve at all the information about how well these predictors are correlated with the response variable. The procedure that does take into account this additional information is called partial least-squares (PLS) regression; see Abdi (2003) for a brief, but comprehensive review. It is reasonable and advisable to apply PLS to the set of optimal predictors determined via PCR cross-validation, rather than to the original, much larger set of predictors.

Similarly to the PCR procedure, the leading PLS predictor is defined as a linear combination of the original predictor time series, but in this case the quantity being maximised is the correlation between this time series and the predictor time series. We found that applying PLS to each response variable, i.e. to each model tendency in (2.13), individually produces better results than the matrix formulation of the PLS algorithm; in the latter, one also considers linear combinations of all response variables and finds two sets of coefficients that define the mode of response and the mode of predictor variables that are maximally correlated (Abdi 2003). In the general multivariate case, the weights of the leading PLS mode are found using singular value decomposition (SVD; Press *et al.* 1994) as the first right singular vector of the matrix $\mathbf{X}^T\mathbf{Y}$, where $\mathbf{X}$ and $\mathbf{Y}$ are the matrices whose columns are the time series of the predictor and response variables, respectively. The right singular vectors of $\mathbf{X}^T\mathbf{Y}$ define the weights for the response variables; in the univariate case, the single such weight is naturally equal to 1.

The time series of the leading PLS mode is obtained by summing the original time series of the predictor variables with the weights obtained as above. The signal associated with the leading PLS mode is then regressed out of the response-variable time series, as well as out of all the predictor time series; once again, we only retain the residual of the linear regression of each of these time series onto the time series associated with the leading PLS mode. The procedure just described is then applied to the 'reduced' response and predictor time series to obtain the next PLS mode, and so on. The optimal number of modes to retain in this PLS procedure is also determined by cross-validation.

Appendix B. Selection of predictor variables

A few regression coefficients found by the application of PCR and PLS regularisation, as described in Appendix A, can be translated by trivial matrix manipulation into the coefficients of the EMR model in the original predictor-variable basis. Many of these coefficients are fairly small and do not contribute much to the predictive capability of the EMR model. The following procedure iteratively fine-tunes the selection of the predictor variables by throwing out original predictors whose corresponding regression coefficients are not significantly different from zero (Kravtsov *et al.* 2008).

This selection procedure is also based on subsampling of the original predictor and response variables. We first obtain 100 sets of regression coefficients by applying PCR-and-PLS regularisation to 100 randomly sampled subsets of the full original time series, each of which includes 80% of the original data points. The regression coefficients so obtained are then translated into the original predictor-variable space.

Now if the interval between the 2nd and 97th percentile of a given regression coefficient obtained as described above contains zero, we exclude the corresponding predictor variable from consideration, thus forming a new, smaller subset of predictor variables. This subset is in turn subsampled 100 times and subjected to PCR and PLS regression to identify coefficients not significantly different from zero, and so on, until all coefficients of the final set of predictors are found to be significant. The final regression coefficients are then found by applying the PCR-and-PLS regularisation to the fully sampled set of optimal predictors.

References

Abdi, H. 2003 Partial least squares (PLS) regression. In *Encyclopedia of Social Sciences Research Methods*, ed. M. Lewis-Beck, A. Bryman, T. Futing, Sage: Thousand Oaks, CA.

Achatz, U. & Branstator, G. W. 1999 A two-layer model with empirical linear corrections and reduced order for studies of internal climate variability. *J. Atmos. Sci.*, **56**, 3140–3160.

Achatz, U. & Opsteegh, J. D. 2003a Primitive-equation-based low-order models with seasonal cycle. Part I: model construction. *J. Atmos. Sci.*, **60**, 466–477.

Achatz, U. & Opsteegh, J. D. 2003b Primitive-equation-based low-order models with seasonal cycle. Part II: application to complexity and nonlinearity of large-scale atmosphere dynamics. *J. Atmos. Sci.*, **60**, 478–490.

Alexander, M. A., Blade, I., Newman, M. *et al.* 2002 The atmospheric bridge: the influence of ENSO teleconnections on air–sea interaction over the global oceans. *J. Climate*, **15**, 2205–2231.

Barnston, A. G., van den Dool, H. M. & Zebiak, S. E. 1994 Long-lead seasonal forecasts – where do we stand? *Bull. Am. Meteorol. Soc.*, **75**, 2097–2114.

Barnston, A. G., Glantz, M. H. & He, Y. 1999 Predictive skill of statistical and dynamical climate models in forecasts of SST during the 1997–98 El Niño episode and 1998 La Niña onset. *Bull. Am. Meteorol. Soc.*, **80**, 217–244.

Berloff, P. & McWilliams, J. C. 2002 Material transport in oceanic gyres. Part II: hierarchy of stochastic models. *J. Phys. Oceanogr.*, **32**, 797–830.

Berner, J. 2005 Linking nonlinearity and non-Gaussianity of planetary wave behavior by the Fokker–Planck equation. *J. Atmos. Sci.*, **62**, 2098–2117.

Box, G. E. P., Jenkins, G. M. & Reinsel, G. C. 1994 *Time Series Analysis, Forecasting, and Control*, 3rd edn. Prentice Hall.

Branstator, G. 1992 The maintenance of low-frequency atmospheric anomalies. *J. Atmos. Sci.*, **49**, 1924–1945.

Branstator, G. 1995 Organization of storm track anomalies by recurring low-frequency circulation anomalies. *J. Atmos. Sci.*, **52**, 207–226.

Branstator, G. & Berner, J. 2005 Linear and nonlinear signatures in the planetary wave dynamics of an AGCM: phase space tendencies. *J. Atmos. Sci.*, **62**, 1792–1811.

Branstator, G. & Haupt, S. E. 1998 An empirical model of barotropic atmospheric dynamics and its response to tropical forcing. *J. Climate*, **11**, 2645–2667.

Bretherton, C. S. & Battisti, D. S. 2000 An interpretation of the results from atmospheric general circulation models forced by the time history of the observed sea surface temperature distribution. *Geophys. Res. Lett.*, **27**, 767–770.

Cheng, X. H. & Wallace, J. M. 1993 Analysis of the northern-hemisphere wintertime 500-hPa height field spatial patterns. *J. Atmos. Sci.*, **50**, 2674–2696.

Crommelin, D. T. & Majda, A. J. 2004 Strategies for model reduction: comparing different optimal bases. *J. Atmos. Sci.*, **61**, 2206–2217.

D'Andrea, F. (2002). Extratropical low-frequency variability as a low-dimensional problem. Part II: stationarity and stability of large-scale equilibria. *Q. J. R. Meteorol. Soc.*, **128**, 1059–1073.

D'Andrea, F. & Vautard, R. 2001 Extratropical low-frequency variability as a low-dimensional problem. Part I: a simplified model. *Q. J. R. Meteorol. Soc.*, **127**, 1357–1374.

Dee, D. P., Cohn, S. E., Dalcher, A. & Ghil, M. 1985 An efficient alogorithm for estimating noise covariances in distributed systems. *IEEE Trans. Automatic Control*, **AC-30**, 1057–1065.

Deloncle, A., Berk, R. D'Andrea, F. & Ghil, M. 2007 Weather regime prediction using statistical learning, *J. Atmos. Sci.*, **64**(5), 1619–1635.

DelSole, T. 1996 Can the quasi-geostrophic turbulence be modeled stochastically? *J. Atmos. Sci.*, **53**, 1617–1633.

DelSole, T. 2000 A fundamental limitation of Markov models. *J. Atmos. Sci.*, **57**, 2158–2168.

DelSole, T. 2001 Optimally persistent patterns in time-varying fields. *J. Atmos. Sci.*, **58**, 1341–1356.

DelSole, T. 2004 Stochastic models of quasi-geostrophic turbulence. *Surveys Geophys.*, **25**, 107–149.
Deser, C. 2000 On the teleconnectivity of the "Arctic Oscillation." *Geophys. Res. Lett.*, **27**, 779–782.
Ehrendorfer, M. 2000 The total energy norm in a quasi-geostrophic model. *J. Atmos. Sci.*, **57**, 3443–3451.
Farrell, B. F. & Ioannou, P. J. 1993 Stochastic forcing of the linearized Navier–Stokes equations. *Phys. Fluids A*, **5**, 2600–2609.
Farrell, B. F. & Ioannou, P. J. 1995 Stochastic dynamics of the midlatitude atmospheric jet. *J. Atmos. Sci.*, **52**, 1642–1656.
Feldstein, S. B. 2000 The time scale, power spectra, and climate noise properties of teleconnection patterns. *J. Climate*, **13**, 4430–4440.
Franzke, C. & Majda, A. J. 2006 Low-order stochastic mode reduction for a prototype atmospheric GCM. *J. Atmos. Sci.*, **63**, 457–479.
Franzke, C., Majda, A. J. & Branstator, G. 2007 The origin of nonlinear signatures of planetary wave dynamics: mean phase space tendencies and contributions from non-Gaussianity. *J. Atmos. Sci.*, **64**, 3987–4003.
Franzke, C., Majda, A. J. & Vanden-Eijnden, E. 2005 Low-order stochastic mode reduction for a realistic barotropic model climate. *J. Atmos. Sci.*, **62**, 1722–1745.
Gardiner, C. W. 1985 *Handbook of Stochastic Methods*. Springer-Verlag.
Ghil, M. 1997 Advances in sequential estimation for atmospheric and oceanic flows. *J. Meteorol. Soc. Japan*, **75**, 289–304.
Ghil, M. 2001 Hilbert problems for the geosciences in the 21st century. *Nonlinear Proc. Geophys.*, **8**, 211–222.
Ghil, M. & Childress, S. 1987 *Topics in Geophysical Fluid Dynamics: Atmospheric Dynamics, Dynamo Theory and Climate Dynamics*. Springer-Verlag.
Ghil, M. & Jiang, N. 1998 Recent forecast skill for the El Niño/Southern Oscillation. *Geophys. Res. Lett.*, **25**, 171–174.
Ghil, M. & Robertson, A. W. 2000 Solving problems with GCMs: general circulation models and their role in the climate modeling hierarchy. In *General Circulation Model Development: Past, Present and Future*, ed. D. Randall, pp. 285–325. Academic Press.
Ghil, M. & Robertson, A. W. 2002 "Waves" vs. "particles" in the atmosphere's phase space: a pathway to long-range forecasting? *Proc. Natl. Acad. Sci.*, **99** (Suppl. 1), 2493–2500.
Ghil, M., Kondrashov, D., Lott, F. & Robertson, A. W. 2003 Intraseasonal oscillations in the mid-latitudes: observations, theory and GCM results. In *Proceedings of the ECMWF/CLIVAR Workshop on Simulation and Prediction of Intra-Seasonal Variability with Emphasis on the MJO, 3–6 November*. The European Centre for Medium-Range Weather Forecasts (ECMWF), pp. 35–53.
Goodman, J. & J. Marshall 1999 A model of decadal middle-latitude atmosphere–ocean coupled modes. *J. Climate*, **12**, 621–641.
Hannachi, A. 1997 Low-frequency variability in a GCM: three-dimensional flow regimes and their dynamics. *J. Climate*, **10**, 1357–1379.
Hannachi, A. & A. O'Neill, A. 2001 Atmospheric multiple equilibria and non-Gaussian behavior in model simulations. *Q. J. R. Meteorol. Soc.*, **127**, 939–958.
Hannachi, A., Stephenson, D. B. & Sperber, K. R. 2003 Probability-based methods for quantifying nonlinearity in the ENSO. *Clim. Dyn.*, **20**, 241–256.
Held, I. M. 2005 The gap between simulation and understanding in climate modeling. *Bull. Am. Meteorol. Soc.*, 1609–1614.

Jiang, N., Neelin, J. D. & Ghil, M. 1995 Quasi-quadrennial and quasi-biennial variability in the equatorial Pacific. *Clim. Dyn.*, **12**, 101–112.

Jin, F.-F., Neelin, J. D. & Ghil, M. 1996 El Niño/Southern Oscillation and the annual cycle: subharmonic frequency-locking and aperiodicity. *Physica D*, **98**, 442–465.

Kaplan, A., Cane, M., Kushnir, Y. *et al.* 1998 Analyses of global sea-surface temperature 1856–1991. *J. Geophys. Res.*, **103**, 18 567–18 589.

Kawanishi, T., Sezai, T., Itô, Y. *et al.* 2003 The Advanced Microwave Scanning Radiometer for the Earth Observing System (AMSR-E), NASDA's contribution to the EOS for Global Energy and Water Cycle Studies. *IEEE Trans. Geosci. Remote Sensing*, **41**, 184–194.

Khasminsky, R. Z. 1963 Principle of averaging for parabolic and elliptic differential equations and for Markov processes with small diffusion. *Theory Prob. Appl.*, **8**, 1–21.

Kimoto, M. & Ghil, M. 1993a Multiple flow regimes in the Northern Hemisphere winter. Part I: methodology and hemispheric regimes. *J. Atmos. Sci.*, **50**, 2625–2643.

Kimoto, M. & Ghil, M. 1993b Multiple flow regimes in the Northern Hemisphere winter. Part II: sectorial regimes and preferred transitions. *J. Atmos. Sci.*, **50**, 2645–2673.

Kondrashov, D., Ide, K. & Ghil, M. 2004 Weather regimes and preferred transition paths in a three-level quasi-geostrophic model. *J. Atmos. Sci.*, **61**, 568–587.

Kondrashov, D., Kravtsov, S. & Ghil, M. 2005 A hierarchy of data-based ENSO models. *J. Climate*, **18**, 4425–4444.

Kondrashov, D., Kravtsov, S. & Ghil M. 2006 Empirical mode reduction in a model of extratropical low-frequency variability. *J. Atmos. Sci.*, **63**, 1859–1877.

Kondrashov, D., Sun, C.-J. & Ghil M. 2008 Data assimilation for a coupled ocean–atmosphere model. Part II: parameter estimation. *Mon. Weather Rev.*, **136**, 5062–5076.

Kravtsov, S., Robertson, A. W. & Ghil M. 2003 Low-frequency variability in a baroclinic beta-channel with land–sea contrast. *J. Atmos. Sci.*, **60**, 2267–2293.

Kravtsov, S., Robertson, A. W. & Ghil M. 2005a Bimodal behavior in the zonal mean flow of a baroclinic beta-channel model. *J. Atmos. Sci.*, **62**,1746–1769.

Kravtsov, S., Kondrashov, D. & Ghil, M. 2005b Multi-level regression modeling of nonlinear processes: derivation and applications to climatic variability. *J. Climate*, **18**, 4404–4424.

Kravtsov, S., Kondrashov, D., Kamenkovich, I. and Ghil, M. 2008 An empirical stochastic model of sea-surface temperature and sea-level wind. Unpublished work.

Kurtz, T. G. 1973 A limit theorem for perturbed operators semigroups with applications to random evolution. *J. Funct. Anal.*, **12**, 55–67.

Kwasniok, F. 1996 The reduction of complex dynamical systems using principal interaction patterns. *Physica D*, **92**, 28–60.

Kwasniok, F. 2004 Empirical low-order models of barotropic flow. *J. Atmos. Sci.*, **61**, 235–245.

Landsea, C. W. & Knaff, J. A. 2000 How much "skill" was there in forecasting the very strong 1997–98 El Niño? *Bull. Am. Meteorol. Soc.*, **81**, 2107–2120.

Legras, B. & Ghil M. 1985 Persistent anomalies, blocking and variations in atmospheric predictability. *J. Atmos. Sci.*, **42**, 433–471.

Liu, W. T. 2002 Progress on scatterometer application. *J. Oceanogr.*, **58**, 121–136.

Lorenz, E. N. 1963 Deterministic nonperiodic flow. *J. Atmos. Sci.*, **20**, 130–141.

Lorenz, D. J. & Hartmann, D. L. 2001 Eddy–zonal flow feedback in the Southern Hemisphere. *J. Atmos. Sci.*, **58**, 3312–3327.

Lorenz, D. J. & Hartmann, D. L. 2003 Eddy–zonal flow feedback in the Northern Hemisphere winter. *J. Climate*, **16**, 1212–1227.

Marshall, J. & Molteni, F. 1993 Toward a dynamical understanding of atmospheric weather regimes. *J. Atmos. Sci.*, **50**, 1792–1818.

Majda, A. J. & Timofeyev, I. 2004 Low dimensional chaotic dynamics versus intrinsic stochastic chaos: a paradigm model. *Physica D*, **199**, 339–368.

Majda, A. J., Timofeyev, I. & Vanden-Eijnden, E. 1999 Models for stochastic climate prediction. *Proc. Natl. Acad. Sci. USA*, **96**, 14687–14691.

Majda, A. J., Timofeyev, I. & Vanden-Eijnden, E. 2001 A mathematical framework for stochastic climate models. *Commun. Pure Appl. Math.*, **54**, 891–974.

Majda, A. J., Timofeyev, I. & Vanden-Eijnden, E. 2002 A priori test of a stochastic mode reduction strategy. *Physica D*, **170**, 206–252.

Majda, A. J., Timofeyev, I. & Vanden-Eijnden, E. 2003 Systematic strategies for stochastic mode reduction in climate. *J. Atmos. Sci.*, **60**, 1705–1722.

Majda, A. J., Timofeyev, I. & Vanden-Eijnden, E. 2006 Stochastic models for selected slow variables in large deterministic systems. *Nonlinearity*, **19**, 769–794.

Majda, A. J., Franzke, C. & Khouider, B. 2008 An applied mathematics perspective on stochastic modelling for climate. *Phil Trans. R. Soc.*, **366**, 2429–2455.

McWilliams, J. C. 2007 Irreducible imprecision in atmospheric and oceanic simulations. *Proc. Nat. Acad. Sci.*, **104**, 8709–8713.

Mo, K. & Ghil, M. 1987 Cluster analysis of multiple planetary flow regimes. *J. Geophys. Res.*, **93D**, 10,927–10,952.

Molteni, F. 2002 Weather regimes and multiple equilibria. In *Encyclopedia of Atmospheric Science*, ed. J. R. Holton, J. Curry & J. Pyle, pp. 2577–2585. Academic Press.

Monahan, A. H. 2004 A simple model for skewness of global sea surface winds. *J. Atmos. Sci.*, **61**, 2037–2049.

Monahan, A. H. 2006a The probability distribution of sea surface wind speeds. Part I: theory and SeaWinds observations. *J. Climate*, **19**, 497–520.

Monahan, A. H. 2006b The probability distribution of sea surface wind speeds. Part II: dataset intercomparison and seasonal variability. *J. Climate*, **19**, 521–534.

Neelin, J. D., Latif, M. & Jin, F.-F. 1994 Dynamics of coupled ocean–atmosphere models: the tropical problem. *Annu. Rev. Fluid Mech.*, **26**, 617–659.

Neelin, J. D., Battisti, D. S., Hirst, A. C. *et al.* 1998 ENSO theory. *J. Geophys. Res.*, **103** (C7), 14 261–14 290.

Penland, C. 1989 Random forcing and forecasting using principal oscillation pattern analysis. *Mon. Weather Rev.*, **117**, 2165–2185.

Penland, C. 1996 A stochastic model of Indo-Pacific sea-surface temperature anomalies. *Physica D*, **98**, 534–558.

Penland, C. & Ghil, M. 1993 Forecasting Northern Hemisphere 700-mb geopotential height anomalies using empirical normal modes. *Mon. Weather Rev.*, **121**, 2355–2372.

Penland, C. & Matrosova, L. 1998 Prediction of tropical Atlantic sea-surface temperatures using linear inverse modeling. *J. Climate*, **11**, 483–496.

Penland, C. & Sardeshmukh, P. D. 1995 The optimal growth of tropical sea-surface temperature anomalies. *J. Climate*, **8**, 1999–2024.

Philander, S. G. H. 1990 *El Niño, La Niña, and the Southern Oscillation*. Academic Press.

Preisendorfer, R. W. 1988 *Principal Component Analysis in Meteorology and Oceanography*. Elsevier.

Press, W. H., Teukolsky, S. A., Vetterling, W. T. & Flannery, B. P. 1994 *Numerical Recipes*, 2nd edn. Cambridge University Press.

Reinhold, B. B. & Pierrehumbert, R. T. 1982 Dynamics of weather regimes: quasistationary waves and blocking. *Mon. Weather Rev.*, **110**, 1105–1145.
Rinne, J. & Karhila, V. 1975 A spectral barotropic model in horizontal empirical orthogonal functions. *Q. J. R. Meteorol. Soc.*, **101**, 365–382.
Robinson, W. 1996 Does eddy feedback sustain variability in the zonal index? *J. Atmos. Sci.*, **53**, 3556–3569.
Robinson, W. 2000 A baroclinic mechanism for the eddy feedback on the zonal index. *J. Atmos. Sci.*, **57**, 415–422.
Schneider, S. H. & Dickinson, R. E. 1974 Climate modeling. *Rev. Geophys. Space Phys.*, **12**, 447–493.
Schubert, S. D. 1985 A statistical-dynamical study of empirically determined modes of atmospheric variability. *J. Atmos. Sci.*, **42**, 3–17.
Seager R., Blumenthal, M. B. & Kushnir, Y. 1995 An advective atmospheric mixed layer model for ocean modeling purposes: global simulation of surface heat fluxes. *J. Climate*, **8**, 1951–1964.
Selten, F. 1993 Toward an optimal description of atmospheric flow. *J. Atmos. Sci.*, **50**, 861–877.
Selten, F. M. 1995 An efficient description of the dynamics of the barotropic flow. *J. Atmos. Sci.*, **52**, 915–936.
Selten, F. M. 1997 Baroclinic empirical orthogonal functions as basis functions in an atmospheric model. *J. Atmos. Sci.*, **54**, 2100–2114.
Selten, F. M. & Branstator, G. 2004 Preferred regime transition routes and evidence for an unstable periodic orbit in a baroclinic model. *J. Atmos. Sci.*, **61**, 2267–2282.
Smyth, P., Ide, K. & Ghil, M. 1999 Multiple regimes in Northern Hemisphere height fields via mixture model clustering. *J. Atmos. Sci.*, **56**, 3704–3723.
Stephenson, D. B., Pavan, V. & Bojariu, R. 2000 Is the North Atlantic Oscillation a random walk? *Int. J. Climatol.*, **20**, 1–18.
Strong, C. M., Jin, F.-F. & Ghil, M. 1995 Intraseasonal oscillations in the barotropic model with annual cycle, and their predictability. *J. Atmos. Sci.*, **52**, 2627–2642.
Strounine, K., Kravtsov, S., Kondrashov, D. & Ghil, M. 2008 Reduced models of extratropical low-frequency variability: parameter estimation and comparative performance. Unpublished work.
Sura, P. 2003 Stochastic analysis of Southern and Pacific Ocean sea surface winds. *J. Atmos. Sci.*, **60**, 654–666.
Sura, P. & Gille, S. 2003 Intepreting wind-driven Southern Ocean variability in a stochastic framework. *J. Mar. Res.*, **61**, 313–334.
Sura, P. & Newman, M. 2008 The impact of rapid wind variability upon air–sea thermal coupling. *J. Climate*, **21**, 621–637.
Sura, P., Newman, M., Penland, C. & Sardeshmukh, P. D. 2005 Multiplicative noise and non-Gaussianity: a paradigm for atmospheric regimes? *J. Atmos. Sci.*, **62**, 1391–1409.
Sura, P. & Sardeshmukh, P. D. 2008 A global view of non-Gaussian SST variability. *J. Phys. Oceanogr.*, **38**, 639–647.
Sura, P., Newman, M. & Alexander, M. A. 2006 Daily to decadal sea-surface temperature variability driven by state-dependent stochastic heat fluxes. *J. Phys. Oceanogr.*, **36**, 1940–1958.
Tennekes, H. & Lumley, J. L. 1972 *A First Course in Turbulence*. MIT Press.
Wallace, J. M. 2000 North Atlantic Oscillation/annular mode: two paradigms – one phenomenon. *Q. J. R. Meterol. Soc.*, **126**, 791–805.
Wallace, M. & Gutzler, D. S. 1981 Teleconnections in the geopotential height field during the Northern Hemisphere winter. *Mon. Weather Rev.*, **109**, 784–812.

Wetherill, G. B. 1986 *Regression Analysis with Applications*. Chapman and Hall.
Winkler, C. R., Newman, M. & Sardeshmukh, P. D. 2001 A linear model of wintertime low-frequency variability. Part I: formulation and forecast skill. *J. Climate*, **14**, 4474–4494.
Zhang, Y. & Held, I. M. 1999 A linear stochastic model of a GCM's midlatitude storm tracks. *J. Atmos. Sci.*, **56**, 3416–3435.

3

An applied mathematics perspective on stochastic modelling for climate

ANDREW J. MAJDA, CHRISTIAN FRANZKE AND BOUALEM KHOUIDER

Systematic strategies from applied mathematics for stochastic modelling in climate are reviewed here. One of the topics discussed is the stochastic modelling of mid-latitude low-frequency variability through a few teleconnection patterns, including the central role and physical mechanisms responsible for multiplicative noise. A new low-dimensional stochastic model is developed here, which mimics key features of atmospheric general circulation models, to test the fidelity of stochastic mode reduction procedures. The second topic discussed here is the systematic design of stochastic lattice models to capture irregular and highly intermittent features that are not resolved by a deterministic parameterisation. A recent applied mathematics design principle for stochastic column modelling with intermittency is illustrated in an idealized setting for deep tropical convection; the practical effect of this stochastic model in both slowing down convectively coupled waves and increasing their fluctuations is presented here.

3.1 Introduction

Stochastic modelling for climate is important for understanding the intrinsic variability of dominant low-frequency teleconnection patterns in climate, to provide cheap low-dimensional computational models for the coupled atmosphere–ocean system and to reduce model error in standard deterministic computer models for extended-range prediction through appropriate stochastic noise (Palmer 2001).

This chapter is a research-expository paper on systematic strategies for stochastic climate modelling from the perspective of modern applied mathematics. In the modern applied mathematics, 'modus operandi' rigorous mathematical analysis, qualitative, asymptotic and numerical modelling are all blended together in a multidisciplinary fashion to provide systematic guidelines to address real-world problems (Majda 2000). For stochastic modelling in climate, the modern applied mathematics tool kit includes stochastic differential equations and discontinuous Markov jump processes (Gardiner 1985), systematic asymptotic reduction techniques, nonlinear dynamical systems theory and ideas from both statistical physics (Majda & Wang 2006) and mathematical statistics (Kravtsov *et al.* 2005; Majda *et al.* 2006a); mathematical rigour provides unambiguous guidelines in idealised models. Another facet of the modern applied mathematics philosophy is the development of qualitative models that represent a Platonic ideal for central issues simultaneously in diverse scientific disciplines such as material science, biomolecular dynamics and climate science.

In Section 3.2, we illustrate and apply this modern applied mathematics philosophy to stochastic modelling of the low-frequency variability of the atmosphere. The systematic mathematical theory (Majda *et al.* 1999, 2001, 2002, 2003, 2006b; collectively referred to as MTV hereafter; Franzke *et al.* 2005; Franzke & Majda 2006) for these problems is briefly reviewed including the central role and physical mechanisms responsible for multiplicative noise in the low-frequency dynamics. In this context, the Platonic ideal from applied mathematics is the truncated Burgers–Hopf model (Majda & Timofeyev 2000). A new simplified low-dimensional stochastic model that reproduces key features of atmospheric general circulation models (GCMs) is used there to test the fidelity of stochastic mode reduction techniques. A recent diagnostic statistical test with firm mathematical underpinning for understanding and interpreting the dynamical sources of the small departures from Gaussianity in low-frequency variables (Franzke *et al.* 2007) is also developed there.

While Section 3.2 deals with applied mathematical modelling through stochastic differential equations and Section 3.3 is devoted to the systematic development of stochastic lattice models to capture unresolved features that are highly intermittent in space and time, such as deep convective clouds, cloud cover in subtropical boundary layers, sub-mesoscale eddies in the ocean and mesoscale sea-ice cover. Here the mathematical tools involve a family of discontinuous Markov jump processes with multiscale behaviour in space time called stochastic spin-flip models. The key mathematical development involves systematic strategies to coarse grain such stochastic spin-flip models to achieve computational efficiency while retaining crucial features of the microscale interactions (Katsoulakis & Vlachos 2003; Katsoulakis *et al.* 2003a,b). The use of such stochastic lattice models to parameterise key features of tropical convection is briefly reviewed (Majda & Khouider 2002; Khouider *et al.* 2003). For the coupling of continuum models like a GCM to a stochastic lattice model as well as in many diverse applications, an applied mathematics Platonic ideal model has recently been introduced and analysed by Katsoulakis *et al.* (2004, 2005, 2006, 2007; hereafter KMS). This model consists of a system of ordinary differential equations (ODEs) for continuum variables $\boldsymbol{X}$,

$$\frac{\mathrm{d}\boldsymbol{X}}{\mathrm{d}t} = \boldsymbol{F}(\boldsymbol{X}, \overline{\sigma}), \tag{3.1.1}$$

two-way coupled to a stochastic spin-flip model written abstractly here as

$$\frac{\mathrm{d}}{\mathrm{d}t}\mathbb{E}f(\sigma) = \mathbb{E}Lf(\sigma), \tag{3.1.2}$$

where $\overline{\sigma}$ denotes the spatial coverage; L is the generator; f is a test function; and $\mathbb{E}$ denotes the expected value. This idealised class of models has been used systematically to analyse the effects of various coarse-graining procedures on processes with intermittency, large-scale bifurcations and microscale phase transitions (KMS 2004, 2005, 2006, 2007). A concrete example for tropical convection in climate is given in Section 3.3. A new application of these stochastic lattice models to capture intermittent features and improve the fidelity of deterministic parameterisations of convection with clear deficiencies is also developed in Section 3.3. First, the systematic design principles for (3.1.1) and (3.1.2) (KMS 2006, 2007) are used to calibrate a stochastic column model for tropical convection with intermittency and then the new results are presented on the practical effect of slowing down convectively coupled waves and increasing their fluctuations through the stochastic lattice models.

3.2 Systematic low-dimensional stochastic mode reduction and atmospheric low-frequency variability

A remarkable fact of Northern Hemisphere low-frequency variability is that it can be efficiently described by only a few teleconnection patterns that explain most of the total variance (e.g. Wallace & Gutzler 1981). These few teleconnection patterns not only exert a strong influence on regional climate and weather but are also related to climate change (Hurrell 1995). These properties of teleconnection patterns make them an attractive choice as basis functions for climate models with a highly reduced number of degrees of freedom. The development of such reduced climate models involves the solution of two major issues: (i) how to properly account for the unresolved modes that are also known as the closure problem, and (ii) how to define a small set of basis functions that optimally represent the dynamics of the major teleconnection patterns. This section addresses primarily the first issue and presents a rigorous strategy of how to systematically account for the unresolved degrees of freedom.

The simplest approach to derive highly truncated models of teleconnection patterns is empirically to fit simple stochastic models (e.g. autoregressive models and fractionally differenced models) to individual scalar teleconnection indices

(Feldstein 2000; Stephenson *et al.* 2000; Percival *et al.* 2001). Statistical tests usually cannot distinguish if short- or long-memory models provide the better fit. A more complex approach, which also tries to capture deterministic interactions between different teleconnection patterns, is to linearise the equations of motion around a climatological mean state. Such models can be determined empirically from data or by using the linearised equations of motion. These models can be forced either by a random forcing (Branstator 1990; Newman *et al.* 1997; Whitaker & Sardeshmukh 1998; Zhang & Held 1999) or by an external forcing representing tropical heating (Branstator & Haupt 1998). To ensure stability of these linear models, damping is added according to various ad hoc principles. There is a recent survey of such modelling strategies (DelSole 2004).

A more powerful method is to empirically fit nonlinear stochastic models with possibly multiplicative (state dependent) noise by using the Fokker–Planck equation (Gardiner 1985; Sura 2003; Berner 2005). To reliably estimate the drift and diffusion coefficients in the Fokker–Planck equation is a subtle inverse problem that requires very long time series, and is further complicated by the need to retain the leading-order eigenvalue structure of the Fokker–Planck operator in order to keep the autocorrelation time scales of the original model (Crommelin & Vanden-Eijnden 2006); the fitting procedure in most of the recent work is the most attractive current regression strategy for low-frequency behaviour. Recently, Kravtsov *et al.* (2005) have developed a simplified nonlinear regression strategy that produces very good results for a three-layer quasi-geostrophic model with a realistic climate. However, order 2000 regression coefficients need to be fitted in a model with order 1000 state variables to achieve these results. Some inherent limitations of this approach in describing the correct physics are discussed briefly below in a simplified model.

All the work presented above derives reduced models by regression fitting of the resolved modes. Another approach is to take advantage of the basis function property of teleconnection patterns. Schubert (1985), Selten (1995), Achatz & Branstator (1999) and Achatz & Opsteegh (2003a,b) developed low-order models with empirical orthogonal functions (EOFs) as basis functions. Truncated EOF models experience climate drift due to the neglected interactions with the unresolved modes. Selten (1995) and Achatz & Branstator (1999) parameterise these neglected interactions by a linear damping, whose strength is determined empirically. A possibly more powerful tool to represent the dynamics of a system is principal interaction patterns (PIPs; Hasselmann 1988; Kwasniok 1996, 2004). The calculation of PIPs takes into account the dynamics of the model for which one tries to find an optimal basis and also often involves ad hoc closure through linear damping and an ansatz for nonlinear interactions. Crommelin & Majda (2004) compare different optimal bases. They find that the models based on PIPs are superior to models based on EOFs. On the other hand, they also point out that the determination of PIPs can show sensitivities regarding the calculation procedure, at least for some low-order atmospheric dynamical systems with regime transitions. Furthermore, PIPs have two more disadvantages: (i) much higher computational cost than EOFs, and (ii) one needs not only data as for EOFs but also the dynamical equations to calculate PIPs. These features can make PIPs possibly a less attractive basis.

Majda *et al.* (1999, 2001, 2002, 2003, 2005, 2006a,b) provide a systematic framework for how to account for the effect of the fast degrees of freedom on the slow modes in combination with using the dominant teleconnection patterns as basis functions. In contrast to the empirical fitting procedures applied in the studies discussed above, the stochastic mode reduction strategy put forward in MTV *predicts* the functional form of all deterministic and stochastic correction terms and provides a *minimal* regression fitting procedure of only the *fast modes* (Franzke *et al.* 2005; Franzke & Majda 2006). In general, only an estimate for the variance and eddy turnover time for each fast mode is needed. It has been applied and tested on a wide variety of simplified models and examples.

3.2.1 Overview of the MTV strategy

We illustrate the ideas for stochastic climate modelling by considering the following prototype equation for geophysical flow:

$$\frac{\partial u}{\partial t} = F + Lu + B(u, u). \tag{3.2.1}$$

The above functional form (3.2.1) is typical of dry dynamical cores of climate models, but the MTV strategy is easily extended to include non-quadratic nonlinearities such as associated with boundary fluxes and moist processes. In stochastic climate modelling, the variable u is decomposed into an orthogonal decomposition through the variables $\tilde{u}$ and u', which are characterised by strongly differing time scales (MTV 1999; Majda *et al.* 2005). The variable $\tilde{u}$ denotes a slow low-frequency mode (also referred to as climate mode) of the system, which evolves slowly in time compared with the u' variables (also referred to as fast mode). By decomposing $u = \tilde{u} + u'$ in terms of some optimal energy norm basis, we can write them as

$$u = \sum_{i=1}^{N} a_i e_i = \sum_{i=1}^{R} \alpha_i e_i + \sum_{j=R+1}^{N} \beta_j e_j, \tag{3.2.2}$$

with $\tilde{u} = \sum_{i=1}^{R} \alpha_i e_i$ and $u' = \sum_{j=R+1}^{N} \beta_j e_j$, where R is the number of climate modes; a_i denotes the expansion coefficients; and α_i and β_j are the expansion coefficients of the slow (fast) modes. The use of the energy norm ensures the conservation of energy by the nonlinear operator (Selten 1995). By properly projecting the energy norm basis, derived from the geophysical model, onto equation (3.2.1), we get two sets of equations for slow α_i and fast β_i modes

$$\dot{\alpha}_i(t) = \varepsilon H_i^{\alpha} + \sum_j L_{ij}^{\alpha\alpha} \alpha_j(t) + \frac{1}{\varepsilon} \sum_j L_{ij}^{\alpha\beta} \beta_j(t) + \sum_{jk} B_{ijk}^{\alpha\alpha\alpha} \alpha_j(t) \alpha_k(t)$$
$$+ \frac{2}{\varepsilon} \sum_{jk} B_{ijk}^{\alpha\alpha\beta} \alpha_j(t) \beta_k(t) + \frac{1}{\varepsilon} \sum_{jk} B_{ijk}^{\alpha\beta\beta} \beta_j(t) \beta_k(t), \tag{3.2.3}$$

$$\dot{\beta}_i(t) = \varepsilon H_i^{\beta} + \frac{1}{\varepsilon} \sum_j L_{ij}^{\beta\alpha} \alpha_j(t) + \frac{1}{\varepsilon} \sum_j L_{ij}^{\beta\beta} \beta_j(t) + \frac{1}{\varepsilon} \sum_{jk} B_{ijk}^{\beta\alpha\alpha} \alpha_j(t) \alpha_k(t)$$
$$+ \frac{2}{\varepsilon} \sum_{jk} B_{ijk}^{\beta\alpha\beta} \alpha_j(t) \beta_k(t) + \frac{1}{\varepsilon^2} \sum_{jk} B_{ijk}^{\beta\beta\beta} \beta_j(t) \beta_k(t), \tag{3.2.4}$$

where the nonlinear operators have been symmeterised, that is, $B_{ijk} = B_{ijk}$ in (3.2.3) and (3.2.4). The upper indices α and β indicate the respective subsets of the full operators in (3.2.1). Here ε is a small positive parameter that controls the separation of time scale between the slow and fast modes and measures the ratio of the correlation time of the slowest non-climate mode u' to the fastest climate mode $\tilde{u}$. In placing the parameter in front of particular terms, we tacitly assume that they evolve on a faster time scale than the terms involving the climate modes alone (see MTV 2001 and Franzke *et al.* 2005 for more details). Ultimately, ε is set to the value $\varepsilon = 1$ in developing all the final results (MTV 2002, 2003), that is, introducing ε is only a technical step in order to carry out the MTV mode reduction strategy. Such a use of ε has been checked on a wide variety of idealised examples where the actual value of ε ranges from quite small to order one (MTV 2002, 2003, 2006b; Majda & Timofeyev 2004). Following MTV (1999, 2001, 2002, 2003, 2006b) and Franzke *et al.* (2005), the mode elimination procedure is based on the assumption that the dynamics of the fast modes alone in (3.2.4), that is, the dynamical system

$$\dot{c}_i = \sum_{jk} B_{ijk}^{\beta\beta\beta} c_j c_k \tag{3.2.5}$$

is ergodic and mixing with integrable decay of correlation. In other words, we assume that for almost all initial conditions, and suitable functions f and g, we have

$$\lim_{T\to\infty} \frac{1}{T} \int_0^T f(c(t))\, \mathrm{d}t = \langle f \rangle, \tag{3.2.6}$$

where $\langle\cdot\rangle$ denotes expectation with respect to some appropriate invariant distribution and

$$G(s) = \lim_{T\to\infty} \frac{1}{T} \int_0^T g(c(t+s),\, c(t))\,\mathrm{d}t - \lim_{T\to\infty} \frac{1}{T^2} \int_0^T \int_0^T g(c(t),\, c(t'))\,\mathrm{d}t\,\mathrm{d}t' \tag{3.2.7}$$

is an integrable function of s, that is, $|\int_0^\infty G(s)\,\mathrm{d}s| < \infty$. Furthermore, we assume that the low-order statistics for the fast modes in (3.2.5) are Gaussian. Under the above assumptions, it can be shown in the limit $\varepsilon \to 0$ (Kurtz 1973; MTV 2001) that the dynamics of the slow modes α_i in (3.2.3) can be written as the following Itô stochastic equation for the slow modes alone:

$$\begin{aligned}
\mathrm{d}\alpha_i(t) = \lambda_{\mathrm{B}} &\left(H_i^{\alpha}\mathrm{d}t + \sum_j L_{ij}^{\alpha\alpha}\alpha_j(t)\mathrm{d}t + \sum_{jk} B_{ijk}^{\alpha\alpha\alpha}\alpha_j(t)\alpha_k(t)\mathrm{d}t \right) \\
&+ \lambda_{\mathrm{A}}^2 \sum_j \tilde{L}_{ij}^{(2)}\alpha_j(t)\mathrm{d}t + \lambda_{\mathrm{A}}\sqrt{2}\sum_j \sigma_{ij}^{(2)}\mathrm{d}W_j^{(2)} \\
&+ \lambda_{\mathrm{M}}^2 \left(\sum_j \tilde{L}_{ij}^{(3)}\alpha_j(t)\mathrm{d}t + \sum_{jkl} \tilde{M}_{ijkl}\alpha_j(t)\alpha_k(t)\alpha_l(t)\mathrm{d}t \right) \\
&+ \lambda_{\mathrm{L}}^2 \left(\sum_j \tilde{L}_{ij}^{(1)}\alpha_j(t)\mathrm{d}t) + \lambda_{\mathrm{M}}\lambda_{\mathrm{L}}(\tilde{H}_j^{(1)}\mathrm{d}t + \sum_{jk} \tilde{B}_{ijk}\alpha_j(t)\alpha_k(t)\mathrm{d}t \right) \\
&+ \lambda_{\mathrm{A}}\lambda_{\mathrm{F}}\tilde{H}_j^{(2)}\mathrm{d}t + \sqrt{2}\sum_j \sigma_{ij}^{(1)}(\alpha(t))\mathrm{d}W_j^{(1)},
\end{aligned} \tag{3.2.8}$$

where the nonlinear noise matrix $\sigma^{(1)}$ satisfies

$$\lambda_{\mathrm{L}}^2 Q_{ij}^{(1)} + \lambda_{\mathrm{L}}\lambda_{\mathrm{M}} \sum_k U_{ijk}\alpha_k(t) + \lambda_{\mathrm{M}}^2 \sum_{kl} V_{ijkl}\alpha_k(t)\alpha_l(t) = \sum_k \sigma_{ik}^{(1)}(\alpha(t))\sigma_{jk}^{(1)}(\alpha(t)). \tag{3.2.9}$$

It is guaranteed (MTV 2001) that the operator on the left-hand side of (3.2.9) is always positive definite ensuring the existence of the nonlinear noise matrix on the right-hand side. All coefficients are defined explicitly in MTV (2001) and Franzke *et al.* (2005). A comprehensive mathematical theory of the stochastic mode reduction strategy for geophysical applications is developed in MTV (2001) with many new mathematical phenomena in the resulting equations explored there.

To see which of these correction terms play a vital role in the integrations of the low-order stochastic model (3.2.8), we grouped the interaction terms between the slow and fast modes according to their physical origin and set a parameter λ_i in front of the corresponding interaction coefficient (see Franzke *et al.* 2005; Franzke & Majda 2006 for more details). The bare truncation is indicated by a λ_B and describes the interaction between the slow modes. The interaction between the triads $B^{\alpha\beta\beta}$ and $B^{\beta\alpha\beta}$ gives rise to additive noise and a linear correction term and arises from the advection of the fast modes by the slow ones; we name these triads 'additive' triads and set a λ_A in front of them (MTV 1999, 2001, 2002, 2003). The other type of triad interaction is between $B^{\alpha\alpha\beta}$ and $B^{\beta\alpha\alpha}$. These interactions create multiplicative noises and cubic nonlinear correction terms (MTV 1999, 2001, 2002, 2003); we call them 'multiplicative' triads hereafter and indicate them by a λ_M. These triad interactions describe the advection of the slow modes by the fast modes that induce tendencies in the slow modes. The linear coupling between the slow and fast modes $L^{\alpha\alpha}$ and $L^{\beta\alpha}$ gives rise to additive noise and a linear correction term (MTV 2001; Franzke *et al.* 2005), which is called the augmented linearity here and is indicated by a λ_L. The augmented linearity describes the effect of the linear interaction between the fast (slow) modes and the climatological mean state onto the slow (fast) modes and is the main interaction captured in the linear stochastic modelling strategy (DelSole 2004). We set a λ_F in front of the last remaining interaction term $L^{\beta\beta}$, the linear coupling of the fast modes. The quadratic nonlinear corrections, a forcing term and a further multiplicative noise contribution are caused by the interaction between the linear coupling terms and the multiplicative triads. Another forcing correction term comes from the interaction between additive triads and the linear coupling of the fast modes.

In chapter 3 of Majda *et al.* (2005), a simplified three-mode elementary 'toy climate model' is discussed and the MTV procedure is applied explicitly to that example. The origin of all the terms in (3.2.8) is developed in a transparent fashion in these examples. Once the low-order stochastic model has been developed from the above procedure, one can assess the importance of the various deterministic and stochastic processes systematically by varying the coefficients λ_B, λ_A, λ_L, λ_M and λ_F systematically in (3.2.8) (Franzke *et al.* 2005) and even develop simple physically motivated regression fitting strategies (Franzke & Majda 2006). In interesting recent work, Sura & Sardeshmukh (2008) have used scalar linear stochastic models with multiplicative and additive noise to explain non-Gaussian sea-surface temperature (SST) variability. If the reduced stochastic models in (3.2.8) from the MTV procedure are linearised at the climate mean state, they automatically produce vector systems of linear stochastic equations with both multiplicative and additive noise with the same structure, with clear sources for the underlying physical contributions to this equation.

3.2.2 Idealised models for stochastic mode reduction

The idealised models, where the procedure has been tested, have order 100 degrees of freedom and include those with trivial climates (MTV 2002), periodic orbits or multiple equilibria (MTV 2003) and heteroclinic chaotic orbits coupled to a deterministic bath of modes satisfying the truncated Burgers–Hopf equation (Majda & Timofeyev 2000, 2004). The truncated Burgers–Hopf equation is a toy model with some remarkable features mimicking behaviour in the real atmosphere; it has a well-defined equipartition spectrum and a simple scaling theory for correlations with the large scales decorrelating more slowly than the small scales, that is, low-frequency variability. Furthermore, these predictions are confirmed with very high precision by numerical simulations (Majda & Timofeyev 2000; Majda & Wang 2006). The MTV procedure has been validated in these examples even when there is little separation of time scales between the slow and fast modes. In the example of a four-dimensional resonant system with chaotic dynamics coupled to the truncated Burgers system, only one empirical regression fitting coefficient is used and complex bifurcation diagrams and probability distribution functions (PDFs) in a climate change scenario are reproduced by the four-dimensional stochastic mode reduction resulting from the MTV procedure applied to the 104 degrees of freedom deterministic system. An especially stringent recent test is the application of this procedure to the first few large-scale modes of the truncated Burgers equations in the turbulent cascade (MTV 2006b).

3.2.3 A-priori stochastic modelling for mountain torque

The ideal barotropic quasi-geostrophic equations with a large-scale zonal mean flow U on a $2\pi \times 2\pi$ periodic domain (Carnevale & Frederiksen 1987) are given by

$$\left.\begin{aligned} \frac{\partial q}{\partial t} + \nabla^{\perp}\psi \cdot \nabla q + U\frac{\partial q}{\partial x} + \beta\frac{\partial \psi}{\partial x} &= 0, \\ q &= \Delta\psi + h \\ \frac{\mathrm{d}U}{\mathrm{d}t} &= \frac{1}{4\pi^2}\int h\frac{\partial \psi}{\partial x}\mathrm{d}x\,\mathrm{d}y, \end{aligned}\right\} \tag{3.2.10}$$

where q is the potential vorticity; U is the large-scale zonal mean flow; ψ is the stream function; and h is the topography. In (3.2.10), the mean flow changes in time through the topographic stress; this effect is the direct analogue for periodic geometry of the change in time of angular momentum due to mountain torque in spherical geometry (Frederiksen *et al.* 1996; Majda & Wang 2006). Here the a-priori stochastic modelling strategy (MTV 1999, 2001) is applied to the stochastic modelling of the topographic stress terms in (3.2.10) as an analogue for mountain torque; thus, the variable U is the slow variable while all the modes ψ_k are fast variables for the MTV procedure.

In this example, the systematic stochastic modelling procedure (MTV 2003) results in the predicted nonlinear reduced equation for U,

$$\frac{dU}{dt} = -\gamma(U)U + \frac{\gamma'(U)}{\alpha\mu} + \sqrt{\frac{2\gamma(U)}{\alpha\mu}}\dot{W}, \tag{3.2.11}$$

where $\gamma'(U) = d\gamma/dU$ and

$$\gamma(U) = 2\sum_k \frac{\mu k_x^2 |H_k|^2 \gamma_k}{\gamma_k^2 + (\Omega_k - k_x U)^2}. \tag{3.2.12}$$

Here, $\Omega_k = k_x\beta/|\boldsymbol{k}|^2 - \overline{U}k_x$ is the Rossby wave frequency, Doppler shifted by the mean flow. Under the additional assumption that $|k_x U|^2 \ll \gamma_k^2 + (\Omega_k)^2$, a standard predicted linear stochastic model for U emerges from (3.2.11) with $\gamma = \gamma(0)$ from (3.2.12) and $\gamma'(0) = 0$ (MTV 1999, 2001; see MTV 2003 for definitions of μ and $\gamma_{\boldsymbol{k}}$). It is shown in MTV (2003) that this nonlinear stochastic equation is superior to the linear one for large-amplitude topography, where $\varepsilon \cong 0.7$. This is the simplest example with multiplicative noise. Egger (2005) has used the systematic strategies from MTV (1999, 2003) to improve regression strategies for analysing observational data for angular momentum. In MTV (2006a), the recent systematic regression fitting strategy mentioned earlier (Crommelin & Vanden-Eijnden 2006) is applied to (3.2.10) and independently confirms the predictions in (3.2.11) and (3.2.12).

3.2.4 Geophysical and climate models

Franzke *et al.* (2005) put the above systematic stochastic mode reduction strategy in a form that makes the practical implementation of the MTV procedure in the complex geophysical models simpler with the same reduced stochastic equations for the fast modes. The EOFs in an appropriate metric are used to distinguish between the slow and fast modes in high-dimensional geophysical systems, such as climate models. Even though EOFs do not strictly decompose the modes according to their autocorrelation time scale, the leading EOFs usually evolve on a slower time scale than the higher EOFs. In the study by Franzke *et al.* (2005), a T21-truncated barotropic model on the sphere with a realistic climate was used to derive low-order stochastic models based on kinetic energy norm EOFs by the MTV strategy. Low-order models with as little as two slow modes succeed in capturing the geographical distributions of the climatological mean field, the variance and the eddy forcing. Furthermore, the envelope of the autocorrelation functions is captured reasonably well.

Recently, Franzke & Majda (2006) applied the systematic stochastic mode reduction strategy to a baroclinic three-layer quasi-geostrophic model on the sphere (Marshall & Molteni 1993), which mimics the climatology of the European Centre for Medium-Range Weather Forecasts reanalysis data. The low-order stochastic climate model consists of climate modes as slow modes defined as the leading total energy norm EOFs and the stochastic mode reduction procedure predicts all forcing, linear, quadratic and cubic correction terms as well as additive and multiplicative noises; these correction terms and noises account for the interaction of the climate modes with the neglected non-climate modes and the self-interaction among the non-climate modes. For the three-layer quasi-geostrophic model, low-order stochastic models with 10 or fewer climate modes reproduce the geographical distributions of the standard deviation and eddy forcing well. They underestimate the standard deviations by at most a factor of approximately 1.5. Furthermore, they reproduce the autocorrelation functions reasonably well. A budget analysis shows that both linear and nonlinear correction terms as well as both additive and multiplicative noises are important. The physical intuition behind the noises as derived from the MTV procedure is as follows: the additive noise stems from the linear interaction between the fast modes and the climatological mean state and the multiplicative noise comes from the advection of the slow modes by the fast modes. All these deterministic correction terms and noises (both additive and multiplicative) are *predicted* by the systematic stochastic mode reduction strategy, whereas previous studies a priori approximate the nonlinear part of the equations by a linear operator and additive noise. This noise is typically white in time but may be spatially correlated. In other words, these studies truncate the dynamics on both the slow and fast modes and add ad hoc damping in order to stabilise the linear model (Whitaker & Sardeshmukh 1998; Zhang & Held 1999). The systematic MTV approach summarised briefly above truncates the dynamics only on the fast modes and predicts the functional form of all necessary nonlinear correction terms and noises; therefore, it also predicts the necessary damping.

The MTV stochastic climate models for this application experience some climate drift. A minimal empirical MTV model without climate drift can be constructed through three-parameter regression fitting by downscaling the bare truncation terms and upscaling the two important MTV processes (augmented linearity and multiplicative triads). These empirical MTV stochastic climate models with minimal regression fitting still capture the geographical distribution of the standard deviation and eddy forcing and the autocorrelation functions reasonably well, while not experiencing climate drift. This surprising result can be interpreted as the fact that the climate modes are predominantly driven by the fast modes and the self-interactions among the slow modes are less important, as can already be seen from the bare-truncation models, which do not capture any feature of the actual dynamics. Furthermore, these empirical MTV stochastic climate models suggest that the bare truncation is probably the cause of the climate drift. Integrations of bare-truncation models (without any MTV correction terms) already produce a big climate drift (Franzke & Majda 2006). The MTV mode reduction procedure is able to reduce the climate drift in most of the slow modes, but is not able to overcome it completely. Previous results with a variety of simplified models show no climate drift in an MTV framework (MTV 1999, 2001, 2002, 2003; Majda & Timofeyev 2004). This is probably because these simplified models are constructed in such a way that they have an optimal basis that captures the dynamics of the climate modes. This gives evidence that total energy norm EOFs are not an adequate dynamical basis in capturing the dynamics of the slow modes. Further details of this application can be found in Franzke & Majda (2006).

3.2.5 A simple stochastic model with key features of atmospheric low-frequency variability

In this section, we present a four-mode stochastic climate model of the kind put forward in Majda *et al.* (2005, chapter 3). This simple stochastic climate model is set up in such a way that it features many of the important dynamical features of comprehensive GCMs but with many fewer degrees of freedom. Such simple toy models allow the efficient exploration of the whole parameter space that is impossible to conduct with GCMs. Thus, we are able to test the predictions of the above framework with direct model experiments by switching on and off certain terms rather than relying on diagnostic methods.

While this model is not rigorously derived from a geophysical flow model (e.g. barotropic vorticity equation), it has the same functional form one would end up with when deriving a reduced stochastic model from a geophysical model. Thus, consistent with geophysical flow models, the toy model has a quadratically nonlinear part that conserves energy, a linear operator and a constant forcing, which in a geophysical model would represent the effects of external forcing such as solar insulation and sea-surface temperature. The linear operator has two contributions: one is a skew-symmetric part formally similar to the Coriolis effect and topographic Rossby wave propagation. The other is a negative-definite symmetric part formally similar to dissipative processes such as surface drag and radiative damping.

The model is constructed in such a way that there are two modes, denoted by $\boldsymbol{x}$, that evolve more slowly than the other two modes, $\boldsymbol{y}$. In the realistic models, there would be very many additional fast modes representing, for example, synoptic weather systems or convection. To mimic their combined effect, we include damping and stochastic forcing, $-(\gamma/\varepsilon)y + (\sigma/\sqrt{\varepsilon})\dot{W}$, in the equations for $\boldsymbol{y}$, where $\boldsymbol{W}$ denotes a Wiener process. The motivation for this approximation is that these fast modes are associated with turbulent energy transfers and strong mixing and that we do not require a more detailed description since we are only interested in their effect on the slow modes. The two fast modes carry most of the variance in this model but as noted earlier, these two modes are surrogates in the model for the entire bath of fast modes so this is very natural. Note that this stochastic Ansatz is different from most of the previous studies (e.g. Newman *et al.* 1997; Whitaker & Sardeshmukh 1998) that truncated the model on *both* the resolved and unresolved modes and replaced all nonlinear terms by a stochastic process. In this study, only the nonlinear interactions among the fast modes are modelled by this stochastic Ansatz. The parameter ε controls the time-scale separation between the slow and fast variables. For testing the predictions of the general framework that is derived in the previous section, we will treat the two slow modes $\boldsymbol{x}$ as the climate modes and the two fast modes as the non-climate modes $\boldsymbol{y}$. Therefore, our toy model has the following form:

$$dx_1 = \{[-x_2(L_{12} + a_1x_1 + a_2x_2) + d_1x_1 + F_1] + L_{13}y_1 + b_{123}x_2y_1\}\,dt, \tag{3.2.13a}$$

$$dx_2 = \{[+x_1(L_{21} + a_1x_1 + a_2x_2) + d_2x_2 + F_2] + L_{24}y_2 + b_{213}x_1y_1\}\,dt, \tag{3.2.13b}$$

$$dy_1 = \left(-L_{13}x_1 + b_{312}x_1x_2 + F_3 - \frac{\gamma_1}{\varepsilon}y_1\right) dt + \frac{\sigma_1}{\sqrt{\varepsilon}}\,dW_1, \tag{3.2.13c}$$

$$dy_2 = \left(-L_{24}x_2 + F_4 - \frac{\gamma_2}{\varepsilon}y_2\right) dt + \frac{\sigma_2}{\sqrt{\varepsilon}}\,dW_2. \tag{3.2.13d}$$

Note that in this case we use a different scaling in powers of ε in order to derive easily a solution by working directly on the stochastic differential equation instead of the Fokker–Planck equation as for equation (3.2.8) (MTV 1999, 2001). To ensure energy conservation of the nonlinear operator, the coefficients have to satisfy $b_{123} + b_{213} + b_{312} = 0$, while the nonlinear bare-truncation terms also conserve energy. In this particular set-up, the slow and the fast modes are coupled through two mechanisms: one is a skew-symmetric linear coupling and the other is nonlinear triad interaction. The nonlinear coupling involving b_{ijk} produces multiplicative noise in the MTV framework (1999, 2003; see also equation (3.2.15a)–(3.2.15d)) so we refer to it as a multiplicative triad. One advantage of the model in (3.2.13a)–(3.2.13d) is that the stochastic mode reduction as $\varepsilon \to 0$ is done explicitly through elementary manipulations (MTV 1999, 2001; Majda *et al.* 2005). The corresponding reduced Itô stochastic differential equation (SDE) for the climate variables alone is given by

$$\begin{aligned} \mathrm{d}x_1(t) = {} & \{-x_2(t)[L_{12} + a_1x_1(t) + a_2x_2(t)] + d_1x_1(t) + F_1\}\,\mathrm{d}t \\ & + \frac{\varepsilon}{\gamma_1}[L_{13}F_3 - L_{13}L_{13}x_1(t) + b_{123}F_3x_2(t) + L_{13}b_{312}x_1(t)x_2(t) \\ & - L_{13}b_{123}x_1(t)x_2(t) + b_{312}b_{123}x_1(t)x_2^2(t)]\,\mathrm{d}t \\ & + \varepsilon\frac{1}{2}\frac{\sigma_1^2}{\gamma_1^2}b_{213}b_{123}x_1(t)\,\mathrm{d}t + \sqrt{\varepsilon}\frac{\sigma_1}{\gamma_1}[L_{13} + b_{123}x_2(t)]\,\mathrm{d}W_1(t), \end{aligned} \tag{3.2.14a}$$

$$\begin{aligned} \mathrm{d}x_2(t) = {} & \{x_1(t)[L_{21} + a_1x_1(t) + a_2x_2(t)] + dx_2(t) + F_2\}\,\mathrm{d}t \\ & + \frac{\varepsilon}{\gamma_2}[L_{24}F_4 - L_{24}L_{24}x_2(t)]\,\mathrm{d}t + \frac{\varepsilon}{\gamma_1}[-b_{213}L_{13}x_1^2(t) \\ & + b_{213}b_{312}x_1^2(t)x_2(t) + b_{213}F_3x_1(t)]\,\mathrm{d}t + \varepsilon\frac{1}{2}\frac{\sigma_1^2}{\gamma_1^2}(b_{213}b_{123}x_2(t) \\ & + L_{13}b_{213})\,\mathrm{d}t + \sqrt{\varepsilon}\frac{\sigma_1}{\gamma_1}b_{213}x_1(t)\,\mathrm{d}W_1(t) + \sqrt{\varepsilon}\frac{\sigma_2}{\gamma_2}L_{24}\,\mathrm{d}W_2(t). \end{aligned} \tag{3.2.14b}$$

Note that coarse-graining time as $t \to t/\varepsilon$ amounts to setting $\varepsilon = 1$, because by coarse-graining time the factors of ε disappear (see Majda *et al.* (2005) for more details).

To evaluate the performance of the reduced dynamics, we calculate the autocorrelation function $\rho(s) = \langle x'(t+s)x'(t)\rangle / \langle x'(t)x'(t)\rangle$ and the third-order two-time moment $K(s) = \langle x'^2(t+s)x'(t)\rangle / \langle x(t)'^2\rangle^{3/2}$, which is a measure of deviations from Gaussianity (MTV 2002; Majda *et al.* 2005). The comparison of the reduced model (3.2.14a) and (3.2.14b) with the full model (3.2.13a)–(3.2.13d) results shows excellent agreement for moderate values of $\varepsilon = 0.1, 0.5$ and still good agreement for $\varepsilon = 1.0$ (Figs. 3.1 and 3.2). Especially, the non-Gaussian features are reproduced with high accuracy for $\varepsilon = 0.1$ and also $\varepsilon = 0.5$, even though slightly less well. For large values of ε, the reduced dynamics get the sign of the non-Gaussianity right. All of the reduced models in (3.2.11)–(3.2.14a) and (3.2.14b) cannot be approximated by the interesting regression strategy of Kravtsov *et al.* (2005) because there is nonlinear multiplicative noise in (3.2.13a)–(3.2.13d), (3.2.14a) and (3.2.14b), nonlinear triad interaction in (3.2.13a)–(3.2.13d) and augmented cubic nonlinearity in (3.2.14a) and (3.2.14b). Thus, these regression strategies necessarily have large model error in this example.

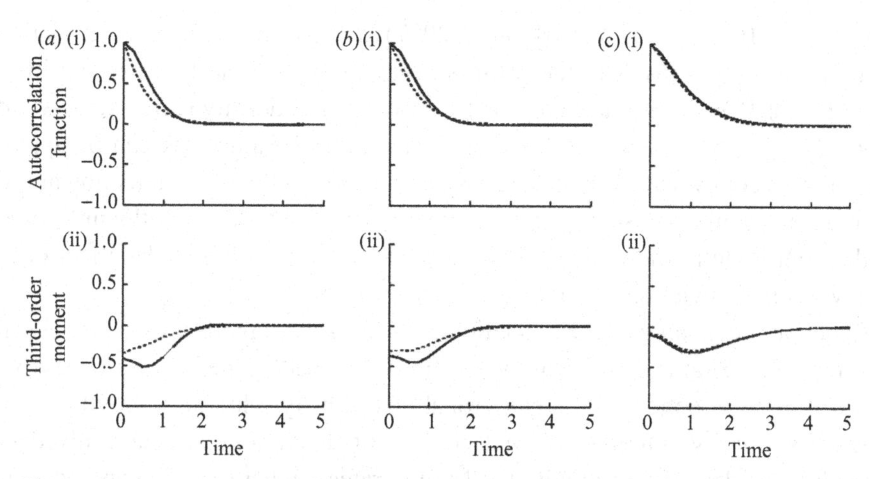

Figure 3.1 Autocorrelation function and third-order moment for different values of ε. (*a*) $\varepsilon = 1.0$, (*b*) $\varepsilon = 0.5$, and (*c*) $\varepsilon = 0.1$. Solid lines, full dynamics (3.2.13a)–(3.2.13d) and dashed lines, reduced dynamics (3.2.14a) and (3.2.14b).

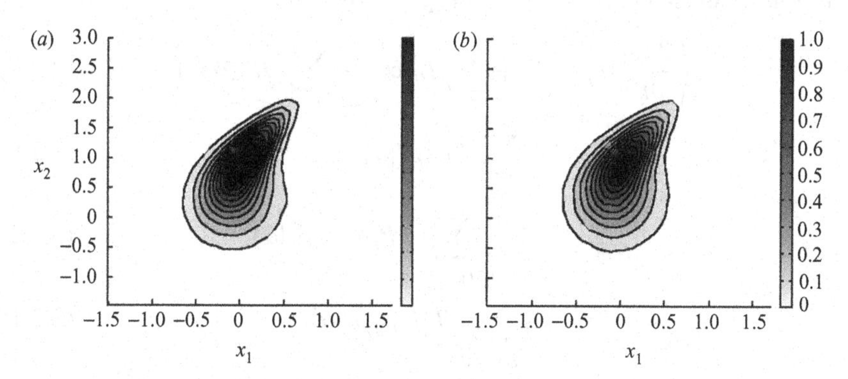

Figure 3.2 Joint PDF of the slow modes x_1 and x_2 for $\varepsilon = 0.1$. (*a*) Full dynamics (3.2.13a)–(3.2.13d) and (*b*) reduced dynamics (3.2.14a) and (3.2.14b).

3.2.6 A mathematical framework for the mean tendency equation as a dynamical diagnostic

In this section, we provide a general framework to estimate the origin of nonlinear signatures of planetary wave dynamics and how important the observed deviations from Gaussianity are for the planetary waves (Franzke *et al.* 2007). This general framework diagnoses contributions to the mean values of state-dependent tendencies in a low-dimensional subspace of complex geophysical systems. The mean phase-space tendencies in some GCMs show distinct nonlinear signatures in certain planes, which are spanned by its leading EOFs (Selten & Branstator 2004;

Branstator & Berner 2005; Franzke *et al.* 2007) that also show only weak deviations from Gaussianity; mostly in the form of weak skewness and kurtosis and in the case of joint PDFs multiple radial ridges of enhanced density (Franzke & Majda 2006; Berner & Branstator 2007). This mathematical framework can be applied to complex geophysical systems and reveals how much the self-interaction among the modes spanning those planes (climate modes) and how much the unresolved modes contribute to these mean phase-space tendencies and also the effect of the observed small deviations from Gaussianity.

To derive the mean tendency equation for the resolved modes, we split the state vector of a quadratic dynamical system into resolved modes α and unresolved modes β and use conditional mean probability density relations. Note that while there is a similarity in motivation in decomposing the system into the resolved and unresolved modes compared with the MTV strategy, where the system is decomposed into the slow and fast modes, the mean tendency equation is more general such that it applies also to systems without time-scale separation. The general formula for the *conditional mean tendencies* for the dynamics in the resolved variables is derived in Franzke *et al.* (2007) and is given by

$$\left\langle \frac{\partial \alpha_i}{\partial t} \middle| \alpha \right\rangle = F_i + \sum_{j \in I_{\mathrm{R}}} L_{ij} \alpha_j + \sum_{j,k \in I_{\mathrm{R}}} B_{ijk} \alpha_j \alpha_k \tag{3.2.15a}$$

$$+ \sum_{j \in I_{\mathrm{U}}} L_{ij} \langle \beta_j | \alpha \rangle \tag{3.2.15b}$$

$$+ 2 \sum_{j \in I_{\mathrm{R}}, k \in I_{\mathrm{U}}} B_{ijk} \alpha_j \langle \beta_k | \alpha \rangle \tag{3.2.15c}$$

$$+ \sum_{j,k \in I_{\mathrm{U}}} B_{ijk} \langle \beta_j \beta_k | \alpha \rangle . \tag{3.2.15d}$$

Note that the right-hand side of (3.2.15a) is the *bare truncation* restricted to interactions among the resolved modes, while (3.2.15b) and (3.2.15c) involve all of the conditional mean statistics $\langle \beta_j | \alpha \rangle$; the terms of (3.2.15c) are associated with *multiplicative triad interactions* [leading to multiplicative noise in an MTV (1999, 2003) framework] of β_j; the terms in (3.2.15d) involve all of the conditional interaction statistics $\langle \beta_j \beta_k | \alpha \rangle$ of second moments and include all of the *additive triad interactions* (leading to additive noise in an MTV framework) of β_j and β_k. Thus, equation (3.2.15) provides a general framework to investigate the mean phase-space tendencies, which allows the decomposition into contributions from interactions among the resolved planetary waves themselves and various contributions involving unresolved degrees of freedom.

Now we simplify the above conditional mean tendency equation for purely Gaussian EOF modes. The PDFs of EOFs of GCM data are nearly Gaussian

(Hsu & Zwiers 2001; Franzke *et al.* 2005; Franzke & Majda 2006; Majda *et al.* 2006a; Berner & Branstator 2007) and there are many geophysical models without damping and forcing that exactly satisfy these assumptions such as barotropic flow on the sphere with topography (Carnevale & Frederiksen 1987; Salmon 1998; Majda & Wang 2006); thus, it is reasonable as a starting point to *assume* that the PDF is *exactly Gaussian*. Thus, in the EOF basis, the PDF factors like

$$p(\alpha, \beta) = \left[\Pi p_i^G(\alpha_i)\right]\left[\Pi p_j^G(\beta_j)\right], \tag{3.2.16}$$

where $p_i^G(\alpha_i)$ and $p_j^G(\beta_j)$ are Gaussian distributions with mean zero. Thus, in the *Gaussian* case, the conditional mean tendency equation simplifies to (see Franzke *et al.* 2007 for more details)

$$\left\langle \frac{\partial \alpha_i}{\partial t} \middle| \alpha \right\rangle = F_i + \sum_j L_{ij}\alpha_j + \sum_{j,k} B_{ijk}\alpha_j\alpha_k \tag{3.2.17a}$$

$$+ \sum_j B_{ijj}\lambda_j. \tag{3.2.17b}$$

Note that contributions from (3.2.15b) and (3.2.15c), that is, linear coupling and multiplicative triad interactions, are identically zero. These terms vanish because the modes are uncorrelated in the EOF basis. Thus, in this Gaussian case, the conditional mean tendency equation recognises bare truncation (3.2.15a) and a constant forcing from additive triad interactions (3.2.15d). Since the leading EOFs of GCM data have PDFs with only small, but significant departures from Gaussianity the behaviour in (3.2.16), (3.2.17a) and (3.2.17b) serves as a 'null hypothesis' for these deviations from Gaussianity in the climate models.

This general framework predicts that in the case of purely Gaussian modes the nonlinear signatures are stemming from the bare truncation (i.e. the self-interaction among the planetary waves resolved in the low-dimensional plane). In Franzke *et al.* (2007), these diagnostics were applied to a plane of two low-frequency EOFs in a three-layer climate model with a nonlinear double swirl (see Fig. 3.5*a* in Franzke *et al.* 2007); the origin of this double swirl is primarily the three contributions in (3.2.15b)–(3.2.15d) from the unresolved modes and not nonlinear effects from the bare truncation. The EOFs of the three-layer model have small but significant departures from Gaussianity (see figure 9 in Franzke *et al.* 2007). Therefore, these deviations account for the contribution of the unresolved onto the resolved modes in the mean phase-space tendencies. Thus, these results show that small deviations from Gaussianity have a substantial impact on the dynamics of the leading EOFs.

3.3 Coarse-grained stochastic lattice models for climate: tropical convection

The current practical models for prediction of both weather and climate involve GCMs where the physical equations for these extremely complex flows are discretised in space and time, and the effects of unresolved processes are parameterised according to various recipes. With the current generation of supercomputers, the smallest possible mesh spacings are approximately 10–50 km for short-term weather simulations and of order 100 km for short-term climate simulations. There are many important physical processes that are unresolved in such simulations such as the mesoscale sea-ice cover, the cloud cover in subtropical boundary layers and deep convective clouds in the tropics. Most of these features are highly intermittent in space and time. An appealing way to represent these unresolved features is through a suitable coarse-grained stochastic model that simultaneously retains crucial physical features of the interaction between the unresolved and resolved scales in a GCM. In work from 2002 and 2003, two of the authors have developed a new systematic stochastic strategy (Majda & Khouider 2002; Khouider *et al.* 2003) to parameterise key features of deep convection in the tropics involving suitable stochastic spin-flip models and also a systematic mathematical strategy to coarse grain such microscopic stochastic models to practical mesoscopic meshes in a computationally efficient manner while retaining crucial physical properties of the interaction.

As regards tropical convection, crucial scientific issues involve the fashion in which a stochastic model affects the climate mean state and the strength and nature of fluctuations about the climate mean. Here the strategy to develop a new family of coarse-grained stochastic models for tropical deep convection is briefly reviewed (Majda & Khouider 2002; Khouider *et al.* 2003) as an illustrative example of the potential use of stochastic lattice models. In Khouider *et al.* (2003), it has been established that in suitable regimes of parameters, the coarse-grained stochastic parameterisations can significantly alter the climatology as well as increase wave fluctuations about the climatology. This was established in Khouider *et al.* (2003) in the simplest scenario for tropical climate involving the Walker circulation, the east–west climatological state that arises from the local region of enhanced surface heat flux, mimicking the Indonesian marine continent. Convectively coupled waves in the tropics such as the Madden–Julian oscillation play an important role in medium-range forecasts, yet the current generation of computer models fails to represent such waves adequately (Lin *et al.* 2006). Palmer (2001) has emphasised the potential of stochastic parameterisation to reduce the model error in a deterministic computer model. Here, in an idealised setting, we show how to develop a stochastic parameterisation to modify and improve the behaviour of convectively coupled waves in a reasonable prototype GCM; this is achieved by following a path guided by the systematic design principle for the idealised model in (3.1.1) (KMS 2006, 2007) to build in suitable intermittency effects.

3.3.1 The microscopic stochastic model for convective inhibition

In a typical GCM, the fluid dynamical and thermodynamical variables denoted here by the generic vector $\boldsymbol{u}$ are regarded as known only over a discrete horizontal mesh with $\boldsymbol{u}_j = \boldsymbol{u}\,(j\Delta x,\, t)$ denoting these discrete values. Throughout the discussion, one horizontal spatial dimension along the equator in the east–west direction is assumed for simplicity in notation and explanation. As mentioned above, the typical mesh spacing in a GCM is coarse with Δx ranging from 50 to 250 km depending on the time duration of the simulation. The stochastic variable used to illustrate the approach is convective inhibition (CIN). Observationally, CIN is known to have significant fluctuations on a horizontal spatial scale on the order of a kilometre, the microscopic scale here, with changes in CIN attributed to different mechanisms in the turbulent boundary layer, such as gust fronts, gravity waves and turbulent fluctuations in equivalent potential temperature (Mapes 2000). In Majda & Khouider (2002) and Khouider *et al.* (2003), it was proposed that all of these different microscopic physical mechanisms, and their interaction, that increase and decrease CIN are too complex to model in detail in a coarse-mesh GCM parameterisation and instead, as in statistical mechanics, should be modelled by a simple-order parameter σ_{I}, taking only two discrete values,

$$\left.\begin{aligned} \sigma_{\mathrm{I}} &= 1 \text{ at a site if convection is inhibited (a CIN site)} \\ \sigma_{\mathrm{I}} &= 0 \text{ at a site if there is potential for deep convection (a PAC site).} \end{aligned}\right\} \qquad (3.3.1)$$

The value of CIN at a given coarse-mesh point is determined by the averaging of CIN over the microscopic states in the vicinity of the given mesh point, i.e.

$$\overline{\sigma}_{\mathrm{I}}(j\Delta x,\, t) = \frac{1}{\Delta x}\int_{(j-1/2)\Delta x}^{(j+1/2)\Delta x} \sigma_{\mathrm{I}}(x,\, t)\,\mathrm{d}x. \qquad (3.3.2)$$

Note that the mesh size Δx is mesoscopic, that is, between the microscale $O(1\,\text{km})$ and the macroscale $O(10\,000\,\text{km})$, and that $\overline{\sigma}_{\mathrm{I}}$ can have any value in the range $0 \le \overline{\sigma}_{\mathrm{I}} \le 1$. Discrete sums over microscopic mesh values (of order 1 km) and continuous integrals are used interchangeably for notational convenience (Majda & Khouider 2002).

3.3.2 The simplest coarse-grained stochastic model

In practical parameterisation, it is desirable for computational feasibility to replace the microscopic dynamics by a process on the coarse mesh, which retains critical dynamical features of the interaction. Following the general procedure developed and tested in Katsoulakis & Vlachos (2003) and Katsoulakis *et al.* (2003a,b), the simplest local version of the systematic coarse-grained stochastic process is developed in Khouider *et al.* (2003) and summarised here.

Each coarse cell Δx_k, $k = 1, \ldots, m$, of the coarse-grained lattice is divided into q microscopic cells, such that $\Delta x_k \equiv 1/q\ 1, 2, \ldots, q$, $k = 1, \ldots, m$. In the coarse-grained procedure, given the coarse-grained sequence of random variables

$$\eta_t(k) = \sum_{y \in \Delta x_k} \sigma_{\mathrm{I},t}(y), \tag{3.3.3}$$

so that the average in (3.3.2) verifies $\overline{\sigma}_{\mathrm{I}}(j\Delta x) = \eta(k)/q$, for $j = k$ in some sense, the microscopic dynamics is replaced by a birth–death Markov process defined on the variables $0, 1, \ldots, q$ for each k, such that $\eta_t(k)$ evolves according to the following probability law:

$$\left.\begin{aligned}
&Prob\{\eta_{t+\Delta t}(k) = n+1 | \eta_t(k) = n\} = C_{\mathrm{a}}(k,\ n)\Delta t + o(\Delta t)\\
&Prob\{\eta_{t+\Delta t}(k) = n-1 | \eta_t(k) = n\} = C_{\mathrm{d}}(k,\ n)\Delta t + o(\Delta t)\\
&Prob\{\eta_{t+\Delta t}(k) = n | \eta_t(k) = n\} = 1 - (C_{\mathrm{a}}(k,\ n) + C_{\mathrm{d}}(k,\ n))\Delta t + o(\Delta t)\\
&Prob\{\eta_{t+\Delta t}(k) \neq n,\ n-1,\ n+1\ | \eta_t(k) = n\} = o(\Delta t).
\end{aligned}\right\} \tag{3.3.4}$$

Here C_{a} and C_{d} are the coarse-grained adsorption and desorption rates representing the rates at which CIN sites are being created and destroyed, respectively, within one coarse-grained cell; that is, the birth–death rates for the Markov process is

$$\left.\begin{aligned}
C_{\mathrm{a}}(k, \eta) &= \tfrac{1}{\tau^{\mathrm{I}}}[q - n(k)]\\
C_{\mathrm{d}}(k, \eta) &= \tfrac{1}{\tau^{\mathrm{I}}}\eta(k)e^{\beta \overline{V}k}
\end{aligned}\right\} \tag{3.3.5}$$

where

$$\overline{V}(k) = \overline{J}(0, 0)[\eta(k) - 1] + h_{\mathrm{ext}}, \tag{3.3.6}$$

with the coarse-grained interaction potential within the coarse cell given by $\overline{J}(0, 0) = 2U_0/(q-1)$, where U_0 is the mean strength of the microscopic interaction potential J between neighbouring sites (Katsoulakis *et al.* 2003a,b). The coarse-grained energy content for CIN is given by the coarse-grained Hamiltonian

$$\overline{H}(\eta) = \frac{U_0}{q-1} \sum_k \eta(k)[\eta(k) - 1] + h_{\mathrm{ext}} \sum_k \eta(k), \tag{3.3.7}$$

where h_{ext} is the external potential that modifies the energy for CIN depending on the large-scale flow variables. The canonical invariant Gibbs measure for the coarse-grained stochastic process is a product measure given by

$$G_{m,q,\beta}(\eta) = (Z_{m,q,\beta})^{-1} \, \mathrm{e}^{\beta H(\eta)} P_{m,q}(\mathrm{d}\eta), \tag{3.3.8}$$

where $P_{m,q}(\mathrm{d}\eta)$ is a given explicit/prior distribution (Katsoulakis *et al.* 2003b). As shown in Katsoulakis *et al.* (2003b), the coarse-grained birth/death process above satisfies detailed balance with respect to the Gibbs measure in (3.3.8) as well as a number of other attractive theoretical features. The simplest coarse-grained approximation given above assumes that the effect of the microscopic interactions on the mesoscopic scales occurs within the mesoscopic coarse-mesh scale Δx, otherwise systematic non-local couplings are needed (Katsoulakis *et al.* 2003b). The accuracy of these approximations is tested for diverse examples from material science elsewhere (Katsoulakis & Vlachos 2003; Katsoulakis *et al.* 2003a,b) and for the instructive idealised coupled models in (3.1.1) (KMS 2004, 2005, 2006, 2007).

The practical implementation of the coarse-grained birth/death process in (3.3.3)–(3.3.6) requires specification of the parameters, τ_{I}, U_0, q and the external potential $h_{ext}(\boldsymbol{u}_{\mathrm{j}})$ as well as the statistical parameter β. The advantages of such a stochastic lattice model are as follows:

- retains systematically the energetics of unresolved features through the coarse-grained Gibb's measure;
- has minimal computational overhead since there are rapid algorithms for updating birth/death processes;
- incorporates feedbacks of the resolved modes on the unresolved modes and their energetics through an external field; and
- includes dynamical coupling through not only sampling the probability distributions of unresolved variables but also their evolving behaviour in time is constrained by the large-scale dynamics.

3.3.3 The model deterministic convective parameterisation

A prototype mass flux parameterisation with crude vertical resolution (Majda & Shefter 2001; Khouider & Majda 2006b) is used to illustrate the fashion in which the coarse-grained stochastic model for CIN can be coupled to a deterministic convective mass flux parameterisation. The prognostic variables (u, θ, θ_{eb}, θ_{em}) are the x-component of the fluid velocity, u, the potential temperature in the middle troposphere, θ, the equivalent potential temperatures, θ_{eb} and θ_{em}, measuring the potential temperatures plus moisture content of the boundary layer and middle troposphere, respectively. The vertical structure is determined by projection on a first baroclinic heating mode (Majda & Shefter 2001; Khouider & Majda 2006b). The dynamic equations for these variables in the parameterisation are given by

$$\left.\begin{aligned}
&\frac{\partial u}{\partial t} - \bar{\bar{\alpha}} = \frac{\partial \theta}{\partial x} = -\left(C_{\mathrm{D}}^{0}\frac{1}{h}\sqrt{u_0^2 + u^2}\right)u - \frac{1}{\tau_{\mathrm{D}}}u \\
&\frac{\partial \theta}{\partial t} - \overline{\alpha}\frac{\partial u}{\partial x} = \mathcal{S} - Q_{\mathrm{R}}^{0} - \frac{\theta}{\tau_{\mathrm{R}}} \\
&h\tfrac{\partial \theta_{\mathrm{eb}}}{\partial t} = -D(\theta_{\mathrm{eb}} - \theta_{\mathrm{em}}) + \left(C_{\theta}\sqrt{u_0^2 + u^2}\right)(\theta_{\mathrm{eb}}^{*} - \theta_{\mathrm{eb}}) \\
&H\tfrac{\partial \theta_{\mathrm{em}}}{\partial t} = D(\theta_{\mathrm{eb}} - \theta_{\mathrm{em}}) - H Q_{\mathrm{R}}^{0} - H\tfrac{\theta_{\mathrm{em}}}{\tau_{\mathrm{R}}}
\end{aligned}\right\} \tag{3.3.9}$$

while the constants Q_{R}^{0} and θ_{eb}^{*} are externally imposed and represent the radiative cooling at equilibrium in the upper troposphere and saturation equivalent potential temperature in the boundary layer. The constants h and H measure the depths of the boundary layer and the troposphere above the boundary layer, respectively. The typical values used here are $h = 500$ m and $H = 16$ km, while $u_0 = 2$ m s^{-1} and $C_{\mathrm{D}}^{0} = 0.001$, corresponding to a momentum spin-up time of approximately three days. The explicit values for the other constants used in (3.3.9) and elsewhere in this section can be found in Majda & Shefter (2001) and Khouider *et al.* (2003).

The crucial quantities in the prototype mass flux parameterisation are the terms $\mathcal{S}$ and D, where $\mathcal{S}$ represents the middle troposphere heating due to deep convection, while D represents the downward mass flux on the boundary layer. The heating term $\mathcal{S}$ is given by

$$\mathcal{S} = M\sigma_{\mathrm{c}}[(\mathrm{CAPE})^{+}]^{1/2}, \tag{3.3.10}$$

where M is a fixed constant; σ_{c} is the area fraction for deep convective mass flux; and CAPE $= R(\theta_{\mathrm{eb}} - \gamma\theta)$ is the convectively available potential energy. Here R is a dimensional constant (Majda & Shefter 2001; Khouider *et al.* 2003). The downward mass flux on the boundary layer D includes the environmental downdrafts m_{e} and the downward mass flux due to convection m_{-}, which are non-negative quantities with explicit formulae described elsewhere (Majda & Shefter 2001; Khouider *et al.* 2003). The notation $(X)^{+}$ denotes the positive part of X. This parameterisation respects conservation of vertically integrated moist static energy.

The equations (3.3.9) and (3.3.10) represent an idealised GCM with crude vertical resolution based on a reasonable design principle for deep convection including basic conservation principles. However, like the current generation of GCMs, the model has major deficiencies as regards convectively coupled waves. First, instabilities for the full model in (3.3.9) arise only through the nonlinearity from $\sqrt{u_0^2 + u^2}$ and are called the wind-induced surface heat exchange (WISHE) instability (Majda

& Shefter 2001, and references therein). Second, if the bulk aerodynamic flux coefficients in (3.3.9), $\sqrt{u_0^2 + u^2}$ are replaced by the constant value $|u_0|$, that is no WISHE, all waves in the system are stable and no fluctuations emerge in an aquaplanet simulation (Majda & Shefter 2001). There is no observational evidence supporting the role of WISHE in driving the large-scale convectively coupled waves in nature (Lin *et al.* 2006 and references therein); here the WISHE term is regarded as a deterministic fix in a GCM parameterisation to generate instabilities. The simulation results reported in Fig. 3.5*a* show that in an aquaplanet model above the equator, two regular periodic convectively coupled waves moving eastward and westward at roughly equal strength are generated by the idealised GCM. Since GCMs often have convectively coupled waves that move too fast and are too regular (Lin *et al.* 2006), the goal here is to see whether the stochastic lattice coupling will slow down the waves in the deterministic parameterisation and simultaneously increase the spatio-temporal fluctuations in those waves. It is important to note here that there are recent deterministic multicloud models (Khouider & Majda 2006a, 2007, 2008) for convectively coupled waves involving the three cloud types in observations above the boundary layer, congestus, deep and stratiform, and their heating structure that reproduces key features of the observational record for convectively coupled waves (Lin *et al.* 2006). The mechanism of instability in these models (Khouider & Majda 2006a) is completely different from WISHE that is not active in the multicloud models. For the idealised setting of flow above the equator, the multicloud models can produce packets of convectively coupled waves moving in one direction at 15–20 m s^{-1} with their low-frequency envelopes moving at 4–7 m s^{-1} in the opposite direction across the warm pool in a fashion like the Madden–Julian oscillation (Khouider & Majda 2007, 2008).

3.3.4 Coupling of the stochastic CIN model into the parameterisation

The equations (3.3.9) and (3.3.10) are regarded here as the prototype deterministic GCM parameterisation when discretised in a standard fashion using central differences on a coarse mesh Δx with Δx ranging from 50 to 250 km. In the simulations from Khouider *et al.* (2003), and those presented below, $\Delta x = 80$ km. The coarse-grained stochastic CIN model is coupled to this basic parameterisation. The area fraction for deep convection σ_c governing the upward mass flux strength is allowed to vary on the coarse mesh and is given by

$$\sigma_c(j\Delta x) = [1 - \overline{\sigma}_I(j\Delta x)]\sigma_c^+, \tag{3.3.11}$$

where $\overline{\sigma}_I$ is the average in (3.3.2) and σ_c^+ is a threshold constant and equal to 0.002 (Majda & Shefter 2001; Khouider *et al.* 2003). When the order parameter σ_I

signifies strong CIN locally so that $\overline{\sigma}_{\mathrm{I}} = 1$ the flux of deep convection is diminished to zero while with PAC locally active $\overline{\sigma}_{\mathrm{I}} = 0$, this flux increases to the maximum allowed by the value σ_c^+. To complete the coupling of the stochastic CIN model into the parameterisation, the coarse-mesh external potential $h_{\mathrm{ext}}(\boldsymbol{u}_j)$, from (3.3.6) and (3.3.7), needs to be specified from the coarse-mesh values $\boldsymbol{u}_{\mathrm{j}}$. There is no unique choice of the external potential but its form can be dictated by simple physical reasoning. In Khouider *et al.* (2003), the plausible physical assumption is made that when the convective downward mass flux m_{-} decreases the energy for CIN decreases. Since the convective downward mass flux results from the evaporative cooling induced by precipitation falling into dry air, it constitutes a mechanism that carries negatively buoyant cool and dry air from the middle troposphere onto the boundary layer, hence tending to reduce CAPE and deep convection.

Another natural external potential is the boundary-layer equivalent potential, since the flux at the boundary is crucial physically. As such, it yields somewhat stronger and more intermittent fluctuations. Thus, the choice

$$h_{\mathrm{ext}}(j\Delta x, t) = \tilde{\gamma}\theta_{\mathrm{eb}}(j\Delta x, t) \tag{3.3.12}$$

is used here with $\tilde{\gamma}$ a calibration factor. The other parameters in the stochastic lattice model are chosen as $\tau_{\mathrm{I}} = 2$ h, $\beta = 1$, $U_0 = 1$ so that CIN sites are favoured in the equilibrium Gibbs measure.

3.3.5 The stochastic single-column model and intermittency

A central issue is how to calibrate the stochastic lattice model to generate intermittent fluctuations with plausible magnitudes as observed in tropical convection. A natural design framework is first to achieve such behaviour in the stochastic single-column model given by the following equations:

$$\left.\begin{aligned}
\frac{\mathrm{d}\theta}{\mathrm{d}t} &= (1-\sigma_{\mathrm{I}})M\sqrt{R(\theta_{\mathrm{eb}}-\gamma\theta)^{+}} - Q_{\mathrm{R}}^{0} - \frac{1}{\tau_{\mathrm{R}}}\theta \\
\frac{\mathrm{d}\theta_{\mathrm{eb}}}{\mathrm{d}t} &= -\frac{1}{\tau_{\mathrm{eb}}}(1-\sigma_{\mathrm{I}})\sqrt{R(\theta_{\mathrm{eb}}-\gamma\theta)^{+}}(\theta_{\mathrm{eb}}-\theta_{\mathrm{em}}) + \frac{1}{\tau_{\mathrm{e}}}(\theta_{\mathrm{eb}}^{*}-\theta_{\mathrm{em}}) \\
\frac{\mathrm{d}\theta_{\mathrm{em}}}{\mathrm{d}t} &= \frac{1}{\tau_{\mathrm{em}}}(1-\sigma_{\mathrm{I}})\sqrt{R(\theta_{\mathrm{eb}}-\gamma\theta)^{+}} - (\theta_{\mathrm{eb}}-\theta_{\mathrm{em}}) - \frac{1}{\tau_{\mathrm{R}}}\theta.
\end{aligned}\right\} \tag{3.3.13}$$

Here σ_{I} is computed from the birth–death stochastic model in (3.3.3)–(3.3.6) with the interaction potential defined through θ_{eb} from (3.3.7) and the definition of h_{ext} in (3.3.12) above. The different time scales were computed for a radiative convective equilibrium consistent with the Jordan tropical sounding and the values $\tau_{\mathrm{eb}} \approx 6$ h, $\tau_{\mathrm{em}} \approx 8$ d, $\tau_{\mathrm{e}} \approx 8$ h, $Q_{\mathrm{R}}^{0} = 1\,\mathrm{K\,d^{-1}}$ and $\tau_{\mathrm{R}} = 50$ d are used here (Khouider *et al.* 2003). This is a specific example of the prototype models from (3.1.1) studied

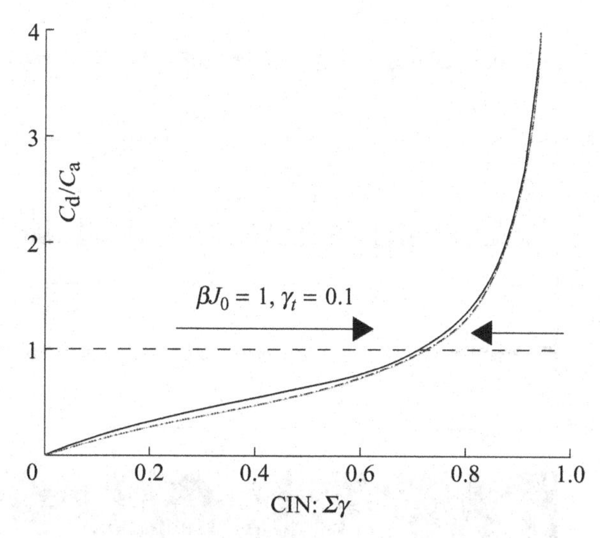

Figure 3.3 Ratio of the adsorption and desorption rates C_d/C_a for the birth–death stochastic lattice model with parameters $\overline{\gamma} = 0.1$, $\beta = U_0 = 1$ and a number of microscopic sites $q = 10$ (——), $q = 100$ (·–·–) and $q = 1000$ (...); the last two plots almost converge. Note that a CIN-dominating equilibrium with large CIN values yield large PAC (adsorption) rates. The balanced equilibrium curve $C_d/C_a = 1$ is shown. Arrows indicate overall tendency of CIN coverage.

extensively in idealised settings (KMS 2004, 2005, 2006). As mentioned earlier, the choices $\beta = 1$, $U_0 = 1$ guarantee that the Gibbs measure in (3.3.8) has a high probability for CIN states; this is a natural requirement for deep convection, where the area fraction of deep convection is much smaller than the area with positive CAPE. In order to generate intermittency and fluctuations in the stochastic column model, it is very natural to require that the ratio of the creation rate of PAC sites to the creation rate of CIN sites, C_d/C_a, becomes large as $\sigma_I \nearrow 1$ (KMS 2006, 2007). In Fig. 3.3, it is shown that this property of the stochastic model is satisfied for the coefficient $\tilde{\gamma} = 0.1$ in the external potential in (3.3.12) with similar behaviour for $q = 10$, 100 and 1000. Recall that q is the number of microscopic states in the birth–death process. A time series of the stochastic column model implemented with these parameters is displayed in Fig. 3.4 with $q = 10$. The strongly intermittent fluctuations in θ_{eb} over several Kelvin as well as similar intermittency in the mass flux is evident. They are consistent with SST variations observed in the Indian Ocean warm pool (e.g. Fasullo & Webster 1999). Also note that the fluctuations in the mid-troposphere potential temperature are much weaker in magnitude as occurs in the actual tropics, where mid-tropospheric temperature anomalies are on the order of 0.1–0.5 °C (e.g. Wheeler *et al.* 2000).

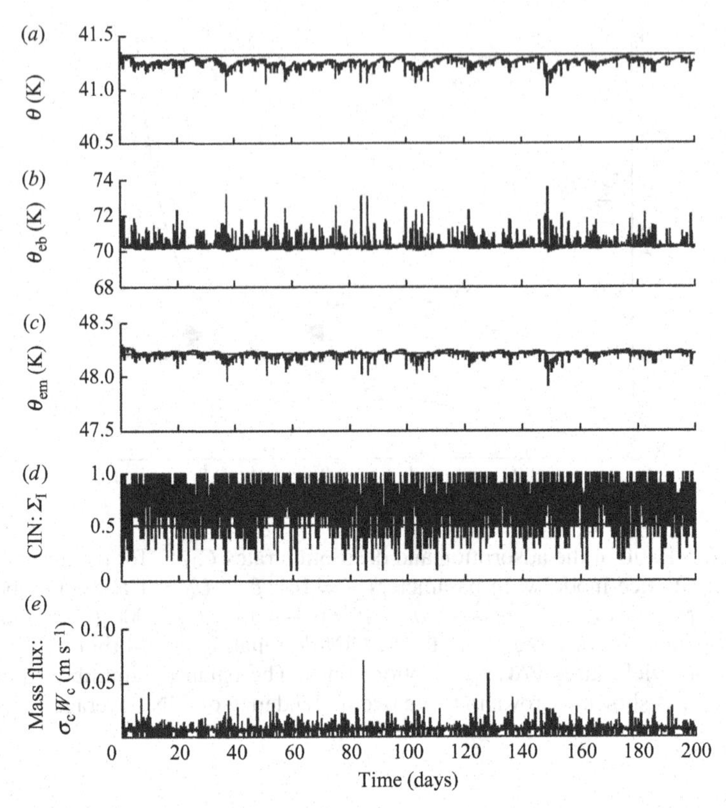

Figure 3.4 (*a–e*) Time series of dynamical variables for the stochastic single-column model with $\tau_{\mathrm{I}} = 2$ h, $\overline{\gamma} = 0.1$, $\beta = U_0 = 1$, $q = 10$. Note the manifestation of strong intermittent bursts beyond the stochastically generated CIN; especially, in θ_{eb} and mass flux $\sigma_{\mathrm{c}} W_{\mathrm{c}} \equiv (1 - \sigma_{\mathrm{I}})\sqrt{R(\theta_{\mathrm{eb}} - \gamma\theta_1)}$. The deterministic equilibrium value is represented by the horizontal line on each panel.

3.3.6 The effects of the stochastic parameterisation on convectively coupled waves

All parameters in the stochastic lattice model have been determined through systematic design principles in the previous section. Here the results of numerical simulations of the stochastic model in statistical steady state are reported for flow above the equator in a standard aquaplanet set-up with uniform SST (Majda & Khouider 2002; Khouider *et al.* 2003). The results of various simulations are reported in Fig. 3.5 for comparison with the deterministic parameterisation with WISHE reported in Fig. 3.5*a*. As shown in Fig. 3.5*b*, the effect of the fluctuations of the stochastic lattice model is simultaneously to create more realistic intermittency in the convectively coupled waves and to slow down their phase speed from 15 to

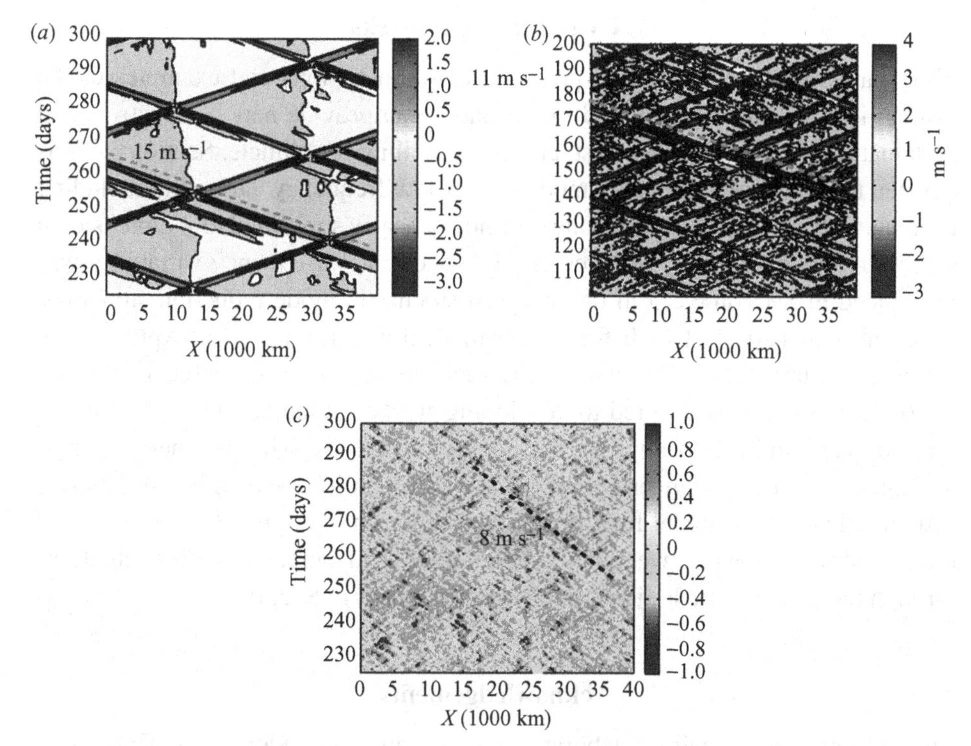

Figure 3.5 Effect of the birth–death stochastic CIN on large-scale convectively coupled waves with the stochastic parameters $\beta = U_0 = 1$, $q = 10$. (*a*) Purely deterministic WISHE waves without the stochastic coupling, (*b*) effect of stochastic CIN on WISHE waves, (*c*) intermittent bursts of convective episodes with apparent tracks of waves maintained by the stochastic CIN effects alone – WISHE is off. The apparent speed of propagation is shown by the dashed lines. Note the different time scales in (*b*) compared with (*a,c*). Space-time velocity contours are shown as indicated.

11 m s^{-1}; as mentioned earlier, these are desirable qualitative features of a stochastic parameterisation. In Fig. 3.5*c*, we report the result of running the model parameterisation in (3.3.9) without WISHE but coupled to the stochastic lattice model; recall that this is the situation where the deterministic model without WISHE produces no waves in the statistical steady state. In Fig. 3.5*c*, there is a clear evidence for stochastic-generated convectively coupled waves with the reduced phase speed of roughly 8 m s^{-1} and roughly half the amplitude; these waves have been created by coupling alone to the intermittent stochastic lattice model and without deterministic instability. This is another attractive feature of the present stochastic lattice models in changing the character of model error for convection (Palmer 2001).

3.4 Concluding remarks

This chapter both reviewed and provided new illustrations and examples of the fashion in which modern applied mathematics can provide new perspectives and systematic design principles for stochastic modelling for climate. Section 3.2 was devoted to systematic stochastic modelling of low-frequency variability including the quantitative sources for multiplicative noise. A new simplified low-dimensional stochastic model with key features of atmospheric low-frequency variability was introduced in Section 3.2.5 in order to test stochastic mode reduction strategies. A recent diagnostic test with firm mathematical underpinning for exploring the subtle departures from Gaussianity and their sources was discussed in Section 3.2.6. Section 3.3 was devoted to developing stochastic lattice models to capture intermittent features and improve the fidelity of deterministic parameterisations. Recent systematic design principles (KMS 2006, 2007) were used to calibrate a stochastic column model for tropical convection in Section 3.3.5; the practical effect of slowing down convectively coupled waves and increasing their fluctuations through these stochastic lattice models was presented in Section 3.3.6.

Acknowledgements

The authors thank their collaborators, Eric Vanden-Eijnden, Ilya Timofeyev, Markos Katsoulakis and Alex Sopasakis, for their explicit and implicit contributions to the work presented here. The research of A. J. M. is partially funded by the National Science Foundation grant no. DMS-0456713, the Office of Naval Research grant no. N00014-05-1-0164 and the Defense Advanced Research Project Agency grant no. N00014-07-1-0750. The research of B. K. is partly supported by a grant from the National Sciences and Engineering Research Council of Canada.

References

Achatz, U. & Branstator, G. 1999 A two-layer model with empirical linear corrections and reduced order for studies of internal climate variability. *J. Atmos. Sci.*, **56**, 3140–3160 (doi:10.1175/1520–0469 (1999)056<3140:ATLMWE>2.0.CO;2).

Achatz, U. & Opsteegh, J. D. 2003a Primitive-equation-based low-order models with seasonal cycle. Part I: model construction. *J. Atmos. Sci.*, **60**, 465–477 (doi:10.1175/1520–0469(2003)060 <0465:PEBLOM>2.0.CO;2).

Achatz, U. & Opsteegh, J. D. 2003b Primitive-equation-based low-order models with seasonal cycle. Part II: application to complexity and nonlinearity of large-scale atmosphere dynamics. *J. Atmos. Sci.*, **60**, 478–490 (doi:10.1175/1520–0469(2003)060<0478:PEBLOM>2.0.CO;2).

Berner, J. 2005 Linking nonlinearity and non-Gaussianity of planetary wave behaviour by the Fokker–Planck equation. *J. Atmos. Sci.*, **62**, 2098–2117 (doi:10.1175/JAS3468.1).

Berner, J. & Branstator, G. 2007 Linear and nonlinear signatures in the planetary wave dynamics of an AGCM: probability density functions. *J. Atmos. Sci.*, **64**, 117–136 (doi:10.1175/JAS3822.1).

Branstator, G. 1990 Low-frequency patterns induced by stationary waves. *J. Atmos. Sci.*, **47**, 629–648 (doi:10.1175/1520–0469(1990) 047<0629:LFPIBS>2.0.CO;2).

Branstator, G. & Berner, J. 2005 Linear and nonlinear signatures in the planetary wave dynamics of an AGCM: phase space tendencies. *J. Atmos. Sci.*, **62**, 1792–1811 (doi:10.1175/JAS3429.1).

Branstator, G. & Haupt, S. E. 1998 An empirical model of barotropic atmospheric dynamics and its response to tropical forcing. *J. Climate*, **11**, 2645–2667 (doi:10.1175/1520–0442(1998)011 <2645:AEMOBA>2.0.CO;2).

Carnevale, G. & Frederiksen, J. S. 1987 Nonlinear stability and statistical mechanics for flow over topography. *J. Fluid Mech.*, **175**, 157–181 (doi:10.1017/S002211208700034X).

Crommelin, D. T. & Majda, A. J. 2004 Strategies for model reduction: comparing different optimal bases. *J. Atmos. Sci.*, **61**, 2206–2217 (doi:10.1175/1520–0469(2004)061<2206:SFMRCD>2.0.CO;2).

Crommelin, D. T. & Vanden-Eijnden, E. 2006 Reconstruction of diffusion using spectral data from timeseries. *Commun. Math. Sci.*, **4**, 651–668.

DelSole, T. 2004 Stochastic models of quasi-geostrophic turbulence. *Surv. Geophys.*, **25**, 107–149 (doi:10.1023/B:GEOP.0000028164.58516.b2).

Egger, J. 2005 A stochastic model for the angular momentum budget of latitude belts. *J. Atmos. Sci.*, **62**, 2592–2601 (doi:10.1175/JAS3480.1).

Fasullo, J. & Webster, P. 1999 Warm pool SST variability in relation to the surface energy balance. *J. Climate*, **12**, 1292–1305 (doi:10.1175/1520–0442(1999)012<1292:WPSVIR>2.0.CO;2).

Feldstein, S. B. 2000 The timescale, power spectra, and climate noise properties of teleconnection patterns. *J. Climate*, **13**, 4430–4440 (doi:10.1175/1520–0442(2000)013<4430:TTPSAC>2.0.CO;2).

Franzke, C. & Majda, A. J. 2006 Low-order stochastic mode reduction for a prototype atmospheric GCM. *J. Atmos. Sci.*, **63**, 457–479 (doi:10.1175/JAS3633.1).

Franzke, C., Majda, A. J. & Vanden-Eijnden, E. 2005 Low-order stochastic mode reduction for a realistic barotropic model climate. *J. Atmos. Sci.*, **62**, 1722–1745 (doi:10.1175/JAS3438.1).

Franzke, C., Majda, A. J. & Branstator, G. 2007 The origin of nonlinear signatures of planetary wave dynamics: mean phase space tendencies and contributions from non-Gaussianity. *J. Atmos. Sci.*, **64**, 3987–4003 (doi:10.1175/2006JAS2221.1).

Frederiksen, J. S., Dix, M. R. & Kepert, S. M. 1996 Systematic energy errors and the tendency toward canonical equilibrium in atmospheric circulation models. *J. Atmos. Sci.*, **53**, 887–904 (doi:10.1175/1520–0469(1996)053<0887:SEEATT>2.0.CO;2).

Gardiner, C. W. 1985 *Handbook of Stochastic Methods*. Springer-Verlag.

Hasselmann, K. 1988 PIPs and POPs: the reduction of complex dynamical systems using principal interaction and oscillation patterns. *J. Geophys. Res.*, **93**, 11015–11021 (doi:10.1029/JD093iD09p11015).

Hsu, C. J. & Zwiers, F. 2001 Climate change in recurrent regimes and modes of northern hemisphere atmospheric variability. *J. Geophys. Res.*, **106**, 20145–20159 (doi:10.1029/2001JD900229).

Hurrell, J. W. 1995 Decadal trends in the North Atlantic Oscillation: regional temperatures and precipitation. *Science*, **269**, 676–679 (doi:10.1126/science.269.5224.676).

Katsoulakis, M. A. & Vlachos, G. D. 2003 Hierarchical kinetic Monte Carlo simulations for diffusion of interacting molecules. *J. Chem. Phys.*, **112**, 9412–9427 (doi:10.1063/1.1616513).

Katsoulakis, M. A., Majda, A. J. & Vlachos, G. D. 2003a Coarse-grained stochastic processes for microscopic lattice systems. *Proc. Natl Acad. Sci. USA*, **100**, 782–787 (doi:10.1073/pnas.242741499).

Katsoulakis, M. A., Majda, A. J. & Vlachos, G. D. 2003b Coarse-grained stochastic processes and Monte Carlo simulations in lattice systems. *J. Comp. Phys.*, **186**, 250–278 (doi:10.1016/S0021–9991(03)00051–2).

Katsoulakis, M. A., Majda, A. J. & Sopasakis, A. 2004 Multiscale coupling in prototype hybrid deterministic/stochastic systems. Part I: deterministic closures. *Commun. Math. Sci.*, **2**, 255–294.

Katsoulakis, M. A., Majda, A. J. & Sopasakis, A. 2005 Multiscale coupling in prototype hybrid deterministic/stochastic systems. Part II: stochastic closures. *Commun. Math. Sci.*, **3**, 453–478.

Katsoulakis, M. A., Majda, A. J. & Sopasakis, A. 2006 Intermittency, metastability, and coarse graining for coupled deterministic–stochastic lattice systems. *Nonlinearity*, **19**, 1021–1047 (doi:10.1088/0951–7715/19/5/002).

Katsoulakis, M. A., Majda, A. J. & Sopasakis, A. 2007 Prototype hybrid couplings of macroscopic deterministic models and microscopic stochastic lattice dynamics. *Contemp. Math.*, **429**, 143–187.

Khouider, B. & Majda, A. J. 2006a A simple multicloud parametrization for convectively coupled tropical waves. Part I: linear analysis. *J. Atmos. Sci.*, **63**, 1308–1323 (doi:10.1175/JAS3677.1).

Khouider, B. & Majda, A. J. 2006b Model multicloud parametrizations with crude vertical structure. *Theor. Comp. Fluid Dyn.*, **20**, 351–375 (doi:10.1007/s00162-006-0013–2).

Khouider, B. & Majda, A. J. 2007 A simple multicloud parametrization for convectively coupled tropical waves. Part II: nonlinear simulations. *J. Atmos. Sci.*, **64**, 381–400 (doi:10.1175/JAS3833.1).

Khouider, B. & Majda, A. J. 2008 Multicloud models for tropical convection: enhanced congestus heating. *J. Atmos. Sci.*, **65**, 895–914 (doi:10.1175/2007JAS2408.1).

Khouider, B., Majda, A. J. & Katsoulakis, M. 2003 Coarse grained stochastic models for tropical convection. *Proc. Natl Acad. Sci. USA*, **100**, 11941–11946 (doi:10.1073/pnas.1634951100).

Kravtsov, S., Kondrashov, D. & Ghil, M. 2005 Multi-level regression modeling of nonlinear processes: derivation and applications to climate variability. *J. Climate*, **62**, 4404–4424 (doi:10.1175/JCLI3544.1).

Kurtz, T. G. 1973 A limit theorem for perturbed operators semigroups with applications to random evolution. *J. Funct. Anal.*, **12**, 55–67 (doi:10.1016/0022–1236(73)90089-X).

Kwasniok, F. 1996 The reduction of complex dynamical systems using principal interaction patterns. *Physica D*, **92**, 28–60 (doi:10.1016/0167–2789(95)00280–4).

Kwasniok, F. 2004 Empirical low-order models of barotropic flow. *J. Atmos. Sci.*, **61**, 235–245 (doi:10.1175/1520–0469(2004)061<0235:ELMOBF>2.0.CO;2).

Lin, J.-L., Kiladis, G. N., Mapes, B. E. *et al.* 2006 Tropical intraseasonal variability in 14 IPCC AR4 climate models. Part 1: convective signals. *J. Climate*, **19**, 2665–2690 (doi:10.1175/JCLI3735.1).

Majda, A. J. 2000 Real world turbulence and modern applied mathematics. *Mathematics Frontiers and Perspectives*, ed. V. Arnold, pp. 137–151. American Mathematical Society.

Majda, A. J. & Khouider, B. 2002 Stochastic and mesoscopic models for tropical convection. *Proc. Natl Acad. Sci. USA*, **99**, 1123–1128 (doi:10.1073/pnas.032663199).

Majda, A. J. & Shefter, M. 2001 Waves and instabilities for model tropical convective parametrizations. *J. Atmos. Sci.*, **58**, 896–914 (doi:10.1175/1520–0469(2001)058<0896:WAIFMT>2.0.CO;2).

Majda, A. J. & Timofeyev, I. 2000 Remarkable statistical behavior for truncated Burgers–Hopf dynamics. *Proc. Natl Acad. Sci. USA*, **97**, 12413–12417 (doi:10.1073/pnas.230433997).

Majda, A. J. & Timofeyev, I. 2004 Low dimensional chaotic dynamics versus intrinsic stochastic chaos: a paradigm model. *Physica D*, **199**, 339–368 (doi:10.1016/j.physd.2004.05.012).

Majda, A. J. and Wang, X. 2006 *Nonlinear Dynamics and Statistical Theories for Basic Geophysical Flows*. Cambridge University Press.

Majda, A. J., Timofeyev, I. & Vanden-Eijnden, E. 1999 Models for stochastic climate prediction. *Proc. Natl Acad. Sci. USA*, **96**, 14687–14691 (doi:10.1073/pnas.96.26.14687).

Majda, A. J., Timofeyev, I. & Vanden-Eijnden, E. 2001 A mathematical framework for stochastic climate models. *Commun. Pure Appl. Math.*, **54**, 891–974 (doi:10.1002/cpa.1014).

Majda, A. J., Timofeyev, I. & Vanden-Eijnden, E. 2002 *A priori* tests of a stochastic mode reduction strategy. *Physica D*, **170**, 206–252 (doi:10.1016/S0167–2789(02)00578-X).

Majda, A. J., Timofeyev, I. & Vanden-Eijnden, E. 2003 Systematic strategies for stochastic mode reduction in climate. *J. Atmos. Sci.*, **60**, 1705–1722 (doi:10.1175/1520–0469(2003)060<1705:SSFSMR>2.0.CO;2).

Majda, A. J., Abramov, R. & Grote, M. 2005 *Information Theory and Stochastics For Multiscale Nonlinear Systems*. Monograph Series of Center for Research in Mathematics, University of Montreal. American Mathematical Society.

Majda, A. J., Franzke, C., Fischer, A. & Crommelin, D. T. 2006a Distinct metastable atmospheric regimes despite nearly Gaussian statistics: a paradigm model. *Proc. Natl Acad. Sci. USA*, **103**, 8309–8314 (doi:10.1073/pnas.0602641103).

Majda, A. J., Timofeyev, I. & Vanden-Eijnden, E. 2006b Stochastic models for selected slow variables in large deterministic systems. *Nonlinearity*, **19**, 769–794 (doi:10.1088/0951–7715/19/4/001).

Mapes, B. E. 2000 Convective inhibition, subgridscale triggering energy, and "stratiform instability" in a toy tropical wave model. *J. Atmos. Sci.*, **57**, 1515–1535 (doi:10.1175/1520–0469(2000)057<1515:CISSTE>2.0.CO;2).

Marshall, J. & Molteni, F. 1993 Toward a dynamical understanding of planetary-scale flow regimes. *J. Atmos. Sci.*, **50**, 1792–1818 (doi:10.1175/1520–0469(1993)050<1792:TADUOP>2.0.CO;2).

Newman, M., Sardeshmukh, P. D. & Penland, C. 1997 Stochastic forcing of the wintertime extratropical flow. *J. Atmos. Sci.*, **54**, 435–455 (doi:10.1175/1520–0469(1997)054<0435:SFOTWE>2.0.CO;2).

Palmer, T. N. 2001 A nonlinear dynamical perspective on model error: a proposal for non-local stochastic-dynamic parametrizations in weather and climate prediction models. *Q. J. R. Meteorol. Soc.*, **127**, 279–304 (doi:10.1002/9j.49712757202).

Percival, D. B., Overland, J. E. & Mofjeld, H. O. 2001 Interpretation of North Pacific variability as a short- and long-memory process. *J. Climate*, **14**, 4545–4559 (doi:10.1175/1520–0442(2001)014<4545:IONPVA>2.0.CO;2).

Salmon, R. 1998 *Lectures on Geophysical Fluid Dynamics*. Oxford University Press.

Schubert, S. D. 1985 A statistical-dynamical study of empirically determined modes of atmospheric variability. *J. Atmos. Sci.*, **42**, 3–17 (doi:10.1175/1520–0469(1985)042<0003:ASDSOE>2.0.CO;2).

Selten, F. M. 1995 An efficient description of the dynamics of barotropic flow. *J. Atmos. Sci.*, **52**, 915–936 (doi:10.1175/1520–0469(1995)052<0915:AEDOTD>2.0.CO;2).

Selten, F. M. & Branstator, G. 2004 Preferred regime transition routes and evidence for an unstable periodic orbit in a baroclinic model. *J. Atmos. Sci.*, **61**, 2267–2282 (doi:10.1175/1520–0469(2004)061<2267:PRTRAE>2.0.CO;2).

Stephenson, D. B., Pavan, V. & Bojariu, R. 2000 Is the North Atlantic Oscillation a random walk? *Int. J. Climatol.*, **20**, 1–18 (doi:10.1002/(SICI)1097–0088(200001)20:1<1::AID-JOC456> 3.0.CO;2-P).

Sura, P. 2003 Stochastic analysis of Southern and Pacific Ocean sea surface winds. *J. Atmos. Sci.*, **60**, 654–666 (doi:10.1175/1520–0469(2003)060<0654:SAOSAP>2.0.CO;2).

Sura, P. & Sardeshmukh, P. D. 2008 A global view of non-Gaussian SST variability. *J. Phys. Oceanogr.*, **38**, 639–647 (doi:10.1175/2007JPO376.1).

Wallace, J. M. & Gutzler, D.S. 1981 Teleconnections in the geopotential height field during the Northern Hemisphere winter. *Mon. Weather Rev.*, **109**, 784–804 (doi:10.1175/1520–0493(1981)109<0784:TITGHF>2.0.CO;2).

Wheeler, M., Kiladis, G. N. & Webster, P. J. 2000 Large scale dynamical fields associated with convectively coupled equatorial waves. *J. Atmos. Sci.*, **57**, 613–640 (doi:10.1175/1520–0469(2000)057<0613:LSDFAW>2.0.CO;2).

Whitaker, J. S. & Sardeshmukh, P. D. 1998 A linear theory of extratropical synoptic eddy statistics. *J. Atmos. Sci.*, **55**, 237–258 (doi:10.1175/1520–0469(1998)055<0237:ALTOES>2.0.CO;2).

Zhang, Y. & Held, I. M. 1999 A linear stochastic model of a GCM's midlatitude storm tracks. *J. Atmos. Sci.*, **56**, 3416–3435 (doi:10.1175/1520–0469(1999)056<3416:ALSMOA>2.0.CO;2).

4

Predictability in nonlinear dynamical systems with model uncertainty

JINQIAO DUAN

Nonlinear systems with model uncertainty are often described by stochastic differential equations. Some techniques from random dynamical systems are discussed. They contribute to a better understanding of solution processes of stochastic differential equations and thus may shed light on predictability in nonlinear systems with model uncertainty.

4.1 Introduction

Nonlinear systems are often influenced by random fluctuations, such as uncertainty in specifying initial conditions or boundary conditions, external random forcing, and fluctuating parameters. In building mathematical models for these nonlinear systems, sometimes, if not often, less-known, less well-understood, or less well-observed processes (e.g. highly fluctuating fast or small-scale processes) are ignored due to limitations in our analytical ability or computational power.

The limitation of predicting dynamical behaviour in nonlinear systems due to uncertainty in initial conditions has been widely investigated (Guckenheimer and Holmes 1983). This chapter discusses model uncertainty in nonlinear systems. This issue has attracted a lot of attention in the geophysical community (Samelson 1989; Griffa & Castellari 1991; Müller 1996; Saravanan & McWilliams 1998; Cessi and Louazel 2001; Orrell *et al.* 2001; Farrell & Ioannou 2002a,b; DelSole 2004; Allen *et al.* 2006; Palmer *et al.* 2006).

The uncertainties in simulation may also be regarded as a kind of model uncertainty. This arises in numerical simulations of multiscale systems that display a wide range of spatial and temporal scales, with no clear scale separation. Due to the limitations of computer power, at present and for the conceivable future, not all scales of variability can be explicitly simulated or resolved. Although these unresolved scales may be very small or very fast, their long time impact on the resolved simulation may be delicate (i.e. may be negligible or may have significant effects; or in other words, uncertain). Thus, to take the effects of unresolved scales on the resolved scales into account, representations or parameterisations of these effects are required (Berselli *et al.* 2005).

Stochastic parameterisation of unresolved scales or unresolved processes leads to stochastic dynamical models in weather and climate prediction (Leith 1975; Hasselmann 1976; Chorin *et al.* 1999; Arnold 2001; Just *et al.* 2001; Imkeller & Monahan 2002; Sura and Penland 2002; Huisinga *et al.* 2003; Berloff 2005; Palmer *et al.* 2005; Wilks 2005; Williams 2005; Palmer & Hagedorn 2006; Duan and Nadiga 2007; Pasquero and Tziperman 2007; Du and Duan 2009).

There has been a recent research focus in the dynamical systems community to better understand the solution orbits of stochastic dynamical models (Horsthemke & Lefever 1984; Rozovskii 1990; Karatzas & Shreve 1991; Da Prato & Zabczyk 1992; Arnold 1998; Freidlin & Wentzell 1998; Garcia-Ojalvo & Sancho 1999; Crauel & Gundlach 1999; Boffetta *et al.* 2002; Oksendal 2003; Waymire & Duan 2005; Wang & Duan 2007). This is relevant to the issue of predictability under uncertainty in nonlinear systems, which is concerned with factors and mechanisms for uncertainties of forecasts, and techniques for quantifying and reducing these uncertainties (North and Cahalan 1981; Smith *et al.* 1999; Boffetta *et al.* 2000; Vannitsem and Toth 2002; Khade and Hansen 2004; Mu *et al.* 2004; Nicolis *et al.* 2004; Palmer and Hagedorn 2006; Blomker and Duan 2007). Various measures have been proposed to quantify predictability (Schneider and Griffies 1999; Kleeman 2002; Majda *et al.* 2002; Mu *et al.* 2003), and the impact of measure selection on prediction results has also been discussed (Orrell 2002).

We consider the following stochastic system defined by Itô stochastic differential equations (SDEs) in $\mathbb{R}^n$:

$$dX_t = b(X_t)dt + \sigma(X_t)dW(t), \quad X(0) = x_0, \tag{4.1}$$

where b and σ are vector and matrix functions, taking values in $\mathbb{R}^n$ and $\mathbb{R}^{n\times m}$, respectively. The standard vector Brownian motion $W(t)$ takes values in $\mathbb{R}^m$. Note that n and m may be equal or different. We treat X, X_t, $X(t)$ or $X_t(\omega)$ as the same random quantity.

The noise term σdW_t may be regarded as model uncertainty or model error. It could be caused by external fluctuations or random influences, or by fluctuating coefficients or parameters in the model. Stochastic parameterisation of unresolved scales or unresolved processes leads to stochastic dynamical systems (Palmer 2001; Palmer *et al.* 2005; Wilks 2005; Williams 2005; Duan & Nadiga 2007). Moreover, numerical simulation of stochastic partial differential equations may also lead to SDEs (Millet & Morien 2005; Roberts 2003–2004).

The Brownian motion $W(t)$, also denoted as W_t, is a Gaussian stochastic process on a underlying probability space $(\Omega, \mathcal{F}, \mathbb{P})$, where Ω is a sample space, $\mathcal{F}$ is a

σ-field composed of measurable subsets of Ω (called 'events') and $\mathbb{P}$ is a probability (also called probability measure). Being a Gaussian process, W_t is characterised by its mean vector (taken to be the zero vector) and its covariance operator, an $n \times n$ symmetric positive definite matrix (taken to be the identity matrix). More specifically, W_t satisfies the following conditions (Oksendal 2003):

(a) $W(0) = 0$, almost surely (a.s.)
(b) W has continuous paths or trajectories, a.s.
(c) W has independent increments,
(d) $W(t) - W(s) \sim N(0, (t - s)I)$, t and $s > 0$ and $t \geq s \geq 0$, where I is the $n \times n$ identity matrix.

Remark 1. (*i*) The covariance operator here is a constant $n \times n$ identity matrix I, i.e. $Q = I$ and the trace of Q is $Tr(Q) = n$.

(*ii*) From now on, we consider two-sided Brownian motion W_t, $t \in \mathbb{R}$, by means of two independent usual Brownian motions W_t^1 and $W_t^2 (t \geq 0)$: For $t \geq 0$, $W_t := W_t^1$, while for $t < 0$, $W_t := W_{-t}^2$.

(*iii*) $W(t) \sim N(0, |t|I)$, i.e. $W(t)$ has probability density function $p_t(x) = \frac{1}{(2\pi t)^{\frac{n}{2}}} e^{-\frac{x_1^2 + \cdots + x_n^2}{2t}}$.

(*iv*) For every $\alpha \in (0, \frac{1}{2})$, for almost every (a.e.) $\omega \in \Omega$, there exists $C(\omega)$ such that

$$|W(t, \omega) - W(s, \omega)| \leq C(\omega)|t - s|^{\alpha},$$

namely, Brownian paths are Hölder continuous with an exponent less than one half.

The Euclidean space $\mathbb{R}^n$ has the usual distance $d(x, y) = \sqrt{\sum_{j=1}^n (x_j - y_j)^2}$, norm $\|x\| = \sqrt{\sum_{j=1}^n x_j^2}$, and the scalar product $x \cdot y = < x, y > = \sum_{j=1}^n x_j y_j$.

This chapter is organised as follows. After reviewing some basics about stochastic differential equations in Section 4.2, random dynamic systems are discussed in Section 4.3. The impact of uncertainty and error growth are considered in Section 4.4, residence time, exit probability and predictability in Section 4.5 and invariant manifolds and predictability in Section 4.6. Finally, nonlinear systems under non-Gaussian noise and coloured noise are discussed in Section 4.7 and 4.8, respectively.

4.2 Stochastic differential equations

4.2.1 Itô and Stratonovich calculus

Note that the Stratonovich stochastic differential $\sigma(X) \circ dW(t)$ and the Itô stochastic differential $\sigma(X)\, dW(t)$ are interpreted through their corresponding definitions of stochastic integrals (Oksendal 2003):

$$\int_0^T \sigma(X) \circ \mathrm{d}W(t) := \text{mean-square} \lim_{\Delta t_j \to 0} \sum_j \sigma \left[X \left(\frac{t_{j+1} + t_j}{2} \right) \right] (W_{t_{j+1}} - W_{t_j}),$$

$$\int_0^T \sigma(X) \mathrm{d}W(t) := \text{mean-square} \lim_{\Delta t_j \to 0} \sum_j \sigma[X(t_j)](W_{t_{j+1}} - W_{t_j}).$$

Note the difference in the sums: in the Stratonovich integral, the integrand is evaluated at the midpoint $\frac{t_{j+1}+t_j}{2}$ of a subinterval (t_j, t_{j+1}), while for the Itô integral, the integrand is evaluated at the left end point t_j. See Oksendal (2003) for a discussion about the difference in physical modelling by these two kinds of stochastic differential equations. There are also dynamical differences for these two types of stochastic equations, even at linear level (Caraballo and Langa 2001).

If the integrand $f(t, \omega)$ is sufficiently smooth in time, e.g. Hölder continuous in time in mean-square norm, with an exponent larger than 1, then both Itô and Stratonovich integrals coincide; see Oksendal (2003), p.39. But in general, these two integrals differ. Note that W_t is only Hölder continuous in time (Klebaner 2005) with the exponent $\alpha < \frac{1}{2}$. So that is why the following stochastic integrals are different:

$$\int_t^T W_t \mathrm{d}W_t = \frac{1}{2}(W_t^2 - W_T^2) - \frac{1}{2}(T - t),$$

$$\int_t^T W_t \circ \mathrm{d}W_t = \frac{1}{2}(W_t^2 - W_T^2).$$

Thus we have the two different kinds of SDEs of Itô and Stratonovich type:

$$\mathrm{d}X = b(X)\mathrm{d}t + \sigma(X)\mathrm{d}W(t), \quad X(0) = x_0, \tag{4.2}$$

$$\mathrm{d}X = b(X)\mathrm{d}t + \sigma(X) \circ \mathrm{d}W(t), \quad X(0) = x_0, \tag{4.3}$$

However, systems of Stratonovich SDEs can be converted to Itô SDEs and vice versa (Kloeden & Platen 1992; Oksendal 2003). In the following we only consider the Itô type of SDEs.

4.2.2 Itô's formula and product rule

Itô's formula in one dimension (scalar case)

Consider a scalar SDE (b, σ and W_t are all scalars),

$$\mathrm{d}X_t = b(X_t)\mathrm{d}t + \sigma(X_t)\mathrm{d}W_t.$$

Let $g(t, x)$ be a given (deterministic) scalar smooth function.

Itô's formula in differential form is:

$$\mathrm{d}g(t, X_t) = [g_t(t, X_t) + g_x(t, X_t)b(X_t) + \frac{1}{2}g_{xx}(t, X_t)\sigma^2(X_t)]\mathrm{d}t$$
$$+ g_x(t, X_t)\sigma(X_t)\mathrm{d}W_t. \quad (4.4)$$

The term $\frac{1}{2}g_{xx}(t, X_t)\sigma^2(X_t)$ is called the Itô correction term. Symbolically, we may use the following rules in manipulating Itô differentials:

$$\mathrm{d}t\mathrm{d}t = \mathrm{d}t\mathrm{d}W_t = 0, \quad \mathrm{d}W_t\mathrm{d}W_t = \mathrm{d}t.$$

Itô's formula in integral form is:

$$g(t, X_t) = g(0, X_0) + \int_0^t [g_t(s, X_s) + g_x(s, X_s) + \frac{1}{2}g_{xx}(s, X_s)\sigma^2(X_s)]\mathrm{d}s$$
$$+ \int_0^t g_x(s, X_s)\sigma(X_s)\mathrm{d}W_s. \quad (4.5)$$

The generator A for this scalar SDE is

$$Ag = g_x b + \frac{1}{2}g_{xx}\sigma^2.$$

Itô's formula in n *dimensions (vector case)*

Consider a SDE system

$$\mathrm{d}X_t = b(X_t)\mathrm{d}t + \sigma(X_t)\mathrm{d}W_t,$$

where b is an n-dimensional vector function, σ is an $n \times m$ matrix function, and $W_t(\omega)$ is an m-dimensional Brownian motion.

Let $g(t, x)$ be a given (deterministic) scalar smooth function for $x \in \mathbb{R}^n$. Itô's formula in differential form is:

$$\mathrm{d}g(t, X_t) = \left\{g_t(t, X_t) + [\nabla g(t, X_t)]^{\mathrm{T}}b + \frac{1}{2}\,\mathrm{trace}[\sigma\sigma^{\mathrm{T}}H(g)](t, X_t)\right\}\mathrm{d}t$$
$$+ [\nabla g(t, X_t)]^{\mathrm{T}}\sigma(X_t)\mathrm{d}W_t, \quad (4.6)$$

where the superscript T denotes transpose matrix and $H(g) = (g_{x_i x_j})$ is the $n \times n$ Hessian matrix, also denoted $D^2(g)$.

The generator A for this SDE system is

$$Ag = (\nabla g)^{\mathrm{T}}b + \frac{1}{2}\,\mathrm{trace}[\sigma\sigma^{\mathrm{T}}D^2(g)], \quad (4.7)$$

where the gradient vector of g is $\nabla g = (g_{x_1}, \ldots, g_{x_n})^{\mathrm{T}}$ and the $n \times n$ Hessian matrix of g is

$$D^2(g) := (g_{x_i x_j}).$$

Symbolically we may also use the rules:

$$\mathrm{d}t\mathrm{d}t = 0, \quad \mathrm{d}t\mathrm{d}W_t = \mathbf{0}, \quad \mathrm{d}W_t \cdot \mathrm{d}W_t = n\mathrm{d}t = \operatorname{trace}(Q)\mathrm{d}t.$$

Note that trace$(Q) = n$ for n-dimensional Brownian motion W_t.

Itô's formula in integral form is:

$$g(t, X_t) = g(0, X_0) + \int_0^t \{g_t(s, X_s) + [\nabla g(s, X_s)]^{\mathrm{T}} b + \frac{1}{2}\operatorname{trace}[\sigma\sigma^{\mathrm{T}} H(g)](s, X_s)\}\mathrm{d}s + \int_0^t [\nabla g(s, X_s)]^{\mathrm{T}} \sigma(X_s)\mathrm{d}W_s. \tag{4.8}$$

Remark 2. For Itô's formula above, a somewhat remote connection is the material derivative of the fluid velocity

$$\frac{\mathrm{d}}{\mathrm{d}t} g(x, t) = \partial_t g + u \cdot \nabla g, \tag{4.9}$$

where $u = \dot{x}$ is the underlying driving flow.

Stochastic product rule

Taking $g = xy$ for a two-dimensional SDE system, we get

$$\mathrm{d}(X_t Y_t) = X_t \mathrm{d}Y_t + (\mathrm{d}X_t)Y_t + \mathrm{d}X_t \mathrm{d}Y_t. \tag{4.10}$$

4.2.3 Estimation of Itô's integrals

Let us now look at some properties and estimations of Itô's integrals.

Itô isometry

$$\mathbb{E}\left(\int_0^T f(t, \omega)\mathrm{d}W_t\right)^2 = \mathbb{E}\int_0^T f^2(t, \omega)\mathrm{d}t \tag{4.11}$$

Generalised Itô isometry

$$\mathbb{E}\left(\int_0^a f(t, \omega)\mathrm{d}W_t \int_0^b g(t, \omega)\mathrm{d}W_t\right) = \mathbb{E}\int_0^{a\wedge b} f(t, \omega)g(t, \omega)\mathrm{d}t, \tag{4.12}$$

where $a \wedge b = \min(a, b)$.

Proof. Look at the case $a = b$.

Denote $I_1 = \int_0^a f(t, \omega)\mathrm{d}W_t$ and $I_2 = \int_0^a g(t, \omega)\mathrm{d}W_t$. Note that $I_1 I_2 = \frac{1}{2}[(I_1 + I_2)^2 - I_1^2 - I_2^2]$ and use the isometry property.

For $a \neq b$, say $a < b$, i.e. $\min(a, b) = a$. Extend f to the time interval $[a, b]$ by setting it zero there. Then apply the above proof. □

Itô isometry in vector case

Let $F(t, \omega)$ and $G(t, \omega)$ be $n \times n$ matrices, and W_t be n-dimensional Brownian motion:

$$\mathbb{E}\left(\int_0^a F(t,\omega)\mathrm{d}W_t\right)\cdot\left(\int_0^b G(t,\omega)\mathrm{d}W_t\right) = \mathbb{E}\int_0^{a\wedge b} \mathrm{trace}(GF^{\mathrm{T}})(t,\omega)\mathrm{d}t, \tag{4.13}$$

where $(\cdot)$ denotes the usual scalar product in $\mathbb{R}^n$, and the trace of the matrix is the sum of diagonal entries.

In particular,

$$\mathbb{E}\left\|\int_0^a F(t,\omega)\mathrm{d}W_t\right\|^2 = \mathbb{E}\int_0^a \mathrm{trace}(FF^{\mathrm{T}})(t,\omega)\mathrm{d}t, \tag{4.14}$$

$$\mathbb{E}\left(\int_0^a F(t,\omega)\mathrm{d}W_t\right)\cdot\left(\int_0^b F(t,\omega)\mathrm{d}W_t\right) = \mathbb{E}\int_0^{a\wedge b} \mathrm{trace}(FF^{\mathrm{T}})(t,\omega)\mathrm{d}t. \tag{4.15}$$

Inequalities involving Itô's integrals

By the Itô isometry and the Doob martingale inequality (Oksendal 2003, p.33), we have, for any constant $\lambda > 0$,

$$\mathbb{P}\left(\sup_{t_0\le t\le T}\left|\int_{t_0}^t f(s,\omega)\mathrm{d}W_s\right| \ge \lambda\right) \le \frac{1}{\lambda^2}\mathbb{E}\int_{t_0}^t |f(s,\omega)|^2\mathrm{d}s. \tag{4.16}$$

According to Arnold (1974), p.81, we have the following two estimates:

$$\mathbb{E}\left(\sup_{t_0\le t\le T}\left|\int_{t_0}^t f(s,\omega)\mathrm{d}W_s\right|^2\right) \le 4\mathbb{E}\int_{t_0}^t |f(s,\omega)|^2\mathrm{d}s, \tag{4.17}$$

and

$$\mathbb{E}\left|\int_{t_0}^t f(s,\omega)\mathrm{d}W_s\right|^{2k} \le (k(2k-1))^{k-1}(t-t_0)^{k-1}\mathbb{E}\int_{t_0}^t |f(s,\omega)|^{2k}\mathrm{d}s. \tag{4.18}$$

4.2.4 Some examples

Example 1. **Langevin equation**

$$\mathrm{d}X_t = -bX_t\mathrm{d}t + a\mathrm{d}W_t, \tag{4.19}$$

where a, b are real parameters, and the initial condition $X_0 \sim \mathbb{N}(0, \sigma^2)$. The solution is

$$X_t = \mathrm{e}^{-bt}X_0 + a\mathrm{e}^{-bt}\int_0^t \mathrm{e}^{bs}\mathrm{d}W_s. \tag{4.20}$$

Note that $\mathbb{E}X_t = 0$ and X_t is a Gaussian process.

$$\mathrm{Cov}(X_s, X_t) = \sigma^2 \mathrm{e}^{-b(s+t)} + \frac{a^2}{2b}[\mathrm{e}^{-b|s-t|} - \mathrm{e}^{-b(s+t)}], \tag{4.21}$$

$$\mathrm{Cov}(X_0, X_t) = \sigma^2 \mathrm{e}^{-bt} + \frac{a^2}{2b}[\mathrm{e}^{-b|t|} - \mathrm{e}^{-bt}], \tag{4.22}$$

$$\mathrm{Var}(X_t) = \sigma^2 \mathrm{e}^{-2bt} + \frac{a^2}{2b}[1 - \mathrm{e}^{-2bt}], \tag{4.23}$$

$$\mathrm{Cor}(X_s, X_t) = \frac{\mathrm{Cov}(X_s, X_t)}{\sqrt{\mathrm{Var}(X_s)}\sqrt{\mathrm{Var}(X_t)}}. \tag{4.24}$$

When $\sigma^2 = \frac{a^2}{2b}$, we have

$$\mathrm{Cov}(X_s, X_t) = \sigma^2 \mathrm{e}^{-b|s-t|}, \tag{4.25}$$

$$\mathrm{Var}(X_t) = \tfrac{a^2}{2b}. \tag{4.26}$$

Namely, in this case, X_t is a stationary process.

Example 2. **Stochastic population model**
Consider the following linear scalar SDE with multiplicative noise:

$$\mathrm{d}X_t = rX_t\mathrm{d}t + \alpha X_t \mathrm{d}W_t, \tag{4.27}$$

where r and α are real constants, and $X_t > 0$, a.s. The SDE is rewritten as

$$\frac{\mathrm{d}X_t}{X_t} = r\mathrm{d}t + \alpha \mathrm{d}W_t. \tag{4.28}$$

The Itô formula is applied to $\ln X_t$ to obtain

$$\mathrm{d}(\ln X_t) = \frac{\mathrm{d}X_t}{X_t} - \frac{1}{2}\alpha^2 \mathrm{d}t. \tag{4.29}$$

That is, $\frac{\mathrm{d}X_t}{X_t} = \mathrm{d}(\ln X_t) + \frac{1}{2}\alpha^2\mathrm{d}t$. Thus (4.28) becomes

$$\mathrm{d}(\ln X_t) = \left(r - \frac{1}{2}\alpha^2\right)\mathrm{d}t + \alpha \mathrm{d}W_t. \tag{4.30}$$

Integrating from 0 to t,

$$\ln \frac{X_t}{X_0} = \left(r - \frac{1}{2}\alpha^2\right)t + \alpha W_t.$$

We hence get the final solution

$$X_t = X_0 \exp\left[\left(r - \frac{1}{2}\alpha^2\right)t + \alpha W_t\right]. \tag{4.31}$$

Example 3. **A linear scalar SDE**

We consider the following scalar SDE (Arnold 1974; Kloeden and Platen 1992)

$$\mathrm{d}X_t = [a_1(t)X_t + a_2(t)]\mathrm{d}t + [b_1(t)X_t + b_2(t)]\mathrm{d}W_t, \quad X_{t_0} \text{ given.} \tag{4.32}$$

The fundamental solution is

$$\Phi_{t,t_0} = \exp\left[\int_{t_0}^{t}\left(a_1(s) - \frac{1}{2}b_1^2(s)\right)\mathrm{d}s + \int_{t_0}^{t} b_1(s)\mathrm{d}W_s\right]. \tag{4.33}$$

Thus the general solution is

$$X_t = \Phi_{t,t_0}\left\{X_{t_0} + \int_{t_0}^{t}[a_2(s) - b_1(s)b_2(s)]\ \Phi_{s,t_0}^{-1}\mathrm{d}s + \int_{t_0}^{t} b_2(s)\ \Phi_{s,t_0}^{-1}\mathrm{d}W_s\right\}. \tag{4.34}$$

Example 4. **A linear system of SDEs**

We consider the following system of SDEs (Oksendal 2003)

$$\mathrm{d}X_t = [AX_t + f(t)]\mathrm{d}t + \sum_{k=1}^{m} g_k(t)\mathrm{d}W_k(t), \quad X_{t_0} \text{ given,} \tag{4.35}$$

where A is a constant $n \times n$ matrix, $X(t)$, $f(t)$ and $g_k(t)$ are n-dimensional vector functions, and W_k $(k = 1, \ldots, m)$ are independent scalar Brownian motions. This is a system with a constant coefficient matrix and additive noise. In this case, we can find out the solution completely with the help of a matrix exponential.

The fundamental solution matrix for the corresponding linear system $\mathrm{d}X_t = AX_t\mathrm{d}t$ is

$$\Phi_{t,t_0} = \mathrm{e}^{A(t-t_0)}. \tag{4.36}$$

The solution for the non-homogeneous linear system with the constant coefficient matrix (4.35) is

$$X_t = \mathrm{e}^{A(t-t_0)}\left\{X_{t_0} + \int_{t_0}^{t}\mathrm{e}^{-A(s-t_0)}f(s)\mathrm{d}s + \sum_{k=1}^{m}\int_{t_0}^{t}\mathrm{e}^{-A(s-t_0)}g_k(s)\mathrm{d}W_k(s)\right\} \tag{4.37}$$

$$= \mathrm{e}^{A(t-t_0)}X_{t_0} + \int_{t_0}^{t}\mathrm{e}^{A(t-s)}f(s)\mathrm{d}s + \sum_{k=1}^{m}\int_{t_0}^{t}\mathrm{e}^{A(t-s)}g_k(s)\mathrm{d}W_k(s). \tag{4.38}$$

Example 5. **Stochastic oscillations**
Let us now look at the following second-order SDE modelling an oscillator under random forcing (Mao 1997; Oksendal 2003):

$$\ddot{x} + a\dot{x} + bx = \sigma \dot{W}_t, \tag{4.39}$$

where a, b, σ are real constants, and W_t is a scalar Brownian motion. This second-order SDE may be rewritten as a first-order SDE system:

$$\dot{x} = y, \tag{4.40}$$

$$\dot{y} = -bx - ay + \sigma \dot{W}_t. \tag{4.41}$$

In matrix form this becomes

$$\dot{X} = AX + K\dot{W}_t, \tag{4.42}$$

where

$$A = \begin{pmatrix} 0 & 1 \\ -b & -a \end{pmatrix}$$

and

$$K = \begin{pmatrix} 0 \\ \sigma \end{pmatrix}.$$

The solution is

$$X(t) = \mathrm{e}^{At} X(0) + \int_0^t \mathrm{e}^{A(t-s)} K \,\mathrm{d}W_s. \tag{4.43}$$

A special case of this model is the stochastic harmonic oscillator:

$$\ddot{x} + kx = h\dot{W}_t, \tag{4.44}$$

where k, h are positive constants. In this case ($a = 0$),

$$A = \begin{pmatrix} 0 & 1 \\ -k & 0 \end{pmatrix}.$$

Noticing that $A^2 = -kI$ with I the 2×2 identity matrix, we have

$$\mathrm{e}^{At} = \begin{pmatrix} \cos(\sqrt{k}t) & \frac{1}{\sqrt{k}}\sin(\sqrt{k}t) \\ -\sqrt{k}\sin(\sqrt{k}t) & \cos(\sqrt{k}t) \end{pmatrix}. \tag{4.45}$$

The solution for the stochastic harmonic oscillator is

$$x(t) = x_0 \cos(\sqrt{k}t) + \frac{y_0}{\sqrt{k}} \sin(\sqrt{k}t) + \frac{h}{\sqrt{k}} \int_0^t \sin[\sqrt{k}(t-s)]\mathrm{d}W_s, \tag{4.46}$$

$$y(t) = -x_0\sqrt{k}\sin(\sqrt{k}t) + y_0\cos(\sqrt{k}t) + h\int_0^t \cos[\sqrt{k}(t-s)]\mathrm{d}W_s. \tag{4.47}$$

4.3 Random dynamical systems

In this section we introduce some definitions in stochastic dynamical systems, as well as recalling some usual notations in probability.

We consider stochastic systems in the state space $\mathbb{R}^n$. All the sample paths or sample orbits and invariant manifolds are in this state space.

Some stochastic processes, such as a Brownian motion, can be described by a canonical (deterministic) dynamical system (see Arnold 1998, Appendix A). A standard Brownian motion (or Wiener process) $W(t)$ in $\mathbb{R}^n$, with two-sided time $t \in \mathbb{R}$, is a stochastic process with $W(0) = 0$ and stationary independent increments satisfying $W(t) - W(s) \sim \mathcal{N}(0, |t - s|I)$. Here I is the $n \times n$ identity matrix. The Brownian motion can be realised in a canonical sample space of continuous paths passing the origin at time 0:

$$\Omega = C_0(\mathbb{R}, \mathbb{R}^n) := \{\omega \in C(\mathbb{R}, \mathbb{R}^n) : \omega(0) = 0\}.$$

We identify $W_t(\omega)$ with $\omega(t)$, namely $W_t(\omega) = \omega(t)$. The convergence concept in this sample space is the uniform convergence on bounded and closed time intervals, induced by the following metric

$$\rho(\omega, \omega') := \sum_{n=1}^{\infty} \frac{1}{2^n} \frac{\|\omega - \omega'\|_n}{1 + \|\omega - \omega'\|_n}, \text{ where } \|\omega - \omega'\|_n := \sup_{-n \le t \le n} \|\omega(t) - \omega'(t)\|.$$

With this metric, we can define events represented by open balls in Ω. For example, a ball centered at zero with radius 1 is $\{\omega\colon \rho(\omega, 0) < 1\}$. We define the Borel σ-algebra $\mathcal{F}$ as the collection of events represented by open balls, As, complements of open balls, A^cs, unions and intersections of As and/or A^cs, together with the empty event, the whole event (the sample space Ω), and all events formed by doing the complements, unions and intersections forever in this collection.

Taking the (incomplete) Borel σ-algebra $\mathcal{F}$ on Ω, together with the corresponding Wiener measure $\mathbb{P}$, we obtain the canonical probability space $(\omega, \mathcal{F}, \mathbb{P})$, also called the Wiener space. This is similar to the game of gambling with a dice, where the canonical sample space is $\Omega_{\text{dice}} = \{1, 2, 3, 4, 5, 6\}$. Moreover, $\mathbb{E}$ denotes the mathematical expectation with respect to probability $\mathbb{P}$.

The canonical *driving* dynamical system describing the Brownian motion is defined as

$$\theta(t) : \Omega \to \Omega, \quad \theta(t)\omega(s) := \omega(t + s) - \omega(t), \; s, t \in \mathbb{R}.$$

Then $\theta(t)$, also denoted as θ_t, is a homeomorphism for each t and $(t, \omega) \rightarrowtail \theta(t)\omega$ is continuous, hence measurable. The Wiener measure $\mathbb{P}$ is invariant and ergodic under this so-called Wiener shift θ_t. In summary, θ_t satisfies the following properties:

- $\theta_0 = id$, where *id* is the identity mapping in the state space $\mathbb{R}^n$
- $\theta_t\theta_s = \theta_{t+s}$, for all $s, t \in \mathbb{R}$,
- the map $(t, \omega) \mapsto \theta_t\omega$ is measurable and $\theta_t\mathbb{P} = \mathbb{P}$ for all $t \in \mathbb{R}$.

We now introduce an important concept. A filtration is an increasing family of information accumulations, called σ-algebras, $\mathcal{F}_t$. For each t, σ-algebra $\mathcal{F}_t$ is a collection of events in sample space Ω. One might observe the Wiener process W_t over time t and use $\mathcal{F}_t$ to represent the information accumulated up to and including time t. More formally, on $(\Omega, \mathcal{F})$, a filtration is a family of σ-algebras $\mathcal{F}_s : 0 \leq s \leq t$ with $\mathcal{F}_s$ contained in $\mathcal{F}$ for each s, and $\mathcal{F}_s \subset \mathcal{F}_\tau$ for $s \leq \tau$. It is also useful to think of $\mathcal{F}_t$ as the σ-algebra generated by infinite union of $\mathcal{F}_s$s, which is contained in $\mathcal{F}_t$. So a filtration is often used to represent the change in the set of events that can be measured, through gain or loss of information.

For understanding stochastic differential equations from a dynamical point of view, the natural filtration is defined as a two-parameter family of σ-algebras generated by increments:

$$\mathcal{F}_s^t := \sigma[\omega(\tau_1) - \omega(\tau_2) : s \leq \tau_1, \tau_2 \leq t], \quad s, t \in \mathbb{R}.$$

This represents the information accumulated from time s up to and including time t. This two-parameter filtration allows us to define forward as well as backward stochastic integrals, and thus we can solve a stochastic differential equation from an initial time forward as well as backward in time (Arnold 1998).

The solution operator for the stochastic system (4.1) with initial condition $X(0) = x_0$ is denoted as $\varphi(t, \omega, x_0)$.

The dynamics of the system on the state space $\mathbb{R}^n$, over the driving flow θ_t is described by a cocycle. A cocycle φ is a mapping:

$$\varphi : \mathbb{R} \times \Omega \times \mathbb{R}^n \to \mathbb{R}^n,$$

which is $[\mathcal{B}(\mathbb{R}) \otimes \mathcal{F} \otimes \mathcal{B}(\mathbb{R}^n), \mathcal{B}(\mathbb{R}^n)]$-measurable such that

$$\varphi(0, \omega, x) = x \in \mathbb{R}^n,$$
$$\varphi(t_1 + t_2, \omega, x) = \varphi(t_2, \theta_{t_1}\omega, \varphi(t_1, \omega, x)),$$

for $t_1, t_2 \in \mathbb{R}$, $\omega \in \Omega$ and $x \in \mathbb{R}^n$. Then φ, together with the driving dynamical system, is called a *random dynamical system*. Sometimes we also use $\varphi(t, \omega)$ to denote this system.

Under very general smoothness conditions on the drift b and diffusion σ, the stochastic differential system (4.1) generates a random dynamical system in $\mathbb{R}^n$; see Kunita (1990) and Arnold (1998). Let us see an example.

Example 6. **A linear random dynamical system**
We consider a simple linear SDE and show that its solution operator defines a random dynamical system:

$$dX_t = X_t dt + dW_t, \quad X_0 = x \in \mathbb{R}, \quad t \in \mathbb{R}.$$

The solution is $X_t(\omega) = e^t x + \int_0^t e^{t-\tau} dW_\tau(\omega)$. Thus the solution operator is

$$\varphi(t, \omega, x) := e^t x + \int_0^t e^{t-\tau} dW_\tau(\omega).$$

Note that

$$\varphi(0, \omega, x) = x. \tag{4.48}$$

Now let us show that

$$\varphi(t + s, \omega, x) = \varphi[t, \theta_s\omega, \varphi(s, \omega, x)]. \tag{4.49}$$

Indeed, on the one hand,

$$\varphi(t + s, \omega, x) = e^{t+s} x + \int_0^{t+s} e^{t+s-\tau} dW_\tau(\omega).$$

On the other hand,

$$\begin{aligned}\varphi(t, \theta_s\omega, \varphi(s, \omega, x)) &= e^t \varphi(s, \omega, x) + \int_0^t e^{t-\tau} dW_\tau(\theta_s\omega) \\ &= e^t [e^s x + \int_0^s e^{s-\tau} dW_\tau] + \int_0^t e^{t-\tau} dW_\tau(\theta_s\omega).\end{aligned}$$

Now we only need to show the following claim: $\int_0^t e^{t-\tau} dW(\theta_s\omega) = \int_s^{t+s} e^{t+s-\tau} dW_\tau(\omega)$. To this end, we can prove that both sides of this claim are identical. In fact, noticing that $dW_\tau(\theta_s\omega) = d(W_{s+\tau} - W_s)$,

$$\textit{left-hand side} = \text{lim mean-square} \sum_j e^{t-\tau_j}(W_{s+\tau_{j+1}} - W_{s+\tau_j}), \tag{4.50}$$

and

$$\begin{aligned}\textit{right-hand side} &= \text{lim mean-square} \sum_j e^{t+s-(s+\tau_j)}(W_{s+\tau_{j+1}} - W_{s+\tau_j}) \\ &= \text{lim mean-square} \sum_j e^{t-\tau_j}(W_{\tau_{j+1}} - W_{\tau_j}).\end{aligned} \tag{4.51}$$

Hence the claim is proved. Therefore, the solution operator $\varphi(t, \omega x)$ satisfies the cocycle property:

$$\varphi(t + s, \omega, x) = \varphi(t, \theta_s\omega, \varphi(s, \omega, x)). \tag{4.52}$$

We recall some concepts in dynamical systems. A *manifold* M is a set, which locally looks like an Euclidean space. Namely, a 'patch' of the manifold M looks like a 'patch' in $\mathbb{R}^n$. For example, curves in $\mathbb{R}^3$ are one-dimensional differentiable manifolds, but tori and spheres in $\mathbb{R}^3$ are two-dimensional manifolds. However, a manifold arising from the study of invariant sets for dynamical systems in $\mathbb{R}^n$, can be very complicated. So we give a formal definition of manifolds. For more discussions on differentiable manifolds, see Abraham *et al.* (1988) and Perko (1990).

Definition 1 (*Differentiable manifold and Lipschitz manifold*)
An n-dimensional differentiable manifold M, is a connected metric space with an open covering $\{U_\alpha\}$, i.e. $M = \bigcup_\alpha U_\alpha$, such that

(i) for all α, U_α is homeomorphic to the open unit ball in $\mathbb{R}^n$, $B = \{x \in \mathbb{R}^n\colon |x| < 1\}$, i.e. for all α there exists a homeomorphism of U_α onto B, $h_\alpha : U_\alpha \to B$ and
(ii) if $U_\alpha \cap U_\beta \neq \emptyset$ and $h_\alpha : U_\alpha \to B$, $h_\beta : U_\beta \to B$ are homeomorphisms, then $h_\alpha(U_\alpha \cap U_\beta)$ and $h_\beta(U_\alpha \cap U_\beta)$ are subsets of $\mathbb{R}^n$ and the map

$$h = h_\alpha \circ h_\beta^{-1} : h_\beta(U_\alpha \cap U_\beta) \to h_\alpha(U_\alpha \cap U_\beta) \tag{4.53}$$

is differentiable, and for all $x \in h_\beta\ (U_\alpha \cap U_\beta)$, the Jacobian determinant $Dh(x) \neq 0$.

If the map (4.53) is only Lipschitz continuous, then we call M an n-dimensional Lipschitz continuous manifold.

Recall that a homeomorphism of A to B is a continuous one-to-one map of A onto B, $h : A \to B$, such that $h^{-1} : B \to A$ is continuous.

Just as invariant sets are important building blocks for deterministic dynamical systems, random invariant sets are basic geometric objects to help understand stochastic dynamics (Arnold 1998). Here we present two different concepts about invariant sets for stochastic systems: random invariant sets and almost sure invariant sets.

Definition 2 (*Random set*)
A collection $M = M(\omega)_{\omega\in\Omega}$, of non-empty closed sets $M(\omega)$, $\omega \in \Omega$, contained in $\mathbb{R}^n$, is called a random set if

$$\omega \mapsto \inf_{y \in M(\omega)} d(x, y)$$

is a random variable for any $x \in \mathbb{R}^n$.

Definition 3 (*Random invariant set*)
A random set $M(\omega)$ is called an invariant set for a random dynamical system φ if

$$\varphi(t, \omega, M(\omega)) \subset M(\theta_t \omega), \quad t \in \mathbb{R} \text{ and } \omega \in \Omega.$$

Random stationary orbits (Arnold 1998) and periodic orbits (Zhao and Zheng 2009) are special invariant sets.

Definition 4 (*Stationary orbit*)
A random variable $y(\omega)$ is called a stationary orbit for a random dynamical system φ if

$$\varphi(t, \omega, y(\omega)) = y(\theta_t \omega), \ \ a.s. \text{ for all } t.$$

Example 7. **Stationary orbit for a linear system**
Let us consider an example of stationary orbits:

$$\mathrm{d}u(t) = -u(t)\mathrm{d}t + \mathrm{d}W(t), \ u(0) = u_0. \tag{4.54}$$

This SDE defines a random dynamical system

$$\varphi(t, \omega, u_0) := u = \mathrm{e}^{-t}u(0) + \int_0^t \mathrm{e}^{-(t-s)}\mathrm{d}W(s). \tag{4.55}$$

A stationary orbit of this random dynamical system is given by

$$Y(\omega) = \int_{-\infty}^{0} \mathrm{e}^{s}\mathrm{d}W_s(\omega). \tag{4.56}$$

Indeed, it follows from (4.55) and (4.56) that

$$\begin{aligned}
\varphi(t, \omega, Y(\omega)) &= \mathrm{e}^{-t}Y(\omega) + \int_0^t \mathrm{e}^{-(t-s)}\mathrm{d}W_s(\omega) \\
&= \mathrm{e}^{-t}\int_{-\infty}^{0} \mathrm{e}^{s}\mathrm{d}W_s(\omega) + \int_0^t \mathrm{e}^{-(t-s)}\mathrm{d}W_s(\omega) \\
&= \int_{-\infty}^{0} \mathrm{e}^{-(t-s)}\mathrm{d}W_s(\omega) + \int_0^t \mathrm{e}^{-(t-s)}\mathrm{d}W_s(\omega) \\
&= \int_{-\infty}^{t} \mathrm{e}^{-(t-s)}\mathrm{d}W_s(\omega).
\end{aligned} \tag{4.57}$$

By (4.55) we also see that

$$\begin{aligned}
Y(\theta_t\omega) &= \int_{-\infty}^{0} \mathrm{e}^{s}\mathrm{d}W_s(\theta_t\omega) \\
&= \int_{-\infty}^{0} \mathrm{e}^{s}\mathrm{d}W_{s+t}(\omega) \\
&= \int_{-\infty}^{t} \mathrm{e}^{-(t-s)}\mathrm{d}W_s(\omega).
\end{aligned} \tag{4.58}$$

Thus $\varphi[t, \omega, Y(\omega)] = Y(\theta_t\omega)$, i.e. $Y(\omega) = \int_{-\infty}^{0} \mathrm{e}^{s}\mathrm{d}W_s(\omega)$ is a stationary orbit for the random dynamical system (4.54).

Definition 5 (*Periodic orbit*)
A random process $y(t, \omega)$ is called an invariant random periodic orbit of period T for a random dynamical system φ if

$$y(t+T, \omega) = y(t, \omega), \ a.s.$$
$$\varphi(t, \omega, y(t_0, \omega)) = y(t + t_0, \theta_t \omega), \ a.s.$$

for all t and t_0.

Definition 6 (*Random invariant manifold*)
If a random invariant set M can be represented by a graph of a Lipschitz mapping

$$\gamma^*(\omega, \cdot) : H^+ \to H^-, \text{ with direct sum decomposition } \quad H^+ \oplus H^- = \mathbb{R}^n$$

such that

$$M(\omega) = \{x^+ + \gamma^*(\omega, x^+), x^+ \in H^+\},$$

then M is called a Lipschitz continuous invariant manifold.

We will also consider deterministic invariant sets or manifolds, while the invariance is in the sense of almost sure (a.s.) (Aubin and Da Prato 1990; Filipovic 2000).

Definition 7 (*Almost sure invariant set and manifold*)
A (deterministic) set M in $\mathbb{R}^n$ is called locally almost surely invariant for (4.1), if for all $(t_0, x_0) \in \mathbb{R} \times M$, there exists a continuous local weak solution $X^{(t_0,x_0)}$ with lifetime $\tau = \tau(t_0, x_0)$, such that

$$X^{(t_0,x_0)}_{t\wedge\tau} \in M, \quad \forall t > t_0, \quad \text{a.s. } \omega \in \Omega,$$

where $t \wedge \tau = (t, \tau)$. When M is a manifold, it is called an almost sure invariant manifold.

4.4 Impact of model uncertainty and error growth

Consider an n-dimensional SDE system:

$$dX_t = b(X_t)dt + \sigma(X_t)dW_t, \tag{4.59}$$

A typical application of the Itô's formula for SDEs is to estimate moments of solutions. For example, for the second moment, by taking $g = \frac{1}{2}\|x\|^2 = \frac{1}{2}x \cdot x$ we obtain

$$\frac{1}{2}d\|X_t\|^2 = dg(X_t) = \left[X_t \cdot b + \frac{1}{2}\text{trace}(\sigma\sigma^{\mathrm{T}})\right] dt + X_t\sigma(X_t)dW_t. \tag{4.60}$$

Taking the mean, we get

$$\frac{1}{2}\frac{\mathrm{d}}{\mathrm{d}t}\mathbb{E}\|X_t\|^2 = \mathbb{E}(X_t \cdot b) + \frac{1}{2}\mathbb{E}\,\mathrm{trace}[\sigma(X_t)\sigma^{\mathrm{T}}(X_t)]. \tag{4.61}$$

This tells us how the fluctuating force affects the evolution of the mean energy of the system. The final term $\mathrm{trace}[\sigma(X_t)\,\sigma^{\mathrm{T}}(X_t)]$ is the effect of noise on mean energy.

Consider the deterministic system without model uncertainty

$$\mathrm{d}Y_t = b(Y_t)\mathrm{d}t. \tag{4.62}$$

Then the solution error $U_t = X_t - Y_t$ satisfies

$$\mathrm{d}U_t = [b(U_t + Y_t) - b(Y_t)]\mathrm{d}t + \sigma(U_t + Y_t)\mathrm{d}W_t. \tag{4.63}$$

Thus

$$\frac{1}{2}\frac{\mathrm{d}}{\mathrm{d}t}\mathbb{E}\|U_t\|^2 = \mathbb{E}(U_t \cdot [b(U_t + Y_t) - b(Y_t)]) + \frac{1}{2}\mathbb{E}\,\mathrm{trace}[\sigma(U_t + Y_t)\sigma^{\mathrm{T}}(U_t + Y_t)]. \tag{4.64}$$

This describes the error growth under uncertainty. The final term $\mathrm{trace}[\sigma(U_t + Y_t)\,\sigma^{\mathrm{T}}(U_t + Y_t)]$ is the effect of noise on error growth.

Let us look at an example.

Example 8. **Lorenz system under uncertainty**

Consider the Lorenz system with multiplicative noise

$$\begin{aligned}
\mathrm{d}x &= (-sx + sy)\mathrm{d}t + \sqrt{\varepsilon}\, x\, \mathrm{d}W_1(t),\\
\mathrm{d}y &= (rx - y - xz)\mathrm{d}t + \sqrt{\varepsilon}\, y\, \mathrm{d}W_2(t),\\
\mathrm{d}z &= (-bz + xy)\mathrm{d}t + \sqrt{\varepsilon}\, z\, \mathrm{d}W_3(t),
\end{aligned}$$

where W_1, W_2 and W_3 are independent scalar Brownian motions, and r, s, b, ε are positive parameters. The classical chaos case is when $r = 28$, $s = 10$ and $b = 8/3$.

Let $X := (x, y, z)^{\mathrm{T}}$. Then by the Itô's formula, we obtain an energy estimate

$$\begin{aligned}
\frac{1}{2}\frac{\mathrm{d}}{\mathrm{d}t}\mathbb{E}\|X\|^2 &= \mathbb{E}\left[-sx^2 - y^2 - bz^2 + (r+s)xy + \frac{1}{2}\varepsilon(x^2 + y^2 + z^2)\right]\\
&\le \left[-\min(s, 1, b) + \frac{1}{2}(r + s + \varepsilon)\right]\mathbb{E}\|X\|^2,
\end{aligned}$$

where we have used the fact that $xy \le \frac{1}{2}(x^2 + y^2) \le \frac{1}{2}(x^2 + y^2 + z^2)$. We can see that in this case, the noisy terms add 'energy' into the system.

Now we consider error growth due to uncertainty. Let $\hat{X} := (\hat{x}, \hat{y}, \hat{z})^{\mathrm{T}}$ be the (deterministic) solution ($\varepsilon = 0$ case), and let $U = (u, v, w)^T := X - \hat{X}$ be the error. Then by Itô's formula, we obtain the error growth estimate

$$\frac{1}{2}\frac{\mathrm{d}}{\mathrm{d}t}\mathbb{E}\|U\|^2 = \mathbb{E}\Big[- su^2 - v^2 - bw^2 + (r+s)uv + \hat{y}uw - \hat{z}uv + \frac{1}{2}\varepsilon(u^2+v^2+w^2)\Big]$$
$$\le \Big[-\min(s,1,b) + \frac{1}{2}(r+s+|\hat{y}|+|\hat{z}|+\varepsilon)\Big]\ \mathbb{E}\|U\|^2,$$

where we have used the fact that $E(\hat{y}uw) \le |\hat{y}|\ \mathbb{E}|uw| \le \frac{1}{2}|\hat{y}|\ \mathbb{E}(u^2+w^2)$. Note that under suitable conditions, this system has a random attractor (Schmalfuss 1997).

4.5 Residence time, exit probability and predictability

We start with an SDE system

$$\mathrm{d}X_t = b(X_t)\mathrm{d}t + \sigma(X_t)\mathrm{d}W_t, \quad X_0 \text{ given} \tag{4.65}$$

where b is an n-dimensional vector function, σ is an $n \times m$ matrix function, and $W_t(\omega)$ is an m-dimensional Brownian motion. The generator for this SDE is a linear second-order differential operator as in Section 4.2:

$$Ag = (\nabla g)^{\mathrm{T}} b + \frac{1}{2}\mathrm{trace}[\sigma\sigma^{\mathrm{T}} D^2(g)], \tag{4.66}$$

where D^2 is the Hessian differential matrix.

For a bounded domain D in $\mathbb{R}^n$, we can consider the exit problem of random solution trajectories of (4.65) from D. To this end, let ∂D denote the boundary of D and let Γ be a part of the boundary ∂D. The escape probability $p(x, y)$ is the probability that the trajectory of a particle starting at (x, y) in D first hits ∂D (or escapes from D) at some point in Γ, and $p(x, y)$ is known to satisfy (Schuss 1980; Lin and Segel 1988; Brannan *et al.* 1999 and references therein):

$$Ap = 0, \tag{4.67}$$
$$p|_{\Gamma} = 1, \tag{4.68}$$
$$p|_{\partial D - \Gamma} = 0. \tag{4.69}$$

Suppose that initial conditions (or initial particles) are uniformly distributed over D. The average escape probability P that a trajectory will leave D along the sub-boundary Γ, before leaving the rest of the boundary, is given by (e.g. Schuss 1980;

Lin and Segel 1988):

$$P = \frac{1}{|D|} \int \int_D p(x, y)\mathrm{d}x\mathrm{d}y, \tag{4.70}$$

where $|D|$ is the area of domain D.

The residence time of a particle initially at (x, y) inside D is the time until the particle first hits ∂D (or escapes from D). The mean residence time $u(x, y)$ is given by (e.g. Schuss 1980; Risken 1984; Naeh *et al.* 1990 and references therein):

$$Au = -1, \tag{4.71}$$

$$u|_{\partial D} = 0. \tag{4.72}$$

Relevance to the predictability problem. For low-dimensional SDE systems, such as the Lagrangian dynamical model for fluid particles in random fluid flows or other truncated models like the Lorenz model, the exit probability and mean residence time may be computed by deterministic partial differential equations solvers (Brannan *et al.* 1999). Be selecting the above domain D appropriately, say corresponding to observational data ('data domain'), we may determine a predictability time window, by monitoring when the system exits the data domain.

4.6 Invariant manifolds and predictability

Invariant manifolds provide geometric structures that describe dynamical behaviour of nonlinear systems. Dynamical reductions to attracting invariant manifolds or dynamical restrictions to other (not necessarily attracting) invariant manifolds are often sought to gain understanding of nonlinear dynamics.

A number of recent studies have been carried out on invariant manifolds for stochastic differential equations (Wanner 1995; Arnold 1998; Duan *et al.* 2003, 2004). Random invariant manifolds in the sense of Definition 6 are difficult to obtain, even locally in state space. But almost sure invariant manifolds in the sense of Definition 7 may be determined locally in state space (which also means for finite time in evolution), for some SDE systems, by a method of solving first-order deterministic partial differential equations (Du and Duan 2007).

We consider the following stochastic system defined by Itô stochastic differential equations in $\mathbb{R}^n$:

$$\mathrm{d}X = b(X)\mathrm{d}t + \sigma(X)\mathrm{d}W(t), \quad X(0) = x_0, \tag{4.73}$$

where again b and σ are vector and matrix functions in $\mathbb{R}^n$ and $\mathbb{R}^{n \times n}$, respectively, and $W(t)$ is standard vector Brownian motion in $\mathbb{R}^n$. We also assume that $b(\cdot) \in C^1(\mathbb{R}^n; \mathbb{R}^n)$ and $\varsigma(\cdot) \in C^1(\mathbb{R}^n; \mathbb{R}^{n \times n})$.

For the nonlinear stochastic system (4.73), we study deterministic almost sure invariant manifolds, which are not necessarily attracting. We reformulate the local invariance condition as invariance equations, i.e. first-order partial differential equations, and then solve these equations by the method of characteristics. Although the local invariant manifold is deterministic, the restriction of the original stochastic system on this deterministic local invariant manifold is still a stochastic system but with reduced dimensions.

We are going to derive representations of invariant finite dimensional manifolds in terms of b and σ, by using the tangency conditions for a deterministic C^2 smooth manifold (a supersurface) M in $\mathbb{R}^n$:

$$\mu(\omega, x) := b(\omega, x) - \frac{1}{2}\sum_j [D\sigma^j(\omega, x)]\, \sigma^j(\omega, x) \in T_x M, \tag{4.74}$$

$$\sigma^j(\omega, x) \in T_x M, \quad j = 1, \ldots, n, \tag{4.75}$$

where D represents the Jacobian operator and σ^j is the j-th column of the matrix σ. The above tangency conditions are shown to be equivalent to almost sure local invariance of manifold M; see e.g. Aubin & Da Prato (1990); Filipovic (2000).

The almost sure invariance conditions (4.74)–(4.75) for manifold M mean that the $n+1$ vectors, μ and $\sigma^j, j = 1, \ldots, n$, are tangent vectors to M. Namely, these $n+1$ vectors are orthogonal to the normal vectors of manifold M.

In other words, if the normal vector for M at x is $N(x)$, then the almost sure invariance conditions (4.74)–(4.75) become the following *invariance equations* for manifold M: for all $x \in M$,

$$< \mu(x), N(x) > = 0, \tag{4.76}$$

$$< \sigma^j(x), N(x) > = 0, \quad j = 1, \ldots, n, \tag{4.77}$$

where, as before, $< \cdot, \cdot >$ denotes the usual scalar product in $\mathbb{R}^n$.

Invariant manifolds are usually represented as graphs of some functions in $\mathbb{R}^n$. By investigating the above invariance equations (4.76)–(4.77), we may be able to find some local invariant manifolds M for the stochastic system (4.73).

The goal of this section is to present a method to find some of these local invariant manifolds. Although the following result and example are stated for a codimension 1 local invariant manifold, the idea extends to other lower-dimensional local invariant manifolds, as long as the normal vectors $N(x)$ (or tangent vectors) may be represented; see tangency conditions (4.76)–(4.77) above and (4.79)–(4.80) below.

Local almost sure invariant manifold

Let the local invariant manifold M for the stochastic dynamical system (4.73) be represented as a graph defined by the algebraic equation

$$M : G(x_1, \ldots, x_n) = 0. \tag{4.78}$$

Then G satisfies a system of first-order (deterministic) partial differential equations and the local invariant manifold M may be found by solving these partial differential equations by the method of characteristics. By restricting the original dynamical system (4.73) on this local invariant manifold M, we obtain a locally valid, reduced lower-dimensional system.

In fact, the normal vector to this graph or surface is, in terms of partial derivatives, $\nabla G(x) = (G_{x_1}, \ldots, G_{x_n})$. Thus the invariance equations (4.76)–(4.77) are now

$$< \mu(x), \nabla G(x) > = 0, \tag{4.79}$$

$$< \sigma^j(x), \nabla G(x) > = 0, \quad j = 1, \ldots, n. \tag{4.80}$$

This is a system of first-order partial differential equations in G. We apply the method of characteristics to solve for G, and therefore obtain the invariant manifold M, represented by a graph in state space $\mathbb{R}^n$: $G(x_1, \ldots, x_n) = 0$.

Method of characteristics. Consider a first-order partial differential equation for the unknown scalar function u of n variables $x_1, \ldots, x_n$,

$$\sum_{j=1}^{n} a_i(x_1, \ldots, x_n)u_{x_i} = c(x_1, \ldots, x_n), \tag{4.81}$$

with continuous coefficients a_i and c.

Note that the solution surface $u = u(x_1, \ldots, x_n, t)$ in $x_1, \ldots x_n$ u-space has normal vectors $N := (u_{x_1}, \ldots, u_{x_n}, -1)$. This partial differential equation implies that the vector $V =: (a_1, \ldots, a_n, c)$ is perpendicular to this normal vector and hence must lie in the tangent plane to the graph of $z = u(x_1, \ldots, x_n)$.

In other words, $(a_1, \ldots, a_n, c)$ defines a vector field in $\mathbb{R}^n$, to which graphs of the solutions must be tangent at each point (McOwen 2003). Surfaces that are tangent at each point to a vector field in $\mathbb{R}^n$ are called *integral surfaces* of the vector field. Thus to find a solution of (4.81), we should try to find integral surfaces.

How can we construct integral surfaces? We can try using the characteristics curves that are the integral curves of the vector field. That is, $X = [x_1(t), \ldots, x_n(t)]$ is a *characteristic* if it satisfies the following system of ordinary differential equations:

$$\frac{dx_1}{dt} = a_1(x_1, \ldots, x_n), \ldots$$
$$\frac{dx_n}{dt} = a_n(x_1, \ldots, x_n),$$
$$\frac{du}{dt} = c(x_1, \ldots, x_n).$$

A smooth union of characteristic curves is an integral surface. There may be many integral surfaces. Usually an integral surface is determined by requiring it to contain (or pass through) a given initial curve or an n–1 dimensional manifold Γ:

$$\begin{aligned} x_i &= f_i(s_1, \ldots, s_{n-1}), i = 1 \ldots n \\ u &= h(s_1, \ldots, s_{n-1}) \end{aligned}$$

This generates an n-dimensional integral manifold M parameterised by $(s_1, \ldots, s_{n-1}, t)$. The solution $u(x_1, \ldots, x_n)$ is obtained by solving for $(s_1, \ldots, s_{n-1}, t)$ in terms of variables $(x_1, \ldots, x_n)$.

Remark 3. If the initial curve Γ is non-characteristic, i.e. it is nowhere tangent to the vector field $V = (a_1, \ldots, a_n, c)$, and $a_1, \ldots, a_n, c$ are C^1 (and thus locally Lipschitz continuous), then there exists a unique integral surface $u = u(x_1, \ldots, x_n)$ containing Γ, defined at least locally near Γ.

Now applying the above method of characteristics to (4.79)–(4.80), we obtain a solution $G = G(x_1, \ldots, x_n)$. However, the local invariant manifold M that we look for is represented by the equation

$$G(x_1, \ldots, x_n) = 0.$$

Therefore, skill is needed to make sure that the solution $G = G(x_1, \ldots, x_n)$ actually penetrates the plane $G = 0$ in the $x_1 \ldots x_n$ G-space; see Fig. 4.1. This needs to be achieved by selecting an appropriate initial curve Γ. The invariant manifold M we thus obtain is defined at least locally near the initial curve Γ.

Relevance to predictability problem. When a SDE system starts to evolve inside a local almost sure invariant manifold M, it remains inside the manifold for a certain time period $0 < t < T$. As determined above, this manifold holds solutions for the system, the time period T may be taken as a lower bound of the predictability time scale.

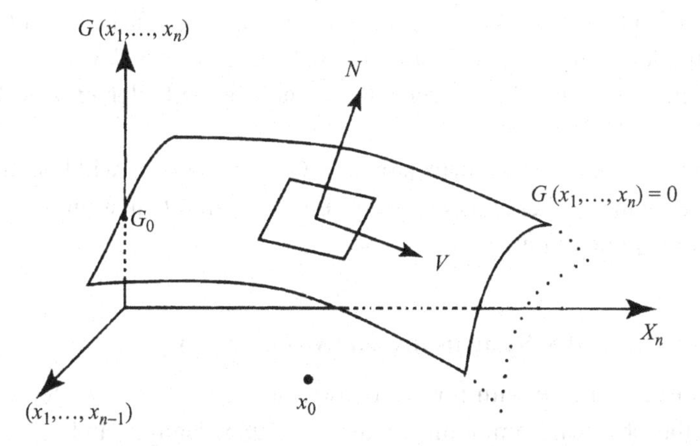

Figure 4.1 Local invariant manifold M is represented by the equation $G(x_1, \ldots, x_n) = 0$ in the $x_1 \ldots x_n$-space. Namely, M is the intersection of the surface $G = G(x_1, \ldots, x_n)$ with the plane $G = 0$ in $x_1 \ldots x_n$ G-space. Here $G(x_1, \ldots, x_n)$ is the solution of (4.79)–(4.80) via the method of characteristics. Note that $N = (u_{x_1}, \ldots, u_{x_n}, -1)$ and $V = (a_1, \ldots, a_n, c)$.

4.7 Systems driven by non-Gaussian noise

Although Gaussian processes like Brownian motion have been widely used in modelling fluctuations in geophysical modelling, it turns out that many physical phenomena are involved with non-Gaussian Lévy motions. For instance, it has been argued that diffusion by geophysical turbulence (Shlesinger *et al.* 1995) corresponds, loosely speaking, to a series of 'pauses', when the particle is trapped by a coherent structure; and 'flights' or 'jumps' or other extreme events, when the particle moves in the jet flow. Palaeoclimatic data (Ditlevsen 1999) also indicate such irregular processes.

Lévy motions are thought to be appropriate models for non-Gaussian processes with jumps (Sato 1999). Let us recall that a Lévy motion $L(t)$ has independent and stationary increments, i.e. increments $\Delta L\ (t, \Delta t) = L(t + \Delta t) - L(t)$ are stationary (therefore ΔL has no statistical dependence on t) and independent for any non-overlapping time lags Δt. Moreover, its sample paths are only continuous in probability, namely, $\mathbb{P}[|L(t) - L(t_0)| \geq \delta] \to 0$ as $t \to t_0$ for any positive δ. This continuity is weaker than the usual continuity in time.

This generalises the Brownian motion $B(t)$, as $B(t)$ satisfies all these three conditions. But additionally, (i) almost every sample path of the Brownian motion is continuous in time in the usual sense and (ii) increments of Brownian motion are Gaussian distributed.

Stochastic differential equations driven by non-Gaussian Lévy noises,

$$dX_t = b(X_t)dt + \sigma(X_t)dL(t), \tag{4.82}$$

have attracted much attention (Janicki and Weron 1994; Schertzer *et al.* 2001; Applebaum 2004) but the investigation of such SDEs is less developed. Recently, mean exit-time estimates have been investigated by Imkeller *et al.* (2006), and Yang and Duan (2008).

Further progresses in SDEs driven by non-Gaussian noises will benefit research in predictability in weather and climate systems with non-Gaussian (which is more common) model uncertainty.

4.8 Systems driven by coloured noise

Coloured noise, or noise with non-zero correlation in time, has been considered or used in the physical community (Gardiner 1985; Hanggi and Jung 1995). A good candidate for modelling coloured noise is the fractional Brownian motion. A fractional Brownian motion (fBm) process B^H, where $H \in (0,1)$ is fixed, is still a Gaussian process. But it is characterised by the stationarity of its increments and a memory property. The increments of the fractional Brownian motion are not independent, except in the standard Brownian case ($H = \frac{1}{2}$). Thus a fractional Brownian motion is not a Markov process except when $H = \frac{1}{2}$. Specifically, $B^H(0) = 0$ and var $[B^H(t) - B^H(s)] = |t - s|^{2H}$. It also exhibits power scaling and path regularity properties with the Hölder parameter H, which are very distinct from Brownian motion. The standard Brownian motion is a special fBm with $H = \frac{1}{2}$.

The stochastic calculus involving fBm is currently being developed; see e.g. Nualart (2003) and Tudor and Viens (2007) and references therein. This will lead to more advances in the study of SDEs driven by coloured fBm noise:

$$dX_t = b(X_t)dt + \sigma(X_t)dB^H(t). \tag{4.83}$$

Since the fBm $B^H(t)$ is not Markov, the solution process X_t is not Markov either. Thus the usual techniques from Markov processes will not be applicable to the study of SDEs driven by fBms. However, the random dynamical systems approach, as described in Section 4.3 above, looks promising (Maslowski and Schmalfuss 2004). The theory of random dynamical systems, developed by Arnold and coworkers (1998), describes the qualitative behaviour of systems of stochastic differential equations in terms of stability, Lyapunov exponents, invariant manifolds and attractors.

Further progresses in SDEs driven by coloured noises will benefit research in the predictability of weather and climate systems with more general (non-white noise) model uncertainty.

Acknowledgements

This work was partly supported by the NSF Grants 0542450 and 0620539.

References

Abraham, R., Marsden, J. E. & Ratiu, T. 1988 *Manifolds, Tensor Analysis, and Applications*, 2nd edn. Springer-Verlag.
Allen, M., Frame, D., Kettleborough J. & Stainforth, D. 2006 Model error in weather and climate forecasting. In *Predictability of Weather and Climate*, ed. T. Palmer & R. Hagedorn, pp. 391–428. Cambridge University Press.
Applebaum, D. 2004 *Lévy Processes and Stochastic Calculus*. Cambridge University Press.
Arnold, L. 1974 *Stochastic Differential Equations*. John Wiley & Sons.
Arnold, L. 1998 *Random Dynamical Systems*. Springer-Verlag.
Arnold, L. 2001 Hasselmann's program visited: the analysis of stochasticity in deterministic climate models. In *Stochastic Climate Models*. ed. J.-S. von Storch and P. Imkeller, pp. 141–158. Birkhäuser.
Aubin, J.-P. & Da Prato, G. 1990 Stochastic viability and invariance. *Scuola Norm. Sup. Pisa*, **127**, 595–694.
Berloff, P. S. 2005 Random-forcing model of the mesoscale oceanic eddies. *J. Fluid Mech.*, **529**, 71–95.
Berselli, L. C., Iliescu, T. & Layton, W. J. 2005. *Mathematics of Large Eddy Simulation of Turbulent Flows*. Springer-Verlag.
Blomker, D. & Duan, J. 2007 Predictability of the Burgers dynamics under model uncertainty. In Boris Rozovsky 60th birthday volume *Stochastic Differential Equations: Theory and Applications*, ed. P. Baxendale and S. Lototsky, pp. 71–90. World Scientific.
Boffetta, G., Celani, A., Cencini, M., Lacorata, G. & Vulpiani, A. 2000 The predictability problem in systems with an uncertainty in the evolution law. *J. Phys. A*, **33**, 1313–1324.
Boffetta, G., Cencini, M., Falcioni, M. & Vulpiani, A. 2002 Predictability: a way to characterize complexity. *Phys. Rep.*, **356**, 367–474.
Brannan, J., Duan, J. & Ervin, V. 1999 Escape probability, mean residence time and geophysical fluid particle dynamics, *Physica D*, **133**, 23–33.
Caraballo, T. & Langa, J. 2001 A comparison of the longtime behavior of linear Itô and Stratonovich partial differential equations. *Stoch. Anal. Appl.*, **19**, 183–195.
Cessi, P. & Louazel, S. 2001 Decadal oceanic response to stochastic wind forcing, *J. Phys. Oceanogr.*, **31**, 3020–3029.
Chorin, A., Kast, A. & Kupferman, R. 1999 Unresolved computation and optimal predictions. *Commun. Pure Appl. Math.*, **52**, 1231–1254.
Crauel, H. & Gundlach, M. (eds.), 1999 *Stochastic Dynamics*. Papers from the Conference on Random Dynamical Systems held in Bremen, April 28–May 2, 1997. Springer-Verlag.
Da Prato, G. & Zabczyk, J. 1992 *Stochastic Equations in Infinite Dimensions*. Cambridge University Press.
DelSole, T. 2004 Stochastic models of quasigeostrophic turbulence. *Surveys Geophys.*, **25**, 107–149.
Ditlevsen, P. D. 1999 Observation of α–stable noise induced millennial climate changes from an ice record. *Geophys. Res. Lett.*, **26**, 1441–1444.
Du, A. & Duan, J. 2007 Invariant manifold reduction for stochastic dynamical systems. *Dyn. Syst. Appl.*, **16**, 681–696.
Du, A. & Duan, J. 2009 A stochastic approach for parameterizing unresolved scales in a system with memory. *J. Algorithms Comput. Technol.*, **3**, 393–405.
Duan, J. & Nadiga, B. 2007 Stochastic parameterization for large eddy simulation of geophysical flows. *Proc. Am. Math. Soc.*, **135**, 1187–1196.

Duan, J., Lu, K. & Schmalfuβ, B. 2003. Invariant manifolds for stochastic partial differential equations. *Ann. Prob.*, **31**, 2109–2135.

Duan, J., Lu, K. & Schmalfuβ, B. 2004 Smooth stable and unstable manifolds for stochastic evolutionary equations. *J. Dyn. Differential Equations*, **16**, 949–972.

Farrell, B. F. & Ioannou, P. J. 2002a Optimal perturbation of uncertain systems. Special issue on stochastic climate models. *Stoch. Dyn.*, **2**, 395–402.

Farrell, B. F. & Ioannou, P. J. 2002b Perturbation growth and structure in uncertain flows. I, II. *J. Atmos. Sci.*, **59**, 2629–2646, 2647–2664.

Filipovic, D. 2000 Invariant manifolds for weak solutions to stochastic equations. *Prob. Theory Rel. Fields*, **118**, 323–341.

Freidlin, M. I. & Wentzell, A. D. 1998 *Random Perturbations of Dynamical Systems*, 2nd edn. Springer-Verlag.

Garcia-Ojalvo, J. & Sancho, J. M. 1999 *Noise in Spatially Extended Systems*. Springer-Verlag.

Gardiner, C. W. 1985 *Handbook of Stochastic Methods*, 2nd edn. Springer-Verlag.

Griffa, A. & Castellari, S. 1991 Nonlinear general circulation of an ocean model driven by wind with a stochastic component. *J. Mar. Res.*, **49**, 53–73.

Guckenheimer, J. & Holmes, P. 1983 *Nonlinear Oscillations, Dynamical Systems and Bifurcations of Vector Fields*. Springer-Verlag.

Hanggi, P. & Jung, P. 1995 Colored noise in dynamical systems. *Adv. Chem. Phys.*, **89**, 239–326.

Hasselmann, K. 1976 Stochastic climate models: Part I. Theory. *Tellus*, **28**, 473–485.

Horsthemke, W. & Lefever, R. 1984 *Noise-Induced Transitions*. Springer-Verlag.

Huisinga, W., Schutte, C. & Stuart, A. M. 2003 Extracting macroscopic stochastic dynamics: model problems. *Commun. Pure Appl. Math.*, **56**, 234–269.

Imkeller, P. & Monahan, A. (eds.) 2002 *Conceptual Stochastic Climate Models*. Special Issue: *Stoch. Dyn.*, **2**.

Imkeller, P. & Pavlyukevich, I. 2006 First exit time of SDEs driven by stable Lévy processes. *Stoch. Processes Appl.*, **116**, 611–642.

Imkeller, P., Pavlyukevich, I. & Wetzel, T. 2007 First exit times for Lévy-driven diffusions with exponentially light jumps. *Ann. Probab.*, **37**, 530–564.

Janicki, A. & Weron, A. 1994 *Simulation and Chaotic Behavior of α–Stable Stochastic Processes*. Marcel Dekker, Inc.

Just, W., Kantz, H., Rodenbeck, C. & Helm, M. 2001 Stochastic modelling: replacing fast degrees of freedom by noise. *J. Phys. A: Math. Gen.*, **34**, 3199–3213.

Karatzas, I. & Shreve, S. E. 1991 *Brownian Motion and Stochastic Calculus*, 2nd edn. Springer-Verlag.

Khade, V. M. & Hansen, J. A. 2004 State dependent predictability: impact of uncertainty dynamics, uncertainty structure and model inadequacies. *Nonlinear Processes Geophys.* **11**, 351–362.

Klebaner, F. C. 2005 *Introduction to Stochastic Calculus with Applications*. Imperial College Press.

Kleeman, R. 2002 Measuring dynamical prediction utility using relative entropy. *J. Atmos. Sci.*, **59**, 2057–2072.

Kloeden, P. E. & Platen, E. 1992 *Numerical Solution of Stochastic Differential Equations*. Springer-Verlag.

Kunita, H. 1990 *Stochastic Flows and Stochastic Differential Equations*. Cambridge University Press.

Leith, C. E. 1975 Climate response and fluctuation dissipation, *J. Atmos. Sci.*, **32**, 2022–2025.

Lin, C. C. & Segel, L. A. 1988 *Mathematics Applied to Deterministic Problems in the Natural Sciences*. SIAM.
Majda, A., Kleeman, R. & Cai, D. 2002 A mathematical framework for quantifying predictability through relative entropy. Special issue dedicated to Daniel W. Stroock and Srinivasa S. R. Varadhan on the occasion of their 60th birthday. *Meth. Appl. Anal.* **9**, 425–444.
Mao, X. 1997 *Stochastic Differenntial Equations and Applications*. Horwood Publishing.
Maslowski, B. & Schmalfuss, B. 2004 Random dynamical systems and stationary solutions of differential equations driven by the fractional Brownian motion. *Stoch. Anal. Appl.*, **22**, 1577–1607.
McOwen, R. C. 2003. *Partial Differential Equations*. Pearson Education.
Millet, A. & Morien, P.-L. 2005 On implicit and explicit discretization schemes for parabolic SPDEs in any dimension. *Stoch. Processes Appl.*, **115**, 1073–1106.
Mu, M., Duan, W. S. & Wang, B. 2003 Conditional nonlinear optimal perturbation and its applications. *Nonlinear Processes Geophys.*, **10**, 493–501.
Mu, M., Duan, W. S. & Chou, J. 2004 Recent advances in predictability studies in China (1999–2002). *Adv. Atmos. Sci.* **21**, 437–443.
Müller, P. 1996 Stochastic forcing of quasi-geostrophic eddies. In *Stochastic Modelling in Physical Oceanography*, ed. R. J. Adler, P. Müller and B. Rozovskii, Birkhäuser.
Naeh, T., Klosek, M. M., Matkowsky, B. J. & Schuss, Z. 1990 A direct approach to the exit problem, *SIAM J. Appl. Math.*, **50**, 595–627.
Nicolis, C. 2004 Dynamics of model error: the role of unresolved scales revisited. *J. Atmos. Sci.*, **61**, 1740–1753.
North, G. R. & Cahalan, R. F. 1981 Predictability in a solvable stochastic climate model. *J. Atmos. Sci.*, **38**, 504–513.
Nualart, D. 2003 Stochastic calculus with respect to the fractional Brownian motion and applications. *Contemp. Math.*, **336**, 3–39.
Oksendal, B. 2003 *Stochastic Differential Equations*, 6th edn. Springer-Verlag.
Orrell, D., Smith, L., Barkmeijer, J. & Palmer, T. N. 2001 Model error in weather forecasting. *Nonlinear Processes Geophys.*, **8**, 357–371.
Palmer, T. N. 2001. A nonlinear dynamical perspective on model error: A proposal for non-local stochastic-dynamic parameterization in weather and climate prediction models. *Q. J. R. Meteorol. Soc.*, **127**, 279–304.
Palmer, T. N. & Hagedorn, R. (eds.) 2006 *Predictability of Weather and Climate*. Cambridge University Press.
Palmer, T. N., Shutts, G. J., Hagedorn, R. *et al.* 2005 Representing model uncertainty in weather and climate prediction. *Annu. Rev. Earth Planet. Sci.*, **33**, 163–193.
Pasquero, C. & Tziperman, E. 2007 Statistical parameterization of heterogeneous oceanic convection. *J. Phys. Oceanogr.*, **37**, 214–229.
Perko, L. 1990 *Differential Equations and Dynamical Systems*. Cambridge University Press.
Risken, H. 1984 *The Fokker–Planck Equation*. Springer-Verlag.
Roberts, A. J. 2003–2004 A step towards holistic discretisation of stochastic partial differential equations. *ANZIAM J.*, **45**, (C), C1–C15.
Rozovskii, B. L. 1990 *Stochastic Evolution Equations*. Kluwer Academic Publishers.
Samelson, R. M. 1989 Stochastically forced current fluctuations in vertical shear and over topography. *J. Geophys. Res.*, **94**, 8207–8215.
Saravanan, R. & McWilliams, J. C. 1998 Advective ocean–atmosphere interaction: an analytical stochastic model with implications for decadal variability. *J. Climate*, **11**, 165–188.

Sato, K.-I. 1999 *Lévy Processes and Infinitely Divisible Distributions*. Cambridge University Press.

Schneider, T. & Griffies, S. M. 1999 A conceptual framework for predictability studies. *J. Climate*, **12**, 3133–3155.

Schertzer, D., Larcheveque, M., Duan, J., Yanovsky, V. & Lovejoy, S. 2001 Fractional Fokker–Planck equation for nonlinear stochastic differential equations driven by non-Gaussian Levy stable noises. *J. Math. Phys.*, **42**, 200–212.

Schmalfuss, B. 1997 The random attractor of the stochastic Lorenz system. *Zeit. Rangew. Math. Phy. (ZAMP)*, **48**, 951–975.

Schuss, Z. 1980 *Theory and Applications of Stochastic Differential Equations*. John Wiley & Sons.

Shlesinger, M. F., Zaslavsky, G. M. & Frisch, U. (eds.) 1995 *Lévy Flights and Related Topics in Physics*. Lecture Notes in Physics, 450. Springer-Verlag.

Smith, L. A., Ziehmann, C. & Fraedrich, K. 1999 Uncertainty dynamics and predictability in chaotic systems. *Q. J. R. Meteorol. Soc.*, **125**, 2855–2886.

Sura, P. & Penland, C. 2002 Sensitivity of a double-gyre model to details of stochastic forcing. *Ocean Modelling*, **4**, 327–345.

Tudor, C. A. & Viens, F. 2007 Statistical aspects of the fractional stochastic calculus. *Ann. Stat.*, **35**, 1183–1212.

Vannitsem, S. & Toth, Z. 2002 Short-term dynamics of model errors. *J. Atmos. Sci.*, **59**, 2594–2604.

Wang, W. & Duan, J. 2007 A dynamical approximation for stochastic partial differential equations. *J. Math. Phys.*, **48**, 102701.

Wanner, T. 1995 Linearization of random dynamical systems. *Dynamics Report*, vol. 4. Spring-Verlag.

Waymire, E. & Duan, J. (eds.) 2005 *Probability and Partial Differential Equations in Modern Applied Mathematics*. Springer-Verlag.

Wilks, D. S. 2005 Effects of stochastic parameterizations in the Lorenz '96 system. *Q. J. R. Meteorol. Soc.*, **131**, 389–407.

Williams, P. D. 2005 Modelling climate change: the role of unresolved processes. *Phil. Trans. R. Soc. A*, **363**, 2931–2946.

Yang, Z. & Duan, J. 2008 An intermediate regime for exit phenomena driven by non-Gaussian Lévy noises. *Stoch. Dyn.*, **8**, 583–591.

Zhao, H. & Zheng, Z. 2009 Random periodic solutions of random dynamical systems. *J. Differential Equations*, **246**, 2020–2038.

5

On modelling physical systems with stochastic models: diffusion versus Lévy processes

CÉCILE PENLAND AND BRIAN D. EWALD

Stochastic descriptions of multiscale interactions are more and more frequently found in numerical models of weather and climate. These descriptions are often made in terms of differential equations with random forcing components. In this chapter, we review the basic properties of stochastic differential equations driven by classical Gaussian white noise and compare with systems described by stable Lévy processes. We also discuss aspects of numerically generating these processes.

5.1 Introduction

Subgrid-scale processes must be treated somehow in numerical weather and climate models, whatever these models' spatial and temporal resolutions. First of all, one could ignore them. Although this is often done, the procedure has not proven particularly successful. A more common, if not *the* most common, approach is to parameterise them deterministically in terms of resolved processes. Some authors even define 'parameterisation' that way, reserving the term 'unparameterisable' to describe what cannot be represented in terms of what a numerical model can resolve.

Modern numerical models (e.g. Buizza *et al.* 1999) employ stochastic parameterisations to account for both mean effects and variability due to dynamical interactions between processes that cannot be explicitly represented in a numerical model and the large-scale systems the model attempts to predict. At the European Centre for Medium-Range Weather Forecasts, for example, efforts are made to parameterise turbulent energy backscatter (e.g. Fjørtoft 1953) from the unresolved scales to the resolved scales in the ensemble prediction system by means of cellular automata (Shutts 2005) or using red-noise processes (Berner *et al.* 2009).

There is both empirical and modelling evidence that such an approach is both practical and physically meaningful. The almost embarrassing similarity of prediction skill in statistical and numerical models of El Niño (e.g. Penland & Sardeshmukh 1995, hereafter PS95; Saha *et al.* 2006) and medium-range weather (Winkler *et al.* 2001) lends credibility to the idea that at least some aspects of dynamical forcing can be treated stochastically (see also Hasselmann 1976).

The fact that forcing may be treated stochastically does not mean that details of the stochastic treatment are arbitrary. Failure to identify the physical origins of the stochastic forcing, which usually results in a mistaken mathematical description of that forcing, can easily translate into a misinterpretation of the dynamical response. For example, PS95 present evidence that El Niño dynamics, as represented by tropical sea-surface temperatures (SSTs), is very well explained in terms of a stable linear process driven by Gaussian stochastic forcing. In fact, the 'tau test' (see their article for details) that is passed by the statistics of tropical SSTs can *only* be passed by such a system. The problem with PS95 is that the authors placed too strong an emphasis on tests for Gaussianity of the SSTs themselves; while a linear dynamical system driven by Gaussian forcing may be Gaussian, that need not be the case if the stochastic forcing is modulated by a linear function of the SSTs (Müller 1987; Sura & Sardeshmukh 2008; Sardeshmukh & Sura 2009). That is, if the amplitudes of rapidly varying wind stress and heat flux depend on SST, which they do, then the distribution of SST cannot be expected to be Gaussian even when the underlying SST dynamics are stable and linear. A result of PS95's misplaced emphasis on Gaussianity was the publication of articles (e.g. An & Jin 2004), which found a small but significant non-Gaussian behaviour in tropical east Pacific SST, and concluded that the dynamics of El Niño must be primarily nonlinear. In other words, scientists were arguing about the resolved dynamics of an extremely important phenomenon without giving due consideration to the nature of the stochastic forcing. There were indeed studies that considered the effect of stochastic forcing in the El Niño system (e.g. Flügel *et al.* 2004), but these did not consider how the basic mathematical description of the stochastic forcing could affect the marginal distribution of SST.

It turns out that the marginal distribution of a linear Gaussian-driven process may not only be non-Gaussian but may also exhibit skew, fat tails and other properties usually associated with more exotic types of stochastic phenomena, such as non-Gaussian α-stable Lévy processes. Even when the dynamical system can be shown to be linear and Gaussian driven, the distribution of that system depends on whether the dynamics describing it respond to the stochastic forcing through 'Itô' or 'Stratonovich' calculus, a differentiation that we explain below. Nevertheless, although these characteristics permit linear dynamics, they differ from the tau test in which they do not require it. Further, the distribution of these different kinds of stochastic dynamics may have many characteristics in common, but the classes of physical causes giving rise to them are quite different. Thus, when using stochastic numerical models of finite resolution to diagnose the physical system, it is not only necessary to choose accurately the stochastic parameterisations appropriate to the class of subgrid-scale processes one wishes to represent but also to ensure that the numerical algorithms built into the models are able to reproduce the correct stochastic response.

The stochastic parameterisations for which the following remarks are appropriate comprise a somewhat limited class of stochastic models. There are certainly other valuable methods, such as random cascades and particle-interaction techniques, which have a valid place in the literature. However, in this chapter, we shall confine ourselves to the rather restricted field of stochastic differential equations (SDEs) driven by either Gaussian white noise, i.e. diffusion processes, or, more generally, by α-stable Lévy processes. The special case of diffusion processes has already been extensively treated in the climate literature (e.g. Hasselmann 1976; Penland 1989; Penland 1996; Majda *et al.* 1999; Sardeshmukh & Sura 2009), but this literature has yet to be widely applied in climate science. At the same time, the ability of Lévy processes to describe properties such as intermittency has attracted the attention of climate scientists, some of whom (e.g. Ditlevsen 1999) have begun to employ a wider class of SDEs (LSDEs) driven by stable Lévy processes to describe observational records in palaeoclimatology. Many scientific studies using LSDEs do not allow the random forcing to depend on the state of the system, i.e. they employ additive rather than multiplicative Lévy noise. We follow this approach as it greatly simplifies both the theoretical and the numerical descriptions of such systems, deferring the more complex issues to a later paper.

The purpose of this article is to present a fairly basic synopsis of properties of diffusion processes and stable Lévy processes. We emphasise what we believe would be useful to scientists who wish to use SDEs in stochastic parameterisations. This exposition is certainly not exhaustive and is not meant to be so. Rather, we hope we present a starting point from which scientists can develop an intuitive feel for the reasoning behind SDEs and an appreciation for the necessity of rigor. Rather than repeat lengthy derivations, we have tried to give a qualitative understanding of the physical processes for which a class of stochastic models is appropriate. When possible, we have tried to provide sufficient quantitative guidance for the numerical generation of that class.

This chapter is organised as follows: Section 5.2 reviews classical SDEs based on Langevin systems with Brownian motion, i.e. Markov diffusion processes. Section 5.3 discusses α-stable Lévy processes and simple additive LSDEs. Section 5.4 provides examples of the stochastic models discussed in Sections 5.2 and 5.3, including a numerical comparison of a simple system driven by Gaussian white noise with that same system driven by a symmetric α-stable Lévy process with identical scale parameter. Section 5.5 concludes the chapter.

5.2 Stochastic systems with Brownian motion

5.2.1 Preliminary discussion

Classical SDEs are valid when there is a clear time-scale separation between 'fast' and 'slow' terms in a dynamical system, although many researchers (Dozier & Tappert 1978a,b; Penland 1985; Majda *et al.* 1999) have found that this restriction can be quite forgiving, and that useful results can be made even when a clear time scale separation is not obtained. Hasselmann (1976) assigns meteorological meaning to the fast and slow systems, which he calls 'weather' and 'climate', respectively. We sometimes use this terminology, although the reader should in no way take this usage as strict definitions of weather and climate.

As an example of a system that might be governed by an SDE, consider a state vector $\boldsymbol{x}(t)$ representing some surface phenomenon (e.g. SST, etc.) in the tropical ocean. Let us also imagine that the evolution of $\boldsymbol{x}(t)$ might be written as a differential equation as follows:

$$\frac{\mathrm{d}x}{\mathrm{d}t} = \boldsymbol{c}(\mathbf{x}, t) + \boldsymbol{w}(\boldsymbol{x}, t). \tag{5.2.1}$$

In equation (5.2.1), the climate variable $\boldsymbol{c}(\boldsymbol{x},t)$ could be modelled easily using a time step of, say, 10 days. This was the time step used in the El Niño model of Zebiak & Cane (1986). By contrast, the weather term $\boldsymbol{w}(\boldsymbol{x}, t)$ may represent tropical convection, so that an accurate model of this term would require a time step on the order of minutes or seconds; the mesoscale model MM5 having a spatial resolution of 1 km uses a time step of 3 s (G. Bryan & T. Hamill 2007, personal communication). It is often assumed that $\boldsymbol{w}(\boldsymbol{x}, t)$ can either be parameterised by processes represented by $\boldsymbol{c}(\boldsymbol{x}, t)$ or even neglected completely on large time scales if it varies rapidly enough. One then assumes that the long time-scale evolution of $\boldsymbol{x}(t)$ in equation (5.2.1) might be well described using only $\boldsymbol{c}(\boldsymbol{x}, t)$ in a model integrated with a 10-day time step. Unfortunately, this is very inaccurate for highly nonlinear systems. On the other hand, time steps small enough to resolve the interactions between $\boldsymbol{c}(\boldsymbol{x}, t)$ and $\boldsymbol{w}(\boldsymbol{x}, t)$ may be impractical. If the time-scale separation between $\boldsymbol{c}(\boldsymbol{x}, t)$ and $\boldsymbol{w}(\boldsymbol{x}, t)$ is sufficiently large, we can still model an approximate version of (5.2.1) using the longer time step, with $\boldsymbol{w}(\boldsymbol{x}, t)$ replaced by a deterministic function of $\boldsymbol{x}$ and t (which may or may not be a constant) multiplying a Gaussian stochastic quantity. That is, $\boldsymbol{w}(\boldsymbol{x}, t)$ varies rapidly enough so that its autocorrelation has decayed to insignificance over the course of the long time step, but the size of $\boldsymbol{w}(\boldsymbol{x}, t)$, though finite, is big enough that its effects on $\boldsymbol{x}$ cannot be neglected.

Thus, the effects of nearly independent values of $w(\boldsymbol{x}, t)$ are combined in such a way that central limit theorem (CLT) behaviour obtains, and Gaussian statistics are introduced into the evolution equation for $\boldsymbol{x}$ on long enough time scales.

5.2.2 An approximation using standard Brownian motion

The Gaussian stochastic quantity introduced in the previous paragraph is not arbitrary, but rather has a variance dependent on the timescales of the system. Quantifying these ideas requires consideration of a 'Wiener process', also called 'Brownian motion', which we denote by $\boldsymbol{W}(t)$. The Brownian motion is also sometimes called a 'continuous random walk'. We shall use these terms interchangeably. Using angle brackets to denote expectation values, we state the following properties of the vector Wiener process $\boldsymbol{W}$.

$\boldsymbol{W}(t)$ is a vector of Gaussian random variables, and

$$\boldsymbol{W}(0) = \boldsymbol{0}, \tag{5.2.2a}$$

$$\langle \boldsymbol{W}(t) \rangle = \boldsymbol{0}, \tag{5.2.2b}$$

$$\langle \boldsymbol{W}(s)\boldsymbol{W}^{\mathrm{T}}(t) \rangle = \mathrm{I} \min(s, t) \tag{5.2.2c}$$

and

$$\langle \mathrm{d}\boldsymbol{W}(s)\mathrm{d}\boldsymbol{W}^{\mathrm{T}}(t) \rangle = \mathrm{I}\, \delta(s - t). \tag{5.2.2d}$$

In (5.2.2), I denotes the identity matrix, and the δ function in (5.2.2d) approaches $\mathrm{d}t$ as s goes to t. The Wiener process is continuous but is only differentiable in a generalised sense,

$$\mathrm{d}W_k = \xi_k \mathrm{d}t. \tag{5.2.3}$$

Equation (5.2.3) defines the kth component ξ_k of 'white noise'. Note that the units of $W_k(t)$ are the square root of time; numerical stochastic models typically involve terms based on deterministic dynamics, which are updated with increments equal to the time step, and other terms involving stochastic terms, which are updated using increments equal to the square root of the time step. More detailed descriptions of Wiener processes and white noise may be found in Arnold (1974, ch. 3).

The Fourier spectrum of $W_k(t)$ varies everywhere as the inverse square of the frequency. Also note that the inverse square spectrum implies that every finite sample of Brownian motion is dominated by an oscillation having a period near to the length of the time series.

Before continuing on to how the Wiener process is used in modelling a physical system with an SDE, it is necessary to mention a crucially important property:

unlike the deterministic Riemann integral, which has a single 'fundamental theorem of integral calculus' (e.g. Purcell 1972), it is possible to define infinitely many different integration rules involving Brownian motion. Two of these calculi are found in nature. The one that primarily interests us corresponds to the continuous multiscale interaction problem with which we began this section. It is called 'Stratonovich calculus' and has the same integration rules as standard Riemannian calculus. The other physically meaningful calculus, called 'Itô calculus', obtains when a system consists of discrete jumps, but the time between jumps is vanishingly small compared with the time scales of interest. That is, there are two limits here that do not commute, the white noise and the continuous limits, and we want to take them both. A computer, of course, requires discretisation with a small enough time step that the system is approximately continuous, but this is yet another approximation and is separate from the two limits that obtain even before we begin to think about the computer program. Now, even though a hydrodynamic system is made up of molecules, we usually base the models of the ocean and the atmosphere on a continuous set of deterministic equations, such as rotating Navier–Stokes. Then, in order to make progress, we make some assumptions about the time scales of the system. This is equation (5.2.1). Finally, we use the dynamic CLT to approximate the fast variables as stochastic terms. These are the conditions leading to Stratonovich calculus, and appropriate numerical schemes are required to approximate numerically the solution to the SDE. We revisit this issue below.

Let us consider a case that is less often used, but may be geophysically relevant in some cases, where a fluid is at an intermediate level of concentration where it is dense enough that continuity equations are still valid on average, but rarified enough that momentum transfer through frequent individual molecular collisions presents a non-negligible stochastic effect. In this case, the physical system might well be described as an Itô SDE, and numerical models of it require different schemes from those modelling Stratonovich systems. Fortunately, although numerical schemes describing the different calculi are not the same, they are often rather similar. Exhaustive discussions may be found in, for example, Kloeden & Platen (1992).

As might be expected, Itô and Stratonovich numerical schemes converge to the same answer in the deterministic limit. This most emphatically does *not* mean that a 'stochastification' of a deterministic numerical scheme is obvious. Computers do not care whether a numerically generated solution to an equation is physical or not, and they can quite happily spit out a perfectly accurate solution to an SDE corresponding to a calculus that does not exist in nature if the numerical scheme is not appropriate to the physical system at hand. We will revisit this issue below, but first we need to set up the problem correctly. This is the subject of Section 5.2.3.

5.2.3 The central limit theorem

Informally, the traditional CLT (e.g. Doob 1953; Wilks 1995) usually employed by geoscientists to justify use of Gaussian distributions, states that the sum of independently sampled quantities having finite variance is approximately Gaussian. As discussed above, we consider here dynamical systems described by a slow time scale and faster time scales. The equations are averaged over a temporal interval large enough that the fast time scales collectively act as Gaussian stochastic forcing of the slow, coarse-grained system. In the mathematical literature (e.g. Feller 1966), the fact that fine details of how the fast processes are distributed do not strongly affect the coarse-grained behaviour of the slower dynamics is often called 'the invariance principle'. As the proof of this theorem is outside the scope of this paper, we state a commonly used form of it and refer the interested reader to the literature for details. Gardiner (1985) gives a heuristic description; we prefer Papanicolaou & Kohler (1974) for a technical statement of the theorem.

A dynamical system consisting of separated time scales may be written in the following manner:

$$\frac{\mathrm{d}\boldsymbol{x}}{\mathrm{d}t} = \varepsilon\, \boldsymbol{G}(\boldsymbol{x}, t) + \varepsilon^2 \boldsymbol{F}(\boldsymbol{x}, t), \tag{5.2.4}$$

where $\boldsymbol{x}$ is an N-dimensional vector. In equation (5.2.4), the smallness parameter ε should not be taken as a measure of importance. It does measure the rapidity with which the terms on the right-hand side of (5.2.4) vary relative to each other; one can think of ε^2 as the ratio of the characteristic time scale of the first term to the characteristic time scale of the second. If we now cast (5.2.4) in terms of a scaled time coordinate

$$\Delta s = \varepsilon^2 \Delta t, \tag{5.2.5}$$

it becomes

$$\frac{\mathrm{d}\boldsymbol{x}}{\mathrm{d}s} = \frac{1}{\varepsilon}\boldsymbol{G}\left(\boldsymbol{x}, \frac{s}{\varepsilon^2}\right) + \boldsymbol{F}\left(\boldsymbol{x}, \frac{s}{\varepsilon^2}\right) \tag{5.2.6}$$

We further assume that the first term in (5.2.4) decays quickly enough in the time interval Δt. In the limit $\varepsilon \to 0$, $\Delta t \to \infty$ with $\varepsilon^2 \Delta t$ remaining finite, the CLT states that (5.2.6) converges weakly to a Stratonovich SDE in the scaled coordinates,

$$\mathrm{d}\boldsymbol{x} = \boldsymbol{F}'(\boldsymbol{x}, s)\mathrm{d}s + \boldsymbol{G}'(\boldsymbol{x}, s) \circ \mathrm{d}\boldsymbol{W}(s). \tag{5.2.7}$$

The primes in (5.2.7) denote only that $\boldsymbol{F}$ and $\boldsymbol{G}$ may be somewhat different from the terms in (5.2.4). The $\boldsymbol{W}$ in (5.2.7) is a K-dimensional vector, each component of which is an independent Wiener process, or Brownian motion, and the '$\circ$' indicates that it is to be interpreted in the sense of Stratonovich. The $\boldsymbol{G}'(\boldsymbol{x}, s)$ is a matrix, the first index of which corresponds to a component of $\boldsymbol{x}$ and the second index of which corresponds to a component of $\mathrm{d}\boldsymbol{W}$.

For details and proof of the CLT, we recommend, for example, the articles by Wong & Zakai (1965), Khasminskii (1966) and by Papanicolaou & Kohler (1974). Examples of geophysical applications may be found in Kohler & Papanicolaou (1977), Penland (1985), Majda *et al.* (1999) and Sardeshmukh *et al.* (2001).

Remark 8 of Papanicolaou & Kohler (1974) is particularly applicable to many problems. In that case, the ith component of the rapidly varying term in equation (5.2.6) can be given as

$$G_i\left(\boldsymbol{x}, \frac{s}{\varepsilon^2}\right) = \sum_{k=1}^{K} G_{ik}(\boldsymbol{x}, s)\eta_k\left(\frac{s}{\varepsilon^2}\right), \tag{5.2.8}$$

where $\eta_k(s/\varepsilon^2)$ is a stationary, centred and bounded random function. The integrated lagged covariance matrix of η is defined to be

$$C_{km} \equiv \int_0^\infty \langle \eta_k(t)\eta_m(t+t')\rangle \mathrm{d}t', \; k, m = 1, 2, \ldots, K. \tag{5.2.9}$$

With these restrictions, the CLT states that in the limit of long times ($t \to \infty$) and small ε ($\varepsilon \to 0$), taken so that $s = \varepsilon^2 t$ remains fixed, the conditional probability density function (PDF) for $\boldsymbol{x}$ at time s given an initial condition $\boldsymbol{x}_0(s_0)$ satisfies the backward Kolmogorov equation (e.g. Horsthemke & Lefever 1984; Bhattacharya & Waymire 1990),

$$\frac{\partial p(\boldsymbol{x}, s|\boldsymbol{x}_0, s_0)}{\partial s_0} = \mathcal{L}p(\boldsymbol{x}, s|\boldsymbol{x}_0, s_0), \tag{5.2.10}$$

where

$$\mathcal{L} = \sum_{i,j=1}^{N} a_{ij}(\boldsymbol{x}_0, s_0)\frac{\partial^2}{\partial x_{0i}\partial \boldsymbol{x}_{0j}} + \sum_{i=1}^{N} b_i(\boldsymbol{x}_0, s_0)\frac{\partial}{\partial x_{0i}} \tag{5.2.11}$$

and

$$a_{ij}(\boldsymbol{x}, s) = \sum_{k,m=1}^{K} C_{km} G_{ik}(\boldsymbol{x}, s) G_{jm}(\boldsymbol{x}, s), \tag{5.2.12a}$$

$$b_i(\boldsymbol{x}, s) = \sum_{k,m=1}^{K} C_{km} \sum_{j=1}^{N} G_{jk}(\boldsymbol{x}, s)\frac{\partial G_{im}(\boldsymbol{x}, s)}{\partial x_j} + F_i(\boldsymbol{x}, s). \tag{5.2.12b}$$

In this limit, the conditional PDF also satisfies a forward Kolmogorov equation (called a 'Fokker–Planck equation' in the scientific literature) in the scaled coordinates

$$\frac{\partial p(\boldsymbol{x}, s|\boldsymbol{x}_0, s_0)}{\partial s} = \mathcal{L}^\dagger p(\boldsymbol{x}, s|\boldsymbol{x}_0, s_0), \tag{5.2.13}$$

where

$$\mathcal{L}^\dagger p = \sum_{i,j=1}^{N} \frac{\partial^2}{\partial x_i \partial x_j}[a_{ij}(\boldsymbol{x}, s)p] - \sum_{i=1}^{N} \frac{\partial}{\partial x_i}[b_i(\boldsymbol{x}, s)p]. \tag{5.2.14}$$

In short, the *moments* of $\boldsymbol{x}$ can be approximated by the moments of the solution to the Stratonovich SDE

$$\mathrm{d}\boldsymbol{x} = \boldsymbol{F}(\boldsymbol{x}, s)\mathrm{d}s + \mathsf{G}(\boldsymbol{x}, s)\mathsf{S} \circ \mathrm{d}\boldsymbol{W}. \tag{5.2.15}$$

In equation (2.15), $\mathsf{G}(\boldsymbol{x}, s)$ is the matrix whose (i, k)th element is $G_{ik}(\boldsymbol{x}, s)$ and S is a matrix where the (k, m)th element of $\mathsf{S}\mathsf{S}^\mathrm{T}$ is C_{km}. Note that S is only unique up to its product with an arbitrary orthogonal matrix. Also note that the usual factor of one-half found in most formulations of the Fokker–Planck equation has been absorbed into the definition of C_{km}.

From the first temporal derivative in (5.2.13) and the second derivative with respect to the state variable in (5.2.14), it can be seen that the Fokker–Planck equation is a type of diffusion equation. For this reason, $\boldsymbol{b}(\boldsymbol{x}, s)$ in (5.2.14) is often called the 'drift', $\boldsymbol{a}(\boldsymbol{x}, s)$ is called the 'diffusion', and the process $\boldsymbol{x}$ is called a 'Markov diffusion process'. This term is used here to make the distinction between systems driven by Gaussian white noise and those driven by non-Gaussian stable Lévy processes, as discussed below.

There do exist Kolmogorov equations (5.2.10) and (5.2.13), for Itô systems, with the modification that in (5.2.12b), $b_i(\boldsymbol{x}, s)$ is simply equal to $F_i(\boldsymbol{x}, s)$. The difference between Itô and Stratonovich calculi is important for scientists to understand because most of the physical phenomena they deal with are Stratonovich, while most mathematical references on stochastic numerical techniques are primarily interested in Itô schemes. There is a formal equivalence between Itô and Stratonovich descriptions of reality, so a theorem about an Itô process can generally be carried over to the corresponding Stratonovich process. More precisely, the Stratonovich SDE (absorbing the matrix S into G)

$$\mathrm{d}x_i = F_i(\boldsymbol{x}, t)\mathrm{d}t + \sum_{\alpha} G_{i\alpha}(x, t) \circ \mathrm{d}W_\alpha \tag{5.2.16a}$$

is equivalent to the Itô SDE

$$\mathrm{d}x_i = \left[F_i(\boldsymbol{x}, t) + \frac{1}{2} \sum_{\alpha, j} G_{j\alpha}(\boldsymbol{x}, t) \frac{\partial G_{i\alpha}(\boldsymbol{x}, t)}{\partial x_j} \right] \mathrm{d}t + \sum_{\alpha} G_{i\alpha}(\boldsymbol{x}, t) \mathrm{d}W_{\alpha}. \quad (2.16\mathrm{b})$$

By 'equivalent' it is meant that solving (5.2.16a) using Stratonovich calculus gives the same solution as solving equation (5.2.16b) using Itô calculus. That is, each equation evaluated for $\boldsymbol{x}$ using its appropriate calculus would describe an experimental outcome equally well as the other; the statistics of $\boldsymbol{x}$ in each case are the same. Note that if $\mathsf{G}(\boldsymbol{x}, s)$ is independent of $\boldsymbol{x}$, Itô and Stratonovich calculi converge to each other. In (5.2.16b), we have omitted '$\circ$' in keeping with standard mathematical notation of an Itô SDE. Unfortunately, the transformation from one description to the other can be prohibitively difficult in physical applications.

For longer discussions on the difference between Itô and Stratonovich systems, we refer the reader to Arnold (1974), Horsthemke & Lefever (1984), Kloeden & Platen (1992) and Ewald & Penland (2008). The important message here is for when one wishes to approximate a continuous real system with short but finite correlation time in a general circulation model (GCM) with a Gaussian stochastic variable. As noted in equation (5.2.9), the correlation structure of the fast variable comes into play. Simply replacing the fast term with a Gaussian random deviate with standard deviation equal to that of the variable to be approximated, and then using deterministic numerical integration schemes, is a recipe for disaster (Sura & Penland 2002; Ewald *et al.* 2004).

5.2.4 Notes on numerical techniques involving SDEs

One can write, and some have written (e.g. Kloeden & Platen 1992), enormous tomes on the theory and practice of numerically integrating SDEs. A review of some of these techniques, including longer discussions of the schemes we present here, may be found in Ewald & Penland (2008). In this subsection, we confine ourselves to discussing the procedure and results of using an explicit stochastic Euler scheme (Rümelin 1982), an explicit stochastic Heun scheme (Rümelin 1982) and an implicit stochastic integration scheme developed by Ewald & Temam (2005; see also Ewald *et al.* 2004) especially for spectral versions of GCMs. To set notation, we denote the time step by Δ, while the $\{z_{\mu i}\}$ denote centred Gaussian random deviates, each with variance Δ, sampled at time t_i. If μ varies from 1 to M, then $z_{\mu i}$ is the μth component of the M-dimensional vector z_i; x_m represents the mth component of an N-dimensional vector $\boldsymbol{x}$ obeying the following SDE:

$$\mathrm{d}\boldsymbol{x} = \boldsymbol{F}(\boldsymbol{x}, s)\mathrm{d}s + \mathsf{G}(\boldsymbol{x}, s)(\circ)\mathrm{d}\boldsymbol{W}. \quad (5.2.17)$$

In equation (5.2.17), $\mathsf{G}(\boldsymbol{x}, s)(\circ)\mathrm{d}\boldsymbol{W}$ equals $\mathsf{G}(\boldsymbol{x}, s)\mathrm{d}\boldsymbol{W}$ if the system is Itô, and $\mathsf{G}(\boldsymbol{x}, s) \circ \mathrm{d}\boldsymbol{W}$ if it is Stratonovich.

The stochastic Euler scheme,

$$x_m(t_{i+1}) = x_m(t_i) + F_m(\boldsymbol{x}, t_i)\Delta + \sum_{\mu} G_{m\mu}(\boldsymbol{x}, t_i) z_{\mu i} \tag{5.2.18}$$

converges to Itô calculus, unless $\mathsf{G}(\boldsymbol{x}, s)$ is not really a function of $\boldsymbol{x}$. In that case, we repeat, the Itô and Stratonovich calculi give the same moments of $\boldsymbol{x}$. It is important that the vector z_i be generated outside the loop performing the summation in (5.2.18); otherwise, random phases are generated, which erroneously eliminate the contribution to $\langle \boldsymbol{x}\boldsymbol{x}^{\mathrm{T}} \rangle$ by any off-diagonal elements of $\mathsf{G}\mathsf{G}^{\mathrm{T}}$. Again, if $\mathsf{G}(\boldsymbol{x}, s)$ is a function of $\boldsymbol{x}$, equation (5.2.18) generates Itô calculus.

For reasons explained elsewhere (e.g. Rümelin 1982; Ewald & Penland 2008), explicit schemes generating Stratonovich calculus are usually predictor–corrector methods. The simplest version is a second-order Runge–Kutta scheme, called the Heun scheme. Here, one generates an intermediate variable using an Euler estimate

$$x'_m(t_{i+1}) = x_m(t_i) + F_m(\boldsymbol{x}, t_i)\Delta + \sum_{\mu} G_{m\mu}(\boldsymbol{x}, t_i) z_{\mu i}, \tag{5.2.19a}$$

which is then updated as follows:

$$\begin{aligned} x_m(t_{i+1}) = x_m(t_i) &+ \frac{1}{2}\{F_m(\boldsymbol{x}, t_i) + F_m(\boldsymbol{x}', t_{i+1})\}\Delta \\ &+ \frac{1}{2}\sum_{\mu}\{G_{m\mu}(\boldsymbol{x}, t_i) + G_{m\mu}(\boldsymbol{x}', t_{i+1})\} z_{\mu i}. \end{aligned} \tag{5.2.19b}$$

Note that the same random vector z_i is used in both (5.2.19a) and (5.2.19b).

Our final example of a numerical scheme designed for SDEs is the implicit scheme of Ewald & Temam (2003, 2005). This algorithm was devised to accommodate the architecture of extant climate models, including barotropic vorticity models (e.g. Sardeshmukh & Hoskins 1988) and full GCMs (e.g. Saha *et al.* 2006). The deterministic climate models usually integrate the state vector first using a leapfrog step, followed by an implicit step. To implement the stochastic analogue of this procedure, we rewrite equation (5.2.17) as

$$\mathrm{d}\boldsymbol{x} = \boldsymbol{F}_1(\boldsymbol{x}, t)\mathrm{d}t + \boldsymbol{F}_2(\boldsymbol{x}, t)\mathrm{d}t + \mathsf{G}(\boldsymbol{x}, t)(\circ)\mathrm{d}\boldsymbol{W}. \tag{5.2.20}$$

In (5.2.20), $\boldsymbol{F}_1(\boldsymbol{x}, t)$ and $\boldsymbol{F}_2(\boldsymbol{x}, t)$ are the explicit and the implicit parts of the model, respectively. The implicit leapfrog scheme of Ewald & Temam (2003, 2005) is as follows:

$$\boldsymbol{x}'(t_{i+2}) = \boldsymbol{x}(t_i) + 2\boldsymbol{F}_1(\boldsymbol{x}(t_{i+1}), t_{i+1})\Delta + \boldsymbol{M}(\boldsymbol{x}(t_i), t_i) + \boldsymbol{M}(\boldsymbol{x}(t_{i+1}), t_{i+1}), \tag{5.2.21a}$$

and

$$\boldsymbol{x}(t_{i+2}) = \boldsymbol{x}'(t_{i+2}) + 2\boldsymbol{F}_2(\boldsymbol{x}(t_{i+2}), t_{i+2})\Delta. \tag{5.2.21b}$$

In the updating expressions, the $\boldsymbol{m}$th component of the vector $\boldsymbol{M}(\boldsymbol{x}(t_i), t_i)$ is

$$M_m(\boldsymbol{x}, t_i) = \sum_{n,\mu,\nu} G_{n\mu}(\boldsymbol{x}, t_i)\frac{\partial G_{m\nu}(\boldsymbol{x}, t_i)}{\partial x_n} I_{(\mu,\nu)} + \sum_{\mu} G_{m\mu}(\boldsymbol{x}, t_i) z_\mu. \tag{5.2.21c}$$

The derivative in (5.2.21c) can be approximated as

$$\frac{\partial G_{m\nu}(\boldsymbol{x}, t_i)}{\partial x_n} = \frac{G_{m\nu}(\boldsymbol{x} + \varepsilon_n\sqrt{\Delta}\hat{\boldsymbol{e}}_n, t_i) - G_{m\nu}(\boldsymbol{x}, t_i)}{\varepsilon_n\sqrt{\Delta}}. \tag{5.2.21d}$$

In (5.2.21d), $\hat{\boldsymbol{e}}_n$ is a unit vector corresponding to the component x_n. The vector ε has components less than or equal to unity, in units of $\boldsymbol{x}/\sqrt{t}$, and allows the modeller to adjust the discretised derivatives to the problem at hand. The difference between Itô and Stratonovich calculi is effected through estimates of the multiple stochastic integral in (5.2.21c)

$$I_{(\mu,\nu)} = \frac{1}{2}(z_\mu z_\nu - \delta_{\mu\nu}\Delta) \qquad \text{(Itô)}, \tag{5.2.22a}$$

$$I_{(\mu,\nu)} = \frac{1}{2} z_\mu z_\nu \qquad \text{(Stratonovich)}. \tag{5.2.22b}$$

In this and in all implicit stochastic algorithms, the stochastic terms enter the problem during the explicit step. When random numbers occur in the denominator of the implicit step, the numerical solution eventually explodes.

5.3 Aspects of modelling with Lévy processes

5.3.1 Preliminary discussion

The fact that a process has infinite variance, or even infinite mean, does not preclude its existence. Consider the game proposed in 1713 by Nicolaus Bernoulli in a letter to Pierre Raymond de Montmort. A player flips a coin repeatedly until he gets 'tails', at which point the game ends. If the tails appears on the first flip, he wins \$1. If it appears on the second flip, he wins double that, or \$2. If he achieves two heads before a tails on the third flip, he wins double *that*, or \$4. If the tails shows up on the $(k+1)$th flip, he wins $\$2^k$. Thus, his expected winnings are a sum of terms, each consisting of the probability that a certain event occurs times the return, or

$1/2 + (1/4) \cdot 2 + (1/8) \cdot 4 + \ldots$, which sums to infinity. Of course, it would take an infinite amount of time to get this infinite return, but that does not mean the game cannot be played to its end. This paradox is called 'the St Petersburg paradox' since it was published in 1738 by Nicolaus' brother Daniel Bernoulli (presumably with Nicolaus' permission) in the *Commentaries of the Imperial Academy of Science of St Petersburg*.

With these ideas in mind, we return to the classic CLT, where independent, identically distributed normalised variables are added together. However, the condition of finite variance is relaxed. The result is a class of distributions called α-stable Lévy distributions, of which the Gaussian distribution is a special case (Lévy 1937; Gnendenko & Kolmogorov 1954). These distributions are characterised by a parameter α, $0 < \alpha \leq 2$, for which moments of order μ diverge for $\mu \geq \alpha$. The exception to this rule is when $\alpha = 2$, which corresponds to the Gaussian distribution. The reason these distributions are called 'stable' is because sums of appropriately centred and normalised (by $n^{1/\alpha}$, with n being the number of terms in the sum) variables sampled from such a distribution belong to the same distribution. For example, the sum of Gaussian variables is also a Gaussian variable.

The class of α-stable Lévy processes are themselves a special case of systems called simply Lévy processes (Appelbaum 2004; Protter 2005). All that is required of a Lévy process $X(t)$ is that (i) it has independent increments, i.e. $X(t) - X(s)$ is independent of $X(r) - X(q)$ at times q, r, s and t for increments that do not overlap; (ii) the increments are stationary, i.e. the pdf of $X(t) - X(s)$ is the same as that of $X(t - s)$; (iii) $X(t)$ is continuous in probability, i.e. for any $\delta > 0$, no matter how small δ is, the probability that $|X(t) - X(s)| > \delta$ is zero in the limit that $t \to s$; and (iv) $X(0) = 0$. Theorems proved for Lévy processes are obviously true for α-stable Lévy processes.

Why do we care about non-Gaussian α-stable Lévy processes? Well, as we saw in Section 5.2, the distance a Brownian particle travels from the origin varies as

$$|W(t)|^2 \sim t. \tag{5.3.1}$$

More generally, the diffusion away from its origin travelled by an α-stable Lévy process $L_\alpha(t)$ is given by

$$|L_\alpha(t)|^2 \sim t^{2/\alpha}. \tag{5.3.2}$$

Thus, for $\alpha < 2$, large excursions from the mean can occur in a much shorter amount of time than for Gaussian-driven processes. This property has its effect on the probability of finding large excursions $P\{L_\alpha(t) > r\}$, which has 'heavy tails', i.e.

$$P\{L_\alpha(t) > r\} \sim r^{-\alpha} \tag{5.3.3}$$

A formalism allowing the existence of heavy tails in the distribution of stochastic forcing allows a much broader class of observed phenomena to be modelled as SDEs of the form

$$\mathrm{d}\boldsymbol{x} = \boldsymbol{F}(\boldsymbol{x}, s)\mathrm{d}s + \mathsf{G}(\boldsymbol{x}, s)(\circ)\mathrm{d}\boldsymbol{L}. \tag{5.3.4}$$

(NB the Itô–Stratonovich quandary persists for $\alpha < 2$.) Indeed, non-Gaussian Lévy models have been applied to atmospheric turbulence (e.g. Viecelli 1998; Ditlevsen 2004) and palaeoclimate (e.g. Ditlevsen 1999), and have been used to explain anomalous diffusion observed in hydrological records (Hurst 1951).

Just as the PDFs of Markov diffusion processes obey a Fokker–Planck equation, so do those of systems involving non-Gaussian stable Lévy processes. However, these Fokker–Planck equations involve fractional derivatives of order α rather than second derivatives with respect to the state variable in the diffusion term (e.g. Chechkin *et al.* 2003; Ditlevsen 2004). It is often more convenient to consider a spectral form of the Fokker–Planck equation so that the fractional derivatives are replaced by conjugate variables raised to the power α. Not surprisingly, much of the theory of Lévy α-stable processes also involves Fourier and Laplace transforms, as we see in the next sections. For clarity, we shall confine our discussion to univariate variables unless the multivariate generalisation is clear.

5.3.2 Poisson processes and compound Poisson processes

This subsection is intended to give the reader an intuitive feel for Lévy processes. To do so, it is extremely useful to employ the fact that Lévy processes can be written in terms of a 'Lévy–Itô decomposition', which is a combination of a drift, a Brownian motion and compound Poisson processes (Appelbaum 2004). We discussed Brownian motions in detail in Section 5.2; we now turn our attention to Poisson and compound Poisson processes.

Consider a rabbit standing (or sitting) at a place we shall designate the origin. After some time t, the bunny may or may not have taken one or more jumps. We shall denote the number of jumps taken by our lapine friend in time t as $N(t)$. Using Protter's (2005) formalism, let us say the rabbit jumps for the nth time at time T_n, with $T_0 = 0$ and $T_{n+1} > T_n$. The counting process $N(t)$ can be written in terms of the indicator function $1_{\{t \geq T_n\}}$, which is one for $t \geq T_n$ and zero for $t < T_n$, i.e.

$$N(t) = \sum_{n \geq 1} 1_{\{t \geq T_n\}}. \tag{5.3.5}$$

If $N(t)$ obeys a Poisson process, then the probability that $N(t)$ equals some non-negative integer n is

$$P\{N(t) = n\} = \frac{(\lambda t)^n \exp(-\lambda t)}{n!}, \qquad (5.3.6)$$

for some $\lambda > 0$. That is, λt is the parameter of the Poisson process and λ is called the *arrival intensity*. The Poisson process has some well-known properties. For example, its mean and variance are time dependent and each is equal to λt. Note that this probability is not only continuous for $t > 0$ but also differentiable with respect to time,

$$\frac{\mathrm{d}P\{N(t) = n\}}{\mathrm{d}t} = \lambda P\{N(t) = n - 1\} - \lambda P\{N(t) = n\}. \qquad (5.3.7)$$

A Poisson process from which the mean λt is subtracted is called a *compensated Poisson process*, and has mean zero. Note further that we confine ourselves to finite λ.

So far, we have not said anything about how big these jumps are. Let us say the kth jump has length Y_k. Then, the distribution of the sum $Y_1 + Y_2 + \cdots + Y_{N(t)}$, with $N(t)$ a Poisson process, is a *compound Poisson process* (Feller 1966; Protter 2005). The compound Poisson processes as just described are Lévy processes.

It is possible to combine the compound Poisson process and Brownian motion into yet another Lévy process: let the combination of a Brownian motion $W(t)$ and a linear drift bt be denoted as $C(t)$, and a compound Poisson process as just described be denoted as $Y(t)$. Then, we may define yet another Lévy process $X(t)$ (Appelbaum 2004) as

$$X(t) = C(t) + Y(t). \qquad (5.3.8)$$

This means that $X(t)$ is simply the Brownian motion with drift until the first jump, say, at time T_1 and height Y_1. After that, $X(t)$ evolves again as a Brownian motion with drift until the next jump, and so on. That is,

$$X(t) = \begin{cases} C(t) & \text{for } 0 \leq t < T_1, \\ C(T_1) + Y_1 & \text{for } t = T_1, \\ X(T_1) + C(t) - C(T_1) & \text{for } T_1 < t < T_2, \\ X(T_1) + C(T_2) - C(T_1) + Y_2 & \text{for } t = T_2, \text{etc.} \end{cases} \qquad (5.3.9)$$

Equation (5.3.9) is an example of a Lévy process represented as the sum of a drift, a Brownian motion and superpositions of compound Poisson processes (Bhattacharya & Waymire 1990; Eliazar & Klafter 2003; Appelbaum 2004). This is what is meant by the qualitative description of Lévy forcing as white noise with jumps, particularly since T_1 and T_2 may be as close together as we like.

5.3.3 Characteristics of Lévy α-stable processes

We do not use the term 'characteristic' lightly. In fact, Lévy α-stable processes are generally defined in terms of their characteristic functions $\phi(u) = \langle \exp(\mathrm{i}uL_\alpha) \rangle$ (e.g. Weron & Weron 1995; Eliazar & Klafter 2003; Dybiec *et al.* 2006; Nolan 2007),

$$\ln \phi(u) = -\sigma^\alpha |u|^\alpha \left\{1 - \mathrm{i}\beta \mathrm{sgn}(u) \tan\left(\frac{\pi\alpha}{2}\right)\right\} + \mathrm{i}\mu u, \ \alpha \neq 1, \quad (5.3.10a)$$

$$\ln \phi(u) = -\sigma^\alpha |u|^\alpha \left\{1 + \mathrm{i}\beta \mathrm{sgn}(u) \frac{2}{\pi} \ln|u|\right\} + \mathrm{i}\mu u, \ \alpha = 1. \quad (5.3.10b)$$

In (5.3.10a) and (5.3.10b), the *location parameter* μ shifts the distribution to the left or right. For $\alpha = 2$, μ represents the Gaussian mean. The *scale parameter* σ, which is greater than zero, represents the width of the distribution about μ. In this notation, the variance of a Gaussian distribution is $2\sigma^2$. This convention is somewhat inconvenient for those of us whose main experience is with Gaussian distributions, but it is the convention used in most of the Lévy process literature, so we may as well get used to it. The *skewness parameter* β lies in the interval $[-1,1]$, and the distribution is symmetric when $\beta = 0$. In fact, when both β and μ are zero, the characteristic function has the form

$$\phi(u) = \exp(-\sigma^\alpha |u|^\alpha). \quad (5.3.11)$$

For the Lévy process corresponding to a Brownian motion $W(t)$, $\sigma^\alpha = \sigma^2 = t/2$. There are two other α-stable Lévy processes for which the PDF is known to have a closed-form solution. One is the Cauchy distribution, where $\alpha = 1$ and $\beta = 0$,

$$p(x) = \frac{1}{\pi} \frac{1}{\sigma^2 + (x - \mu)^2}. \quad (5.3.12)$$

The other is the Lévy distribution, where $\alpha = 0.5$ and $\beta = 1$,

$$p(x) = \sqrt{\frac{\sigma}{2\pi}} \frac{\exp(-\frac{\sigma}{2x})}{x^{3/2}}. \quad (5.3.13)$$

For more general cases, one may evaluate the Fourier transform of (5.3.10a) and (5.3.10b) numerically.

5.3.4 Additive α-stable Lévy-driven SDEs

Consider differential equations of the form

$$\mathrm{d}\boldsymbol{x} = \boldsymbol{F}(\boldsymbol{x}, t)\mathrm{d}t + \mathsf{G}\mathrm{d}\boldsymbol{L}_\alpha, \quad (5.3.14)$$

where the coefficient G of the Lévy noise vector increment $d\boldsymbol{L}_\alpha$ is simply a constant matrix. Although (5.3.14) is far from general, there are certainly plenty of applications for which it is applicable. In these cases, the standard numerical approach (e.g. Dybiec *et al.* 2006) is to use an Euler scheme

$$x_m(t_{i+1}) = x_m(t_i) + F_m(\boldsymbol{x}(t_i), t_i)\Delta + \sum_\mu G_{m\mu}\Delta^{1/\alpha} z_{\mu i}, \qquad (5.3.15)$$

where Δ is the time step and $\{z_{\mu i}\}$ are random variables sampled at time t_i from a centred α-stable Lévy distribution. Note that the time step in (5.3.15) is raised to the power $1/\alpha$. For $\alpha = 2$, (5.3.15) is equivalent to (5.2.18) with $\Delta^{1/2}$ $z_{\mu i}$ representing the Gaussian random deviate of variance Δ.

Techniques for approximating random variables $\{z_{\mu i}\}$ from α-stable Lévy distributions may be found in Dybiec *et al.* (2006) as well as Weron & Weron (1995). As of this writing, mathematician John Nolan, a professor at American University, has a website: http://academic2.american.edu/~jpnolan/stable/stable.html from which one may download Lévy-variable generators for use with proper acknowledgement. An example of an additive LSDE is given in Section 5.4.

5.4 Diffusion and α-stable Lévy processes: some examples

5.4.1 Ornstein–Uhlenbeck processes

To compare the qualitative properties of a system driven by Gaussian white noise with that driven by an α-stable Lévy noise, we consider a simple Ornstein–Uhlenbeck (OU) process as follows:

$$dx = -\gamma x dt + d\zeta. \qquad (5.4.1)$$

In (5.4.1), ζ is either a symmetric α-stable Lévy variable with $\alpha = 1.5$ or Brownian motion ($\alpha = 2$). The decay parameter γ is equal to 0.1/tu (1 tu = 1 time unit) and the scale parameter σ for each noise process is unity. Recall that $\sigma = 1$ for a Gaussian process is equivalent to a variance of 2. Equation (5.4.1) is integrated numerically using an Euler integration scheme [(5.3.15); see also Protter & Talay 1997] with a time step of $\Delta = 0.1$ tu. Random numbers in the interval (0,1) are generated using a Mersenne Twister (Matsumoto & Nishimura 1998), which are then used to generate either Gaussian random variables using a Box–Mueller scheme (Press *et al.* 1992) or Lévy variables (Weron & Weron 1995).

The time series of random variables used to integrate (5.4.1) are shown in Fig. 5.1. Note that the ordinate of the graph showing the Lévy process is an order of magnitude larger than that showing the Gaussian process, though both processes are characterised by the same scale parameter σ. The first 10 tu, comprising 100 time steps, of the time series are shown in Fig. 5.2. There it can be seen that

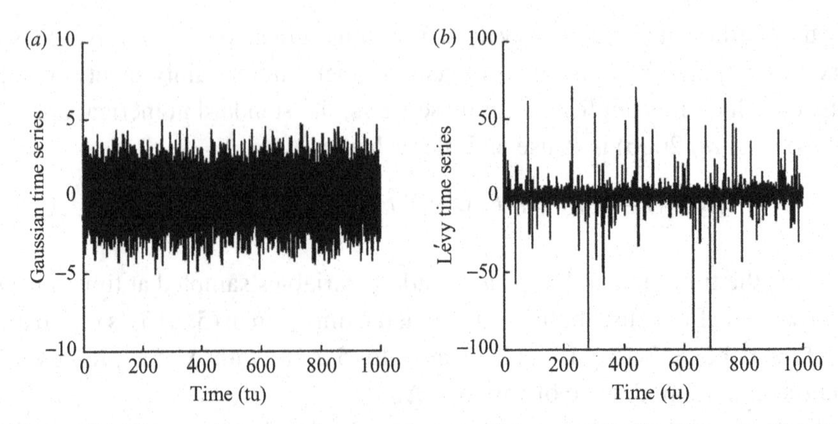

Figure 5.1 Time series of random numbers used to numerically integrate (5.4.1). (*a*) Gaussian random variables and (*b*) Lévy-distributed variables.

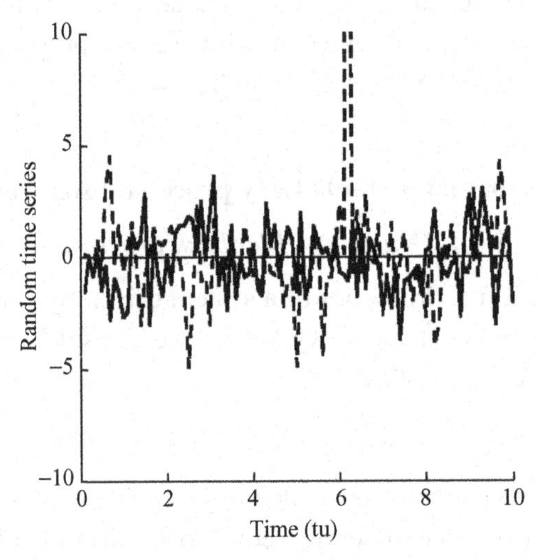

Figure 5.2 First 10 tu of time series as shown in Fig. 5.1. Solid line, Gaussian random noise; dashed line, Lévy-distributed noise.

the two time series are qualitatively similar, except in the region of the occasional jump.

The Gaussian-driven and the Lévy-driven OU processes are shown in Fig. 5.3*a*,*b*, respectively. Also shown (Fig. 5.4) is a close-up of the two OU processes; note the effect of the jump that occurred around 6 tu persists for approximately 15 time steps (1.5 tu) in the Lévy-driven system. The global effect of the jumps is seen in the cumulative distributions of the two OU processes (Fig. 5.5), where the fat tails of the α-stable Lévy distributed forcing are translated into fat tails of the Lévy-driven OU process.

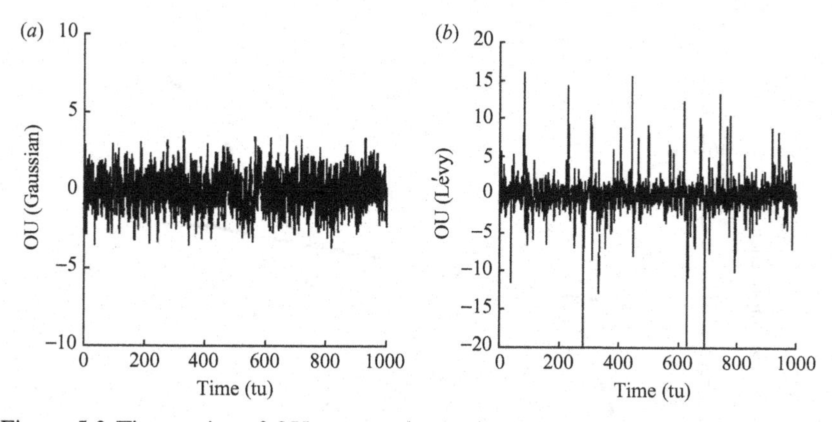

Figure 5.3 Time series of OU process forced by (*a*) Gaussian random noise and (*b*) Lévy-distributed noise.

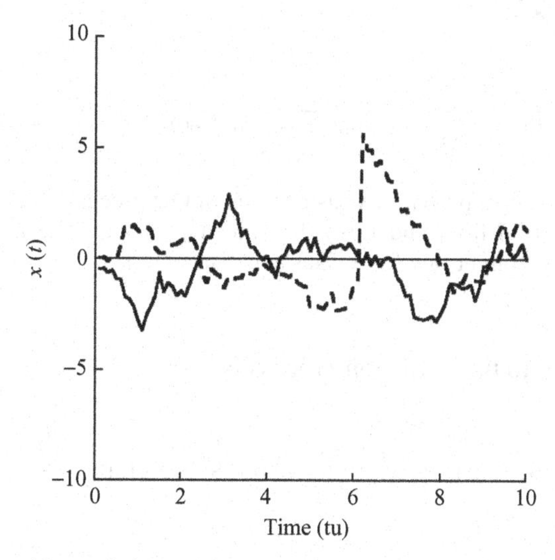

Figure 5.4 First 10 tu of time series as shown in Fig. 5.3. Solid line, Gaussian random noise; dashed line, Lévy-distributed noise.

5.4.2 Diffusion and Lévy processes with some similarities

Sura & Sardeshmukh (2008) have shown that daily departures of SSTs from the annual cycle are well described by the Stratonovich SDE

$$\mathrm{d}T = \left(AT - \frac{1}{2}Eg\right)\mathrm{d}t + b \circ \mathrm{d}W_1 + (ET + g) \circ \mathrm{d}W_2, \qquad (5.4.2)$$

where W_1 and W_2 are independent Wiener processes and A, E, g and b are constants. Sardeshmukh & Sura (2009) similarly showed that daily departures of 300 mb vorticity from the annual cycle obey an equation of the same form. The marginal

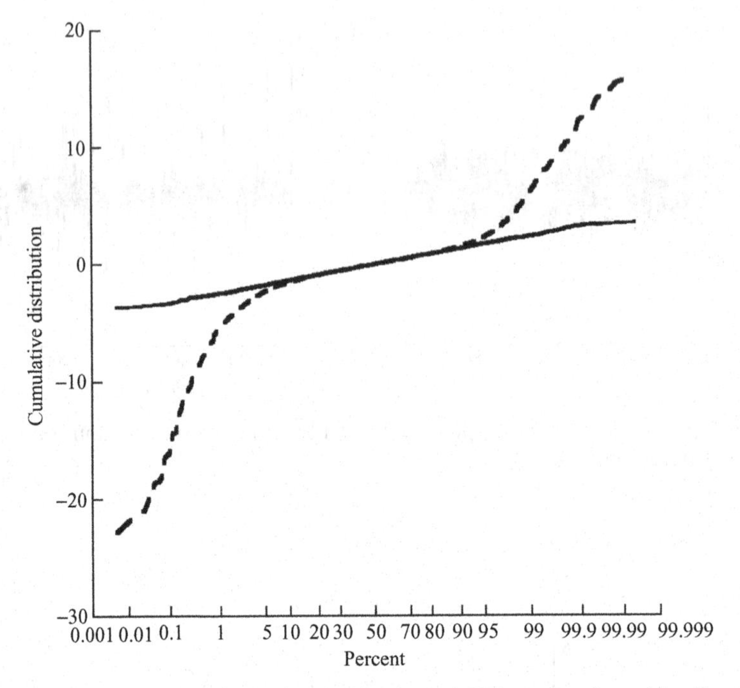

Figure 5.5 Cumulative probability distribution of OU process forced by Gaussian random noise (solid line) and Lévy-distributed noise (dashed line). Nonlinear abscissa yields a straight line for a Gaussian distribution.

PDF corresponding to this diffusion process is

$$p(T) = \frac{1}{\Re}[(ET+g)^2 + b^2]^{-(\gamma+1)} \exp\left[\frac{2\gamma g}{b} \arctan\left(\frac{ET+g}{b}\right)\right], \quad (5.4.3)$$

where $\Re$ is the normalisation constant. The quantity γ is defined as $-(A+0.5E^2)/E^2$ and is positive for physically allowable systems, i.e. A is strictly negative and $|A|$ is larger than $0.5E^2$. Thus, $0 < \gamma < \infty$ and, for large T_0, $P\{T > T_0\} \sim T_0^{-(2\gamma+1)}$. Thus, as pointed out by Sardeshmukh & Sura (2009), moments of order higher than $2\gamma + 1$ do not exist. The distribution of this diffusion process allows both skew and heavy tails, depending on the level of stochastic forcing and the level of correlation between the additive and multiplicative noises. The question arises, then, whether (5.4.3) is the PDF of a diffusion that is also an α-stable Lévy process. In general, the answer is 'No'. While (5.4.3) does converge to a Cauchy distribution for $\gamma \to 0$ and to a Gaussian for $\gamma \to \infty$, it is straightforward to show that the characteristic function of $T(t)$ cannot be written in the form of (5.3.10a) or (5.3.10b) for finite $\gamma > 0$.

5.5 Conclusions

With the increasing popularity of stochastic parameterisations in numerical weather and climate models, it is necessary for us to remind ourselves of how the physical and mathematical bases for such approximations are related. In this chapter, we have tried to summarise some of this theory without pretending to be exhaustive. In particular, we have not discussed the newly developed theory of integrating SDEs driven by state-dependent Lévy processes. The procedures for doing so involve what are called 'Marcus integrals', and are beyond the scope of this chapter. The most accessible reference we have found on the subject is in the book by Appelbaum (2004).

Even the numerical simulation of Gaussian-driven processes is not trivial, but the theory for these systems is quite well developed. Obviously, it is not sufficient for the mere existence of power law tails or skew in a probability distribution to preclude a valid approximation of a physical system as forced by Gaussian white noise (see also Sardeshmukh & Sura 2009). Thus, unless there are physical reasons for doubting the validity of such an approximation, we recommend using the theoretical arsenal developed around classical SDEs if at all possible.

We conclude with a reminder that the difference between Itô and Stratonovich calculi is important. For mathematicians, it may suffice to know that there is a transformation between them; scientists have to know what that transformation is and how to evaluate it. A thermometer gives only one reading at a time, and knowing that an isomorphism exists between the output of a numerical weather prediction model and the temperature that will eventually be observed does not help the farmer or fisherman unless scientists can apply that isomorphism in advance. Of course, if the stochastic forcing in a numerical model can be argued on physical grounds to exist independently of the system being modelled, the issue of multiple calculi becomes moot and one may use the Euler scheme to model an SDE driven by an α-stable Lévy random variable, whether or not $\alpha = 2$. The key phrase here involves argument on physical grounds; concern for the fidelity of any type of parameterisation to the physical system being modelled is always our first priority.

Acknowledgements

The authors are pleased to acknowledge valuable conversations with P. D. Sardeshmukh, P. Imkeller, T. Hamill, M. Charnotskii, P. Sura and I. Pavlyukevich. Particular gratitude is expressed to an anonymous reviewer for extremely valuable advice.

References

An, S.-I. & Jin, F.-F. 2004 Nonlinearity and asymmetry of ENSO. *J. Climate*, **17**, 2399–2412 (doi:10.1175/1520–0442(2004)017<2399:NAAOE>2.0.CO;2).

Appelbaum, D. 2004 *Lévy Processes and Stochastic Calculus*, vol. 93. Cambridge Studies in Advanced Mathematics. Cambridge University Press.

Arnold, L. 1974 *Stochastic Differential Equations: Theory and Applications*. John Wiley & Sons.

Berner, J., Shutts, G. J., Leutbecher, M. & Palmer, T. N. 2009 A spectral stochastic kinetic energy backscatter scheme and its impact on flow-dependent predictability in the ECMWF ensemble prediction system. *J. Atmos. Sci.*, **66**, 603–626.

Bhattacharya, R. N. & Waymire, E. C. 1990 *Stochastic Processes with Applications*. John Wiley & Sons.

Buizza, R., Miller, M. & Palmer, T. N. 1999 Stochastic representation of model uncertainties in the ECMWF ensemble prediction system. *Q. J. R. Meteorol. Soc.*, **125**, 2887–2908 (doi:10.1256/smsqj.56005).

Chechkin, A.V., Klafter, J., Gonchar *et al.* 2003 Bifurcation, bimodality, and finite variance in confined Lévy flights. *Phys. Rev. E*, **67**, 010102 (doi:10.1103/PhysRevE.67.010102).

Ditlevsen, P. D. 1999 Observation of α-stable noise induced millenial climate changes from an ice-core record. *Geophys. Res. Lett.*, **26**, 1441–1444 (doi:10.1029/1999 GL900252).

Ditlevsen, P. D. 2004 *Turbulence and Climate Dynamics*. J. & R. Frydenberg A/S.

Doob, J. L. 1953 *Stochastic Processes*. John Wiley & Sons.

Dozier, L. B. & Tappert, F. D. 1978a Statistics of normal mode amplitudes in a random ocean. 1. Theory. *J. Acoust. Soc. Am.*, **63**, 353–365 (doi:10.1121/1.381746).

Dozier, L. B. & Tappert, F. D. 1978b Statistics of normal mode amplitudes in a random ocean. 2. Computations. *J. Acoust. Soc. Am.*, **64**, 533–547 (doi:10.1121/1.382005).

Dybiec, B., Gudowska-Nowak, E. & Hanggi, R. 2006 Lévy–Brownian motion on finite intervals: mean first passage time analysis. *Phys. Rev. E*, **73**, 046104 (doi:10.1103/PhysRevE.73.046104).

Eliazar, I. & Klafter, J. 2003 Lévy-driven Langevin systems: targeted stochasticity. *J. Stat. Phys.*, **111**, 739–767 (doi:10.1023/A:1022894030773).

Ewald, B. D. & Penland, C. 2008 Numerical generation of stochastic differential equations in climate models. In *Handbook of Numerical Analysis: Computational Methods for the Atmosphere and the Oceans*, ed. R. Temam & J. Tribbia. Elsevier.

Ewald, B. D. & Temam, R. 2003 Analysis of stochastic numerical schemes for the evolution equations of geophysics. *Appl. Math. Lett.*, **16**, 1223–1229 (doi:10.1016/S0893–9659(03)90121–2).

Ewald, B. D. & Temam, R. 2005 Numerical analysis of stochastic schemes in geophysics. *SIAM J. Numer. Anal.*, **42**, 2257–2276 (doi:10.1137/S0036142902418333).

Ewald, B. D., Penland, C. & Temam, R. 2004 Accurate integration of stochastic climate models. *Mon. Weather Rev.*, **132**, 154–164 (doi:10.1175/1520–0493(2004)132<0154:AIOSCM>2.0.CO;2).

Feller, W. 1966 *An Introduction to Probability Theory and Its applications*, vol. II. John Wiley & Sons.

Fjørtoft, R. 1953 On the changes in the spectral distribution of kinetic energy for two-dimensional nondivergent flow. *Tellus*, **5**, 225–230.

Flügel, M., Chang, P. & Penland, C. 2004 The role of stochastic forcing in modulating ENSO predictability. *J. Climate*, **17**, 3125–3140 (doi:10.1175/1520–0442(2004)017<3125:TROSFI>2.0.CO;2).

Gardiner, C. W. 1985 *Handbook of Stochastic Methods*. Berlin, Springer-Verlag.

Gnendenko, B. V. & Kolmogorov, A. N. 1954 *Limit Distributions for Sums of Independent Random Variables*. Addison-Wesley.

Hasselmann, K. 1976 Stochastic climate models. 1. Theory. *Tellus*, **28**, 473–485.

Horsthemke, W. & Lefever, R. 1984 *Noise-Induced Transitions: Theory and Applications in Physics, Chemistry, and Biology*. Springer-Verlag.

Hurst, H. 1951 Long-term storage capacity of reservoirs. *Proc. Inst. Civil Eng.*, **5**, 519–577.

Khasminskii, R. Z. 1966 A limit theorem for solutions of differential equations with random right-hand side. *Theory Prob. Appl.*, **11**, 390–406 (doi:10.1137/1111038).

Kloeden, P. E. & Platen, E. 1992 *Numerical Solution of Stochastic Differential Equations*. Springer-Verlag.

Kohler, W. & Papanicolaou, F. C. 1977 Wave propagation in a randomly inhomogeneous ocean. In *Wave Propagation and Underwater Acoustics*, ed. J. B. Keller & J. S. Papadakis. Lecture Notes in Physics, vol. 70, ch. IV, pp. 153–223. Springer-Verlag.

Lévy, P. 1937 *Theorie de l'addition des variables aleatoires*. Gauthier-Villars.

Majda, A. J., Timofeyev, I. & Vanden-Eijnden, E. 1999 Models for stochastic climate prediction. *Proc. Natl. Acad. Sci. USA*, **96**, 14687–14691 (doi:10.1073/pnas.96.26.14687).

Matsumoto, M. & Nishimura, T. 1998 Mersenne twister: a 623-dimensionally equidistributed uniform pseudorandom number generator. *ACM Trans. Model. Comput. Simul.*, **8**, 3–30 (doi:10.1145/272991.272995).

Müller, D. 1987 Bispectra of sea surface temperature anomalies. *J. Phys. Oceanogr.*, **17**, 26–36 (doi:10.1175/1520–0485(1987)017<0026:BOSSTA>2.0.CO;2).

Nolan, J. P. 2007 *Stable Distributions – Models for Heavy Tailed Data*. Birkhauser.

Papanicolaou, G. C. & Kohler, W. 1974 Asymptotic theory of mixing stochastic differential equations. *Commun. Pure Appl. Math.*, **27**, 641–668.

Penland, C. 1985 Acoustic normal mode propagation through a three-dimensional internal wave field. *J. Acoust. Soc. Am.*, **78**, 1356–1365 (doi:10.1121/1.392906).

Penland, C. 1989 Random forcing and forecasting using principal oscillation pattern analysis. *Mon. Weather Rev.*, **117**, 2165–2185 (doi:10.1175/1520–0493(1989)117<2165:RFAFUP>2.0.CO;2).

Penland, C. 1996 A stochastic model of IndoPacific sea surface temperature anomalies. *Physica D*, **98**, 534–558 (doi:10.1016/0167–2789(96)00124–8).

Penland, C. & Sardeshmukh, P. D. 1995 The optimal growth of tropical sea surface temperature anomalies. *J. Climate*, **8**, 1999–2024 (doi:10.1175/1520–0442(1995) 008<1999:TOGOTS>2.0.CO;2).

Press, W. H., Teukolsky, S. A., Vetterling, W. T. & Flannery, B. P. 1992 *Numerical Recipes in Fortran: the Art of Scientific Computing*. Cambridge University Press.

Protter, P. 2005 *Stochastic Integration and Differential equations*. Springer-Verlag.

Protter, P. & Talay, D. 1997 The Euler scheme for Lévy-driven stochastic differential equations. *Ann. Prob.*, **25**, 393–423 (doi:10.1214/aop/1024404293).

Purcell, E. J. 1972 *Calculus with Analytic Geometry*. Meredith Corporation.

Rümelin, W. 1982 Numerical treatment of stochastic differential equations. *SIAM J. Numer. Anal.*, **19**, 605–613.

Saha, S. *et al.* 2006 The NCEP climate forecast system. *J. Climate*, **19**, 3483–3517 (doi:10.1175/JCLI3812.1).

Sardeshmukh, P. D. & Hoskins, B. J. 1988 The generation of global rotational flow by steady idealized tropical divergence. *J. Atmos. Sci.*, **45**, 1228–1251 (doi:10.1175/1520–0469(1988)045<1228:TGOGRF>2.0.CO;2).
Sardeshmukh, P. D. & Sura, P. 2009. Reconciling non-Gaussian climate statistics with linear dynamics. *J. Climate*, **22**, 1193–1207.
Sardeshmukh, P. D., Penland, C. & Newman, M. 2001 Rossby waves in a fluctuating medium. *Progress in Probability Stochastic Climate Models*, vol. 49, ed. P. Imkeller & von J.-S. Storch. Progress in Probability. Birkhaueser.
Shutts, G. 2005 Kinetic energy backscatter for NWP models and its calibration. In Proceedings of the ECMWF workshop on representation of subscale processes using stochastic–dynamic models, Reading, UK, 6–8 June.
Sura, P. & Penland, C. 2002 Sensitivity of a double-gyre ocean model to details of stochastic forcing. *Ocean Modelling*, **4**, 327–345 (doi:10.1016/S1463–5003(02)00008–2).
Sura, P. & Sardeshmukh, P. D. 2008 A global view of non-Gaussian SST variability. *J. Phys. Oceanogr.*, **38**, 639–647 (doi:10.1175/2007JPO3761.1).
Viecelli, J. A. 1998 On the possibility of singular low-frequency spectra and Lévy law persistence in the planetary-scale turbulent circulation. *J. Atmos. Sci.*, **55**, 677–687 (doi:10.1175/1520–0469(1998)055<0677:OTPOSL>2.0.CO;2).
Weron, A. & Weron, R. 1995 *Computer Simulation of Lévy α-Stable Variables and Processes*. Lecture Notes in Physics, vol. 457, pp. 379–392. Springer-Verlag.
Wilks, D. S. 1995 *Statistical Methods in the Atmospheric Sciences*. Academic Press.
Winkler, C. R., Newman, M. & Sardeshmukh, P. D. 2001 A linear model of wintertime low-frequency variability. Part I: formulation and forecast skill. *J. Climate*, **14**, 4474–4494 (doi:10.1175/1520–0442(2001)014<4474:ALMOWL>2.0.CO;2).
Wong, E. & Zakai, M. 1965 On the convergence of ordinary integrals to stochastic integrals. *Ann. Math. Stat.*, **36**, 1560–1564 (doi:10.1214/aoms/1177699916).
Zebiak, S. & Cane, M. 1986 A model El Niño-Southern Oscillation. *Mon. Weather Rev.*, **115**, 2262–2278 (doi:10.1175/1520–0493(1987)115<2262:AMENO>2.0.CO;2).

6

First passage time analysis for climate prediction

PETER C. CHU

Climate prediction is subject to various input uncertainties such as initial condition errors (predictability of the first kind), boundary condition (or model parameter) errors (predictability of the second kind), and combined errors (predictability of the third kind). Quantification of model predictability due to input uncertainties is a key issue leading to successful climate prediction. The first passage time (FPT), defined as the time period when the prediction error first exceeds a predetermined criterion (i.e. the tolerance level) can be used to quantify the model predictability. A theoretical framework on the basis of the backward Fokker–Planck equation is developed to determine the probability distribution function of FPT. Furthermore, the FPT analysis can also be used for climate index prediction.

6.1 Introduction

Complexity in climate systems makes prediction difficult. Numerical modelling and climate index prediction are often used. For numerical climate modelling, a practical question is commonly asked: How long is a climate model valid since being integrated from its initial state? To answer this question, the instantaneous error (IE, defined as the difference between the prediction and reality), must be investigated. It is widely recognised that the IE is caused by the three types of model uncertainties (Lorenz 1984): (a) measurement errors, (b) model errors such as discretisation and uncertain model parameters and (c) chaotic dynamics. Measurement errors cause uncertainty in initial and/or boundary conditions. Such a method using the IE evolution with respect to uncertain model input is called the forward approach.

For the forward approach, the small amplitude stability analysis (linear error dynamics) is used with the IE growth rate and the corresponding e-folding time scale as the measures for evaluating the model predictability due to initial condition error (the first kind of predictability). The IE growth rate is usually estimated by either the leading (largest) Lyapunov exponent or the amplification factors calculated from the leading singular vectors (SVs) (Dalcher & Kalnay 1987). However, the errors may grow to finite amplitudes such as in medium-range prediction (Vukicevic 1991), in forecasting using 'imperfect' models (Bofetta *et al.* 1998) and in models with open boundaries (Chu 1999). The linear assumption is no longer applicable and nonlinear effects should be considered.

One way to simplify the climate system is to represent low-frequency variability of atmospheric circulations by climate indices, i.e. teleconnection patterns, such as the Arctic Oscillation (AO), Antarctic Oscillation (AAO), North Atlantic Oscillation (NAO), Pacific/North American Pattern (PNA) and Southern Oscillation (SO). Temporally varying indices, $s(t)$, were calculated for these patterns. Here, t denotes time. Among them, the SO index (SOI) was first to show an equivalent barotropic seesaw in atmospheric pressure between the southeastern tropical Pacific and the Australian–Indonesian regions (Walker & Bliss 1937). A popular formula for calculating the monthly SOI is proposed by the Australian Bureau of Meteorology,

$$s(t) = 10 \times \frac{p_{\rm diff}(t) - \langle p_{\rm diff} \rangle}{SD(p_{\rm diff})}. \tag{6.1}$$

Here, $p_{\rm diff}$ is the mean sea-level pressure of Tahiti minus that of Darwin for that month; $\langle p_{\rm diff} \rangle$ is the long-term average of $p_{\rm diff}$ for the month in question; and $SD(p_{\rm diff})$ is the long-term standard deviation of $p_{\rm diff}$ for the month in question. Detailed information can be found at http://www.cpc.noaa.gov/.

How to predict the climate indices effectively has practical significance because of their connections to large-scale atmospheric circulation. Usually, these indices are treated as time series and statistical predictions are conducted (forward method). For example, singular spectrum analysis (Keppenne & Ghil 1992), wavelet analysis (Torrence & Campo 1998), and nonlinear analogue analysis (Drosdowsky 1994) were used to obtain the dominant frequencies of the SOI time series. The power-law correlations were found for the self-affine properties of SOI (Ausloos & Ivanova 2001). Such methods may be called the forward index prediction.

An alternative 'backward' method may be used for numerical climate modelling, index prediction and model predictability detection. This method predicts a typical time span (τ) needed for the IE to first exceed a pre-determined criterion (i.e. the tolerance level) in numerical modelling or for a climate index to exceed a given increment. Such a time span (τ) is defined as the first passage time (FPT), which is widely used in many disciplines such as nonlinear dynamical systems (e.g. Ivanov *et al.* 1994), physics and chemistry (e.g. Rangarajan & Ding 2000), biology and economics (e.g. Chu 2006), etc., but not in meteorology and climatology until recently when the FPT was used to evaluate ocean–atmospheric model predictability (Chu *et al.* 2002a,b,c) and to predict the climate indices (Chu 2008). In this chapter, we will introduce the FPT concept and its broad application in climate studies. The rest of the chapter is outlined as follows. Sections 6.2 and 6.3 discuss the forward and backward approaches. Sections 6.4 and 6.5 describe the backward Fokker–Planck equation (for FPT) and the moments of FPT. Section 6.6 presents the stochastic stability and Section 6.7 shows the power decay law of FPT and its moments. Section 6.8 depicts identification of the Lagrangian stability using FPT analysis and Section 6.9 presents the conclusions.

6.2 Forward approach

The forward approach is commonly used in climate studies. Here, two examples are listed for comparison to the backward approach.

6.2.1 Climate index prediction

Monthly varying climate indices from the National Oceanic and Atmospheric Administration (NOAA) Climate Prediction Center show randomness (Fig. 6.1) with poor predictability because their phases and amplitudes are rather unpredictable; both involve many (time and space) scales which are often intrinsic to chaotic behaviour. The forward approach predicts the change of the index (z) at time t with a given temporal increment τ from analysing single or multiple time series. Due to the stochastic nature, the probability density function (PDF), $p[z(t), \tau]$, should first be constructed. Collette & Ausloos (2004) analysed the NAO monthly index (single time series) from 1825 to 2002 and found that the long-range time correlations are similar to Brownian fluctuations.

The distribution functions of the NAO monthly index fluctuations have a form close to a Gaussian, for all time lags. This indicates the lack of predictive power of the present NAO monthly index. Lind *et al.* (2005) used the standard Markov analysis to get the Chapman–Kolmogorov equation for the conditional PDF of the increments z of the NAO index over different time intervals τ and to compute the diffusion and drift coefficients ($D^{(1)}$, $D^{(2)}$) from the first two moments of such a probability distribution. The random variable $z(t)$ is found to satisfy the Langevin equation

$$\frac{\mathrm{d}z(t)}{\mathrm{d}t} = D^{(1)}\left[z(t), t\right] + \eta(t)\sqrt{D^{(2)}\left[z(t), t\right]} \tag{6.2}$$

where $\eta(t)$ is a fluctuating δ-correlated force with Gaussian statistics, i.e.

$$\langle \eta(t) \rangle = 0, \ \langle \eta(t)\eta(t') \rangle = q^2 \delta(t - t'). \tag{6.3}$$

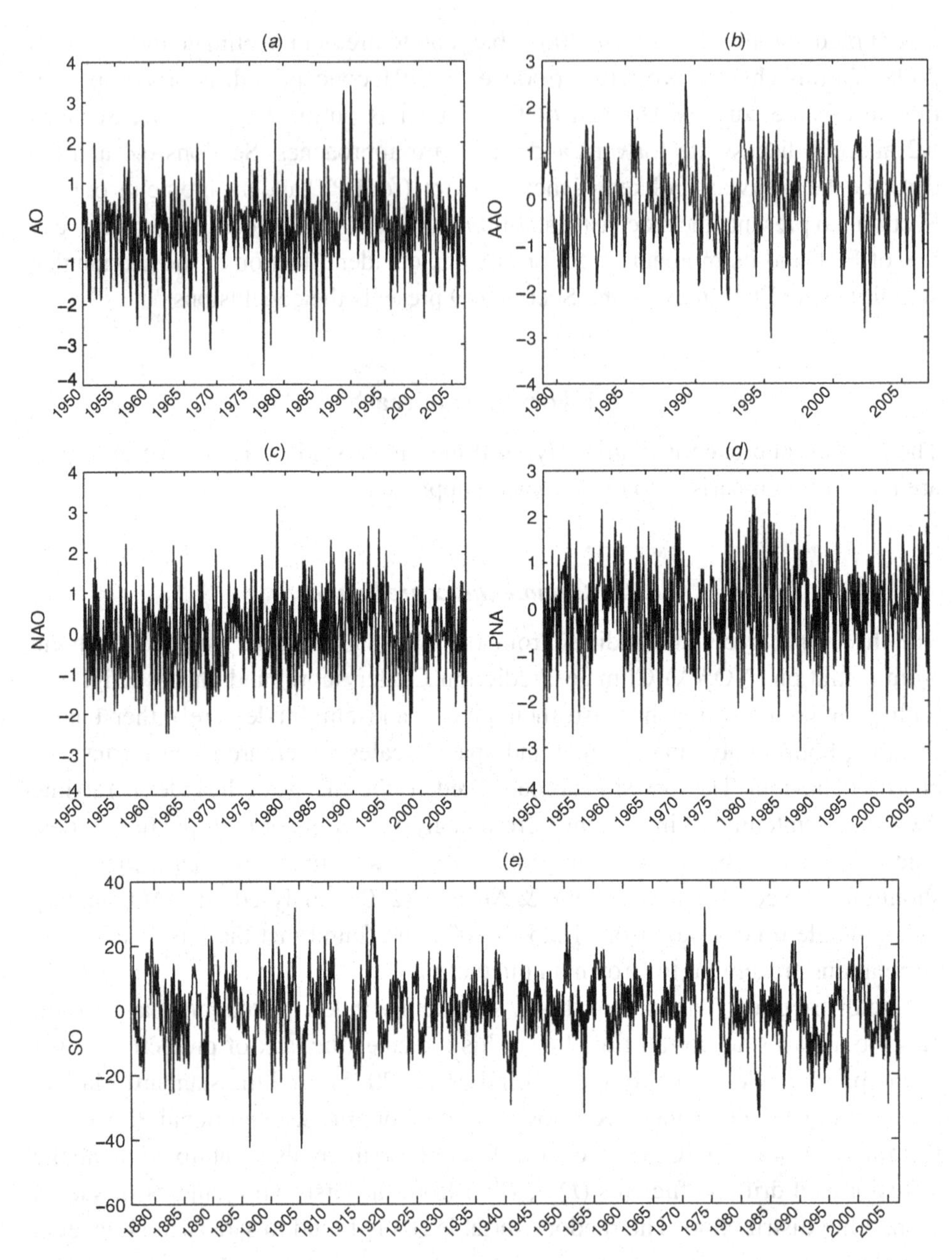

Figure 6.1 The monthly mean climate index data obtained from NOAA Climate Prediction Center for: (*a*) Arctic Oscillation (AO), (*b*) Antarctic Oscillation (AAO), (*c*) North Atlantic Oscillation (NAO), (*d*) Pacific/North American Pattern (PNA) and (*e*) Southern Oscillation (SO). Detailed information can be found at http://www.cpc.noaa.gov/.

Maharaj and Wheeler (2005) predict the daily bi-variate index (multiple time series) of the Madden–Julian oscillation using seasonally varying vector autoregressive models. Although this model shows no strong skill advantage over a lagged regression technique, it has the convenience of employing only a single set of equations to make predictions for multiple forecast horizons.

6.2.2 Model predictability

Let climate evolution in domain Ω be described by the dynamical state vector $\mathbf{X}(\mathbf{x}, t)$ composed from variables such as pressure, velocity, temperature, humidity, salinity and so on, with $(\mathbf{x}, t)$ denoting the spatial and temporal coordinates. In general this vector may only be approximately estimated from the observations. In order to overcome this complexity the estimation of model forecast skill is reduced into the sensitivity analysis of the so-called 'reference solution' $\mathbf{X}_{\mathrm{ref}}(\mathbf{x}, t)$ to perturbations of model input (Lacarra & Talagrand 1988). The numerical climate model is usually represented by

$$\mathrm{d}\mathbf{Y}/\mathrm{d}t = \mathbf{F}(\mathbf{Y}, t), \tag{6.4}$$

with the initial condition

$$\mathbf{Y}(\mathbf{x}, t_0) = \mathbf{Y}_0, \tag{6.5}$$

and the boundary conditions

$$\mathbf{G}(\mathbf{Y}(\mathbf{x}, t)) = \mathbf{Q}, \tag{6.6}$$

where $\mathbf{F}(\mathbf{Y}, t)$ is a nonlinear operator describing forcing, nonlinear interactions and subgrid parameterisations, $\mathbf{Y}_0$ is the initial vector, and $\mathbf{G}$ determines fluxes across the boundary of the calculation domain Ω. When the IE is defined by

$$\mathbf{Z}(\mathbf{x}, t) = \mathbf{Y}(\mathbf{x}, t) - \mathbf{X}_{\mathrm{ref}}(\mathbf{x}, t), \quad \mathbf{Z}(\mathbf{x}, t_0) = \mathbf{Z}_0,$$

the model predictability is estimated by a dimensionless functional

$$J(\mathbf{Z}, \mathbf{W}, t) = \left\{\mathbf{Z}^{\mathrm{T}}\mathbf{W}\mathbf{Z}\right\}, \tag{6.7}$$

where $\mathbf{W}(\mathbf{x}, t)$ is the weight matrix, the superscript 'T' denotes the transpose operator, and the brackets denote the volume integration $\{\ldots\} = \oiiint_{\Omega} \ldots \mathrm{d}V$.

From the physical point of view, the functional J satisfies the two following inequalities,

$$\begin{aligned} \langle J(\mathbf{Z}, \mathbf{W}, t)\rangle &\geq J_{\mathrm{noise}} = \left\langle\left\{\mathbf{Z}_{\mathrm{noise}}^{\mathrm{T}}\mathbf{Z}_{\mathrm{noise}}\right\}\right\rangle, \\ \langle J(\mathbf{Z}, \mathbf{W}, t)\rangle &\leq \varepsilon_0 \left\{\mathbf{X}_{\mathrm{ref}}^{\mathrm{T}}\mathbf{X}_{\mathrm{ref}}\right\}, \end{aligned}$$

where the angular brackets show the ensemble averaging; and $\mathbf{Z}_{\text{noise}}$ characterises the level of non-removed noise existing in the model such as the natural heating noise. In general the level of the intrinsic model noise is determined by the chosen parameterisation schemes for unresolved scales (Chu *et al.* 2002b); and $\varepsilon = \varepsilon_0\{\mathbf{X}_{\text{ref}}^{\text{T}}\mathbf{X}_{\text{ref}}\}$ is the allowed model uncertainty (tolerance level) for the prediction.

The forward approach is to find the temporal evolution of the functional $J(\mathbf{Z}, \mathbf{W}, t)$. The model evaluation becomes a stability analysis on small-amplitude errors in terms of either the leading (largest) Lyapunov exponent or can be calculated from the leading singular vectors (Farrell & Ioannou 1996). The faster the IE grows, the shorter the e-folding scale is. For finite-amplitude IE, however, the linear stability analysis becomes invalid. The statistical analysis of the instantaneous error (both small- and finite-amplitudes) growth (Nicolis 1992), the information–theoretical principles for the predictability power (Schneider & Griffies 1999) and the ensembles for forecast skill identification (Toth *et al.* 2001) become useful. Such methods may be called the forward model predictability.

The linear tangent model of the IE dynamics is commonly used to estimate the prediction time scale through leading (largest) Lyapunov exponents or the amplification factors calculated from the leading singular vectors (the finite-time Lyapunov exponents). Linearisation of the governing equations is realistic when the energy of IE is considerably less than the energy of the reference solution (Palmer 2001). With the linear tangent model, the energy is transferred from the (basic) flow to the perturbations (error field), not vice versa. However, many studies show the existence of upscale processes when the energy is transferred the small scale to large scale perturbations. In this case the nonlinear IE dynamics should be considered even if formally the energy of the reference solution dominates. Thus, approaches that automatically incorporate both linear and nonlinear perspectives of IE should be considered in ocean models, especially in regional and coastal oceans.

Both linear and nonlinear perspectives of IE and corresponding prediction timescales are computed through the probabilistic approach based on the PDF of IE. In general, such a PDF satisfies the Liouville or Fokker–Planck equation (Nicolis 1992). In practice, because the dimension of error subspace is high for large models, the low-order moments of PDF, such as mean and variance, can be estimated only through ensemble averaging techniques (Toth & Kalnay 1997), i.e. practically the error distribution is assumed to be approximately Gaussian. That simplifies the mathematics but often does not agree with the physics of circulation in small semi-closed seas. Using the Gulf of Lion as an example, Auclair *et al.* (2003) pointed out that in coastal ocean modelling, the IE statistics remain largely unknown (they are probably not even exactly Gaussian) and such a well-behaved homogeneous (or even regionally homogeneous) statistical description could seem rather over-simplistic.

6.3 Backward approach and FPT

6.3.1 Climate index prediction

The climate indices change either positively or negatively at a given time (Fig. 6.1). Then, of course, it is of interest to predict the exact change at a point in time; however, this is not possible. Therefore, the best one can do, from a statistical point of view, is to predict the time that is probabilistically favourable for the given index change. This optimal time, as we will see, is determined by the maximum of the first passage time distribution, i.e. the optimal FPT. Given a fixed value of an index reduction (z), the corresponding time span (positive) is estimated for which the index reduction (similar for index enhancement),

$$\gamma_{\Delta t}(t) = s(t + \Delta t) - s(t), \tag{6.8}$$

reaches the level ρ for the first time (Chu 2008),

$$\tau_\rho(t) = \inf\{\Delta t > 0 | \gamma_{\Delta t}(t) \leq -\rho\}, \tag{6.9}$$

which is a random variable and called the FPT. As the index data run through the past history, the FPT can be computed and the PDF of FPT, $p(\tau_\rho)$, can be constructed. Fig. 6.2 shows $p(\tau_\rho)$ with $\rho = 1.9$ for AO, AAO, NAO, PNA indices and with $\rho = 25$ for SOI. The solid curve is the theoretical PDF, which will be discussed in Section 6.4.

The PDF of FPT exhibits a rather well defined and pronounced maximum, followed by an extended tail for very long FPTs indicating a non-zero and important probability of large passage times (note that the τ_ρ-axis is logarithmic). These long (toward El Niño for SOI) FPTs reflect periods where the tropical Pacific is in a strong El Niño phase and needs a long period of time before finally coming to an even stronger El Niño. The short FPTs on the other hand – those around the maximum – are in La Niña periods, which appear to be the most common scenario. A cumulative distribution function (CDF) is introduced for transition times larger than τ_ρ, i.e.

$$P(\tau_\rho) = \int_{\tau_\rho}^{\infty} p(\tau)\mathrm{d}\tau. \tag{6.10}$$

To better understand the tail of this distribution, various values of ρ are considered. If this level is small enough, it is likely that the index reduction will break through the level after the first month, while larger FPTs will become more and more unlikely. However, the probability for a large FPT value will not be zero; if, say, there is a small-level ρ, then a period of strong El Niño will result in a $\tau_\rho(t)$ that might be considerably larger than one month since it takes time to recover from the El Niño event. For instance, after the 1992 El Niño event, it took five years to reach a new El Niño event in 1997.

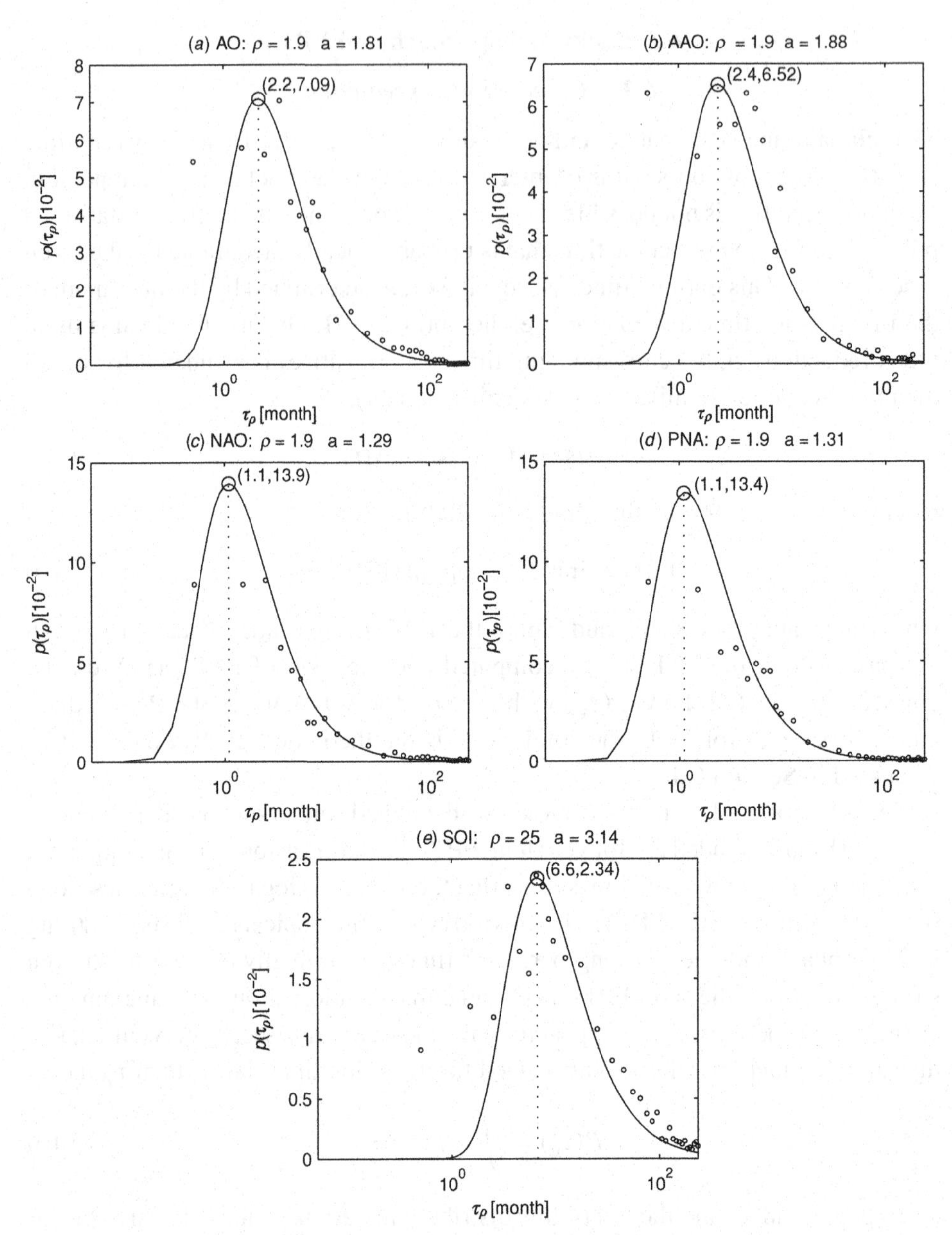

Figure 6.2 First passage time density functions with particular index reduction ρ for (*a*) Arctic Oscillation (AO), (*b*) Antarctic Oscillation (AAO), (*c*) North Atlantic Oscillation (NAO), (*d*) Pacific/North American Pattern (PNA) and (*e*) Southern Oscillation (SO) (after Chu 2008). Here, the parameter a represents the most favourable FPT, $\tau_\rho^{(\max)} = 2a^2/3$, which is the first value in parentheses near the curve maxima. The second value is its probability density.

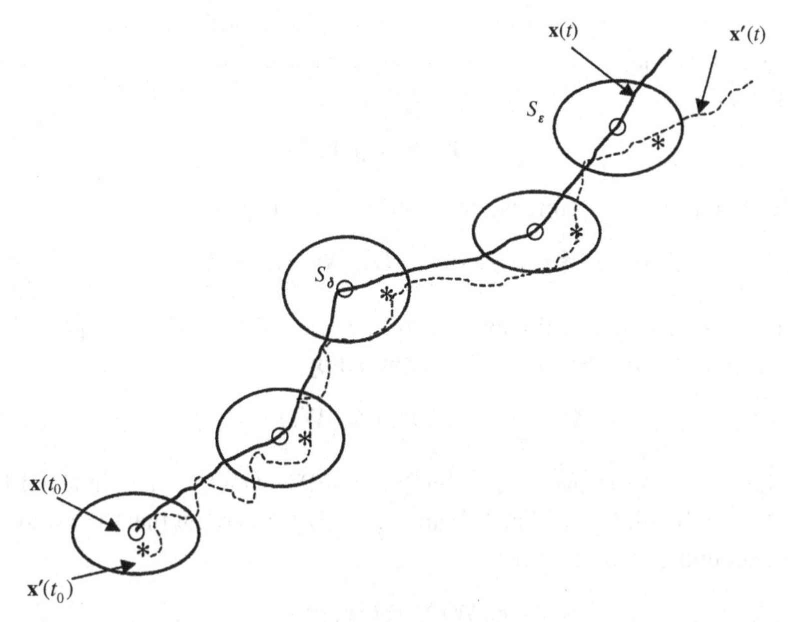

Figure 6.3 Trajectories of model prediction **y** (solid curve) and reality **x** (dashed curve) and error ellipsoid $S_\varepsilon(t)$ centred at y. The reality and prediction trajectories are denoted by '$*$' and 'o', respectively. A valid prediction is represented by a time period $(t - t_0)$ at which the error first goes out of the ellipsoid $S_\varepsilon(t)$ (after Chu *et al.* 2002b).

6.3.2 Model predictability

The backward approach is to find a time period τ (i.e. FPT) at which the functional $J(\mathbf{Z}_0, \mathbf{W}, t)$ reaches a predetermined tolerance level (ε) for the first time,

$$\tau\,(\mathbf{Z}_0, \mathbf{W}, \varepsilon) = \inf_{t \geq 0}(t\,|J(\mathbf{Z}_0, \mathbf{W}, t) > \varepsilon^2). \tag{6.11}$$

The prediction is valid if the state point $\mathbf{x}(t)$ is situated inside the ellipsoid (S_ε, called tolerance ellipsoid) with centre at $\mathbf{y}(t)$ (i.e. the prediction) and size ε. When $\mathbf{y}(t)$ coincides with $\mathbf{x}(t)$, the model has perfect prediction. The prediction is invalid if the state point $\mathbf{x}(t)$ first touches the boundary of the tolerance ellipsoid; the FPT for prediction is the time period from the initial state (Fig. 6.3).

The FPT is a random variable when the model has stochastic forcing or the initial condition has a random error. Its statistics such as the probability density function, mean and variance can represent how long the model can predict. For simplicity and without loss of generality, a one-dimensional population model is used for illustration. Clearly, the FPT defines the model predictability on the condition that any returns of model predictability (i.e. model error smaller than the tolerance level ε after passing the FPT) do not contribute to the prediction skill (the shaded zones in Fig. 6.4*a*). The mean FPT differs from the e-folding or the doubling time when J oscillates or is random. To compare the e-folding and the FPT, we need to assume in (6.11) that

$$J_{\text{norm}} = (\mathbf{Z}_0, \mathbf{Z}_0) \text{ and } \hat{\varepsilon}^2 = e^2.$$

The e-folding time is the time when $\langle J \rangle$ crosses e^2 (Fig. 6.4*b*),

$$\tau_{\text{e}}(\mathbf{W}) = \max_{t \geq 0}(t \,|\, \langle J(\mathbf{Z}_0, \mathbf{W}, t) \rangle \leq e^2), \tag{6.12}$$

where the brackets denote the average over the ensemble of initial perturbations $\mathbf{Z}_0$. The mean FPT for the same e is computed by

$$\langle \tau(\mathbf{W}, e) \rangle = \langle \inf_{t \geq 0}(t | J(\mathbf{Z}_0, \mathbf{W}, t) > e^2) \rangle, \tag{6.13}$$

where the averaging is over the ensemble of FPTs $(\tau_1, \ldots, \tau_N)$ induced by the ensemble of $\mathbf{Z}_0$ (Fig. 6.4*c*). Chu & Ivanov (2005) pointed out that the mean FPT is the lower bound of e-folding time:

$$\tau_{\text{e}}(\mathbf{W}) \geq \langle \tau(\mathbf{W}, e) \rangle.$$

6.4 Backward Fokker–Planck equation

Despite the linearity or nonlinearity of the original dynamical system, the PDF of FPT always satisfies a linear equation called the backward Fokker–Planck equation. For a random variable z satisfying the Langevin equation (6.2), the corresponding backward Fokker–Planck equation (Gardiner 1985; Chu *et al.* 2002a; Chu 2008) is given by

$$\frac{\partial p}{\partial t} - [D^{(1)}(z,t)]\frac{\partial p}{\partial z} - \frac{1}{2}\eta^{(2)} D^{(2)}(z,t)\frac{\partial^2 p}{\partial z^2} = 0, \tag{6.14}$$

which has an analytical solution,

$$p(\tau) = \frac{1}{\sqrt{\pi}}\frac{a}{\tau^{3/2}}\exp\left(-\frac{a^2}{\tau}\right), a > 0, \tau > 0. \tag{6.15}$$

The Lorenz system (1984) is the simplest possible model capable of representing an unmodified or modified Hadley circulation, determining its stability, and, if it is unstable, representing a stationary or migratory disturbance. Projection of the three-dimensional forecast error vector onto the unstable manifold leads to a self-consistent model (Nicolis 1992),

$$\frac{\mathrm{d}\xi}{\mathrm{d}t} = (\sigma - g\xi)\xi + \eta(t)\xi, \quad \xi|_{\mathrm{t=t_0}} = \xi_0, \quad \xi \in (0, \infty), \tag{6.16}$$

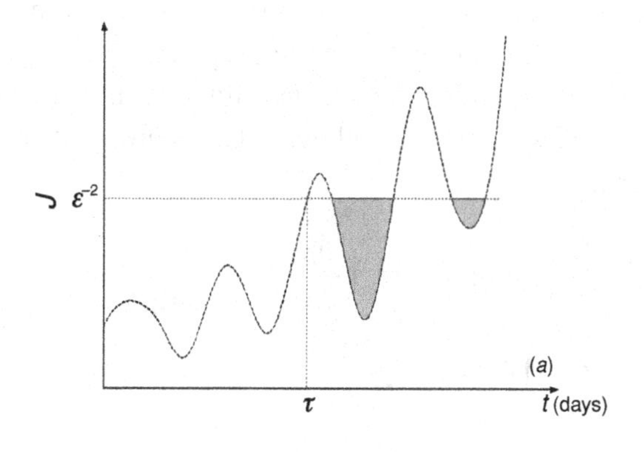

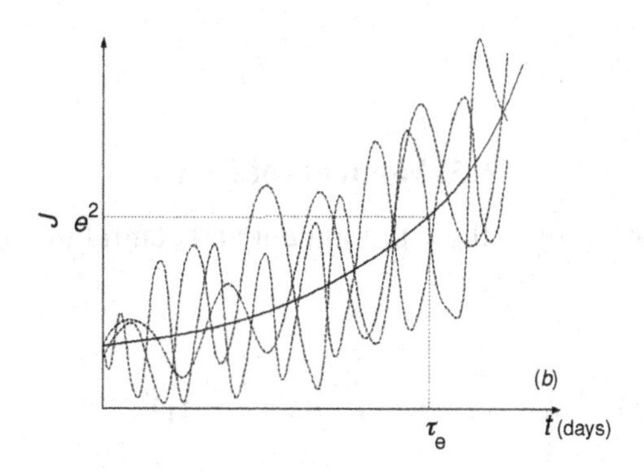

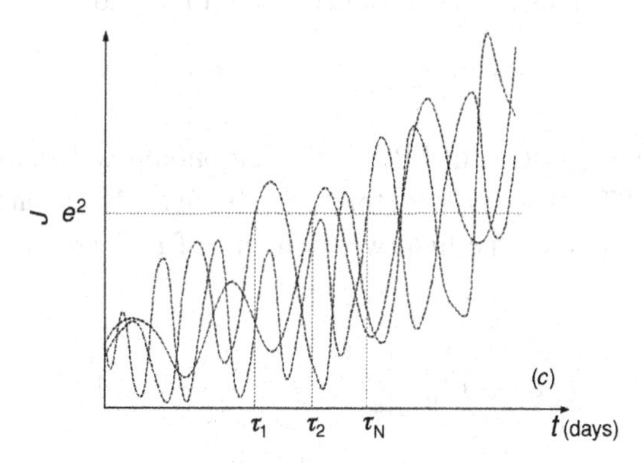

Figure 6.4 The FPT and e-folding time for an oscillating prediction error: (*a*) FPT (τ) computed in an individual forecast (shaded zones show returns of model predictability), (*b*) an ensemble of J (dashed curves), the ensemble averaged J (solid curve) and the e-folding time (τ_e); and (*c*) the ensemble of J (dashed curves) and appropriate ensemble of FPT ($\tau_1, \tau_2, \ldots, \tau_N$) (after Ivanov & Chu 2007a,b).

where ξ is non-dimensional amplitude of error, g is a non-negative, generally time-independent nonlinear parameter whose properties depend on the underlying attractor; and $\eta(t)$ is a fluctuating δ-correlated force with Gaussian statistics [see (6.3)]. The PDF of FPT satisfies the following backward Fokker–Planck equation (Chu *et al.* 2002a,b),

$$\frac{\partial p}{\partial t} - \left(\sigma\xi_0 - g\xi_0^2\right)\frac{\partial p}{\partial \xi^0} - \frac{1}{2}q^2\frac{\partial^2 p}{\partial \xi_0^2} = 0, \tag{6.17}$$

with the initial error (ξ_0) bounded by,

$$\xi_{\text{noise}} \le \xi_0 \le \varepsilon. \tag{6.18}$$

6.5 Moments of FPT

With the known PDF of FPT, the k-th moment and central moment ($k = 1, 2, \ldots$) are calculated by

$$\tau_k = k\int_0^\infty p(\tau)(\tau)^{k-1}\mathrm{d}\tau, \quad \hat{\tau}_k = k\int_0^\infty p(\tau)(\tau - \tau_1)^{k-1}\mathrm{d}\tau, \quad k = 2, \ldots, \infty, \tag{6.19}$$

where τ_1 and $\hat{\tau}_2$ are the mean and variance of the FPT, and

$$\mathrm{SK} \equiv \hat{\tau}_3/\hat{\tau}_2^{3/2}, \mathrm{KU} \equiv \hat{\tau}_4/\hat{\tau}_2^2, \tag{6.20}$$

are the skewness and kurtosis of FPT. For an autonomous dynamical system, $\mathbf{f} = \mathbf{f}(\mathbf{z}_0)$, the PDF of FPT still varies with time [satisfying the Backward Fokker–Planck equation (6.17)]. However, the first two moments of FPT are given by (Chu *et al.* 2002a,b),

$$\left(\sigma\xi_0 - g\xi_0^2\right)\frac{\mathrm{d}\tau_1}{\mathrm{d}\xi_0} + \frac{q^2\xi_0^2}{2}\frac{\mathrm{d}^2\tau_1}{\mathrm{d}\xi_0^2} = -1, \tag{6.21}$$

$$\left(\sigma\xi_0 - g\xi_0^2\right)\frac{\mathrm{d}\tau_2}{\mathrm{d}\xi_0} + \frac{q^2\xi_0^2}{2}\frac{\mathrm{d}^2\tau_2}{\mathrm{d}\xi_0^2} = -2\tau_1, \tag{6.22}$$

with the boundary conditions,

$$\tau_1 = 0, \ \tau_2 = 0 \qquad \text{for } \xi_0 = \varepsilon; \tag{6.23}$$

$$\mathrm{d}\tau_1/\mathrm{d}\xi_0 = 0, \ \mathrm{d}\tau_2/\mathrm{d}\xi_0 = 0 \quad \text{for } \xi_0 = \xi_{\text{noise}}. \tag{6.24}$$

The equations for the moments of FPT (6.21) and (6.22) with the boundary conditions (6.23) and (6.24) have analytical solutions,

$$\tau_1(\bar{\xi}_0, \bar{\xi}_{noise}, \varepsilon) = \frac{2}{q^2} \int_{\bar{\xi}_0}^{1} y^{-\frac{2\sigma}{q^2}} \exp\left(\frac{2\varepsilon g}{q^2} y\right) \left[\int_{\bar{\xi}_{noise}}^{y} x^{\frac{2\sigma}{q^2}-2} \exp\left(-\frac{2\varepsilon g}{q^2} x\right) dx \right] dy, \tag{6.25}$$

$$\tau_2(\bar{\xi}_0, \bar{\xi}_{noise}, \varepsilon) = \frac{4}{q^2} \int_{\bar{\xi}_0}^{1} y^{-\frac{2\sigma}{q^2}} \exp\left(\frac{2\varepsilon g}{q^2} y\right) \left[\int_{\bar{\xi}_{noise}}^{y} \tau_1(x) x^{\frac{2\sigma}{q^2}-2} \exp\left(-\frac{2\varepsilon g}{q^2} x\right) dx \right] dy, \tag{6.26}$$

where

$$\bar{\xi}_0 = \xi_0/\varepsilon, \quad \bar{\xi}_{noise} = \xi_{noise}/\varepsilon$$

are non-dimensional initial-condition error and noise level scaled by the tolerance level ε, respectively. For given tolerance and noise levels (or user input), the mean and variance of FPT can be calculated using (6.25) and (6.26). For the Nicolis error model (6.16) with the same parameter values as in Nicolis (1992),

$$\sigma = 0.64, \quad g = 0.3, \quad q^2 = 0.2. \tag{6.27}$$

Figures 6.5 and 6.6 show the curve plots of $\tau_1(\bar{\xi}_0, \bar{\xi}_{noise}, \varepsilon)$ and $\tau_2(\bar{\xi}_0, \bar{\xi}_{noise}, \varepsilon)$ versus $\bar{\xi}_0$ for four different values of tolerance level, ε (0.01, 0.1, 1 and 2) and four different values of random noise $\bar{\xi}_{noise}$ (0.1, 0.2, 0.4 and 0.6). The following features are obtained: (a) τ_1 and τ_2 decrease with increasing $\bar{\xi}_0$, which implies that the higher the initial error, the lower the predictability (or FPT) is; (b) τ_1 and τ_2 decrease with increasing noise level $\bar{\xi}_{noise}$, which implies that the higher the noise level, the lower the predictability (or FPT) is; and (c) τ_1 and τ_2 increase with the increasing ε, which implies that the higher the tolerance level, the longer the FPT is. Note that the results presented in this subsection are for a given value of stochastic forcing ($q^2 = 0.2$) only.

6.6 Stochastic stability

The stochastic stability addresses the effects of random perturbations on trajectories of a dynamical system and estimates its stability in terms of probabilistic measures, such as expected values or distribution functions. In general, the stochastic stability differs from the predictability. However, if a time scale quantifies the model predictability, and if this scale indicates the time when the forecast uncertainty exceeds some boundary or when information on the initial condition is lost, the

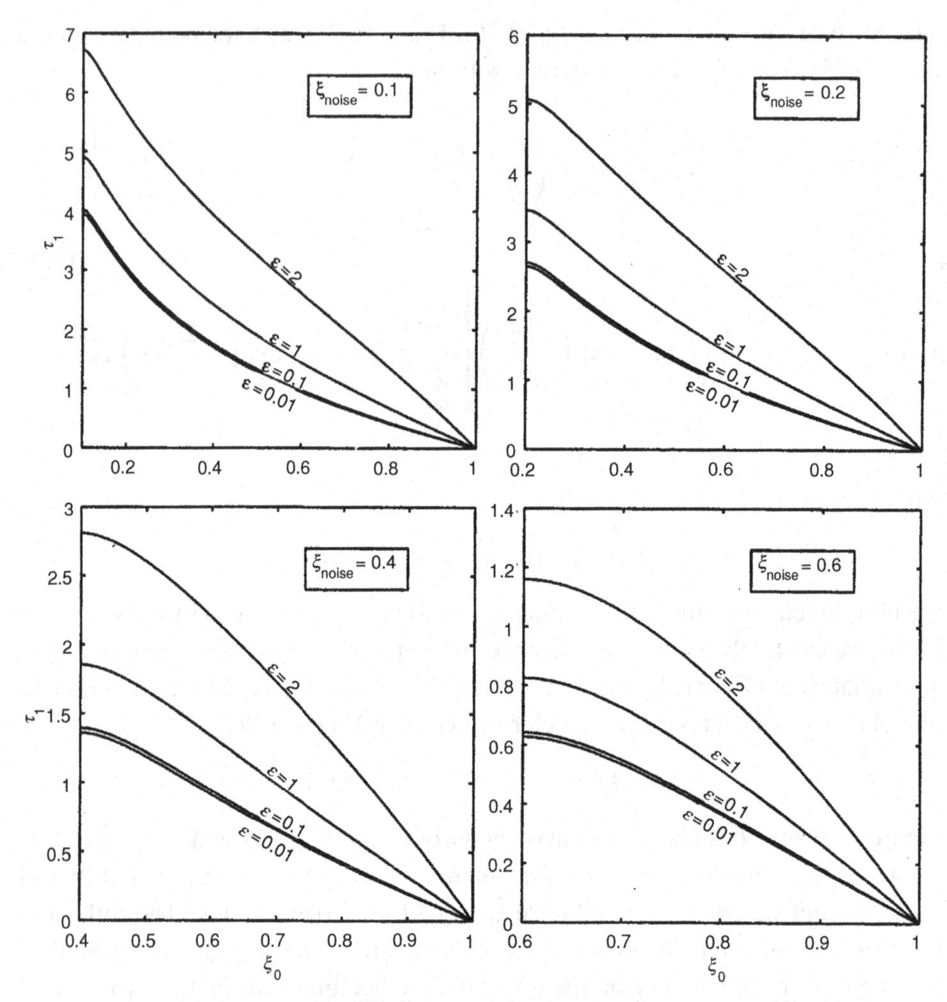

Figure 6.5 Dependence of $\tau_1(\bar{\xi}_0, \bar{\xi}_{\text{noise}}, \varepsilon)$ on the initial condition error $\bar{\xi}_0$ for four different values of ε (0.01, 0.1, 1 and 2) and four different values of random noise $\bar{\xi}_{\text{noise}}$ (0.1, 0.2, 0.4 and 0.6) using the Nicolis (1992) model with stochastic forcing $q^2 = 0.2$ (after Chu *et al.* 2002a).

stochastic stability and predictability are interchangeable. Since these time scales are widely used in meteorology and oceanography, the stochastic stability concept seems to be a useful tool for the predictability analysis of large hydrodynamic models.

Loss of superposition and extreme inhomogeneity are common in nonlinear hydrodynamic models that require local measures of predictability and corresponding time scales (Lorenz 1984). For small-amplitude initial perturbations, the time scales are related to the inverse of the largest Lyapunov exponent estimated by the tangent linear models. The linear approach gives reasonable estimations of

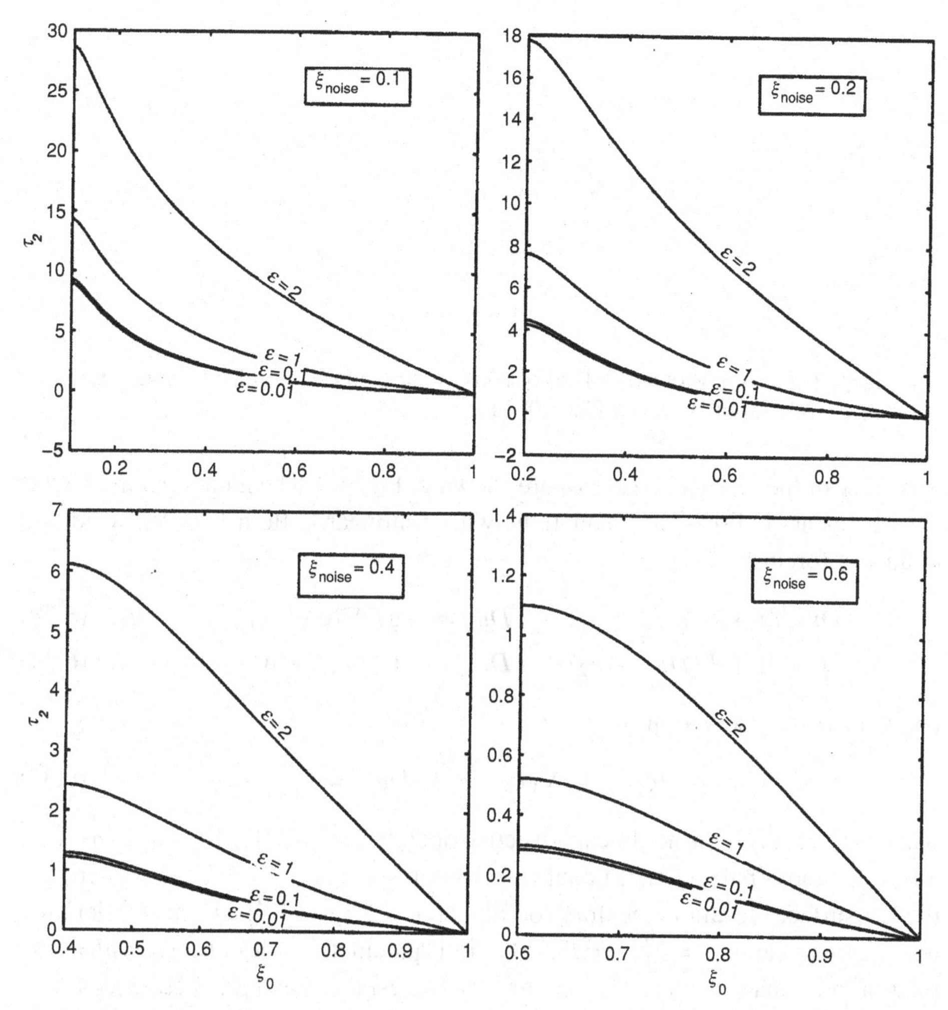

Figure 6.6 Dependence of $\tau_2(\bar{\xi}_0, \bar{\xi}_{\text{noise}}, \varepsilon)$ on the initial condition error $\bar{\xi}_0$ for four different values of ε(0.01, 0.1, 1 and 2) and four different values of random noise $\bar{\xi}_{\text{noise}}$ (0.1, 0.2, 0.4 and 0.6) using the Nicolis (1992) model with stochastic forcing $q^2 = 0.2$ (after Chu *et al.* 2002a).

model predictability in many practical cases. However, it cannot provide critical boundaries on finite-amplitude stability especially in ocean circulation models.

6.6.1 Reference solution

We consider a rectangular semi-closed basin with the horizontal dimensions, $L_1 = 1050$ km and $L_2 = 1000$ km, and with constant depth $H = 2$ km, which is situated on a midlatitude β-plane. The basin has rigid (Γ) and open (Γ') boundaries. The

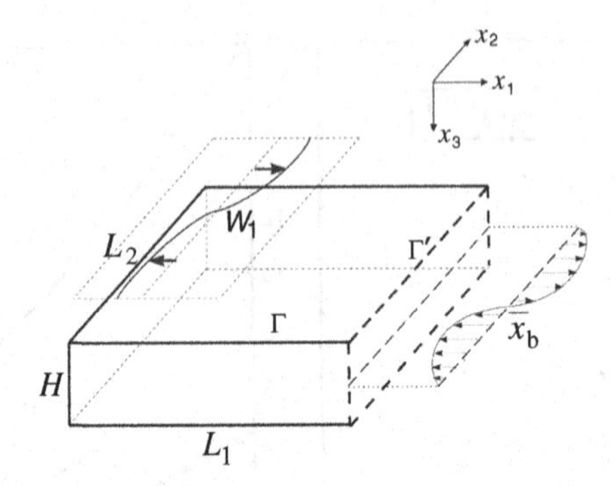

Figure 6.7 Basin geometry. The x_1 and x_2 axes point toward east and north, respectively (after Ivanov & Chu 2007a).

geometry of the basin and its sizes are shown in Fig. 6.7. The numerical model has the nonlinear shallow-water equations with nonlinear bottom friction, wind and boundary forcing

$$\partial Du_1/\partial t + L(Du_2, Du_1) - fDu_2 = -gD\nabla_1\zeta + W_1 - \alpha E^{1/2}u_1, \quad (6.28)$$

$$\partial Du_2/\partial t + L(Du_1, Du_2) + fDu_1 = -gD\nabla_2\zeta + W_2 - \alpha E^{1/2}u_2, \quad (6.29)$$

and the mass conservation equation

$$\partial\zeta/\partial t + (\nabla_1 Du_1 + \nabla_2 Du_2) = 0, \quad (6.30)$$

where $L(\ldots,\ldots)$ is the nonlinear advective operator; $[\nabla_1, \nabla_2] = [\partial/\partial x_1, \partial/\partial x_2]$; u_1 and u_2 are the zonal and meridional velocities, respectively; $D = H + \zeta$, where ζ is the sea-surface elevation; the drag coefficient $\alpha = 2.5 \times 10^{-3}$; g is the acceleration due to gravity; and $E = u_1^2 + u_2^2$. The Coriolis parameter varies linearly with a beta plane approximation $f = f_0 + \beta x_2$, where $f_0 = 2\Omega\sin(\varphi_0)$ and $\beta = (2\Omega/a)\cos(\varphi_0)$. Here, Ω and a are the rate of rotation and the radius of the Earth, respectively; $\varphi_0 = 35°$; for the chosen model parameters: $f_0 = 7.3 \times 10^{-5}$ s^{-1}, $\beta = 2.0 \times 10^{-11}$ m^{-1} s^{-1}. A flow in the semi-closed basin bounded by $\Gamma \cup \Gamma'$ is forced by both the zonal wind forcing W_1 ($W_2 = 0$) varying with latitude as

$$W_1 = -\frac{w_s}{\rho_w}\cos\left(\frac{\pi x_2}{L_2}\right), \quad (6.31)$$

where $\rho_w = 1025$ kg m^{-3}, w_s is the wind stress and $w_s/\rho_w = 1.0 \times 10^{-3}$ m^2 s^{-2}, and by a prescribed net flux [characterised by the normal velocity $\bar{u}_b(x_2, t)$ and surface elevation $\bar{\zeta}_b(x_2, t)$ along the boundary Γ'].

Zero normal velocity and zero Neumann conditions for the surface elevation are imposed on the rigid boundary Γ. The chosen model configuration is suitable for the analysis of ocean model predictability affected by different kinds of stochastic

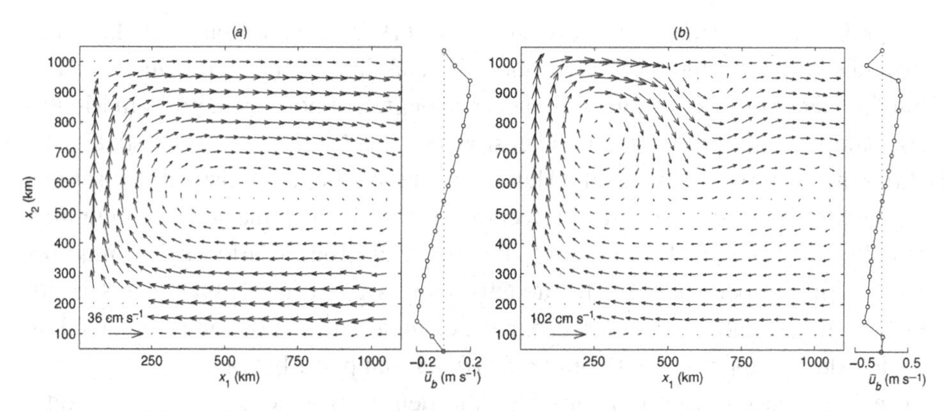

Figure 6.8 Spatial structure of the reference solution at the initial state (*a*) and after integration for 60 days (*b*). Open boundary conditions ($\bar{u}_b$) corresponding to the reference solution are shown to the right of the circulation patterns (after Ivanov & Chu 2007a,b).

uncertainties: errors inserted in initial conditions (Ivanov & Chu 2007a,b), wind (the present study) and open boundary conditions. Cross-correlations between these errors can also be studied. Model (6.28)–(6.30) is similar to that used by Veronis (1966) for the analysis of the nonlinear wind-driven circulation in a closed basin. But in contrast to Veronis (1966) we parameterise bottom friction by the quadratic drag law.

The prescribed non-stationary net flux across the open boundary is computed as explained in Chu *et al.* (1997). The structure of open boundary conditions on day-0 and day-60 is demonstrated in Figs. 6.8*a* and 6.8*b*, respectively. The initial condition represents a non-closed anticyclonic gyre shown in Fig. 6.8*a*. The corresponding initial surface elevation is not shown because its structure is obvious. After 30 days of integration the model reaches a spin up when the spatially averaged kinetic energy oscillates with a period of 120 days. The amplitude of this oscillation reduces with time exponentially with rate of 1000 d^{-1}. The circulation pattern formed after day 30 presents a multigyre structure with maximum velocities up to 0.9 to 1.0 m s^{-1} (Fig. 6.8*b*).

6.6.2 Ensemble prediction

For simplicity, Ivanov & Chu (2007a,b) sampled perturbations from specified, multivariate normal distributions. First, due to simplicity, our hydrodynamic model is able to produce forecast ensembles containing up to $\sim O(10^5)$ perturbations. Second, the Latin hypercube (LHC) design (Latin Hypercube; see http://www.mathworks.com/matlabcentral/fileexchange/4352) was applied to simulate a highly uniform distribution of an initial error in the phase space. Using pure probabilistic arguments, Downing *et al.* (1985) pointed out that the Latin hypercube design is more effective than the classical Monte-Carlo method. For obtaining dense error coverage with the same degree of homogeneity from Monte-Carlo samples and through the LHC design, N^{M_0} and $N(2M_0 + 2)$ statistical realisations are required, respectively. Here, N is the number of statistical realisations necessary to simulate one degree of freedom, and M_0 is the truncated mode number. Typically, $N \sim O(10^3)$. For $M_0 = 15$, the classical Monte-Carlo method requires $\sim O(10^{45})$ initial perturbations for a statistically significant estimate. This is not feasible with available computer resources. Comparable results can be obtained by the LHC design approach with only $\sim O(10^4)$ initial perturbations.

The LHC design approach provides the dense error coverage of the model phase space for $\sim O(10^4)$ initial perturbations. This number of initial perturbations is a trade-off between the ensemble ability to reproduce the main features of IE statistics, and the computational cost. However, the optimal number of initial perturbations (N_{opt}) should be specified for the concrete ocean model. We estimated this number through the Kullback–Leibler (KL) distance (the relative entropy; see White 1994).

The KL distance is a natural distance function from a 'true' probability density, F_∞, to a 'target' probability density, F_N. For continuous density functions, the KL distance is defined as

$$KL_N\,(F_\infty\,|F_N) = \int_0^\infty \mathrm{d}\tau\, F_N(\tau) \log\left[\frac{F_\infty(\tau)}{F_N(\tau)}\right], \tag{6.32}$$

where $F_N(\tau)$ and $F_\infty(\tau)$ are τ-PDFs computed for an N sample ensemble and a hypothetic ensemble with infinite sampling, respectively. In practice, a difference between two distributions is negligible if $KL_N \leq 5.0 \times 10^{-3}$ (White 1994).

To calculate the KL distance we suppose $F_\infty = F_{100000}$ because only small differences in τ-statistics estimated from ensembles of 5.0×10^3, 1.0×10^4, 2.0×10^4, 5.0×10^4 and 1.0×10^5 samples have been observed. The KL_N rapidly reduces with N from 2.0×10^{-1} ($N = 20$) to 4.0×10^{-3} ($N = 10^3$). Therefore, N_{opt} was chosen as 10^3.

6.6.3 Stochastic perturbations

Two types of (stochastic) initial perturbations are introduced: two-dimensional isotropic Gaussian white noise [white-noise-like perturbations (WNLPs)] with the two-point correlation function

$$\langle u'_i(\mathbf{x})u'_j(\mathbf{x}')\rangle = I^2\delta_{ij}\delta(\mathbf{x} - \mathbf{x}'), \tag{6.33}$$

or two-dimensional isotropic Gaussian spatially correlated noise [red-noise-like perturbations (RNLPs)] with the two-point correlation function

$$\langle u_i'(\mathbf{x})u_j'(\mathbf{x}')\rangle = I^2\delta_{ij}\exp\left[-\frac{(\mathbf{x}-\mathbf{x}')^2}{2R_\perp{}^2}\right], \tag{6.34}$$

where $\mathbf{x} = (x_1, x_2)$ and $R_\perp$ and I^2 are correlation radius and noise variance (intensity of perturbations), respectively. These perturbations are directly added to the initial conditions. The technical details of generating Gaussian noises with correlation functions (6.33) and (6.34) can be found in Sabel'feld (1991). The non-dimensional intensity of the initial perturbations $\bar{I}^2 = (I/I_0)^2$ ($I_0 = 1\ \mathrm{m\,s^{-1}}$) will be used below. Noise with correlation functions (6.33) and (6.34) is the popular model of errors for optimal interpolation or spline fitting. Both of these procedures are applied to construct initial conditions for ocean models from irregularly spaced data (Brasseur *et al.* 1996).

The stochastic wind perturbation $\mathbf{U}$ is represented by (Sura *et al.* 2001)

$$\mathbf{U} = [U_1(x_1, x_2, t), U_2(x_1, x_2, t),] = \boldsymbol{\mu}(t)\sigma, \quad G^{1/2}(x_1, x_2), \tag{6.35}$$

where $\boldsymbol{\mu}(t) = [\mu_1(t), \mu_2(t)]$ are white Gaussian vector processes with zero mean and unit variance; σ^2 is the wind variance. The spatial structure function G characterises a degree of spatial inhomogeneity of wind perturbations above an area of interest. Two different structure functions G are used. The first one is given by

$$G_1(x_1, x_2) = \cos\left(\frac{\pi x_2}{L_2}\right). \tag{6.36}$$

In this case only the amplitude of wind stress (6.33) is distorted by the non-Gaussian white noise. The second one is chosen as

$$G_2(x_1, x_2) = \alpha_s\left[\pi\beta_1\beta_2\mathrm{erf}\left(\frac{L_1}{2\beta_1}\right)\mathrm{erf}\left(\frac{L_2}{2\beta_2}\right)\right]^{-1/2}$$

$$\times\exp\left(-\frac{(x_1 - L_1/2)^2}{2\beta_1^2} - \frac{(x_2 - L_2/2)^2}{2\beta_2^2}\right). \tag{6.37}$$

Here, erf is the error function; α_s is a scaling parameter; (β_1, β_2) are the decorrelation scales; and G_2 shows the impact of the localised atmospheric eddy activity near the point $(L_1/2, L_2/2)$ on the surface wind perturbations with $\beta_c = \beta_1 = \beta_2 =$ 600 km (Sura *et al.* 2001). In most numerical experiments the scaling constant α_s is chosen to adjust the weight function in (6.13) to 1. However, a number of computations use β_c between 100 km and 600 km. The noise in the surface wind with $\sigma^2 = 28.0\ \mathrm{m^2\,s^{-2}}$ corresponds to typical observed atmospheric conditions in

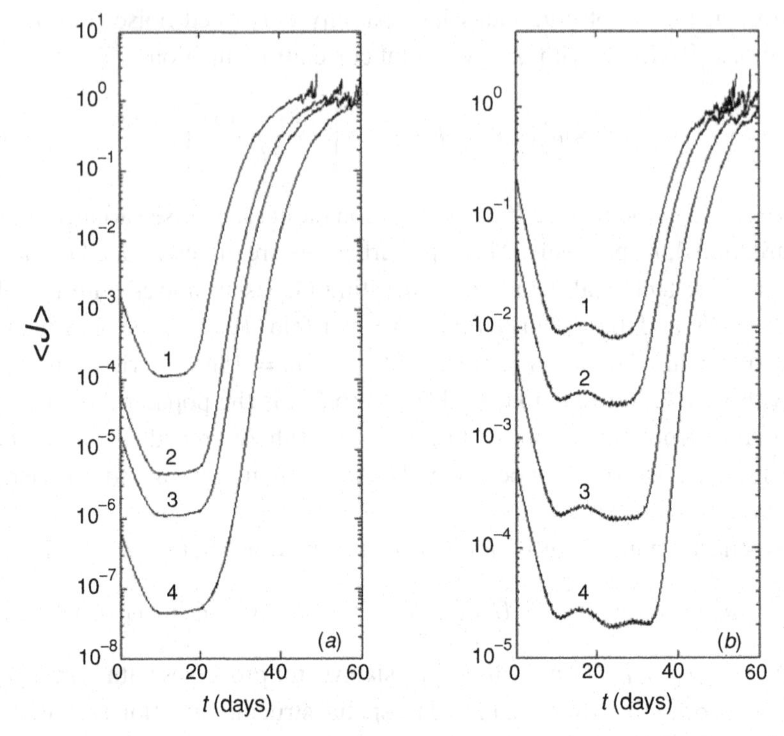

Figure 6.9 Temporal evolution of the root-mean-square error <J> for various initial perturbations: (a) WNLPs with different noise intensities: 0.05, 0.01, 0.005 and 0.001 (denoted by '1', '2', '3', '4'); and (b) RNLPs with the correlation radius of 70 km and different noise intensities: 0.02, 0.01, 0.003 and 0.001 (denoted by '1', '2', '3', '4') (after Ivanov & Chu 2007a).

the North Atlantic region (Wright 1988). Therefore, the stochastic forcing (6.35)–(6.37) is a conceptual tool to study the effect of noise on simple and more complex wind-driven regional ocean models.

6.6.4 Stability due to initial uncertainty

At least four stages for the IE evolution are identified for both WNLPs (Fig. 6.9*a*) and RNLPs (Fig. 6.9*b*). All these stages are clearly identified by the IE growth rate. Initial error decay is observed for the first ten days of IE evolution (Figs. 6.9*a*,*b*) where the IE growth rate, $Q \equiv \mathrm{d}\ln\langle J\rangle/\mathrm{d}t$, evolves as

$$Q \approx Q_0[\exp(\alpha_0 t) - \alpha_1], \tag{6.38}$$

where α_0 is the decay exponent, $Q_0 \approx -0.45$, $\alpha_0 = \ln \alpha_1/t_1$ and $t_1 = 10$ d.

Non-exponential initial error decay corresponding to (6.38) differs from a quasi-exponential decay obtained in Wirth & Ghil's (2000) model with the dissipative operator $\nu\Delta_\perp$ and quite large horizontal viscosity ν. In contrast to this model the zero horizontal viscosity is used in our model. Therefore, the non-exponential decay seems to be caused by nonlinear bottom friction. Then $\langle J(t)\rangle$ has quasi-stationary low values during 10 and 20 days for WNLPs (Fig. 6.9*a*) and RNLPs (Fig. 6.9b), respectively. During the third stage (after day 20 and day 30 for WNLPs and RNLPs, respectively) the IE grows faster than exponentially (Figs. 6.9*a*,*b*).

Linear theory suggests that the larger the amplitude of the initial perturbations, the higher the probability of obtaining low model predictability. The IE should steadily increase with a prediction time scale in the linear predictability regime. In contrast to this, forecast skill may decay more slowly when the amplitude of initial perturbations increases. The growing perturbations rapidly adopt a horizontal scale comparable to that of the reference state (linear predictability regime), and further growth is limited by interactions with this state and among them (nonlinear predictability regime). In the nonlinear predictability regime the IE demonstrates clear contributions from the cumulative effects of flow scales, and the predictability time is no longer measured by the inverse of the leading Lyapunov exponent. Moreover, model predictability is enhanced with the growth of the correlation radius, $R_\perp$, and is less sensitive to the choice of the intensity, $\bar{I}^2$.

Ivanov and Chu (2007a) demonstrated the existence of the linear predictability regime identified by non-Gaussian statistics and (quasi)-exponential growth of prediction error, for a small correlation radius ($R_\perp \ll 50$ km, in the case of WNLPs) and large noise intensities ($\bar{I}^2 \leq 0.2$). For WNLPs, the typical τ-PDF was close to Gaussian if $\bar{I}^2 \sim 0.01$–0.05. The growth of $\bar{I}^2$ up to 0.1–0.2 resulted in a weak asymmetry for the τ-PDF (SK $\rightarrow$ 0.15) and departs from non-Gaussian behaviour (KU $\rightarrow$ 3.10). However, although such a τ-PDF has a short tail (labelled '1' in Fig. 6.10*a*), it was still close to Gaussian and the mean τ-FPT reduced with the growth in amplitude of initial perturbations. The τ-PDF quickly departs from Gaussian with the growth of $\bar{I}^2$ after 0.2. However, from a physical point of view, such initial perturbations seem to be too large to exist in reality.

The nonlinear predictability regime appears as $R_\perp$ grows. Both highly non-Gaussian τ-PDFs, and the mean FPT that grows with increasing $R_\perp$, indicate that the prediction error becomes nonlinear. A typical non-Gaussian τ-PDF (SK $\approx$ 0.8, KU $\approx$ 4.0) computed for a finite correlation radius is demonstrated in Fig. 6.10*b*. The long PDF tail (labelled '2' in Fig. 6.10*b*) is clearly seen in this figure. The tail is formed by rare individual forecasts (FPT up to 60 days), each of which is longer than the mean ensemble forecasting (FPT of about 44 days).

The first four parameters of FPT were computed for various values of correlation radius $R_\perp$. The asymmetry of τ-PDFs becomes higher for the larger values of correlation radius $R_\perp$ (Fig. 6.11). The SK value, which is a measure of asymmetry, increases up to 0.8 when $R_\perp$ tends to 100 km. Larger values of mean FPT

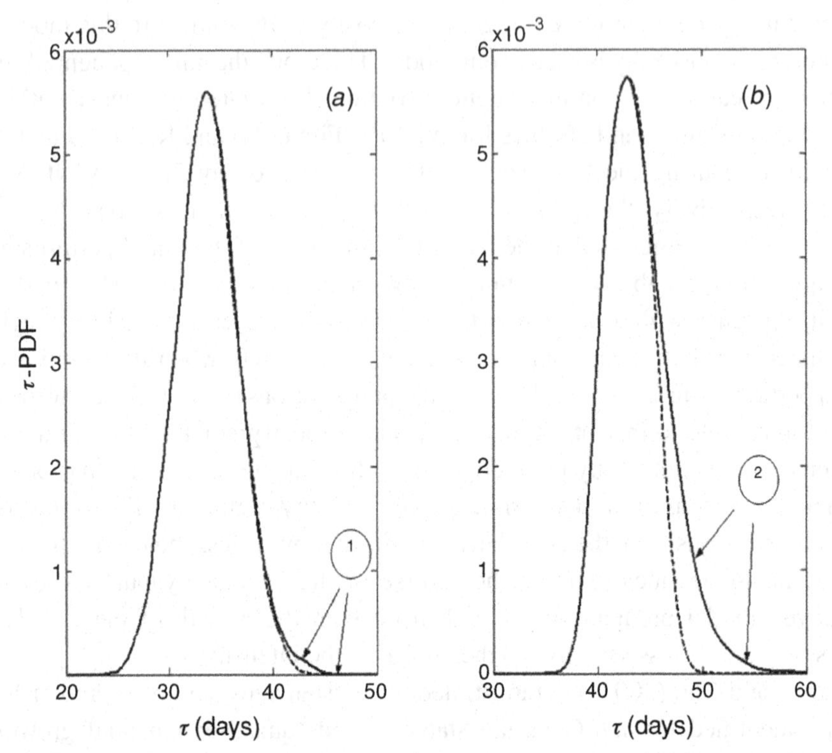

Figure 6.10 The τ-PDFs for (*a*) WNLPs with $\bar{I}^2 = 0.1$, $R_\perp = 30$ km and $\bar{\varepsilon}^2 = 0.5$; and (*b*) RNLPs with $\bar{I}^2 = 0.1$, $R_\perp = 125$ km and $\bar{\varepsilon}^2 = 0.2$. Skewness and kurtosis are 0.15 and 3.09 for (*a*), and 0.77 and 3.95 for (*b*). Dashed lines indicate mirror reflections of the left-hand-side tails of τ-PDFs (from Ivanov & Chu 2007a).

(Fig. 6.11*a*) and τ-variances (Fig. 6.11*b*) correspond to more asymmetric PDFs. Highly non-Gaussian (KU $\approx$ 4, Fig. 6.11*d*) and sharp τ-PDFs with long tails stretching to large prediction times accompany this nonlinear predictability regime. The explicit growth of mean predictability time observed with the growth of correlation radius $R_\perp$ is strong evidence of the nonlinear predictability regimes caused by the inhomogeneous morphology of the model phase space (Kaneko 1998).

6.6.5 Stability due to surface wind uncertainty

Perturbations excited by uncertain winds grow at all scales and during the whole 50–60 day period. This is in contrast to the case when there is uncertainty in the initial condition only. Therefore, the presence of the spatio-temporal noise (6.34) in wind forcing (6.33) causes the monotonic error growth. At least four predictability regimes are identified from Fig. 6.12. In all the cases the IE grows in a monotonic manner but with different speeds. More accurately, these regimes can be identified

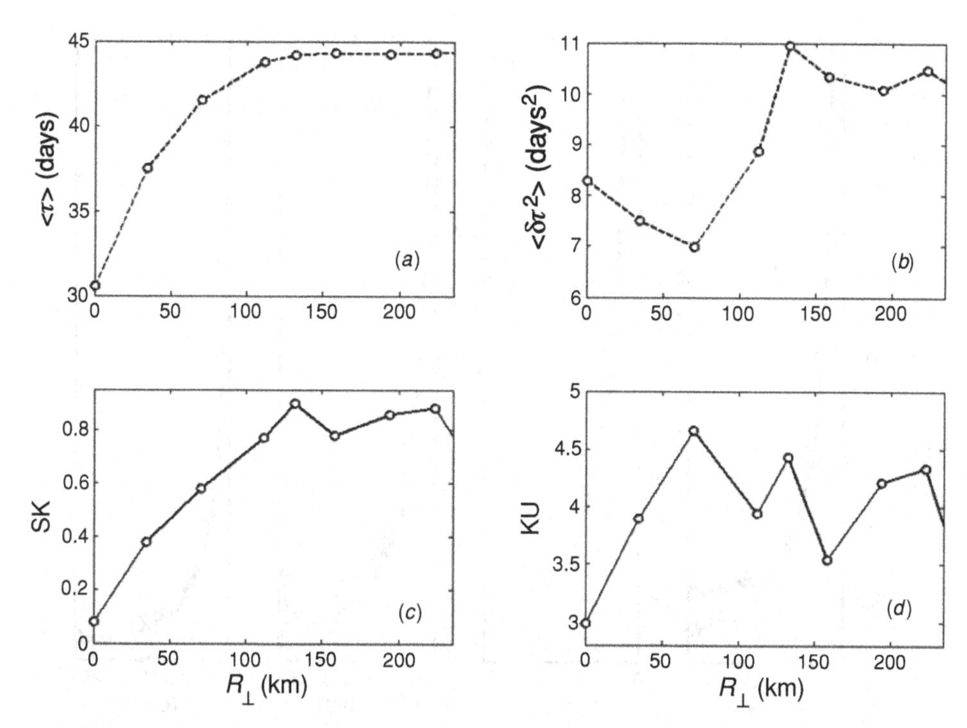

Figure 6.11 Dependence of τ-statistics on the correlation radius for RNLPs with the noise intensity of 0.1, and tolerance level 0.2: (*a*) τ-mean, (*b*) τ-variance, (*c*) τ-skewness, (*d*) τ-kurtosis (after Ivanov & Chu 2007a).

using the error growth rate (Q): linear growth, power growth and super-exponential growth.

The error dynamics strongly depends on the intensity and spatial inhomogeneity of wind uncertainty ($G, \bar{\sigma}^2$) as shown in Fig. 6.12. During the linear error growth, mean square IE is represented by,

$$\langle J \rangle \approx D_{\text{eff}} t, \tag{6.39}$$

Duration of the linear IE growth regime is typically up to 4–5 days if $\bar{\sigma}^2 \sim 0.1$–1.0. The effective coefficient D_{eff} is determined by summation of contributions from the error source term at all wavenumbers (Ivanov & Chu 2007b). The linear law (6.39) was earlier documented in a number of studies (see, for example, Vannitsem & Toth 2002, personal communication). However, Ivanov & Chu (2007b) analytically determined the error sources with strong dependence on the effective coefficient on the variance of wind as well as on the degree of spatial inhomogeneity of the wind forcing.

For moderate but inhomogeneous winds, Ivanov & Chu (2007b) found the power growth of perturbations after the transient phase (for example, see Fig. 6.12).

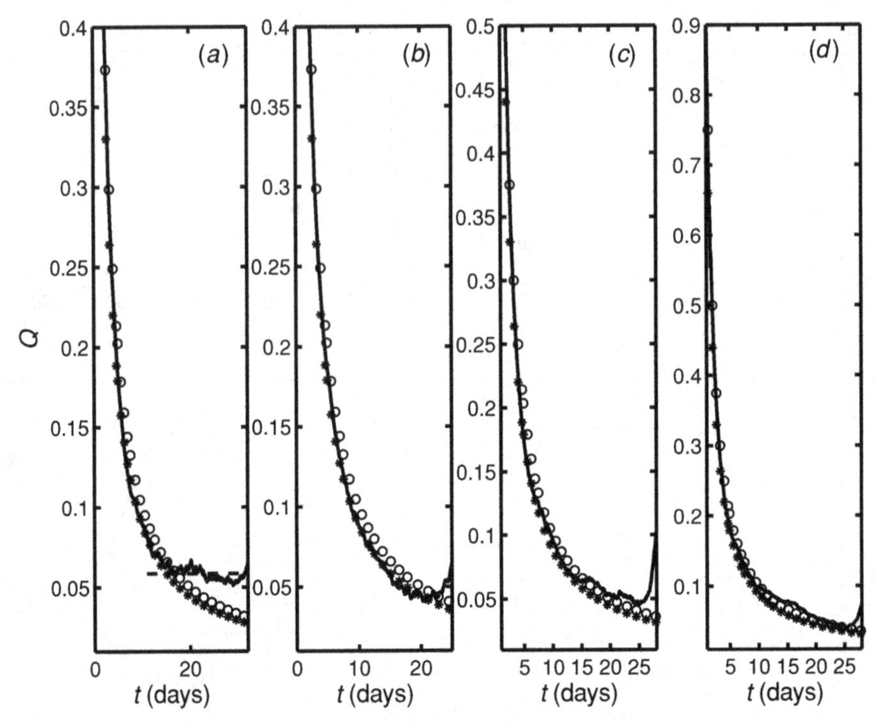

Figure 6.12 The growth rate Q (solid curve) for different $\bar{\sigma}^2$ and G: (*a*) $\bar{\sigma}^2 = 0.1$, $G = G_2$; (*b*) $\bar{\sigma}^2 = 1.0$, $G = G_1$; (*c*) $\bar{\sigma}^2 = 1.0$, $G = G_2$, and (*d*) $\bar{\sigma}^2 = 2.0$, $G = G_2$. The solid curve, circles and asterisks show exponential, linear and power (with scaling exponent of 8.8×10^{-1}) laws, respectively (after Ivanov & Chu 2007b).

For small values of $\bar{\sigma}^2 \ll 1.0$ the power growth is replaced by exponential growth (shown by the dashed line in Fig. 6.12*a*).

If $\bar{\sigma}^2$ exceeds 1, there is no exponential growth and the IE grows with the power law the power exponent being about 8.8×10^{-1}. This regime exists between day-5 and day-15 in Fig. 6.12*a*, between day-7 and day-23 in Fig. 6.12*b*, between day-4 and day-14 in Fig. 6.12*c*, but there is no power-law regime in Fig. 6.12*d* when the stochastic wind uncertainty is too large ($\bar{\sigma}^2 > 2.0$). For such a variance the linear growth of IE dominates. Smaller scales affected by the stochastic wind are subject to strong viscous damping due to the increasing drag coefficient α with the growth of kinetic energy of large-scale perturbations. Therefore, the smaller scales grow more slowly than the unstable large scales. The growing perturbations rapidly adopt the horizontal scales comparable to those of the reference state. Alternatively, stronger stochastic wind ($\bar{\sigma}^2 \geq 1.0$) excites more modes at smaller scales than weak wind, and the coherent behaviour of the modes is clearly observed in this case (for example see Fig. 6.12*b*).

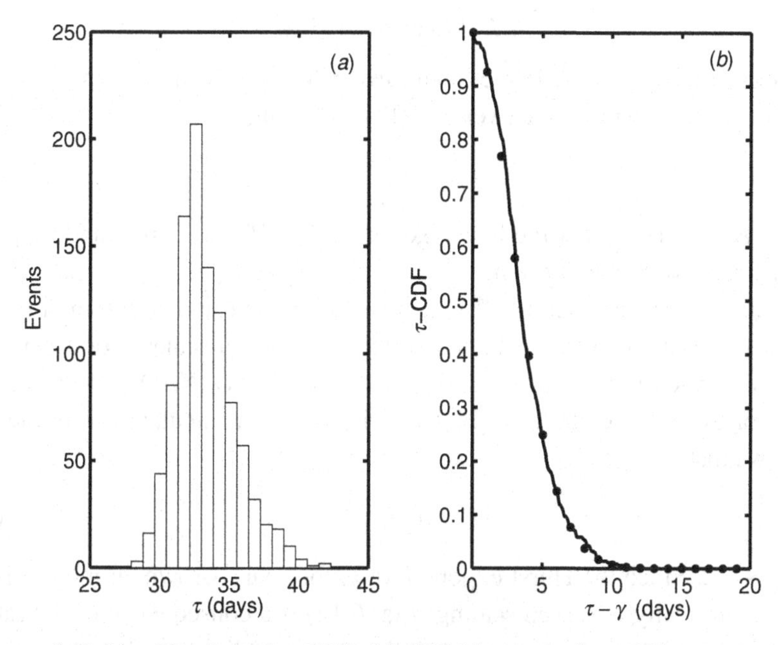

Figure 6.13 Identification of τ-CDF: (a) FPT histogram for a 103-term ensemble; (*b*) CDF computed directly from the ensemble (solid curve) and using the developed method (black dots), $\gamma = 30$ days (after Ivanov & Chu 2007b).

For finite-amplitude IEs, stochastic wind forcing (6.10), in general, induces highly non-Gaussian τ-PDFs. The following question arises: what kind of statistics can be used to represent such τ-PDFs? If an appropriate distribution function is found, it would be possible to identify the ensemble-generated PDFs from a limited observation series and small forecast ensembles, and in turn to estimate the model predictability horizon. Ivanov & Chu (2007b) found that the FPT satisfies the three-parameter Weibull distribution (Fig. 6.13*a*),

$$f(\tau) = \frac{\beta}{\eta}\left(\frac{\tau - \gamma}{\eta}\right)^{\beta-1} \exp\left[-\left(\frac{\tau - \gamma}{\eta}\right)^{\beta}\right], \tag{6.40}$$

with (η, γ, β) the scale, shape and location parameters, which were estimated by an algorithm on the basis of the Kullback–Leibler distance (White 1994). The τ-CDFs computed by the non-parametrical technique (solid curve) and our method (black dots) are compared in Fig. 6.13*b*. The parameters (η, γ, β) have the values 3.71, 30.0 and 1.67 days, respectively. The parameter β affects the length of the PDF tail formed by rare forecasts, which are longer than the mean ensemble forecast $\langle\tau\rangle$. A small value of β indicates an enhanced probability for the realisation of abnormally long (in our case up to 50 days) model forecasts.

6.7 Power decay law

A random process is called fractional Brownian motion if its cumulative FPT density function satisfies the power law (Ding & Yang 1995),

$$P(\tau) \sim \tau^{H-1}, \tag{6.41}$$

with $0 < H < 1$. Here H is the Hurst exponent. For $H = 1/2$, the random process is the ordinary Brownian motion.

The cumulative distribution, $P(\tau_\rho)$, for each climate index datum (Fig. 6.14) shows a power-law feature in the tail of the distribution scales. For a very small value of the index reduction, ρ_0 (0.01 for the AO, AAO, NAO and PNA indices and 0.1 for SOI), the cumulative FPT density function calculated from the index data shows that

$$P(\tau_0) \sim \tau_0^{-\alpha_0}, \tag{6.42}$$

with α_0 ~$1/2$. Since the Hurst exponent of an ordinary Brownian motion is $H = 1/2$, the empirically observed scaling (Fig. 6.14) is a consequence of the (at least close to) Brownian motion behaviour of the climate indices.

This argument of an unbiased Brownian motion is also strengthened by observing that (Fig. 6.14)

$$P(\tau_0 = 1) \sim 1/2. \tag{6.43}$$

This indicates that the climate index has a 50% chance of increase and decrease at each time step (one month). Figure 6.14 also shows the cumulative distributions $P(\tau_\rho)$ for different values of ρ, i.e. $\rho = 0.01$, 0.5, 1.0 and 1.5 for the (AO, AAO, NAO, PNA) indices and $\rho = 0.1$, 10, 20 and 30 for SOI. From this figure it is seen that the tail exponent, α_ρ, is rather insensitive to the value of ρ. In particular one finds that $\alpha_\rho \sim 1/2$ over a broad range of values for ρ, a value that is consistent with the Brownian motion hypothesis.

Such a power decay law in FPT is also found in model predictability. The Gulf of Mexico real-time nowcast/forecast system is taken as an example to show the existence of the power decay law of FPT in the predictability skill. This system (Chu *et al.* 2002c) was built based on the Princeton Ocean Model (POM) with $1/12° \times 1/12°$ horizontal resolution. Real-time sea-surface height (SSH) anomalies derived from NASA/CNES TOPEX and ESA ERS-2 altimeters, and composite SST data derived from the NOAA Advanced Very High Resolution Radiometer (AVHRR) in a continuous data assimilation mode, are used to produce a nowcast and four-week forecast. It was found that the forecast retains considerable skill to about one–two weeks, beyond which it begins to deviate rapidly from reality.

The Gulf of Mexico velocity data at 50 m depth are archived every six hours from the nowcast/forecast system for six months beginning on 9 July 1998. The

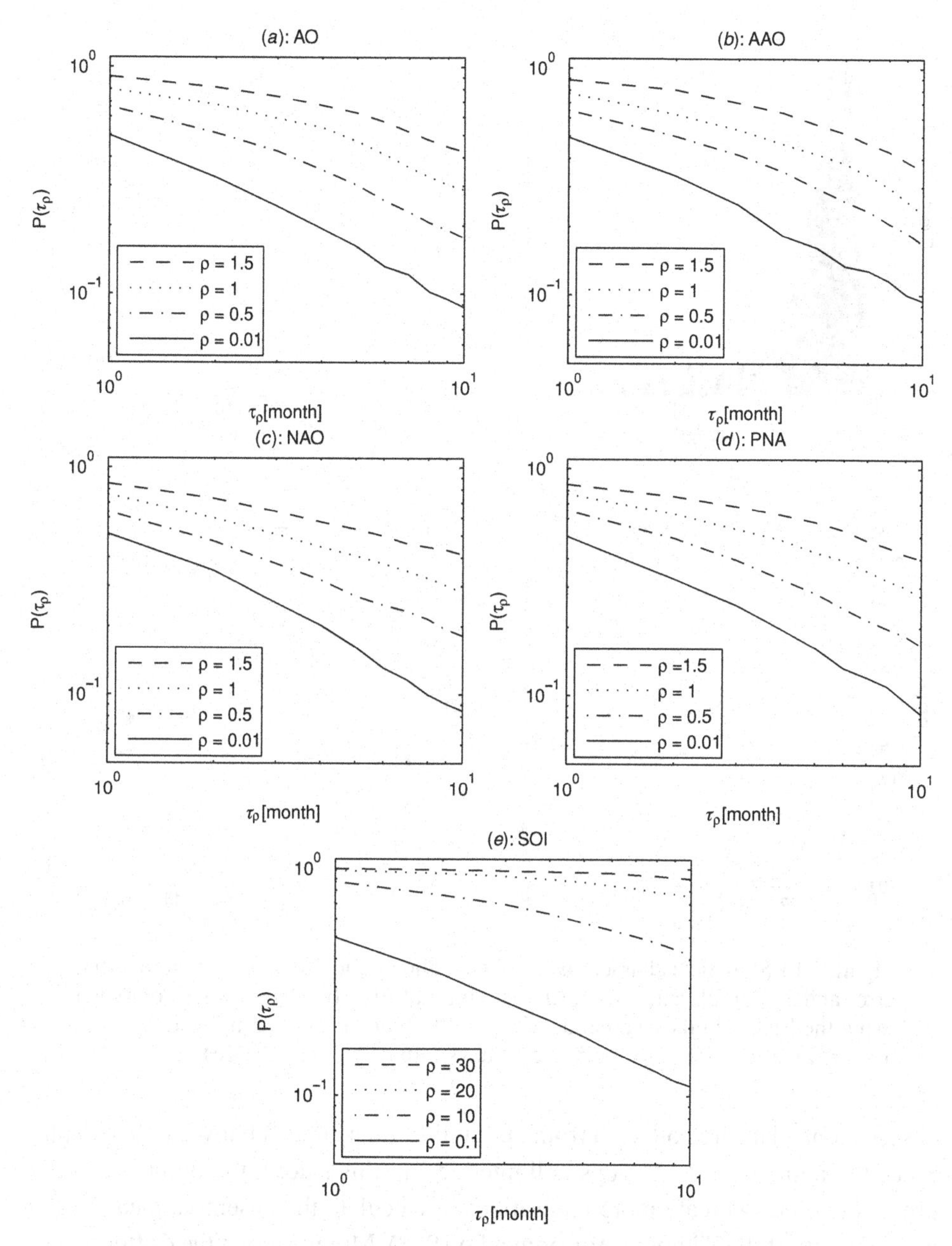

Figure 6.14 The empirical cumulative density functions, $P(\tau_\rho)$, for different values of the index reduction for (*a*) Arctic Oscillation (AO), (*b*) Antarctic Oscillation (AAO), (*c*) North Atlantic Oscillation (NAO), (*d*) Pacific–North American Pattern (PNA) and (*e*) Southern Oscillation (SO). It shows the power-law features (after Chu 2008).

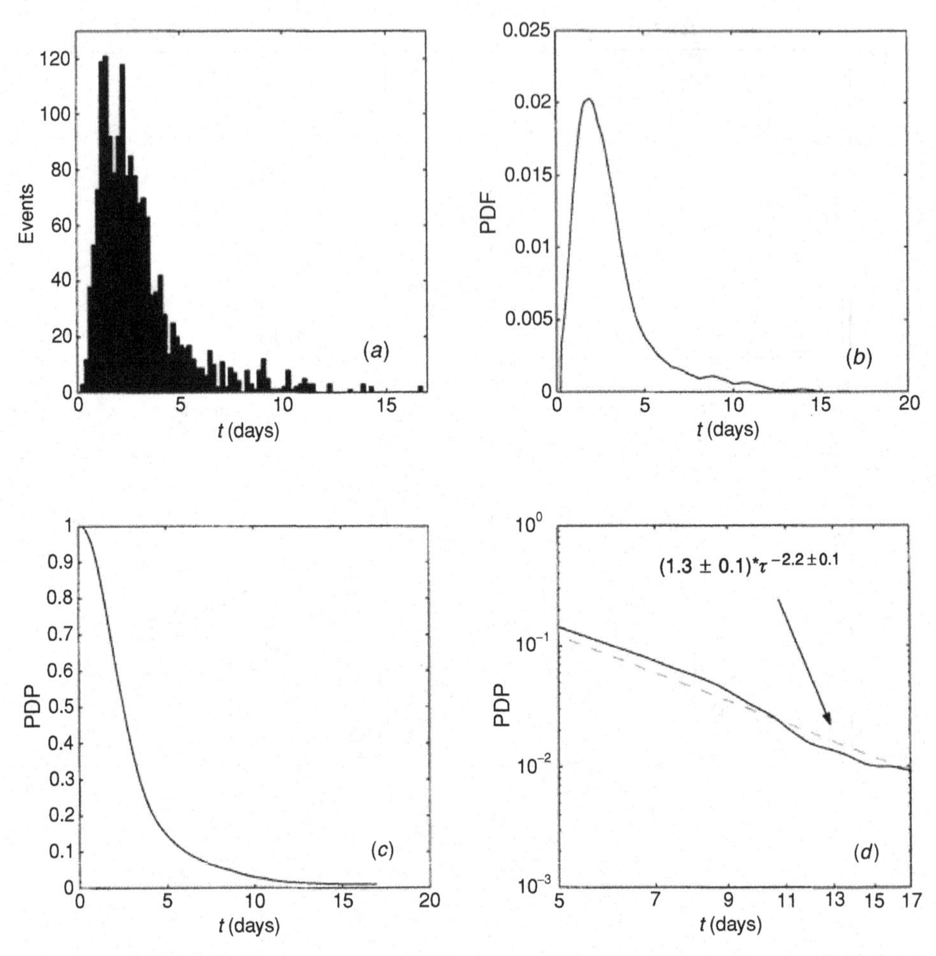

Figure 6.15 Statistical characteristics of the valid prediction period for zero initial error and 55 km tolerance level: (*a*) histogram of FPT, (*b*) PDF of FPT computed using the Epanechnikov kernel density, (*c*) PDP for FPT between 0 and 20 days, and (*d*) PDP for FPT between 5 and 17 days (after Chu *et al.* 2002c).

observational data are collected from 50 satellite-tracked drift buoys at 50 m depth. Since the number of drift buoys is limited (50 in this study), the point stochastic process is used to create more realisations to calculate the ensemble mean of the predictability skill (Tikhonov & Khomenko 1998). Moving along the drifter trajectory $\mathbf{y}(t)$, points are randomly selected and the predictability skill is calculated for each of these points. Such a process is called a point stochastic process. The average over these points is equivalent to the ensemble average due to the ergodic feature of the trajectories (Dymnikov & Filatov 1987; Tikhonov & Khomenko 1998).

The FPT is computed for non-ensemble prediction with a tolerance level (ε) of 55 km and without initial error ($\mathbf{z}_0 = 0$). The PDF of FPT (Fig. 6.15), calculated using the Epanechnikov kernel density (Good 1996) from the histogram (Fig. 6.15*a*), shows a non-Gaussian distribution with a narrow peak and a long tail in the domain of long-term prediction. Successful 10–15 day predictions are abnormally long (called extreme long predictions) compared to the mean FPT (3.2 days) and the most probable FPT (2.4 day). The tail of the probability density of prediction (PDP) follows the power law with the power exponent, $\gamma = 2.2$ (Figs. 6.15*c* and *d*).

6.8 Lagrangian predictability

During the World Ocean Circulation Experiment (WOCE), the ocean velocity observation has been significantly advanced with extensive spatial and temporal coverage using near-surface Lagrangian drifters, RAFOS floats and Autonomous Lagrangian Circulation Explorers (ALACEs). Trajectories of these quasi-Lagrangian drifters reflect the whole spectrum of ocean motions, including meso- and submeso-scale eddies, various waves, inertial and semi-diurnal motions, and provide invaluable resources to estimate the forecast skill. However, direct comparison between model and Lagrangian observational data is difficult since neither the dynamics of numerical models nor their forcing data (such as surface forcing functions) are identical to reality. It needs to be determined if the model–data difference comes from a deficiency in modelling ocean physics, from some unessential imperfection or from unrepresentative data. It is clear that the high-resolution ocean model and the Lagrangian drifter data are compared only in the statistical sense. Since the mesoscale (10–50 km) movements of Lagrangain drifters represent the eddy dynamics, it is reasonable to ask if the high-resolution ocean model reproduces the mesoscale Lagrangian drifter movements. Such a model capability, defined as the 'Lagrangian predictability' (Mariano *et al.* 2002) as distinct from the 'Eulerian predictability' has the capability to reproduce patterns and topological details of the circulation attractor.

The Lagrangian data collected from 15 satellite-tracked sonobuoys, manufactured by the Applied Technology Associates and deployed by Horizon Marine, provided the Lagrangian data used for the study. A nylon drogue at 50 m depth is tethered to the buoy hull of which 0.64 m is in the water and 0.33 m in the air. The drifter trajectories used in the present study are given in Fig. 6.16. The synthetic particle trajectories are obtained from the time integration of the modelled horizontal velocity field at 50 m depth with bi-cubic spline interpolation in space and second-order interpolation in time.

The PDF of FPT for zero initial errors is calculated from the drifter and synthetic particle trajectories with four tolerance levels (ε) (0.25°, 0.5°, 0.75° and 1.25°). It clearly shows non-Gaussian distribution (Fig. 6.17). The long tail stretching

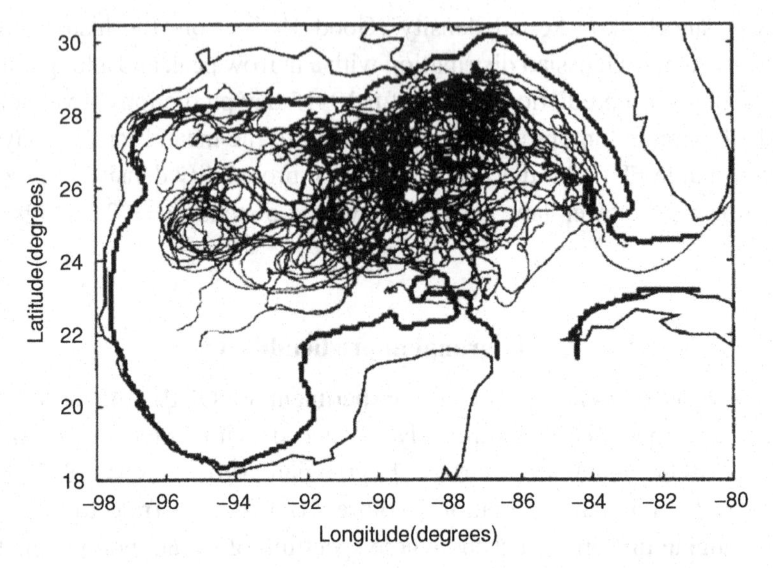

Figure 6.16 Spaghetti of 15 drifting buoys used in the analysis.

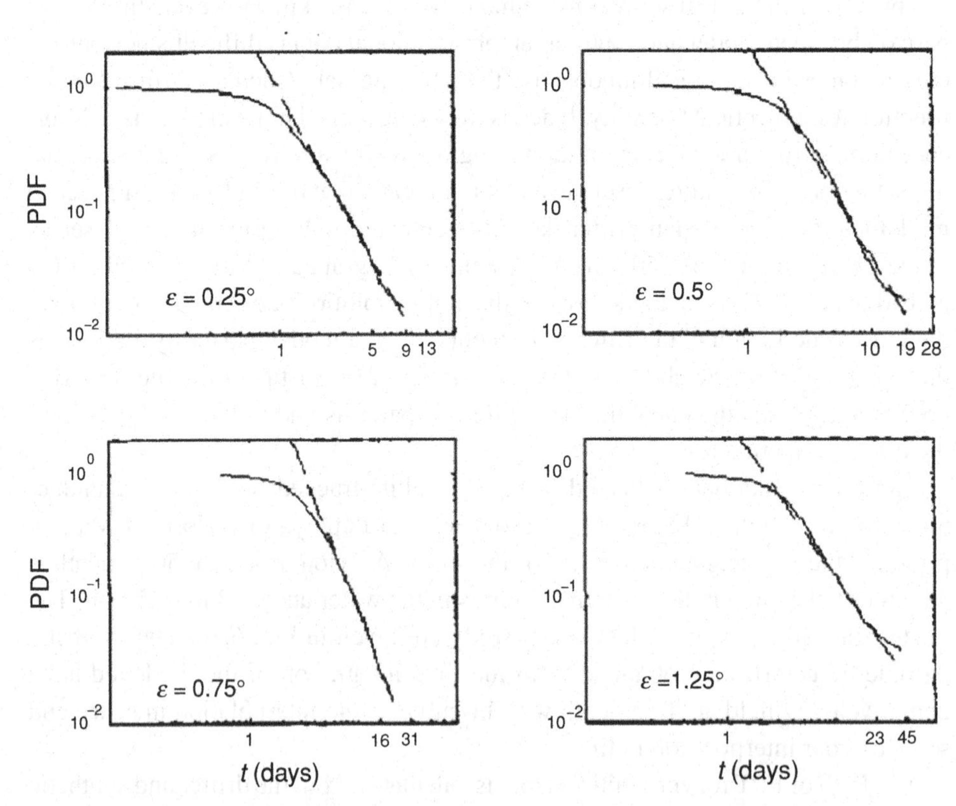

Figure 6.17 PDF of FPT calculated for different levels of tolerance ε. Solid and dashed lines are the PSP and the proposed theoretical power laws to approximate it for large times (from Ivanov and Chu 2007b).

into the long FPT period (t larger than 1 day) demonstrates the existence of a long-term correlation between the drifter and synthetic particle trajectories. Such a phenomenon is called 'extremely successful prediction' (Chu *et al.* 2002a). This long tail has the power behaviour for a long FPT period with power exponents of 2.17, 1.98, 2.08 and 1.77 for ε values of 0.25°, 0.5°, 0.75° and 1.25°, respectively. For a power exponent larger than 2, there is no intermittency in the prediction skill.

Dependence of the first two FPT moments on the tolerance (ε) follows the power law

$$\langle \tau \rangle \sim \varepsilon^{0.99\pm0.08\ldots}, \quad \hat{\tau}_2 \sim \varepsilon^{1.76\pm0.07\ldots}, \tag{6.44}$$

which indicates that the Lagrangian prediction skill of the Gulf of Mexico model is non-Gaussian and non-intermittent. If the FPT statistics is Gaussian, the power exponents of (6.44) should be

$$\alpha = 1, \quad \beta = 2,$$

rather than

$$\alpha = 0.99 \pm 0.08, \quad \beta = 1.76 \pm 0.07.$$

6.9 Conclusions

First passage time (FPT) presents a new way to detect climate model predictability and temporal variability of the climate indices. It is a random variable, the probability density function of which satisfies the backward Fokker–Planck equation. In solving this equation, it is easy to obtain the ensemble mean and variance of the FPT for climate models and climate indices. The solution of the backward Fokker–Planck equation is usually non-Gaussian. The PDF of FPT for a numerical ocean model to uncertain initial conditions satisfies the three-parameter Weibull distribution.

The FPTs for the five climate indices show Brownian fluctuations. This means that the early results on the Brownian fluctuations for the NAO index are also valid for the other indices (AO, AAO, PNA and SO). With $\delta \to 0$ as a limit case (i.e. no index reduction), the FPT density function tends to the δ function.

The power law of FPT is found for climate model predictability. The probability density function of FPT for the Gulf of Mexico nowcast system is asymmetric with a broader and long tail for higher values which indicates long-term predictability. This long tail corresponds to the power law for error growth. For FPT with small mean and mode values, its individual model prediction can be valid for a long period (long-term prediction); this is represented by a long tail for higher FPT values. The extreme long and short predictions share the same statistics.

The power law of FPT is also found in climate index prediction. For very small values of the index reduction, ρ_0 (0.01 for the AO, AAO, NAO, and PNA indices and 0.1 for SOI), the cumulative FPT density function shows a power-law dependence on τ_ρ with an exponent of approximately $-1/2$. Another well-known method for checking power-law dependence is examination of the autocorrelation function. This also confirms that the climate indices have Brownian-type fluctuations.

The FPT is found useful to analyse stochastic stability of numerical models. It is expected that FPT will be applied to many areas such as stochastic climate prediction and stochastic dynamical systems.

Acknowledgements

The author thanks the Banff International Research Station (BIRS) for hosting the Workshop on Stochastic Dynamical Systems and Climate Modeling and the organisers, Jinqiao Duan, Boualem Khouider, Richard Kleeman and Adam Monahan. The author also thanks Tim Palmer and Paul Williams for their work in editing this book.

References

Auclair, F., Marsaleix, P. & De Mey, P. 2003 Space–time structure and dynamics of the forecast error in a coastal circulation model of the Gulf of Lions. *Dyn. Atmos. Ocean*, **36**, 309–346.

Ausloos, M. & Ivanova, K. 2001 Power-law correlations in the Southern-Oscillation-index fluctuations characterizing El Niño. *Phys. Rev. E*, **63**, 047201.

Boffetta, G., Giuliani, P., Paladin, G. & Vulpiani, A. 1998 An extension of the Lyapunov analysis for the predictability problem. *J. Atmos. Sci.*, **55**(23) 3409–3416.

Brasseur, P., Blayo, E. & Verron, J. 1996 Predictability experiments in the North Atlantic Ocean: outcome of a quasi-geostrophic model with assimilation of TOPEX/POSEIDON altimeter data. *J. Geophys. Res.*, **101**, 14161–14173.

Chu, P. C. 1999 Two kinds of predictability in the Lorenz System. *J. Atmos. Sci.*, **56**, 1427–1432.

Chu, P. C. 2006 First-passage time for stability analysis of the Kaldor model. *Chaos, Solitons & Fractals*, **27**, 1355–1368.

Chu, P. C. 2008 First passage time analysis for climate indices. *J. Atmos. Oceanic Technol.*, **25**, 258–270.

Chu, P. C., Fan, C. W. & Ehret, L. L. 1997 Determination of open boundary conditions with an optimization method. *J. Atmos. Oceanic Technol.*, **14**, 723–734.

Chu, P. C., Ivanov, L. M. & Fan, C. W. 2002a Backward Fokker–Planck equation for determining model valid prediction period. *J. Geophys. Res.*, **107**, C6, 10.1029/2001JC000879.

Chu, P. C., Ivanov, L. M., Margolina, T. M. & Melnichenko, O. V. 2002b On probabilistic stability of an atmospheric model to various amplitude perturbations. *J. Atmos. Sci.*, **59**, 2860–2873.

Chu, P. C., Ivanov, L., Kantha, L., Melnichenko, O. & Poberezhny, Y. 2002c Power law decay in model predictability skill. *Geophys. Res. Lett.*, **29**(15), 10.1029/2002GLO14891.

Chu, P. C., Ivanov, L. M., Kantha, L. H. *et al.* 2004 Lagrangian predictabilty of high-resolution regional ocean models. *Nonlinear Proc. Geophys.*, **11**, 47–66.
Chu, P. C. & Ivanov, L. M. 2005 Statistical characteristics of irreversible predictability time in regional ocean models. *Nonlinear Proc. Geophys.*, **12**, 1–10.
Collette, C. & Ausloos, M. 2004 Scaling analysis and evolution equation of the North Atlantic oscillation index fluctuations. *Int. J. Mod. Phys. C*, **15**, 1353–1366.
Dalcher, A. & Kalnay, E. 1987 Error growth and predictability in operational ECMWF forecasts. *Tellus*, **39A**, 474–491.
Ding, M. & Yang, W. 1995 Distribution of the first return time in fractional Brownian motion and its application to the study of on–off intermittency. *Phys. Rev. E*, **52**, 207–213.
Downing, D. J., Gardner, R. H. & Hoffman, F. O. 1985. An examination of response-surface methodologies for uncertainty analysis in assessment of models. *Technometrics*, **27**, 2, 151–163.
Drosdowsky, W. 1994 Analog (nonlinear) forecasts of the Southern Oscillation index time series. *Weather Forecast.*, **9**, 78–84.
Dymnikov, V. P. & Filatov, A. N. 1997 *Mathematics of Climate Modeling*. Birkhauser Publishing Co.
Farrell, B. F. & Ioannou, P. J. 1996 Generalized stability theory. Part 1. Autonomous operations. *J. Atmos. Sci.*, **53**, 14,2025–14,2040.
Gardiner, C. W. 1985 *Handbook of Stochastic Methods for Physics, Chemistry and the Natural Sciences*. Springer-Verlag.
Good, P. I. 1996 *Re-Sampling Methods. A Practical Guide to Data Analysis*. Birkhauser Publishing.
Ivanov, L. M., Kirwan, A. D. Jr. & Melnichenko, O. V. 1994 Prediction of the stochastic behavior of nonlinear systems by deterministic models as a classical time-passage probabilistic problem, *Nonlinear Proc. Geophys.*, **1**, 224–233.
Ivanov, L. M. & Chu, P. C. 2007a On stochastic stability of regional ocean models to finite-amplitude perturbations of initial conditions. *Dyn. Atmos. Oceans*, **43** (3–4), 199–225.
Ivanov, L. M. & Chu, P. C. 2007b On stochastic stability of regional ocean models with uncertainty in wind forcing. *Nonlinear Proc. Geophys.*, **14**, 655–670.
Kaneko, K. 1998 On the strength of attractors in a high-dimensional system. Milnor attractor: Milnor attractor network, robust global attraction, and noise-induced selection. *Physica* D, **124**, 322–344.
Keppenne, C. L. & Ghil, M. 1992 Adaptive spectral analysis and prediction of the Southern Oscillation index. *J. Geophys. Res.*, **97**, 20449–20454.
Lacarra, J. F. & Talagrand, O. 1988 Short-range evolution of small perturbations in a barotropic model. *Tellus*, **40A**, 81–95.
Latin hypercube sampling tool, 2001: http://www.mathepicomepitoos/hs/nrpage.html.
Lind, P. G., Mora, A., Gallas, J. A. C. & Haase, M. 2005 Reducing stochasticity in the North Atlantic Oscillation index with coupled Langevin equations. *Phys. Rev. E*, **72**, 056706.
Lorenz, E. N. 1984 Irregularity. A fundamental property of the atmosphere. *Tellus*, **36A**, 98–110.
Maharaj, E. A. & Wheeler, M. J. 2005 Forecasting an index of the Madden–Julian Oscillation. *Int. J. Climatol.*, **25**, 1611–1618.
Mariano, A. J., Griffa, A., Ozgokmen, T. M. & Zambiancji, E. 2002 Lagrangian analysis and predictability of coastal and ocean dynamics 2000. *J. Atmos. Oceanic Technol.*, **19**, 1114–1126.

Nicolis, C. 1992 Probabilistic aspects of error growth in atmospheric dynamics. *Q. J. R. Meteorol. Soc.*, **118**, 553–568.
Palmer, T. N. 2001 A nonlinear dynamical perspective on model error: a proposal for non-local stochastic-dynamic parameterisation in weather and climate prediction models. *Q. J. R. Meteorol. Soc.*, **127**, 279–304.
Rangarajan, G. & Ding, M. 2000 First passage time distribution for anomalous diffusion. *Phys. Lett. A*, **273**, 322–330.
Sabel'feld, K. 1991 *Monte-Carlo Methods in Boundary Value Problems*. Springer-Verlag.
Schneider, T. & Griffies, S. H. 1999 A conceptual framework for predictability studies. *J. Climate*, **12**, 3133–3155.
Sura, P., Fraedrich, K. & Lunkeit, F. 2001 Regime transitions in a stochastically forced double-gyre model. *J. Phys. Oceanogr.*, **31**, 411–426.
Tikhonov, V. I. & Khimenko, V. I. 1998 Level-crossing problems for stochastic processes in physics and radio engineering: a survey. *J. Comput. Tech. Electr.*, **43**, 457–477.
Torrence, C. & Campo, G. P. 1998 A practical guide to wavelet analysis. *Bull. Am. Meteor. Soc.*, **79**, 62–78.
Toth, Z. & Kalnay, E. 1997 Ensemble forecasting at NCEP: the breeding method, *Mon. Weather Rev.*, **125**, 3297–3318.
Toth, Z., Szunyogh, I., Bishop, C. *et al.* 2001 On the use of targeted observations in operational numerical weather predictions. Preprints, Fifth Symposium on Integrated Observing Systems, American Meteorological Society, January 15–19, Albaquerque, NM, pp. 72–79.
Veronis, G. 1966 Wind-driven ocean circulation, II: numerical solutions of the nonlinear problem. *Deep-Sea Res.*, **13**, 31–55.
Vukicevic, T. 1991 Nonlinear and linear evolution of initial forecast errors. *Mon. Weather Rev.*, **119**, 1602–1611.
Walker, G. T. & Bliss, E. W. 1937 World weather VI. *Mem. R. Meteorol. Soc.*, **4**, 119–139.
White, H. 1994 *Estimation, Inference and Specification Analysis*. Cambridge University Press.
Wirth, A. & Ghil, M. 2000 Error evolution in the dynamics of an ocean general circulation model. *Dyn. Atmos. Ocean*, **32**, 419–431.
Wright, P. B. 1988 An atlas based on the COADS dataset: fields of mean wind, cloudiness and humidity at the surface of global ocean. Technical Report 14, Max-Plank-Institute for Meteorology.

7

Effects of stochastic parameterisation on conceptual climate models

DANIEL S. WILKS

Conceptual climate models are very simple mathematical representations of climate processes, which are especially useful because their workings can be readily understood. The usual procedure of representing effects of unresolved processes in such models using functions of the prognostic variables (parameterisations) that include no randomness generally results in these models exhibiting substantially less variability than do the phenomena they are intended to simulate. A viable yet still simple alternative is to replace the conventional deterministic parameterisations with stochastic parameterisations, which can be justified theoretically through the central limit theorem. The result is that the model equations are stochastic differential equations. In addition to greatly increasing the magnitude of variability exhibited by these models, and their qualitative fidelity to the corresponding real climate system, representation of unresolved influences by random processes can allow these models to exhibit surprisingly rich new behaviours of which their deterministic counterparts are incapable.

7.1 Introduction: model simplification and parameterisation

Mathematical models of the climate system vary enormously in their levels of detail and complexity. The simplest are zero- or one-dimensional representations of the planet, and they calculate explicitly the dynamics of only a small number of (perhaps only one) prognostic variables (Schneider & Dickinson 1974; Saltzman 1978). At the other extreme are modern general circulation models (GCMs) of the coupled atmosphere–ocean system, which include a comprehensive three-dimensional representation of the geophysical fluid dynamics in terms of millions or tens of millions of prognostic variables. Clearly the more complex models resemble the actual physical system more closely, whereas the more simplified models represent the climate

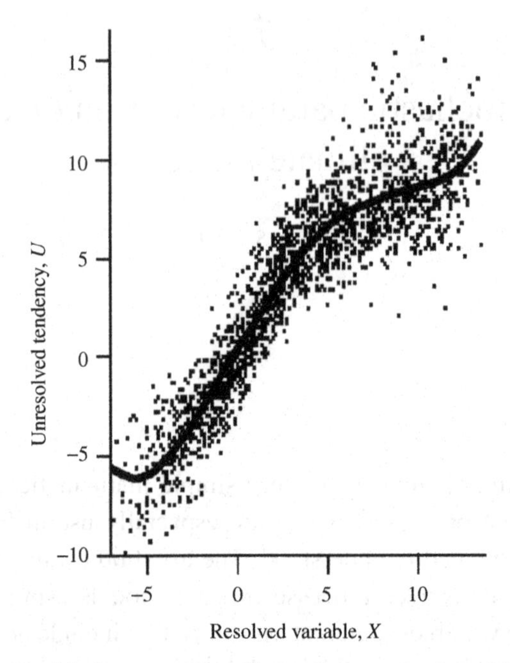

Figure 7.1 Relationship between the effects of unresolved variables on the time tendency of a resolved variable X, as a function of X, in the Lorenz (1996) model. The smooth curve through the individual points represents the mean effect of the unresolved variables conditional on particular values of the resolved variable. Reproduced with permission from Wilks (2005).

system in only a generalised, or conceptual, way. However, because these conceptual models are sufficiently simple for their workings to be readily understood, they can be useful for gaining qualitative understanding of both the climate system, and the relationship between the models and the real system being modelled.

Any mathematical model of the climate requires simplifications relative to the real climate system. Of course, these simplifications are more numerous and severe for the simpler models. But regardless of the level of modelling detail, there are inevitably relevant processes occurring on time and space scales that are too small to be represented explicitly, or which are not fully or quantitatively understood, or both. The effects of such unresolved climate variables and climate processes are included in models in an approximate way, through functions – called parameterisations – of the resolved prognostic variables that summarise the effects of the unresolved processes. Fundamentally, these parameterisations are meant to represent the statistical characteristics of the effects of the unresolved processes on the model's prognostic variables.

By far the most common form for these parameterisations is representation of only the mean effects of the unresolved processes, with their other statistical properties being ignored. For example, Fig. 7.1 shows the relationship between the effects of unresolved processes on a resolved variable X in the idealised Lorenz (1996) model. The cloud of points represents individual 'data' values for the effects of unresolved variables on the time tendency of the resolved variable X, and the smooth curve is a regression function specifying (parameterising) their mean effect, conditional on particular values of X. Such a parameterisation could be implemented using an equation of the form

$$\frac{\mathrm{d}X}{\mathrm{d}t} = R(X) + U(X), \tag{7.1.1}$$

where $R(X)$ represents the explicitly resolved dynamics and $U(X)$ would be the smooth function of X indicated in Fig. 7.1, representing unresolved processes.

A conventional mean-response parameterisation function such as $U(X)$ in (7.1.1), which specifies a single conditional mean effect due to the unresolved scales, is deterministic; i.e. (7.1.1) contains no randomness, even though (according to the data scatter in Fig. 7.1) there is no unique or constant response from the unresolved scales for a given value of the resolved variable. By contrast, nature would provide a realisation from the point cloud around $U(X)$, rather than the value $U(X)$ itself. Deterministic parameterisations have been justified heuristically by analogy to the macroscopic ('resolved scale') temperature response (of, say, the sensing surface of a thermometer) to a very large number of ('unresolved') interactions of the molecules of the medium. Because the number of microscopic interacting molecules is very large, the temperature is practically constant in time, even though the individual molecular kinetic energies, whose average effect yields the temperature, are highly variable. This conventional deterministic approach to parameterisation has been called the 'Reynolds/Richardson paradigm' (Palmer *et al.* 2005), reflecting its historical genesis at the very beginning of modern fluid dynamical weather prediction.

This concept of a 'thermodynamic limit' and its applicability to the parameterisation problem can be viewed in terms of the well-known central limit theorem from statistics (e.g. Lindgren 1968). One consequence of the central limit theorem is that the frequency distribution of the average of n independent random quantities (regardless of the distribution from which they have been drawn) tends towards the Gaussian ('bell curve') distribution as the number of averaged random effects increases. Furthermore, the variance of this distribution for the average decreases in proportion to $1/n$. Craig & Cohen (2006) derived these same properties using a statistical mechanics approach, in which the unresolved small-scale influences are mass fluxes produced by independent (non-interacting) convective clouds in equilibrium with their large-scale forcing.

The solid curves in Fig. 7.2 illustrate these properties for the case in which the individual independent quantities to be averaged are drawn from an exponential

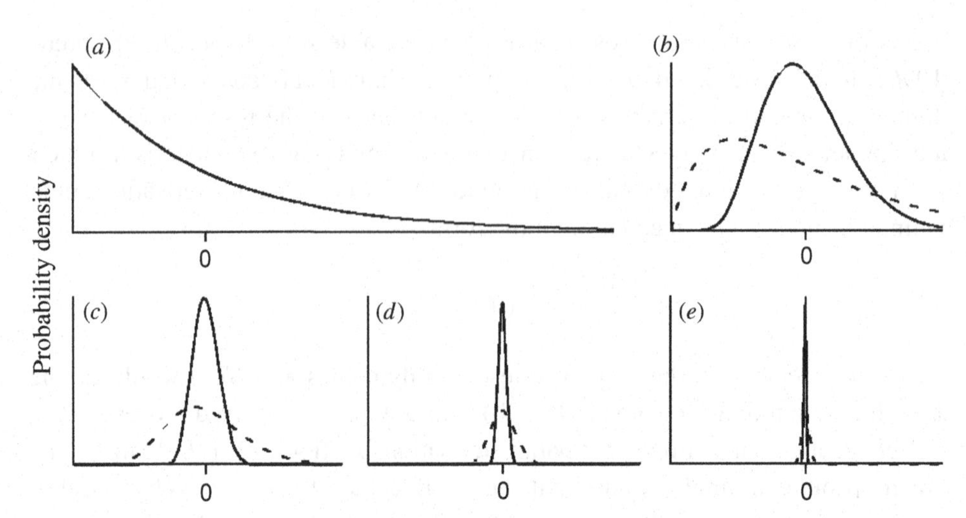

Figure 7.2 Frequency distributions for the average of (*a*) $n = 1$, (*b*) $n = 10$, (*c*) $n = 100$, (*d*) $n = 1000$ and (*e*) $n = 10000$ random variables drawn from the exponential distribution in (*a*). Solid curves show distributions resulting when the individual values being averaged are mutually independent and dashed curves show distributions for averages of moderately strongly correlated values. Horizontal scales are equal and vertical scales increase from (*a*) to (*e*).

distribution, shifted to the left so as to have zero mean, as shown in Fig. 7.2*a*. In this illustration, the probability density function in Fig. 7.2*a* represents the distribution for the effect of a single small-scale, unresolved event on a large-scale time tendency; Fig. 7.2*b* represents this distribution for the average effect of 10 such small-scale events. The dispersion of the probability distributions in Fig. 7.2 around their mean of zero corresponds to the scatter of points in Fig. 7.1 around the deterministic function $U(X)$, for a particular value of X.

As progressively larger numbers of independent draws from the probability distribution in Fig. 7.2*a* are averaged (Fig. 7.2*b–e*), the shapes of the frequency distributions for their average effect approaches the Gaussian relatively quickly, even though the original distribution in Fig. 7.2*a* is very far from the Gaussian. On the other hand, extremely large numbers of these small-scale events must be averaged in order for the uncertainty regarding their net effect to vanish for practical purposes, which condition would be required for a deterministic (mean effect, only) parameterisation of their influence to be justified. In practice, this zero-variance limit for parameterisations is not reached in climate models: if it could be, then the small-scale influences would be resolved, in effect, requiring no approximation of their contribution to the dynamics.

Note that independence of the small-scale variables is not necessary in order for their average to approach the Gaussian, nor for the variance of that distribution to tend towards zero as the average is taken over progressively larger numbers of small-scale values. The probability densities in Fig. 7.2 (dashed curves) indicate corresponding results for moderately strongly correlated ($\rho = 0.8$ for consecutive values in the n-member series) small-scale variables (Kotz & Adams 1964). These show that the distributions for the average of correlated values also tend towards the Gaussian, with decreasing variance as more values are averaged, but for both of these features the convergence is slower than in the case of independent small-scale effects. If the average effect of mutually correlated small-scale events on a large-scale tendency is to be represented by a parameterisation, use of a deterministic (i.e. zero variance) parameterisation is even less well justified.

7.2 Realistic variability in simple models: Hasselmann's paradigm

In a landmark paper, Hasselmann (1976) proposed that conventional mean-response parameterisations be extended to include also the variance associated with unresolved processes.

Integration of full GCMs of the climate system was only just beginning to become computationally feasible at that time, so most mathematical climate studies employed relatively simple conceptual models (Schneider & Dickinson 1974; Saltzman 1978). Among other attributes, these conceptual climate models highlight clearly the extreme simplification that may result from the use of deterministic parameterisations. Because most of these simple models do not exhibit chaotic dynamics, their variability is limited to adjustment towards their equilibrium solutions from non-equilibrium initial conditions, execution of fixed limit cycles (i.e. deterministic periodic solutions) or responses to variable external forcing. Attempts to simulate observed climate variability with deterministically parameterised conceptual models using realistic levels of external (e.g. orbitally varying solar) forcing had not been successful.

Hasselmann's insight was to realise that if the time scales of the resolved ('climate') variables and the to-be-parameterised unresolved ('weather') variables are well separated, then the central limit theorem implies that the effects of the unresolved scales on the resolved prognostic variables in a climate model can be represented as Gaussian white noise. In this paradigm, the governing equations for the climate model become stochastic differential equations, explicitly including randomness that represents effects of unresolved processes. Furthermore, this randomness may be specified using only its first two statistical moments because a Gaussian distribution is fully defined by its mean and variance. Hasselmann's paradigm extends the deterministically parameterised equation (7.1.1) to the stochastic differential equation

$$\frac{\mathrm{d}X}{\mathrm{d}t} = R(X) + U(X) + \sigma z(t), \tag{7.2.1}$$

where σ is the square root of the Gaussian variance characterising the uncertainty of the parameterised effects and z represents (unit variance) Gaussian white noise. When numerically integrated in discrete time, the effect of the term $\sigma z(t)$ is to choose randomly a value from a Gaussian distribution centred on $U(X)$. This random variate corresponds to an individual value from the point cloud vertically above or below the relevant point on $U(X)$ in Fig. 7.1. Although others had been thinking along similar lines at the time (Leith 1975; Lorenz 1975), Hasselmann (1976) was the first to provide a quantitative paradigm for stochastic parameterisation in climate modelling.

Not surprisingly, a stochastically parameterised model of the sort shown in equation (7.2.1) can exhibit much more variability than its deterministic counterpart, equation (7.1.1). Hasselmann (1976) points out that, both conceptually and mathematically, a simple climate model with stochastic parameterisation is analogous to Brownian motion, in which pollen grains exhibit (microscopic) motions that are the integral responses to their interactions with the molecules of the fluid in which they are suspended. In this analogy, the resulting random motions of the pollen grains might correspond to the variations in the average Earth temperature on the time scales of centuries or millennia, in response to the integrated effects of random synoptic (weather) disturbances in the general circulation of the atmosphere. Integration of a GCM of the atmosphere–ocean system would correspond to simulation of the dynamics of the individual fluid molecules, given fixed or only slightly perturbed positions of the pollen grains. Climate simulation using a simplified model having only non-stochastic parameterisations would correspond (in Hasselmann's words) 'to determining the large-particle paths by considering only the interactions between the large particles themselves and the *mean* pressure and stress fields set up by the small-particle motions . . .'. Of course, the dynamics in this last analogy are drastically impoverished.

In one of the first studies to implement Hasselmann's paradigm, Lemke (1977) augmented a simple and well-studied zonally (i.e. around latitude circles) and annually averaged conceptual climate model (Budyko 1969; Sellers 1969) with a stochastic component. The conventional deterministic version of this model computes, for each resolved latitude, the energy balance as the sum of incoming and outgoing radiation (both as functions of the prognostic temperature), and an averaged meridional (i.e. north–south) heat flux convergence due to the net effect of advection by synoptic waves (parameterised as being proportional to the difference between the prognostic temperature at the latitude in question and the globally

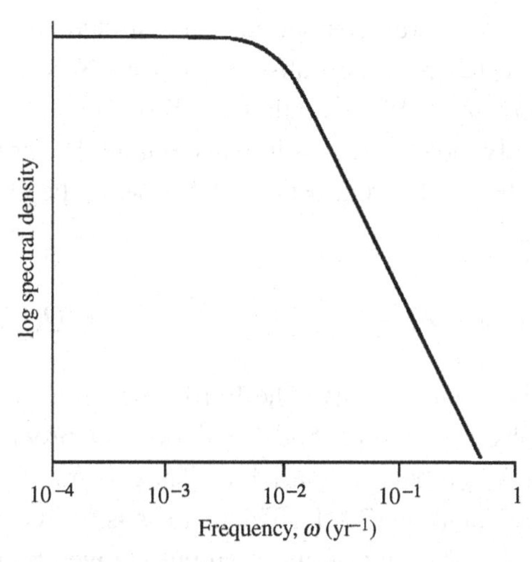

Figure 7.3 Generalised midlatitude temperature spectrum from the stochastic energy balance climate model of Lemke (1977).

averaged temperature). These radiative and advective processes correspond to the terms $R(T)$ and $U(T)$, respectively, in equations (7.1.1) and (7.2.1). In addition, unresolved deviations from the average parameterised meridional heat flux divergence were represented by white-noise forcing [$\sigma z(t)$ in equation (7.2.1)]. Subsequently, Lorenz (1979) justified stochastic, non-diffusive meridional heat flux parameterisations on the basis of atmospheric observations.

A key quantity in (7.2.1) is the standard deviation ('amplitude') of the parameterised random synoptic forcing, σ. Setting this standard deviation equal to the average midlatitude meridional heat flux (matching to the order of magnitude the available observations of heat transport by synoptic waves), Lemke (1977) obtained power spectra for the prognostic temperatures having the form shown in Fig. 7.3. Broadly consistent with observed climate spectra, apart from peaks attributed to external (prominently, orbital) variations in solar input, these spectra are flat at low frequencies and exhibit approximately an ω^{-2} dependence at higher frequencies. Depending on the choice of the magnitude of the modelled albedo dependence on temperature, Lemke (1977) generally found the transition between these two portions of the spectra to occur between 10^{-2} and 10^{-3} per year.

7.3 Stochastic parameterisation and regime dynamics

Besides providing additional and realistic variability in simple models, stochastic parameterisations can also produce surprising new behaviours that are qualitatively different from those of their deterministically parameterised counterparts. This characteristic of stochastic parameterisations is illustrated here using a conceptual model for ice ages, which has been abstracted from Nicolis & Nicolis (1981), Sutera (1981), Benzi *et al.* (1982) and Nicolis (1982).

The model is a highly idealised zero-dimensional model for the global radiative energy balance, with the single prognostic variable being the globally and annually averaged temperature T, satisfying

$$C\frac{\mathrm{d}T}{\mathrm{d}t} = -E_{\mathrm{out}} + E_{\mathrm{in}} = -\varepsilon(5.67 \times 10^{-8})T^4 + Q[1 - \alpha(T)], \qquad (7.3.1)$$

where C is an effective heat capacity. The Earth loses energy to space according to the well-known physics of blackbody radiation, in proportion to the fourth power of the prognostic temperature variable. Thus, the rate of energy loss $-E_{\mathrm{out}}$ corresponds to $R(T)$ in equation (7.1.1). The factor ε is an average effective global emissivity (approx. 0.6). The rate of energy input is given by the proportion $1 - \alpha(T)$ of the annually and globally averaged solar flux Q (approx. 342 W m^{-2}) that is absorbed, where $\alpha(T)$ is the average planetary albedo parameterised as a function of the prognostic temperature. Thus, the energy gain term E_{in} might be associated with the deterministic parameterisation $U(T)$ in equation (7.1.1).

One view of the workings of this simple deterministic model is shown in Fig. 7.4*a*, where E_{in} (dashed curve) and E_{out} (solid curve) are plotted as functions of temperature. The parameterised albedo function $\alpha(T)$ has been chosen in a way that E_{in} crosses the E_{out} curve in three places, yielding the three equilibrium solutions T_-, T_0 and T_+. The solutions T_- and T_+ are stable equilibria because $\mathrm{d}T/\mathrm{d}t > 0$ for $T < T_-$ and $T_0 < T < T_+$, and $\mathrm{d}T/\mathrm{d}t < 0$ for $T_- < T < T_0$ and $T > T_+$. Thus, for any initial $T_{\mathrm{init}} \neq T_0$, the dynamics in (7.3.1) will ultimately arrive at one or the other of the stable solutions T_- or T_+. Although $\mathrm{d}T/\mathrm{d}t = 0$ at $T = T_0$, the slightest perturbation from this unstable solution will also tend towards one of the two stable solutions.

Another view of the dynamics and stability of this simple model is shown by the potential function Ψ in Fig. 7.4*b*, which is related to equation (7.3.1) through

$$-\frac{\partial \Psi}{\partial T} = \frac{\mathrm{d}T}{\mathrm{d}t}. \qquad (7.3.2)$$

The dynamics of (7.3.1) can be visualised by imagining a (two-dimensional) laboratory apparatus having the shape of the potential function in Fig. 7.4*b*, on which a steel ball can roll under the influence of gravity. Because the density of the steel is high, any effects of air turbulence in the laboratory have negligible effect on its motions. Unless the ball is initially balanced rather precisely at T_0, it will promptly roll into one or the other of the potential wells (or 'basins of attraction') and stop at either T_- or T_+. From that point forward, $\mathrm{d}T/\mathrm{d}t = 0$, and the model climate

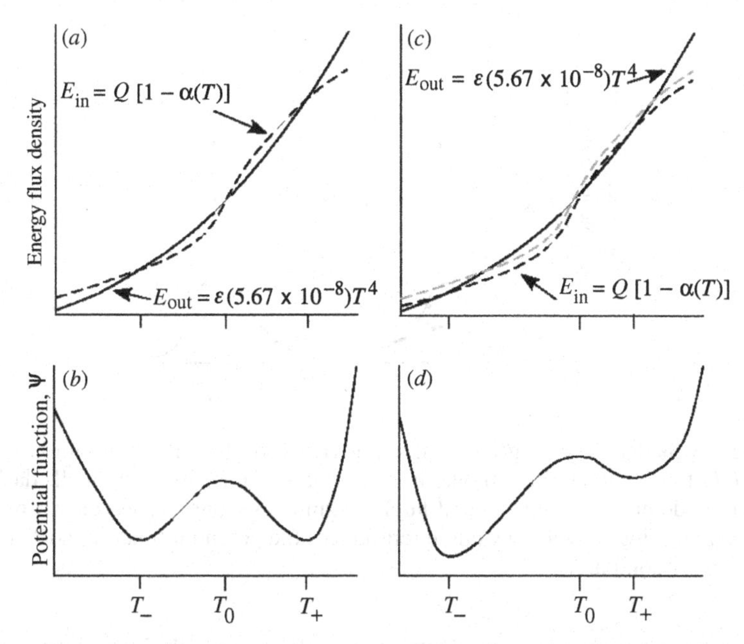

Figure 7.4 (*a*, *c*) Energy inputs and outputs, and (*b*, *d*) sketches of the corresponding potential functions, for the simple zero-dimensional global energy balance model with parameterised albedo feedback function $\alpha(T)$ [equation (7.3.1)]. The energy input (black dashed curve) in (*c*) reflects decreased solar output relative to (*a*), yielding displacements of the equilibrium solutions T_-, T_0 and T_+, and the changes in the potential function in (*d*) relative to (*b*). The energy input function from (*a*) is reproduced as the grey dashed curve in (*c*) for comparison.

exhibits no further temperature variability, even though two temperature regimes, corresponding to the two potential wells, are physically possible.

Consider now this same simple climate model that has been extended with a stochastic parameterisation to include variations in the global energy balance, which are not related to the global albedo according to the function $1 - \alpha(T)$ shown in Fig. 7.4*a*,

$$C\frac{\mathrm{d}T}{\mathrm{d}t} = -E_{\mathrm{out}} + E_{\mathrm{in}} + \sigma z(t) = -\varepsilon(5.67 \times 10^{-8})T^4 + Q[1 - \alpha(T)] + \sigma z(t). \quad (7.3.3)$$

Because the Gaussian noise $z(t)$ is additive, its effect does not depend on T, so that the potential function [equation (7.3.2)] for both (7.3.1) and (7.3.3) is the same. The dynamics expressed by (7.3.3) correspond to an air-filled, low-density ball rolling on the apparatus illustrated in Fig. 7.4*b*. Furthermore, the laboratory windows are open, so the motions of the ball are subject to turbulent accelerations that increase in intensity with increasing σ.

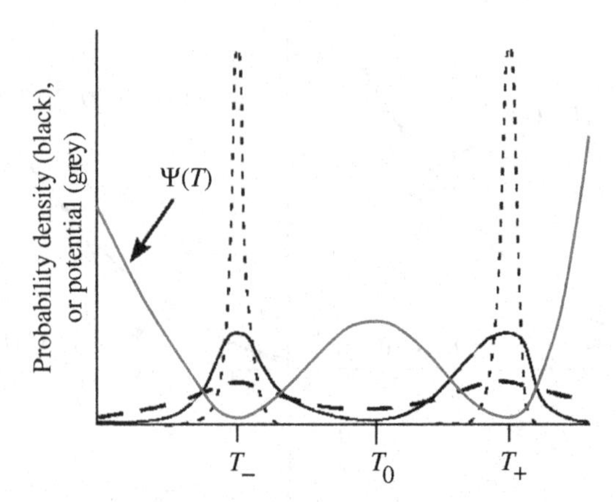

Figure 7.5 Sketches of long-run ('climatological') probability distribution functions $f(T)$ for the stochastic climate model of equation (7.3.3), for small (dashed curve), moderate (solid curve) and large magnitudes (long-dashed curve) of the stochastic forcing σ. The grey curve reproduces the potential function $\Psi(T)$ from Fig. 7.4b for comparison.

In this new situation, the ball still tends to roll towards the two stable solutions T_- and T_+, but nearly always remains in motion due to the erratic turbulent accelerations. If the wind is light (small σ), the random accelerations are small compared with the gravitational acceleration propelling the ball towards the bottoms of the wells, and the ball will usually reside very near T_- or T_+. In this small-σ situation, the ball will almost never switch from one well to the other. For moderate σ, the ball will move fairly frequently between the two wells, but will reside relatively near their centres more often than between or outside of them. If the wind is very strong (large σ), then the topography of the potential function apparatus makes little difference to the position of the ball, so the probabilities for its location are almost independent of position.

Figure 7.5 shows the probability density functions $f(T)$ for the location of the ball in these three situations. In terms of the climate model, these correspond to the climatological distributions for globally averaged temperature, for the three levels of random forcing, σ. Quantitatively, these probability density functions could be computed according to (Nicolis & Nicolis 1981)

$$f(T) \propto \exp\left[-\frac{2}{\sigma^2}\Psi(T)\right], \tag{7.3.4}$$

which becomes progressively flatter and more independent of the potential function as the variance of the stochastic forcing increases.

The double-peaked probability function for small σ in Fig. 7.5 may seem surprising since it implies occasional jumps between the two regimes, each of which requires a relatively large stochastic perturbation or sequence of perturbations. Such events become progressively rarer as σ decreases, but are not impossible unless $\sigma = 0$ [i.e. for the deterministic climate model in equation (7.3.1)]. It can be useful to characterise this tendency of the modelled system to jump between regimes using the average, or typical, time to achieve such a jump (Nicolis & Nicolis 1981; Sutera 1981),

$$\tau \propto \exp\left[\frac{2}{\sigma^2}\Delta\Psi\right], \tag{7.3.5}$$

where $\Delta\Psi$ is the difference between the bottom of the potential well currently occupied and the local maximum $\Psi(T_0)$, specifying the height of the 'barrier'. For sufficiently small σ^2, this characteristic time could be large enough (in comparison with, say, the age of the Earth) that the transitions are theoretical possibilities only.

Simple climate models of the sort exemplified by equation (7.3.3) and Fig. 7.4*a*,*b* were originally constructed for conceptual investigation of the approximately 10^5-year ice-age cycle. In this context, $T_+ \approx 288$ K might be associated with interglacials and $T_- \approx 278$ K with ice-age conditions. Jumps between the two regimes would then correspond to oscillations between these two climate states. However, working with a stochastically parameterised model qualitatively similar to (7.3.3), Benzi *et al.* (1982) found a purely red temperature spectrum, with no hint of ice-age-like periodicities.

Consider now a modification to this simple climate model, in which the incoming solar radiation, Q, in equations (7.3.1) and (7.3.3), varies periodically in a way that mimics the Milankovich eccentricity cycle,

$$Q = Q_0[1 + 0.001\sin(2\pi\omega t)], \tag{7.3.6}$$

where Q_0 is the long-term average of the incoming solar flux and $\omega = 10^{-5}\ \text{y}^{-1}$. Figure 7.4*c*,*d* indicates the effect on equations (7.3.1) and (7.3.3) of a reduction of solar flux that might occur for $t \approx 75000$ y. The effect is to lower the dashed curve for E_{in} in Fig. 7.4*c* relative to its level in Fig. 7.4*a* (which is reproduced in Fig. 7.4*c* with the grey dashed curve, for comparison). Intersections of the modified E_{in} function with E_{out} now define stable equilibria T_- and T_+ that are colder than their counterparts in Fig. 7.4*a*, while the unstable equilibrium solution T_0 increases towards T_+. Concurrently, the cold potential well $\Psi(T_-)$ becomes deeper relative to the well for the warm regime $\Psi(T_+)$. To see why this change in the shape of the potential function will occur, consider the situation where E_{in} is reduced to the point that it only just touches E_{out} at a warm equilibrium T_+. At that point T_0 merges with T_+ and the previously existing 'warm' potential well shallows sufficiently for that local minimum in Ψ to disappear. The reverse effects occur for increases in solar input, e.g. at $t \approx 25000$ y.

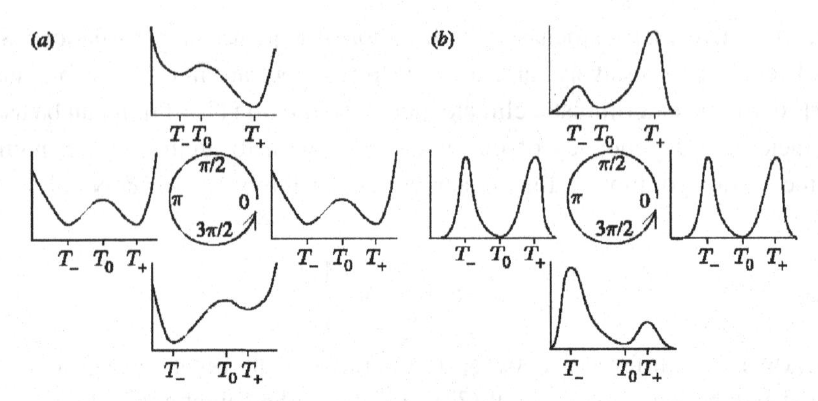

Figure 7.6 (*a*) Schematic cycle of the potential function $\Psi(T)$ for equations (7.3.1) and (7.3.3), resulting from periodic variation in the solar input Q according to equation (7.3.6). (*b*) Corresponding climatological probability distributions $f(T)$ for the stochastically parameterised version of the model [equation (6.3.3)].

Figure 7.6*a* illustrates the cycle of potential functions that follows from periodic variation in the external solar forcing specified by equation (7.3.6). As solar input increases, the stable equilibrium temperatures T_- and T_+ increase, while the depth of the well representing the warm regime increases at the expense of the cold regime well ($\pi/2$). As the solar input subsequently decreases, the topography of the potential functions passes through its initial configuration (π), following which the stable equilibria cool further ($3\pi/2$), with the cold regime potential well deepening at the expense of the warm regime. The overall effect is that the nature of the model dynamics is being affected by the changes in the external solar input.

Imagine the deterministic dynamics of equation (7.3.1) as influenced by the periodic solar forcing, again in terms of the laboratory analogue. The track along which the steel ball may roll changes shape periodically, as shown by the cycle in Fig. 7.6*a*. If the ball begins at $T = T_+$ at $t = 0$, it will execute regular periodic excursions around this initial temperature, with period $1/\omega = 10^5$ y, as the location of the maximum depth of the warm well periodically increases and decreases. However, the steel ball remains trapped in whichever well it begins, and its maximum temperature excursions are substantially smaller than the range of temperature variation between ice ages and interglacials, $T_+ - T_-$. The most prominent climatic temperature excursions in recent geological history have occurred with a period of approximately 10^5 y, but the roughly 0.1% amplitude of the eccentricity variation in solar radiation is the smallest of the three Milankovich cycles. It appears to be inadequate to explain the ice ages, not only in this very simple model but also in other more realistic models.

Qualitatively different behaviour occurs when the stochastically parameterised model [equation (7.3.3)] is subjected to the periodic solar forcing of equation

(7.3.6). In particular, the stochastic forcing $\sigma z(t)$ now allows solutions to make transitions between the interglacial and ice-age regimes, resulting in probability distributions for T that range over a broad range of possibilities. Furthermore, according to equation (7.3.4) and as shown in Fig. 7.6*b*, these climatological temperature distributions $f(T)$ execute a periodic cycle that follows the 10^5-y cycle of the potential function in Fig. 7.6*a*. At times of higher solar forcing ($\pi/2$), the warmer climate (around T_+) becomes the more favoured state. Similarly, the colder climate (around T_-) becomes more likely at times of lower solar forcing ($3\pi/2$).

The existence of these periodically shifting probability distributions is not by itself sufficient for the solutions $T(t)$ of equations (7.3.3) and (7.3.6) to exhibit a realistic behaviour (e.g. a prominent spectral peak near $\omega = 10^{-5}$ y^{-1}). In addition, the variance of the stochastic forcing in equation (7.3.3) must be of an appropriate magnitude, exemplifying a phenomenon that has come to be known as 'stochastic resonance' (Benzi *et al.* 1982). This is not a resonance in the usual sense of a coupling between two preferred frequencies, because the spectrum of the white-noise forcing $z(t)$ is completely flat: it has no preferred frequencies. Rather, stochastic resonance occurs when the variance σ^2 of the stochastic forcing yields a characteristic transition time τ [equation (7.3.5)] that matches approximately the period ω^{-1} of the external forcing. Stochastic forcing that is too weak will result in transitions that occur too rarely, yielding solutions similar to those for the deterministic dynamics [equation (7.3.1)], although with a small amount of superimposed variability. Stochastic forcing that is too strong will tend to produce many more than two regime transitions over the course of a single period of ω^{-1} y.

When τ and ω^{-1} are well matched, the random transitions will tend to occur with the progression of the solar forcing cycle as the potential barrier $\Delta\Psi$ decreases, while the simultaneous deepening of the preferred potential well will tend to retain the solution in the (new) preferred regime until the topography of the potential function changes again. The resulting time series for modelled temperature qualitatively resemble transitions between ice ages and interglacials (Benzi *et al.* 1982; Imkeller 2001), with the temperature response lagging the solar forcing by approximately $\pi/4$ (Nicolis 1982). This capacity for appropriately scaled random perturbations to amplify weak periodic forcings through stochastic resonance was first noted by Benzi *et al.* (1982) in the context of this conceptual climate model and has since seen numerous applications in various areas of physics (e.g. Gammaitoni *et al.* 1998).

The issue of regime switching among multiple climate equilibria is broader than the particular application to ice ages presented in this section, and conceptual models with stochastic parameterisations have been applied to regime behaviour in other settings as well. These include simple models of multiple equilibria in the North Atlantic thermohaline circulation (Stommel & Young 1993; Cessi 1994; Monahan 2002; Monahan *et al.* 2008) and in the large-scale atmospheric circulation (Egger 1981; Benzi *et al.* 1984; De Swart & Grassman 1987). Stochastic parameterisation has also been found to improve the resemblance to the well-known three-dimensional Lorenz (1963) model of a two-dimensional principal components truncation of it (Selten 1995; Palmer 2001), in terms of both the individual trajectories and the overall bimodal 'climatological distribution'.

7.4 Conclusion

This chapter has reviewed a modest selection of the earlier and conceptually simpler work on stochastic parameterisation of climate models, both for a historical perspective and as a pedagogical device. Not surprisingly, a prominent effect of including stochastic variability in parameterisations is to increase the variability exhibited by the models overall, to more realistic levels. This effect has been found also for stochastically parameterised versions of more complete and complex climate models (e.g. Lin & Neelin 2000; Sardeshmukh & Sura 2007). More recent work on stochastic parameterisation in conceptual models has been summarised by Imkeller & von Storch (2001) and Imkeller & Monahan (2002).

The simplest approach to stochastic parameterisation is through additive white (i.e. temporally independent) noise. A prominent time scale separation between the resolved and parameterised processes implies that white noise will be a reasonable representation because successive averages of an autocorrelated process over sufficiently long periods (although still shorter than the resolved time scale) will be approximately independent of each other. When this condition is not met, representation of the unresolved parameterised effects using an autocorrelated stochastic process presents no problem in simulation (e.g. Egger 1981; Benzi *et al.* 1984; De Swart & Grassman 1987; Monahan 2002; Wilks 2005), although it does substantially complicate the calculation of analytical mathematical results (e.g. Egger 1981; Stommel & Young 1993). Similarly, in spatially explicit models, it has been found that spatially correlated random forcing may be appropriate (e.g. Lemke 1977; Penland & Sardeshmukh 1995; Buizza *et al.* 1999).

An important conclusion to be drawn from the very simple studies summarised here is that stochastic influences may fundamentally affect the simulation of regime behaviour in climate models, and in particular that rates and probabilities for regime transitions depend on the presence and magnitude of the stochastic input. Realistic modelling of climate changes may depend critically on realistic modelling of these regime changes. Palmer (1999) has argued that changes in the climate (including those due to anthropogenic influences) may manifest themselves in large part through changes of residence frequencies in its preferred regimes, noting that models with insufficient internal variability tend to exhibit overpopulation of the more stable regimes (Molteni & Tibaldi 1990). A more unsettling view is that climate change may manifest itself through an abrupt shift to a qualitatively different regime (e.g. Broecker 1997).

References

Benzi, R., Parisi, G., Sutera, A. & Vulpiani, A. 1982 Stochastic resonance in climatic change. *Tellus*, **34**, 10–16.

Benzi, R., Hansen, A. R. & Sutera, A. 1984 On stochastic perturbation of simple blocking models. *Q. J. R. Meteorol. Soc.*, **110**, 393–409 (doi:10.1002/qj.49711046406).

Broecker, W. S. 1997 Thermohaline circulation, the Achilles heel of our climate system: will man-made CO_2 upset the current balance? *Science*, **278**, 1582–1588 (doi:10.1126/science.278.5343.1582).

Budyko, M. I. 1969 The effect of solar radiation variations on the climate of the Earth. *Tellus*, **21**, 611–619.

Buizza, R., Miller, M. J. & Palmer, T. N. 1999 Stochastic simulation of model uncertainties in the ECMWF ensemble prediction system. *Q. J. R. Meteorol. Soc.*, **125**, 2887–2908 (doi:10.1256/smsqj.56005).

Cessi, P. 1994 Simple box model of stochastically forced thermohaline flow. *J. Phys. Oceanogr.*, **24**, 1911–1920 (doi:10.1175/1520–0485(1994)024<1911:ASBMOS>2.0.CO;2).

Craig, G. C. & Cohen, B. G. 2006 Fluctuations in an equilibrium convective ensemble. Part I: theoretical formulation. *J. Atmos. Sci.*, **63**, 1996–2004 (doi:10.1175/JAS3709.1).

De Swart, H. E. & Grassman, J. 1987 Effect of stochastic perturbations on a low-order spectral model of the atmospheric circulation. *Tellus*, **39A**, 10–24.

Egger, J. 1981 Stochastically driven large-scale circulations with multiple equilibria. *J. Atmos. Sci.*, **38**, 2606–2618 (doi:10.1175/1520–0469(1981)038<2606:SDLSCW>2.0.CO;2).

Gammaitoni, L., Hanggi, P., Jung, P. & Marchesoni, F. 1998 Stochastic resonance. *Rev. Mod. Phys.*, **70**, 223–287 (doi:10.1103/RevModPhys.70.223).

Hasselmann, K. 1976 Stochastic climate models. Part I. Theory. *Tellus*, **28**, 473–485.

Imkeller, P. 2001 Energy balance models – viewed from stochastic dynamics. In *Stochastic Climate Models*, ed. P. Imkeller & von J.-S. Storch, pp. 213–240. Birkhauser.

Imkeller, P. & Monahan, A. H. 2002 Conceptual stochastic climate models. *Stoch. Dyn.*, **2**, 311–326 (doi:10.1142/S0219493702000443).

Imkeller, P. & von Storch, J.-S. (eds.) 2001 *Stochastic Climate Models*, Birkhauser.

Kotz, S. & Adams, J. W. 1964 Distribution of sums of identically distributed exponentially correlated gamma-variables. *Ann. Math. Stat.*, **35**, 277–283 (doi:10.1214/aoms/1177703750).

Leith, C. 1975 The design of a statistical-dynamical climate model and statistical constraints on the predictability of climate. In *The Physical Basis of Climate and Climate Modeling*. WMO-GARP Publication Series, no. 16, pp. 137–141.

Lemke, P. 1977 Stochastic climate models. Part 3. Application to zonally averaged energy models. *Tellus*, **29**, 385–392.

Lin, J. W.-B. & Neelin, J. D. 2000 Influence of a stochastic moist convective parameterisation on tropical climate variability. *Geophys. Res. Lett.*, **27**, 3691–3694 (doi:10.1029/2000GL011964).

Lindgren, B. W. 1968 *Statistical Theory*. Macmillan.

Lorenz, E. N. 1963 Deterministic non-periodic flow. *J. Atmos. Sci.*, **20**, 130–141 (doi:10.1175/1520–0469(1963)020<0130:DNF>2.0.CO;2).

Lorenz, E. N. 1975 Climatic predictability. In *The Physical Basis of Climate and Climate Modeling*. WMO-GARP Publication Series, no. 16, pp. 133–136.

Lorenz, E. N. 1979 Forced and free variations of weather and climate. *J. Atmos. Sci.*, **36**, 1367 (doi:10.1175/1520–0469(1979)036<1367:FAFVOW>2.0.CO;2).

Lorenz, E. N. 1996 Predictability – a problem partly solved. In *Proceedings of the Seminar on Predictability*, vol. 1, pp. 1–18. ECMWF.

Molteni, F. & Tibaldi, S. 1990 Regimes in the wintertime circulation over northern extratropics. II. Consequences for dynamical predictability. *Q. J. R. Meteorol. Soc.*, **116**, 1263–1288 (doi:10.1002/qj.49711649602).

Monahan, A. H. 2002 Correlation effects in a simple stochastic model of the thermohaline circulation. *Stochastics Dyn.*, **2**, 437–462 (doi:10.1142/S0219493702000510).

Monahan, A. H., Alexander, J. & Weaver, A. J. 2008 Stochastic models of the meridional overturning circulation: time scales and patterns of variability. *Phil. Trans. R. Soc. A*, **366**, 2527–2544 (doi:10.1098/rsta.2008.0045).

Nicolis, C. 1982 Stochastic aspects of climatic transitions – response to a periodic forcing. *Telllus*, **34**, 1–9.

Nicolis, C. & Nicolis, G. 1981 Stochastic aspects of climatic transitions – additive fluctuations. *Tellus*, **33**, 225–234.

Palmer, T. N. 1999 A nonlinear dynamical perspective on climate prediction. *J. Climate*, **12**, 575–591 (doi:10.1175/1520–0442(1999)012<0575:ANDPOC>2.0.CO;2).

Palmer, T. N. 2001 A nonlinear dynamical perspective on model error: a proposal for non-local stochastic-dynamic parmetrization in weather and climate prediction models. *Q. J. R. Meteorol. Soc.*, **127**, 279–304 (doi:10.1002/qj.49712757202).

Palmer, T. N., Shutts, G. J., Hagedorn, R. *et al.* 2005 Representing model uncertainty in weather and climate prediction. *Annu. Rev. Earth Planet. Sci.*, **33**, 163–193 (doi:10.1146/annurev. earth.33.092203.122552).

Penland, C. & Sardeshmukh, P. 1995 The optimal growth of tropical sea surface temperature anomalies. *J. Climate*, **8**, 1999–2024 (doi:10.1175/1520–0442(1995)008<1999:TOGOTS>2.0.CO;2).

Saltzman, B. 1978 A survey of statistical-dynamical models of the terrestrial climate. *Adv. Geophys.*, **20**, 183–304.

Sardeshmukh, P. D. & Sura, P. 2007 Multiscale impacts of variable heating in climate. *J. Climate*, **20**, 5677–5695 (doi:10.1175/2007JCLI1411.1).

Schneider, S. H. & Dickinson, R. E. 1974 Climate modeling. *Rev. Geophys. Space Phys.*, **12**, 447–493 (doi:10.1029/RG012i003p00447).

Sellers, W. D. 1969 A new global climate model based on the energy balance of the earth-atmosphere system. *J. Appl. Meteorol.*, **8**, 392–400 (doi:10.1175/1520–0450(1969)008<0392:AGCMBO>2.0.CO;2).

Selten, F. M. 1995 An efficient empirical description of large-scale atmospheric dynamics. PhD thesis, p. 169. Vrije Universiteit te Amsterdam.

Stommel, H. M. & Young, W. R. 1993 The average T–S relation of a stochastically forced box model. *J. Phys. Oceanogr.*, **23**, 151–158 (doi:10.1175/1520–0485(1993)023<0151:TAROAS>2.0.CO;2).

Sutera, A. 1981 On stochastic perturbation and long-term climate behaviour. *Q. J. R. Meteorol. Soc.*, **107**, 137–151 (doi:10.1256/smsqj.45108).

Wilks, D. S. 2005 Effects of stochastic parameterisations in the Lorenz '96 system. *Q. J. R. Meteorol. Soc.*, **131**, 389–407 (doi:10.1256/qj.04.03).

8

Challenges in stochastic modelling of quasi-geostrophic turbulence

TIMOTHY DELSOLE

This chapter reviews challenges in developing stochastic models of quasi-geostrophic turbulence. These challenges include reconciling the observed atmospheric energy spectrum with classical inertial range theory, rigorously justifying the parameterisations used to construct stochastic models, formulating closure theories for the values of the parameters in stochastic models, and incorporating moisture effects in the stochastic models. In addition, a turbulence closure theory is proposed for specifying the parameters in stochastic models of inhomogeneous turbulence. The central hypothesis is that the stochastic forcing in a region is a monotonic function of the local eddy variance. This hypothesis serves to specify the complete space-lag correlations of the stochastic forcing. Two further assumptions are made in the theory, namely (1) the parameterised nonlinear terms approximately conserve energy, and (2) the energy spectra of the eddies resolved by the stochastic model blend smoothly with the energy spectra predicted from inertial range theory. These assumptions eliminate all but a single parameter of the stochastic model. Comparisons of the predictions from this turbulence theory to fully nonlinear simulations are promising, although the stochastic model substantially overestimates the amount of energy at the largest scales. This last problem is characteristic of many other stochastic models, and its explanation constitutes another challenge in developing stochastic turbulence models.

8.1 Introduction

A major goal in climate science is to understand changes in climate that result from anthropogenic forcing, and to distinguish these changes from natural variations. Accomplishing this goal requires estimating the characteristics of climate variations under natural and anthropogenic forcings. The most convincing estimates are derived from comprehensive climate models that simulate individual atmospheric fluctuations explicitly by solving a set of partial differential equations numerically using time steps on the order of an hour. Unfortunately, confident conclusions require multiple climate simulations of hundreds to thousands of years. If, instead, the statistics of atmospheric fluctuations could be computed directly and more quickly than simulating individual realisations, then faster and computationally cheaper climate models could be constructed for climate study. This paper discusses a class of models that might serve this purpose.

Large-scale atmospheric fluctuations, such as those associated with weather with horizontal length scales of 1000 km or more, play a significant role in maintaining the climate balance by transporting significant amounts of heat, momentum and moisture (Peixoto & Oort 1992). These fluctuations also play a significant role in determining the variability of the climate system by acting as stochastic forcing of the ocean–land–ice system (Hasselmann 1976). For the most part, these fluctuations are believed to be manifestations of quasi-geostrophic turbulence (hereafter called QG turbulence) – that is, a type of turbulence characteristic of rapidly rotating, vertically stratified fluid systems. Ideally, the statistics of turbulent fluctuations should be predicted from well-established physical laws. Unfortunately, this goal encounters barriers that are well known and appreciated: dissipation plays a non-trivial role in turbulence and hence precludes the use of conventional statistical mechanics to derive the statistics directly, and nonlinearity of the governing equations precludes a closed representation of the eddy statistics. Two basic strategies for predicting the statistics of turbulent eddies have emerged: (1) find approximate statistics of the exact governing equations, or (2) find exact statistics of approximate governing equations. A major advantage of (2) is that the statistics are guaranteed to be realisable (e.g. energy is not negative, covariance matrices are positive semi-definite). In the context of turbulence models, the second strategy involves replacing the original nonlinear dynamical system by a model in which the nonlinear interaction between two components is parameterised by a stochastic term and a dissipation term. The damping and forcing are chosen to induce the correct statistical dynamics while also being computable – goals that are often at odds with each other. The resulting model is called a stochastic turbulence model.

Farrell & Ioannou (1993) proposed a class of stochastic models for predicting the statistics of quasi-geostrophic turbulence. This type of model differed dramatically from previous models in that it predicted the statistics of large-scale eddies responsible for most of the turbulent transport. These models have been reviewed comprehensively by DelSole (2004). The purpose of this chapter is to draw attention to challenges in developing these models further. Our goal is to stimulate research rather than to criticise. To begin with, in Section 8.2 the challenges in modelling turbulence in the inertial range are discussed. Specifically, several challenges in reconciling the connection between two-dimensional and quasi-geostrophic turbulence are considered, and the connection of inertial range models to understanding the observed spectrum of the atmosphere. Next, in Section 8.3, the basis of the models introduced by Farrell and Ioannou (1993) is discussed. A new closure theory is proposed in Section 8.4, which shows some promising results as well as some problems with the theory. The chapter closes with a list of challenges for which we have little to offer in the way of advice, but are worth stating in the hope of drawing attention to their existence.

8.2 Challenges in modelling inertial range turbulence

A review of QG turbulence would not be complete without some reference to two-dimensional turbulence (hereafter referred to as 2D turbulence), since the underlying systems share an important similarity. Specifically, 2D and QG inviscid flow are each governed by the conservation of a single scalar invariant, namely vorticity and potential vorticity, respectively (Charney 1971). Conservation of vorticity or potential vorticity profoundly constrains energy transfer in the respective systems. In three-dimensional turbulence, Kolmogorov (1941) proposed that at sufficiently large Reynolds numbers there could exist a range of length scales, called the inertial range, in which eddies transfer energy primarily locally in wavenumber space and in which the direct effects of dissipation and forcing could be ignored. In the inertial range, large eddies break up into smaller eddies, which in turn break up into smaller eddies, effectively transferring energy to small scales. In 2D turbulence, however, simultaneous conservation of energy and vorticity implies that the energy transfer cannot be local in wavenumber space, and that more energy flows upscale than downscale in statistical equilibrium (Kraichnan 1967; Eyink 1996; Gkioulekas & Tung 2007). It should be recognised that some confusion about these points exist in the literature, especially regarding some overreaching statements by Fjortoft (1953); see Merilees & Warn (1975) and Tung & Orlando (2003a).

According to arguments originally put forth by Kraichnan (1967), Leith (1968) and Batchelor (1969), two types of energy spectra can exist in the inertial range of 2D turbulence. The first depends only on total wavenumber k and energy transfer rate ε, and therefore by dimensional arguments must be of the form

$$e(k) = C\varepsilon^{2/3}k^{-5/3} \tag{8.2.1}$$

where C is a universal constant. The energy spectrum $e(k)$ is defined such that the total energy can be written as

$$E = \int_{k_{\min}}^{k_{\max}} e(k)dk, \tag{8.2.2}$$

where $k_{\min}$ and $k_{\max}$ are the minimum and maximum total wavenumbers in the inertial range. The second energy spectrum depends only on total wavenumber k and enstrophy transfer rate η, and therefore by dimensional arguments must be of the form

$$e(k) = C'\eta^{2/3}k^{-3} \tag{8.2.3}$$

where C' is a universal constant. In the case of two-dimensional turbulence, Kraichnan (1967) argued that if the inertial interactions act to spread energy among different wavenumbers, then the $-5/3$ power law given by (8.2.1) is characterised by an upscale energy transfer, the -3 power law given by (8.2.3) is characterised by a downscale enstrophy transfer, the two transfers cannot be self-similar in the same range of wavenumbers, and the enstrophy cascade cannot be local in wavenumber space. Furthermore, Kraichnan conjectured that if an infinite fluid were excited by isotropic stirring forces in a narrow band of wavenumbers, then the $-5/3$ energy-transfer range would carry energy to large scales while the -3 enstrophy-transfer range would carry vorticity to small scales. In other words, both power spectra could exist simultaneously in non-overlapping wavenumber ranges.

The extension of 2D inertial range theory to QG systems has been discussed from interesting perspectives by Salmon (1978, 1998), Hoyer & Sadourny (1982) and Larichev and Held (1995). A new element that emerges is that both energy and enstrophy injected by the large-scale *baroclinic* forcing are transferred downscale toward deformation scales, but the energy transferred down to deformation scales returns to large scales in *barotropic* form to be dissipated, whereas the enstrophy continues cascading to scales shorter than the deformation scale.

A variety of stochastic models have been proposed to describe inertial range turbulence. We will not discuss the details of these models here, as they have been thoroughly documented elsewhere (Leslie 1973; Rose & Sulem 1978; Kraichnan & Montgomery 1980; Lesieur 1990; Frisch 1995). Here we discuss certain recent results that challenge our understanding of inertial range turbulence and its relevance to the observed atmosphere.

First, Tran & Shepherd (2002) effectively show that *neither* the $-5/3$ *nor* -3 power laws can occur in 2D turbulence with forcing at intermediate scales and dissipation by viscosity. More precisely, Tran & Shepherd show that if a doubly periodic, 2D fluid were excited in a band of wavenumbers, and if the dissipation were molecular viscosity alone, then after a sufficiently long time the dissipation of enstrophy takes place on scales no smaller than the forcing scale, in which case there can be no direct enstrophy cascade. Furthermore, while an inverse energy cascade is not precluded, its energy spectrum must be steeper than a -3 power law, which precludes the $-5/3$ power law in this range. This result challenges the idea that the dissipation scale can be made arbitrarily small simply by reducing the dissipation rate. Instead, the result suggests that, regardless of viscosity or model resolution, the turbulence adjusts itself such that enstrophy dissipates at scales no smaller than the forcing scale. If Ekman drag is added to molecular viscosity, then enstrophy dissipation can take place on scales smaller than the forcing scale, but the -3 power law (with or without log correction) is not permitted. These results explain why the $-5/3$ and -3 power laws have not been obtained in numerical simulations employing these types of dissipation, despite tremendous advances in computational ability. Finally, if the dissipation occurs through a 'hypoviscosity', which dissipates large scales more than small scales, together with viscosity, then the $-5/3$ and -3 power laws are possible. This is consistent with the fact that numerical simulations that are claimed to exhibit both scalings, such as those of Borue (1993) and Lindborg & Alvelius (2000), employ an inverse viscosity. Thus, the work of Tran & Shepherd (2002), as well extensions by Tran & Bowman (2003, 2004), clarifies that the choice of dissipation profoundly constrains the scalings of the energy spectrum, and that past failures to produce the scalings may not be due to limited resolution, as is often claimed.

Second, there exists spectacular inconsistencies between inertial range theory and the observed atmosphere. Specifically, Nastrom & Gage (1985) show that the observed energy spectra in the atmosphere has an approximate -3 power law at *low* wavenumbers and an approximate $-5/3$ power law at *high* wavenumbers – precisely opposite to the classical dual cascade picture based on 2D turbulence. This finding has been supported consistently by subsequent observational studies (Hogstrom *et al.* 1999; Lacorata *et al.* 2004). Furthermore, Cho & Lindborg (2001) use the third-order structure function estimated from aircraft and balloon measurements to infer that the actual energy transfer is downscale in the range 10 km to 1000 km, in contrast to upscale energy cascade in the $-5/3$ power-law regime expected from classical inertial range theory. Straus & Ditlevsen (1999) also find a downscale energy transfer based on reanalysis data, although these data exhibit neither the -3 nor $-5/3$ power laws. Straus & Ditlevsen (1999) further show that the atmosphere satisfies few of the assumptions underlying inertial range theory: for example, the energy and enstrophy fluxes are not constant over a significant range of scales, and the energy and enstrophy forcing are not localised in wavenumber space.

Koshyk *et al.* (1999) and Koshyk & Hamilton (2001) claim to have reproduced the observed -3 and $-5/3$ ranges with a high-resolution global model. They find that divergent motions contribute significantly to the kinetic energy spectrum in the $-5/3$ range, suggesting that this regime is probably not well explained by two-dimensional (non-divergent) inertial range theory. However, the robustness of these results are unclear because the $-5/3$ range was simulated only over about a decade of length scales, and the spectra for divergent motions were sensitive to resolution. Lindborg (2005, 2006) argues that the $-5/3$ power law arises from a 'zig-zag' instability that causes eddies in strongly stratified fluids to break up into layers, which in turn break into smaller layers, thereby mimicking a forward cascade and its associated power law. Tung & Orlando (2003b) claim to have simulated the -3 and $-5/3$ power laws in a high-resolution QG model. They argue that the $-5/3$ power law emerges because a small downscale energy flux eventually dominates the inertial range, in which case dimensional arguments predict the $-5/3$ power law. On the other hand, Smith (2004) reproduces the -3 and $-5/3$ power laws in a forced 2D system, but shows that the $-5/3$ power law disappears if the resolution is doubled, suggesting that the $-5/3$ power law is an artifact of inadequate resolution. This latter suggestion raises another curious question: is it a coincidence that an apparent model artifact produces the same $-5/3$ power law as observations?

8.3 Challenges in modelling inhomogeneous turbulence

Inertial range theories are conspicuously silent about the fluxes induced by turbulence eddies. The reason for this is simple: isotropic, homogeneous eddies have, by definition, vanishing eddy fluxes. Dropping the homogeneous assumption in stochastic models of inertial range turbulence usually leads to such a large increase in numerical complexity that they cannot compete with direct numerical simulation. Clearly, a different approach to modelling turbulent eddy fluxes is required.

A characteristic feature of all inhomogeneous turbulent flows is the presence of significant strain rates in the flow. It has been recognised for at least 100 years (Orr 1907) that a properly configured perturbation to a shear flow can grow transiently to large amplitude by extracting energy from the flow through downgradient momentum fluxes. Farrell (1985, 1987, 1989) has shown that this mechanism also operates in baroclinic systems, and hence in QG systems. Since flows with significant strain rates preferentially amplify certain structures, the possibility exists that, under random excitation, the eddies that eventually dominate the eddy energy will have characteristic spatial structures that depend on the flow. In this sense, the eddy dynamics of shear flows differ fundamentally from that of isotropic turbulence, and this difference holds the key to new methods.

There are at least two reasons to expect the atmosphere to have significant large-scale, slowly varying zonal jets with significant strain rates. First, the climate system is subjected to large-scale radiative forcing from the Sun. This solar radiation produces a large-scale pole-to-equator temperature difference, which by thermal wind balance produces significant baroclinic jets. Second, there is a strong tendency for QG turbulence on a beta plane to evolve to large-scale, slowly varying, zonal jets (Rhines 1975; Vallis & Maltrud 1993; Manfroi & Young 1999; Smith & Waleffe 1999; Vallis 2006). These two mechanisms imply the presence of significant, large-scale, slowly varying strain rates in the atmosphere, which motivates a separation of motions into 'eddy' and 'background' fields. The precise basis for separation, however, is unclear. In practice, the background field is identified by a variety of statistical techniques, including principal component analysis (Majda *et al.* 2002), zonal averaging (Farrell & Ioannou 1993) and time averaging (Whitaker & Sardeshmukh, 1998). If the background flow varies on time scales much longer than the available sample, then the background flow will be indistinguishable from the time mean. Furthermore, if the system is non-stationary, the background and eddy components may change with time. An interesting question is whether a well-defined background flow can be derived directly from the equations of motion, rather than derived empirically from observations or simulations.

One attractive basis for distinguishing background and eddy flows is time scale. In particular, if there exists a strong separation in time scales between variables, then reduced models involving just slow variables can be derived rigorously using such methods as adiabatic elimination (Gardiner 1990), slaving (Haken 2004) and stochastic mode reduction (Majda *et al.* 2001). The question arises as to whether there exists an optimal method for distinguishing slow and fast variables. If the time scale of a process is measured by the integral time scale, defined as

$$T_1 = \int_0^\infty \rho_\tau \mathrm{d}\tau, \tag{8.3.1}$$

where ρ_τ is the autocorrelation function of the time series, then a procedure known as optimal persistence analysis (OPA) can find the optimal set of slow variables. More precisely, OPA determines a complete set of uncorrelated components such that the first has the maximum integral time scale, the second has the maximum integral time scale subject to being uncorrelated with the first, and so on (DelSole 2001a). Although OPA is not typically used to identify slow components, the simple point is made that no other technique can produce a stronger separation between slow and fast components. Note further that alternative methods for separating fast and slow components, such as EOF (empirical orthogonal function) analysis, do not preclude the possibility that some linear combination of the 'slow EOFs' has a fast time scale, or vice versa. In contrast, if OPA is used to identify slow and fast variables, then no linear combination of the fast variables can produce a slow variable, and vice versa.

A major step in the construction of stochastic turbulence models is discretisation of the governing equations. This step raises challenges as to whether solutions to the discrete equations relate to solutions of the continuous system. In addition, it requires parameterising the interaction between resolved scales and subgrid scales. These challenges often are circumvented by dispensing with the continuous system altogether and defining the true system to be the discrete system, including its nonlinear interactions, and then to develop turbulence models for the discrete system directly.

Let the state vector for the discrete system be denoted by the time-dependent vector $\mathbf{q}(t)$. Discretising the governing equations for QG flow leads to linear and quadratically nonlinear terms, which we represent symbolically as

$$\mathbf{q}_t = \mathbf{Lq} + \mathbf{N}(\mathbf{q}, \mathbf{q}) + \mathbf{F}, \tag{8.3.2}$$

where subscript t denotes a derivative with respect to time, $\mathbf{L}$ is a linear operator, $\mathbf{N}$ is a bilinear operator and $\mathbf{F}$ specifies the forcing and dissipation of the system. If the background and eddy basis vectors form a complete set, then the state vector $\mathbf{q}$ can be transformed into background–eddy space by a suitable linear transformation. Let this transformation be expressed as

$$\mathbf{q} = \mathbf{G}\begin{pmatrix}\mathbf{b}\\ \mathbf{e}\end{pmatrix}, \tag{8.3.3}$$

where the vectors $\mathbf{b}$ and $\mathbf{e}$ specify the amplitudes of the background and eddy components, respectively, and $\mathbf{G}$ is a non-singular transformation matrix. Since the transformation matrix $\mathbf{G}$ is invertible, any element of $\mathbf{b}$ and $\mathbf{e}$ can be obtained by projecting $\mathbf{q}$ with a suitable projection vector. Applying this projection to the governing equations (8.3.2) yields separate dynamical equations for the two components, which may be written symbolically as:

$$\mathbf{b}_t = \mathbf{L}_{bb}\mathbf{b} + \mathbf{L}_{be}\mathbf{e} + \mathbf{N}^b_{bb}(\mathbf{b}, \mathbf{b}) + \mathbf{N}^b_{be}(\mathbf{b}, \mathbf{e}) + \mathbf{N}^b_{ee}(\mathbf{e}, \mathbf{e}) + \mathbf{F}_b \tag{8.3.4}$$

$$\mathbf{e}_t = \mathbf{L}_{eb}\mathbf{b} + \mathbf{L}_{ee}\mathbf{e} + \mathbf{N}^e_{bb}(\mathbf{b}, \mathbf{b}) + \mathbf{N}^e_{be}(\mathbf{b}, \mathbf{e}) + \mathbf{N}^e_{ee}(\mathbf{e}, \mathbf{e}) + \mathbf{F}_e. \tag{8.3.5}$$

The above procedure does nothing more than change variables. The key step is to *parameterise the eddy–eddy nonlinear interactions by a Langevin model of the form*

$$\mathbf{N}^e_{ee}(\mathbf{e}, \mathbf{e}) = \mathbf{De} + \mathbf{w}, \tag{8.3.6}$$

where $\mathbf{D}$ is a dissipation operator and $\mathbf{w}$ is a stochastic forcing. Substituting this parameterisation in the original eddy equation (8.3.5) gives the stochastic turbulence model

$$\mathbf{e}_t = \mathbf{Ae} + \mathbf{w} + \hat{\mathbf{F}}, \tag{8.3.7}$$

where

$$\mathbf{A} = \mathbf{N}^e_{be}(\mathbf{b}, \cdot) + \mathbf{L}_{ee} + \mathbf{D} \tag{8.3.8}$$
$$\hat{\mathbf{F}} = \mathbf{L}_{eb}\mathbf{b} + \mathbf{N}^e_{bb}(\mathbf{b}, \mathbf{b}) + \mathbf{F}_e. \tag{8.3.9}$$

The dot notation denotes the operation $\mathbf{N}^e_{be}(\mathbf{b}, \cdot)\mathbf{e} = \mathbf{N}^e_{be}(\mathbf{b}, \mathbf{e})$. A few comments are in order. First, the term $\mathbf{N}^e_{be}(\mathbf{b}, \mathbf{e}) + \mathbf{L}_{ee}\mathbf{e}$ can be interpreted as the equations of motion linearised about the background flow. It follows that the dynamical operator $\mathbf{A}$ contains more than the linearised equations, specifically, it also contains an extra damping term that parameterises the dissipative effects of eddy–eddy nonlinear interactions. Second, the forcing term $\hat{\mathbf{F}}$ contains an eddy forcing $\mathbf{F}_e$ which may be a linear function of $\mathbf{e}$, in which case the linear component of $\mathbf{F}_e$ can be absorbed in the dynamical operator $\mathbf{A}$. Third, the forcing term $\hat{\mathbf{F}}$ contains terms that depend on the background variables $\mathbf{b}$, which seemingly implies that the eddy equations are forced by slowly varing terms which in turn will induce a slowly varying signal in the eddy variables. However, if the background flow and forcing are zonally symmetric, then the forcing $\hat{\mathbf{F}}$ vanishes identically. In other cases, the eddy variables often have approximately zero mean when averaged over time intervals long compared to the time scale of eddy fluctuations but short compared to the time scale of background fluctuations; i.e. $\langle \mathbf{e}_t \rangle \approx 0$, and $\langle \mathbf{e} \rangle \approx 0$. In this case, (8.3.7) implies that $\mathbf{w} + \hat{\mathbf{F}}$ has approximately zero mean, suggesting that we can, without loss of generality, drop the forcing term $\hat{\mathbf{F}}$ and assume the stochastic forcing $\mathbf{w}$ has zero mean. The full assumptions underlying this parameterisation are discussed in Majda *et al.* (2001, 2006, 2008).

It should be recognised that the above formulation allows the background flow to change in time, although these changes typically occur on time scales much longer than that of the eddies. Accordingly, one can envisage solving the equations stepwise – first the stochastic model is solved by neglecting time variations in the background flow, then the resulting eddy statistics are used to update the background flow, then the stochastic model is solved again but with an updated background flow. This procedure is repeated to derive a 'quasi-linear' equilibration. In the limit of large time scale separation between eddies and background flow, a reduced stochastic model in which the eddy equations do not appear explicitly can be derived (Majda *et al.* 2001). The quasi-linear model produces solutions that are consistent with the fully nonlinear solutions (DelSole & Farrell 1996), and has been used to model midlatitude jet vacillation, emergence of multiple jets, reorganisation of storm tracks and the quasi-biennial oscillation (Farrell & Ioannou 2003).

It is perhaps worthwhile to show the stochastic model explicitly for a simple case of quasi-geostrophic turbulence. The governing equation for quasi-geostrophic flow is

$$q_t + J(\psi, q) = F, \tag{8.3.10}$$

where the subscript denotes a partial derivative, q denotes potential vorticity and is related to the streamfunction ψ as $q = \psi_{xx} + \psi_{yy} + \psi_{zz} + \beta y$, β is the meridional gradient of the Coriolis parameter, $J(\psi, q) = \psi_x q_y - \psi_y q_x$ and F denotes forcing and dissipation (we consider a simplified set of equations in which density and static stability are constant and certain background state quantities are absorbed in the z-variable). Suppose the background flow is identified with the time mean flow of a fully nonlinear solution. In this case, the equations do not need to be projected into background–eddy space because the decomposition is simple: the background potential vorticity q^b is a single state vector and the eddy component is $q^e = q - q^b$. If background and eddy quantities are denoted by superscript b and e, respectively, then the eddy equations take the form

$$q_t^e + J(\psi^b, q^e) + J(\psi^e, q^b) = N^e, \tag{8.3.11}$$

where N^e contains all remaining terms in the eddy equation. All terms on the left-hand side are linear in ψ^e, therefore the discretised version of these equations can be written in the form

$$\mathbf{q}_t = \mathbf{L}\mathbf{q} + \mathbf{N}. \tag{8.3.12}$$

We now invoke the parameterisation (8.3.6),

$$\mathbf{N} = \mathbf{D}\mathbf{q} + \mathbf{w}, \tag{8.3.13}$$

where $\mathbf{D}$ is a dissipation operator and $\mathbf{w}$ is a stochastic forcing. Substituting this parameterisation into the eddy equation (8.3.12) puts the QG stochastic model into the form (8.3.7). The stochastic forcing is assumed to be white noise in time with zero mean and spatial covariance matrix $\mathbf{Q}$:

$$\langle \mathbf{w}(t + \tau)\mathbf{w}^{\mathrm{T}}(t) \rangle = \alpha \mathbf{Q} \delta(\tau), \tag{8.3.14}$$

where α is a parameter that specifies the overall amplitude of the forcing and superscript T denotes the transpose operation. Also, the dissipation operator is assumed to be of the general form

$$\mathbf{D} = \sum_{j=0}^{J} (-1)^{j+1} \nu_j \nabla^{2j}, \tag{8.3.15}$$

where $\nu_0, \nu_1, \ldots, \nu_J, J$ are (unknown) parameters. Substituting the parameterisation (8.3.13) into the eddy equation (8.3.12) yields the stochastic model

$$\mathbf{q}_t = \mathbf{A}\mathbf{q} + \mathbf{w}, \tag{8.3.16}$$

where $\mathbf{A} = \mathbf{L} + \mathbf{D}$. Solutions to the stochastic model (8.3.16), or equivalently (8.3.13), are well known (see Gardiner 1990; DelSole 2004). In particular, the stationary covariance matrix of $\mathbf{q}$ is given by

$$\mathbf{C} = \int_0^\infty e^{\mathbf{A}s}\mathbf{Q}e^{\mathbf{A}^{\mathrm{T}}s}\mathrm{d}s. \tag{8.3.17}$$

The above model has the following unspecified parameters: $\alpha, \nu_0, \ldots, \nu_J, J$, and $\mathbf{Q}$. The matrix $\mathbf{Q}$ could contain thousands of parameters, depending on the dimension of the system. The solution to this model cannot be obtained without specifying these parameters.

8.4 Closure theory for inhomogeneous turbulence

The key step in constructing a stochastic turbulence model, embodied in (8.3.6), is the parameterisation of the nonlinear interaction between eddies by a stochastic term and dissipation term. However, no satisfactory theory exists for the form of the dissipation operator or for the statistics of the stochastic forcing. One of the most significant discoveries in stochastic turbulence modelling is that very simple assumptions for the dissipation and forcing can yield realistic eddy statistics in background flows with significant strain rates. Specifically, Farrell & Ioannou (1993, 1994, 1995, 1996) and DelSole & Farrell (1995, 1996) showed that a quasi-geostrophic model linearised about the time-mean, zonal-mean flow, forced by white noise with a white energy spectrum on large scales, and dissipated at a uniform rate, can produce realistic structures for the heat fluxes, momentum fluxes and variances. These models contain essentially three empirical parameters: (1) an eddy dissipation rate ν_0, typically 0.1 to 0.5 d^{-1}, (2) an energy injection rate due to random forcing α and (3) a cutoff wavenumber above which the random forcing vanishes. These parameters are chosen empirically to fit observations. DelSole (2004) reviews the extent to which these models can simulate the observed statistics.

The above stochastic models show a great deal of promise in elucidating the physics of turbulent eddies and the dynamics of low-frequency fluctuations in the background flow (see Chapter 3). Nevertheless, the models are limited because they contain empirical parameters that need to be tuned on a case-by-case basis. As such, these models cannot give insight into certain questions. For example, the models cannot explain what determines the amplitude of the eddies, because these amplitudes are determined by the noise variance, which is chosen empirically. Similarly, these models cannot explain what controls the time scale of eddies, because this time scale is controlled primarily by the potential vorticity damping, which also is chosen empirically. To address these kinds of questions, a closure theory for the parameters of the model needs to be formulated. In the remainder of this section, a new closure theory is proposed. Part of this theory is based on that proposed by DelSole (2001b), which essentially specifies the space-lag correlations of the stochastic forcing, thereby eliminating hundreds of parameters. The remainder of the theory specifies the remaining parameters, namely the dissipation parameters and the overall forcing amplitude.

Perhaps the most unrealistic assumption in the above models is that the stochastic forcing is statistically homogeneous even though the turbulence is inhomogeneous. If the variance of turbulent eddies is non-uniform in space, then the stochastic forcing by nonlinear interactions probably is not uniform in space. In fact, since the nonlinear terms scale quadratically with amplitude, a more plausible assumption is that the stochastic forcing varies monotonically with the local eddy variance. The simplest assumption is that the local eddy variance is linearly proportional to the local stochastic forcing variance, which is expressed mathematically as

$$\left\langle (\mathbf{q} - \langle \mathbf{q} \rangle)_i^2 \right\rangle = \gamma \left\langle (\mathbf{w})_i^2 \right\rangle, \tag{8.4.1}$$

where γ is a proportionality constant and i is an index specifying the local region. In terms of the stochastic model (8.3.16), this relation is

$$(\mathbf{C})_{ii} = \gamma \, (\mathbf{Q})_{ii} \tag{8.4.2}$$

which holds for each i individually (no implied summation over i). Note that these relations depend on the coordinate system used to represent the system, or equivalently, the measure used to define variance. For QG systems, enstrophy appears to be the measure of variance that works well (DelSole 2001b). Since these relations imply that the eddy variance has the same structure as the noise variance, we call it the *noise–response similarity hypothesis*.

We now solve the noise–response similarity hypothesis, following DelSole (2001b). Without loss of generality, assume that the stochastic forcing $\mathbf{w}$ is a linear combination of independent white-noise processes with zero mean and standard deviations $a_1, a_2, \ldots, a_N$, each multiplied respectively by the vectors $\mathbf{f}_1, \mathbf{f}_2, \ldots, \mathbf{f}_N$. In this case, the total noise covariance matrix is

$$\mathbf{Q} = a_1^2 \mathbf{f}_1 \mathbf{f}_1^{\mathrm{T}} + a_2^2 \mathbf{f}_2 \mathbf{f}_2^{\mathrm{T}} + \cdots + a_N^2 \mathbf{f}_N \mathbf{f}_N^{\mathrm{T}}. \tag{8.4.3}$$

For the moment, assume that the basis vectors $\mathbf{f}_i$ are given – it will turn out that only one choice allows a realisable solution. Since the solution covariance (8.3.17) is linear in $\mathbf{Q}$, the total solution covariance equals the sum of the covariances due to each term in (8.4.3). Thus, let $\mathbf{C}_k$ denote the asymptotic response of the stochastic model to forcing $\mathbf{f}_k$ with unit variance:

$$\mathbf{C}_k = \int_0^\infty e^{\mathbf{A}s}\mathbf{f}_k\mathbf{f}_k^{\mathrm{T}}e^{\mathbf{A}^{\mathrm{T}}s}\,ds. \tag{8.4.4}$$

Then the total covariance is given by

$$\mathbf{C} = a_1^2\mathbf{C}_1 + a_2^2\mathbf{C}_2 + \cdots + a_N^2\mathbf{C}_N. \tag{8.4.5}$$

The covariances $\mathbf{C}_1, \ldots, \mathbf{C}_N$ describe the responses of the stochastic model to the individual forcing functions $\mathbf{f}_1, \mathbf{f}_2, \ldots, \mathbf{f}_N$, which can be summed because the responses are independent. Substituting (8.4.3) and (8.4.5) into the similarity hypothesis (8.4.2) gives

$$\left(a_1^2\mathbf{C}_1 + a_2^2\mathbf{C}_2 + \cdots + a_N^2\mathbf{C}_N\right)_{ii} = \gamma\left(a_1^2\mathbf{f}_1\mathbf{f}_1^{\mathrm{T}} + a_2^2\mathbf{f}_2\mathbf{f}_2^{\mathrm{T}} + \cdots + a_N^2\mathbf{f}_N\mathbf{f}_N^{\mathrm{T}}\right)_{ii}. \tag{8.4.6}$$

These equations for $i = 1, 2, \ldots, N$ can be written compactly as

$$\mathbf{Hx} = \gamma\mathbf{Rx} \tag{8.4.7}$$

where

$$\mathbf{H}_{ij} = (\mathbf{C}_j)_{ii}, \quad \mathbf{R}_{ij} = \left(\mathbf{f}_j\mathbf{f}_j^{\mathrm{T}}\right)_{ii}, \quad (\mathbf{x})_j = a_j^2. \tag{8.4.8}$$

Each of these quantities has a physical interpretation: $\mathbf{H}_{ij}$ gives the response variance at point i due to the jth forcing function, $\mathbf{R}_{ij}$ gives the variance production rate at point i due to the jth forcing function and $(\boldsymbol{x})_j$ gives the variance with which the jth forcing function is excited.

The above considerations reduce the noise–response similarity hypothesis to the generalised eigenvalue problem (8.4.7). As such, there appears to be not one solution, but several solutions. However, not all eigenvector solutions of (8.4.7) are permissible because all elements of $\mathbf{x}$ must be positive, according to (8.4.8). The condition that all elements of $\mathbf{x}$ be positive may be called the *realisability constraint*, since its violation implies a non-positive definite noise covariance matrix $\mathbf{Q}$. The condition of realisability is then essentially a constraint on the forcing functions f_i. Astonishingly, a sufficient condition for realisability can be found very simply. A theorem in linear algebra known as the Perron–Frobenius theorem states that a square matrix with strictly positive elements has precisely one eigenvector with all positive elements (Horn & Johnson 1985). This theorem can be invoked only after the generalised eigenvalue problem (8.4.7) is put into standard form by multiplying both sides by $\mathbf{R}^{-1}$. Brown *et al.* (1969) show that if $\mathbf{R}$ is non-negative, then $\mathbf{R}^{-1}$ is non-negative if and only if the entries of $\mathbf{R}$ are all zero except for a single positive entry in each row and column. Since the order of the forcing functions is arbitrary, we may order them so that the positive elements occur on the diagonal. Thus, if $\mathbf{R}$ cannot be made diagonal, then $\mathbf{R}^{-1}$ must contain negative elements and the product $\mathbf{R}^{-1}\mathbf{H}$ may contain negative elements. It follows that a realisable solution is guaranteed by imposing the condition that $\mathbf{R}$ be diagonal. Thus, if $\mathbf{H}$ has strictly positive elements, which implies that any forcing leads to a response everywhere (i.e. there are no 'shadow zones'), and $\mathbf{R}$ is diagonal, then the Perron–Frobenius theorem guarantees that there will be only one positive eigenvector of $\mathbf{R}^{-1}\mathbf{H}$, and hence only one realisable solution to the eigenvalue problem (8.4.7). Note that these arguments provide only sufficient conditions for solution. Also, the eigenvector solution is unique up to a multiplicative constant, implying that α is not determined by this solution. Importantly, diagonal $\mathbf{R}$ implies $(\mathbf{f}_k)_i = \delta_{ik}$, which implies that $\mathbf{Q}$ is diagonal and hence the stochastic forcing is locally uncorrelated in space. However, the stochastic forcing is neither isotropic nor homogeneous, since the noise amplitudes vary in space.

The above solution to the noise–response similarity hypothesis completely specifies the noise covariance matrix $\mathbf{Q}$ and the proportionality constant γ, given the dissipation operator $\mathbf{D}$ and overall noise amplitude α. In this way, the structure of the random forcing becomes a property of the background flow, in contrast to previous formulations of this stochastic model.

The stochastic models discussed in this section make no use of the fact that the inviscid QG equations conserve energy and potential enstrophy. These conservation properties follow from the fact that, under suitable boundary conditions,

$$[\psi J(\psi, q)] = 0, \qquad [q J(\psi, q)] = 0 \tag{8.4.9}$$

where the brackets [ ] denote an area integral. Whether these balance constraints hold in the discretised system depends on how the subgrid scales are parameterised. To be consistent with the inertial range theories discussed in Section 8.2, we suggest that energy should be approximately conserved by the resolved scales (since little energy is assumed to transfer downscale) whereas enstrophy should be continually lost by the resolved scales (since the enstrophy cascade is assumed to go downscale). We consider the implications of these assumptions in turn.

Let the energy and potential enstrophy in the discretised system be denoted respectively as

$$\phi_1 = \mathbf{q}^{\mathrm{T}}\mathbf{E}_1\mathbf{q}, \qquad \phi_2 = \mathbf{q}^{\mathrm{T}}\mathbf{E}_2\mathbf{q}, \tag{8.4.10}$$

where $\mathbf{E}_1$ and $\mathbf{E}_2$ are appropriate kernel matrices. For simplicity, let us identify the background flow with the zonally symmetric component of the flow, in which case the eddy flow has vanishing zonal mean. In this case, the nonlinear terms in the eddy equation conserve energy by virtue of the fact that $[\psi^e J(\psi^e, q^e)] = 0$, which for the discretised system becomes

$$0 \approx \mathbf{e}^{\mathrm{T}}\mathbf{E}_1\mathbf{N}^e_{ee}(\mathbf{e}, \mathbf{e}). \tag{8.4.11}$$

It follows that if we parameterise the eddy–eddy nonlinear interactions as in (8.3.6), then the forcing and dissipation should be chosen to conserve energy, which may be expressed mathematically as

$$0 \approx \mathbf{e}^{\mathrm{T}}\mathbf{E}_1\mathbf{N}^{e}_{ee}(\mathbf{e}, \mathbf{e}) = \mathbf{e}^{\mathrm{T}}\mathbf{E}_1\mathbf{D}\mathbf{e} + \mathbf{e}^{\mathrm{T}}\mathbf{E}_1\mathbf{w}. \tag{8.4.12}$$

The last two terms cannot balance each other at every instant because one of them is independent white noise. At most, this balance can be maintained only in a mean sense. Accordingly, we impose the constraint that the forcing and dissipation conserve energy in the ensemble mean sense.

The final assumption in our closure theory is that the resolved scales should transfer enstrophy to subgrid scales, to support the downscale enstrophy cascade in the (unresolved) inertial range. Transfer of enstrophy to subgrid scales implies that the parameterised nonlinear terms destroy enstrophy (because they neglect the transfer to subgrid scales). To the extent that the statistics of subgrid scales are determined by the enstrophy transfer rate to unresolved scales, the energy spectrum should be of the form (8.2.3). A reasonable constraint, then, is that the energy spectrum for the resolved scales should blend smoothly with the subgrid energy spectrum (8.2.3). This can happen only if the stochastic model produces a -3 power law at large wavenumbers. The conditions under which this occurs can be seen as follows. On small scales, the QG equations become dominated by terms with the highest-order derivatives, in which case the QG potential vorticity equations can be approximated as

$$\frac{\partial q}{\partial t} \approx w(t) + (-1)^{J+1}\nu_J \nabla^{2J} q. \tag{8.4.13}$$

For a particular total wavenumber K the stochastic model becomes

$$\frac{\partial q_K}{\partial t} = w_K(t) - \nu_J K^{2J} q_K. \tag{8.4.14}$$

This is a standard Langevin model with stationary variance

$$\langle q_K^2 \rangle = \frac{\langle w_K^2 \rangle}{2\nu_J K^{2J}}. \tag{8.4.15}$$

The isotropic energy spectrum therefore is

$$e(K) = K\frac{\langle q_K^2 \rangle}{K^2} = \frac{\langle w_K^2 \rangle}{2\nu_J K^{2J+1}}. \tag{8.4.16}$$

It turns out that the noise–response similarity hypothesis (8.4.2) produces noise with an approximately white enstrophy spectrum at large wavenumber. In this case, $\langle w_K^2 \rangle \approx$ constant and the -3 power law is obtained for $J = 1$. Thus, the dissipation operator cannot include diffusion operators greater than first order,

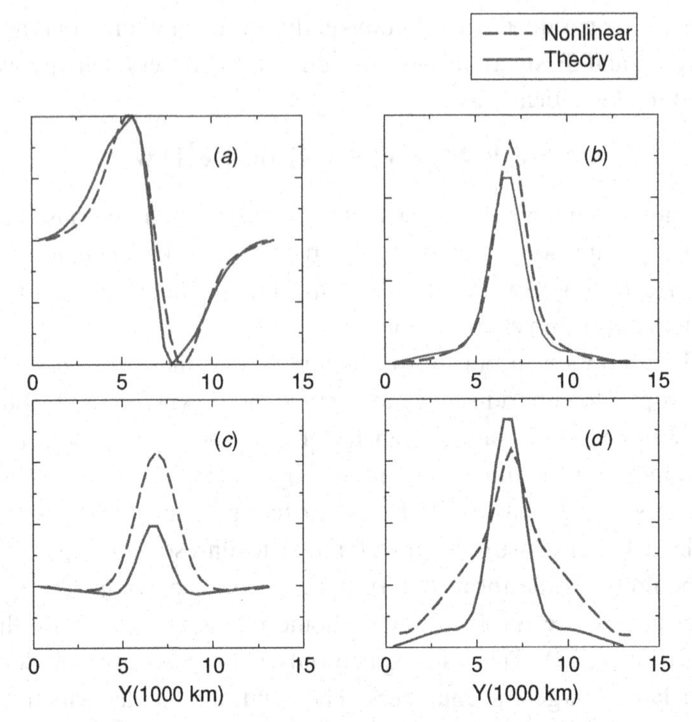

Figure 8.1 Comparison between second-order eddy moments in a fully nonlinear quasi-geostrophic model (dashed) and those predicted by a stochastic model (solid). The statistics being compared are the zonal momentum flux (*a*), temperature variance (*b*), temperature flux (*c*) and enstrophy (*d*), as indicated in the respective figures. The horizontal axis corresponds to meridional distance.

leaving ν_0 and ν_1 as the only dissipation parameters in the theory. Furthermore, non-zero ν_1 guarantees the -3 power law (provided the noise has a white enstrophy spectrum). Importantly, the enstrophy transfer rate η appearing in the inertial range law (8.2.3) is derived from the stochastic model solution as

$$\eta = -\left\langle \mathbf{e}^{\mathrm{T}}\mathbf{E}_2\mathbf{N}^{e}_{ee}(\mathbf{e},\mathbf{e})\right\rangle = -\left\langle \mathbf{e}^{\mathrm{T}}\mathbf{E}_2\mathbf{D}\mathbf{e} + \mathbf{e}^{\mathrm{T}}\mathbf{E}_2\mathbf{w}\right\rangle. \tag{8.4.17}$$

The above constraints eliminate all but one degree of freedom. This remaining degree of freedom is effectively the diffusion coefficient ν_1. It turns out, however, that the results are not very sensitive to the choice of this parameter, as long as it has reasonable values (e.g. 10^{-8}–10^{-7} m^2 s^{-1}).

To test the closure theory, let us consider a two-layer QG model that is thermally relaxed toward a baroclinic jet and contains Ekman damping in the bottom layer. All parameters of the model are the same as in Delsole (2001b). A comparison of various second-order eddy statistics between the fully nonlinear model and the stochastic model is shown in Fig. 8.1. We see that the stochastic model captures

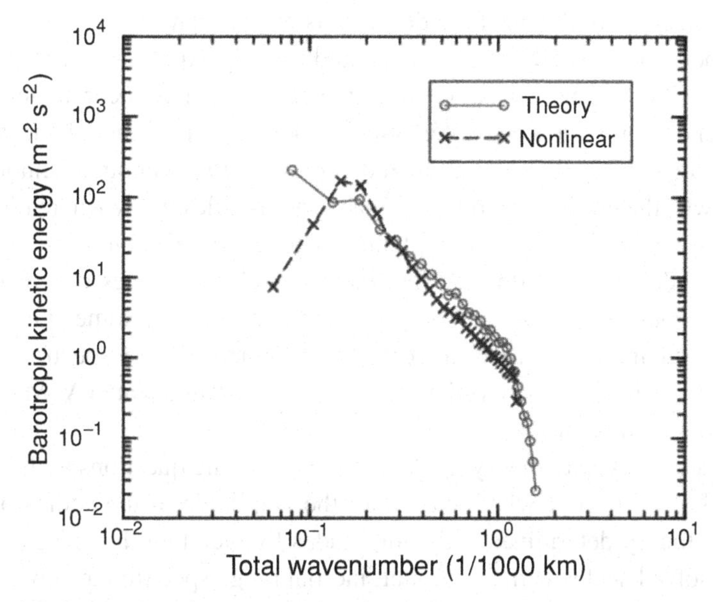

Figure 8.2 Comparison between the barotropic kinetic energy spectrum as a function of total wavenumber in a fully nonlinear quasi-geostrophic model (dash–x) and as predicted by a stochastic model (solid–open circle).

the general structure of the statistics, but the temperature flux is underestimated by a factor of two and the enstrophy tends to be concentrated too much in the centre compared to the nonlinear solution. Since the temperature variance is accurately predicted, the fact that the temperature flux is underpredicted by a factor of two indicates that the correlation between the velocity and temperature is incorrectly predicted by the stochastic model. Part of the problem is that the model grossly overestimates the kinetic energy on large scales, as will become clear shortly. Nevertheless, it should be emphasised that the stochastic model has essentially no tunable parameters and thus constitutes a closure theory for turbulence given the zonal-mean flow and external parameters.

The corresponding barotropic kinetic energy spectrum for the nonlinear model and stochastic model are shown in Fig. 8.2. We see that the stochastic model recovers the spectrum in the nonlinear model in the range 500 km to 1000 km, which in turn is close to a -3 power law. However, for length scales exceeding 1000 km, the stochastic model predicts increasing energy density while the nonlinear model predicts decreasing energy density. We draw attention to the fact that vertical scale is logarithmic and hence the stochastic model overestimates the kinetic energy spectrum by over an order of magnitude. This overestimation of kinetic energy at large length scales is typical in these types of stochastic models, although this fact is not always made clear in the published literature. Furthermore, given the tendency for the linear model to transfer enstrophy downscale and energy upscale when stochastically forced, we should actually expect energy to 'pile up' at large scales. This tendency raises the question of how the nonlinear model manages to reduce energy at large scales. We note that both the linear model and nonlinear model contain exactly the same Ekman damping, so this damping cannot be used to explain why the nonlinear model dissipates large-scale energy whereas the linear model does not. We believe that explaining why energy does not 'pile up' in the nonlinear model, while it does in the linear model, constitutes one of the challenges in stochastic models of quasi-geostrophic turbulence. Some evidence from other stochastic model research suggests that the kinetic energy spectrum could be improved by including zonal variations in the background state (A. Majda 2008, personal communication).

What does the closure theory imply about the opening questions to this section? Consider the question of what determines the amplitude of the eddies. In effect, the closure theory determines eddy amplitudes by blending the energy spectrum at small resolved scales with an implicit inertial range spectrum at subgrid scales. This implies that the eddy amplitudes adjust to whatever level is needed to be consistent with inertial range spectra at small scales. That is, the statistical steady equilibrium depends on how enstrophy is dissipated at subgrid scales. In regard to the question of what determines the time scale of the eddies, the proposed closure theory determines the overall eddy time scale through the energy balance constraint, since this constraint strongly constrains the potential vorticity damping coefficient. This result implies that the gross time scale depends on how quickly the eddy–eddy nonlinear interactions decorrelate in response to perturbations – if the parameterised damping is too weak, then the stochastic forcing overwhelms the damping to produce a net energy source for the eddy–eddy terms; if the parameterised damping is too strong, then the damping overwhelms the stochastic forcing to produce a net energy sink for the eddy–eddy terms. The interesting connection here is that the time scale depends on the degree to which the parameterised forcing and dissipation energetically balance each other.

8.5 Other challenges

We conclude by discussing challenges for which we have little insight to offer, but are worthwhile to mention in the hope of bringing them to the attention of others. A grand challenge in developing stochastic turbulence models suitable for climate modelling is the incorporation of moisture effects. Not only do large-scale atmospheric eddies transport significant amounts of moisture, but they interact with the moisture through latent-heat release and radiative effects of water vapour. In addition, condensation rates are strongly nonlinear functions of moisture content and hence difficult to parameterise with linear models. Recently, Lapeyre & Held (2004) proposed incorporating latent heating effects in the QG framework by treating moisture as a perturbation relative to a mean background field, analogous to temperature. The resulting model cannot be expected to mimic the true effects of moisture on large-scale eddies, but its simplicity may elucidate some important dynamics, and construction of the corresponding stochastic models may be revealing. (Lapeyre & Held note that the model becomes dominated by vortices when latent heating effects become strong. However, we do not expect the stochastic models developed here to be useful in this regime since the vortices are not well captured by linear transient growth mechanisms, nor can they captured in a slowly varying background flow. On the other hand, it is not clear how much these vortices contribute to the net transport, so failure to capture them may not be problematic.) Pierrehumbert *et al.* (2007) and O'Gorman & Schneider (2006) developed random walk models for the moisture flux and mean condensation rate as a function of the eddy length and time scales. Perhaps models of this type could be coupled to stochastic turbulence models such that the stochastic model predicts the eddy length and time scales required to solve the random walk model, and the random walk model predicts the moisture flux and condensation rates that modify the thermodynamics of the eddies.

Recent studies with idealised general circulation models reveal that the eddy energies and fluxes obey scaling laws that hold over several decades (Schneider & Walker 2006, 2008). These scaling laws have been derived from isentropic–coordinate balance conditions in which the intersection of isentropes with the surface play a critical role. Since quasi-geostrophic theory does not adequately represent such intersecting isentropes, these studies call into question the relevance of QG stochastic models for explaining the eddy statistics in the atmosphere. Therefore, it would be of great interest to determine the conditions under which QG stochastic models reproduce these scaling laws (which they presumably can by suitable tuning of the forcing and dissipation). One barrier to making a clean comparison with these scaling laws is that QG models are not easily extended to spherical domains. Specifically, these models require inverting potential vorticity in a spherical domain to obtain the velocity and temperature fields, a task that is difficult owing to the absence of clear balance constraints in the tropics. On the other hand, DelSole & Hou (1999) and Zhang & Held (1999) have shown that stochastic models based on the primitive equations can be constructed that also produce realistic fluxes and variances, though more details of the forcing and dissipation need to be specified. Recent studies also have revealed robust diffusive relations between potential vorticity fluxes and local background potential vorticity gradients in two-layer quasi-geostrophic models, especially in the limit of large-scale baroclinic jets (Pavan & Held 1996). It would be of interest to determine the conditions under which stochastic models can reproduce these relations, and whether the closure theory developed here is consistent with the conditions.

Another challenge in developing stochastic models of shear-flow turbulence is finding ways to speed up the computations. For instance, the closure theory discussed in the previous section is more computationally demanding than the fully nonlinear problem, although this does not diminish the importance of the fact that the stochastic model gives useful predictions of the eddy statistics given only the background state, and therefore gives insight into how the eddy statistics are determined by the background flow. Finding more efficient calculation methods is probably more a technical problem than a scientific problem. One approach is to exploit ray tracing techniques to find the linear response to an impulse of a system with spatially varying background flow – that is, to find the Green function for the system. Lighthill (1978) suggested using stationary phase approximations to solve for the response in the early stages of dispersion, assuming homogeneous background states, and then using the theory of rays in inhomogeneous systems to solve for the far-field response. The asymptotic response to an impulse in a beta-plane has been discussed by Rossby (1945) and Lighthill (1967), and ray tracing theory has been applied to spherical atmospheres by Hoskins *et al.* (1977) and Hoskins & Karoly (1981). Buhler & McIntyre (1999), Buhler *et al.* (1999) and Broutman *et al.* (2001) developed Lighthill's hybrid approach in the context of gravity waves, and perhaps a similar approach can be adapted to quasi-geostrophic systems to derive approximate covariances due to random forcing.

Perhaps the grandest challenge in all of stochastic model theory is to justify rigorously the models themselves. Although Majda *et al.* (2001) rigorously derive stochastic models for the low-frequency variability of multidimensional models, these models still rely on parameterising the nonlinear eddy–eddy interactions by a stochastic forcing and linear dissipation, the parameters of which are chosen empirically. Majda *et al.* (2001) derive the stochastic model in the limit in which the ratio of the correlation time of the slowest eddy variable to the fastest background variable is small. While an approach based on time scale separation is theoretically satisfying for the development of climate models, this may not be the best basis for justifying stochastic models of inhomogeneous turbulence. For instance, the success of these models appears to derive from the fact that background straining fields preferentially amplify certain perturbations, and that these amplification mechanisms are captured by linear operators. This connection raises the question as to whether stochastic turbulence models can be justified on the basis of the degree of transient growth or the degree of non-normality of the dynamical operator.

We end with perhaps the most curious challenge: why do relatively simple assumptions for the forcing and dissipation lead to impressive predictions for the structure of the eddy variances and fluxes? Although we have argued that this success derives from transient growth mechanisms inherent in background straining fields, this argument is based on linear theory and does not explain why the nonlinear system behaves in this way.

Acknowledgements

Some of the ideas presented in this paper arose from stimulating discussions I have had with many colleagues over several years, including Brian Farrell, Petros Ioannou, Andy Majda, David Nolan, David Straus, Michael Tippett and Ka-Kit Tung. This research was supported by the National Science Foundation (ATM0332910, EAR-0233320), National Aeronautics and Space Administration (NNG04GG46G) and the National Oceanic and Atmospheric Administration (NA04OAR4310034).

References

Batchelor, G. K. 1969 Computation of the energy spectrum in homogeneous, two-dimensional turbulence. *Phys. Fluids*, **12**, 233–239.

Borue, V. 1993 Spectral exponents of enstrophy cascade in stationary two-dimensional homogeneous turbulence. *Phys. Rev. Lett.*, **71**, 3967–3970.

Broutman, D., Rottman, J. W. & Eckermann, S. D. 2001 A hybrid method for wave propagation from a localized source, with appliction to mountain waves. *Q. J. R. Meteorol. Soc.*, **127**, 129–146.

Brown, T. A., Juncosa, M. & Klee, V. 1969: Invertibly positive linear operators on spaces of continuous functions. *Math. Ann.*, **183**, 105–114.

Buhler, O. & McIntyre, M. E. 1999 On shear-generated gravity waves that reach the mesosphere. Part II: wave propgation. *J. Atmos. Sci.*, **56**, 3764–3773.

Buhler, O., McIntyre, M. E. & Scinocca, J. F. 1999 On shear-generated gravity waves that reach the mesosphere. Part I: wave generation. *J. Atmos. Sci.*, **56**, 3749–3763.

Charney, J. G. 1971 Geostrophic turbulence. *J. Atmos. Sci.*, **28**, 1087–1095.

Cho, J. Y. N. & Lindborg, E. 2001 Horizontal velocity structure functions in the upper troposphere and lower stratosphere 1. Observations. *J. Geophys. Res.*, **106**(D10), 10223–10232.

DelSole, T. 2001a Optimally persistent patterns in time-varying fields. *J. Atmos. Sci*, 1341–1356.

DelSole, T. 2001b A theory for the forcing and dissipation in stochastic turbulence models. *J. Atmos. Sci.*, **58**, 3762–3775.

DelSole, T. 2004 Stochastic models of quasigeostrophic turbulence. *Surveys Geophys.*, **25**, 107–149.

DelSole, T. & Farrell, B. F. 1995 A stochastically excited linear system as a model for quasigeostrophic turbulence: analytic results for one- and two-layer fluids. *J. Atmos. Sci.*, **52**, 2531–2547.

DelSole, T. & Farrell, B. F. 1996 The quasi-linear equilibrium of a thermally maintained, stochastically excited jet in a quasigeostrophic model. *J. Atmos. Sci.*, **53**, 1781–1797.

DelSole, T. & Hou, A. Y. 1999 Empirical stochastic models for the dominant climate statistics of a general circulation model. *J. Atmos. Sci.*, **56**, 3436–3456.

Eyink, G. 1996 Exact results on stationary turbulence in two dimensions: consequences of vorticity conservation. *Physica D*, **91**, 97–142.

Farrell, B. F. 1985 Transient growth of damped baroclinic waves. *J. Atmos. Sci.*, **42**, 2718–2727.

Farrell, B. F. 1987 Developing disturbances in shear. *J. Atmos. Sci.*, **44**, 2191–2199.

Farrell, B. F. 1989 Optimal excitation of baroclinic waves. *J. Atmos. Sci.*, **46**, 1193–1206.

Farrell, B. F. & Ioannou, P. J. 1993 Stochastic dynamics of baroclinic waves. *J. Atmos. Sci.*, **50**, 4044–4057.

Farrell, B. F. & Ioannou, P. J. 1994 Theory for the statistical equilibrium energy spectrum and heat flux produced by transient baroclinic waves. *J. Atmos. Sci.*, **51**, 2685–2698.

Farrell, B. F. & Ioannou, P. J. 1995 Stochastic dynamics of the midlatitude atmospheric jet. *J. Atmos. Sci.*, **52**, 1642–1656.

Farrell, B. F. & Ioannou, P. J. 1996 Generalized stability theory. Part I: autonomous operators. *J. Atmos. Sci.*, **53**, 2025–2040.

Farrell, B. F. & Ioannou, P. J. 2003 Structural stability of turbulent jets. *J. Atmos. Sci.*, **60**, 2101–2118.

Fjortoft, R. 1953 On the changes in the spectral distribution of kinetic energy for two-dimensional non-divergent flow. *Tellus*, **5**, 225–230.

Frisch, U. 1995 *Turbulence*. Cambridge University Press.

Gardiner, C. W. 1990 *Handbook of Stochastic Models*, 2nd edn. Springer-Verlag.

Gkioulekas, E. & Tung, K. K. 2007 A new proof on net upscale energy cascade in two-dimensional and quasi-geostrophic turbulence. *J. Fluid Mech.*, **576**, 173–189.

Haken, H. 2004 *Synergetics*, 3rd edn. Springer-Verlag.

Hasselmann, K. 1976 Stochastic climate models I. Theory. *Tellus*, **28**, 473–485.

Hogstrom, U., Smedman, A.-S. & Bergstrom, H. 1999 A case study of two-dimensional stratified turbulence. *J. Atmos. Sci.*, **56**(7), 959–976.

Horn, R. A. & Johnson, C. R. 1985 *Matrix Analysis*. Cambridge University Press, New York.

Hoskins, B. J. & Karoly, D. J. 1981 The steady linear response of a spherical atmosphere to thermal and orographic forcing. *J. Atmos. Sci.*, **38**, 1179–1196.

Hoskins, B. J., Simmons, A. J. & Andrews, D. G. 1977 Energy dispersion in a barotropic atmosphere. *Q. J. R. Meteorol. Soc.*, **103**, 553–567.

Hoyer, J.-M. & Sadourny, R. 1982 Closure modeling of fully developed baroclinic instability. *J. Atmos. Sci.*, **39**, 707–721.

Kolmogorov, A. N. 1941 The local structure of turbulence in incompressible viscous fluid for very large Reynolds numbers. *Dokl. Acad. Sci., USSR*, **30**, 299–303.

Koshyk, J. N. & Hamilton, K. 2001 The horizontal kinetic energy spectrum and spectral budget simulated by a high-resolution troposphere–stratosphere–mesosphere GCM. *J. Atmos. Sci.*, **58(4)**, 329–348.

Koshyk, J. N., Hamilton, K. & Mahlman, J. D. 1999 Simulation of the $k^{-5/3}$ mesoscale spectral regime in the GFDL SKYHI general circulation model. *Geophys. Res. Lett.*, **26(7)**, 843–846.

Kraichnan, R. H. 1967 Inertial ranges in two-dimensional turbulence. *Phys. Fluids*, **10**, 1417–1423.

Kraichnan, R. H. & Montgomery, D. 1980 Two-dimensional turbulence. *Reports Progr. Phys.*, **43**, 547–619.

Lacorata, G., Aurell, E., Legras, B. & Vulpiani, A. 2004 Evidence for a $k^{-5/3}$ spectrum from the EOLE Lagrangian balloons in the low stratosphere. *J. Atmos. Sci.*, **61**, 2936–2942.

Lapeyre, G. & Held, I. M. 2004 The role of moisture in the dynamics and energetics of turbulent baroclinic eddies. *J. Atmos. Sci.*, **61**, 1693–1710.

Larichev, V. D. & Held, I. M. 1995 Eddy amplitudes and fluxes in a homogeneous model of fully developed baroclinic instability. *J. Phys. Oceanogr.*, **25**, 2285–2297.

Leith, C. E. 1968 Diffusion approximation for two-dimensional turbulence. *Phys. Fluids*, **11**, 671–673.

Lesieur, M., 1990 *Turbulence in Fluids*, 2nd edn. Kluwer.

Leslie, D. C. 1973 *Developments in the Theory of Turbulence*. Clarendon Press.

Lighthill, M. J. 1967 On waves generated in dispersive systems by travelling forcing effects, wth applications to the dynamics of rotating fluids. *J. Fluid Mech.*, **27**, 725–752.

Lighthill, M. J. 1978 *Waves in Fluids*. Cambridge University Press.

Lindborg, E. 2005 The effect of rotation on the mesoscale energy cascade in the free atmosphere. *Geophys. Res. Lett.*, **32**, L018090.

Lindborg, E. 2006 The energy cascade in a strongly stratified fluid. *J. Fluid Mech.*, **550**, 207–242.

Lindborg, E. & Alvelius, K. 2000 The kinetic energy spectrum of the two-dimensional enstrophy turbulence cascade. *Phys. Fluids*, **12**, 945–947.

Majda, A., Timofeyev, J. & Vanden-Eijnden, E. 2006 Stochastic models for selected slow variables in large deterministic systems. *Nonlinearity*, **19**, 769–794.

Majda, A. J., Franzke, C. & Khouider, B. 2008 An applied mathematics perspective on stochastic modelling for climate. *Phil. Trans. R. Soc.*, **A366**, 2429–2455.

Majda, A. J., Kleeman, R. & Cai, D. 2002 A mathematical framework for quantifying predictability through relative entropy. *Meth. Appl. Anal.*, **9**, 425–444.

Majda, A. J., Timofeyev, I. & Vanden-Eijnden E. 2001 A mathematical framework for stochastic climate models. *Commun. Pure Appl. Math.*, **54**, 891–974.

Manfroi, A. J. & Young, W. R. 1999 Slow evolution of zonal jets on the beta-plane. *J. Atmos. Sci.*, **56**, 784–800.

Merilees, P. E. & Warn, H. 1975 On energy and enstrophy exchanges in two-dimensional non-divergent flow. *J. Fluid Mech.*, **69**, 625–630.

Nastrom, G. D. & Gage, K. S. 1985 A climatology of atmospheric wavenumber spectra of wind and temperature observed by commerical aircraft. *J. Atmos. Sci.*, **42**, 950–960.

O'Gorman, P. A. & Schneider, T. 2006 Stochastic models for the kinematics of moisture transport and condensation in homogeneous turbulent flows. *J. Atmos. Sci.*, **63**, 2992–3005.

Orr, W. M. 1907 Stability of instability of the steady motions of a perfect liquid. *Proc. R Irish Acad.*, **27**, 9–69.

Pavan, V. & Held, I. M. 1996 The diffusive approximation for eddy fluxes in baroclinically unstable jets. *J. Atmos. Sci.*, **53**, 1262–1272.

Peixoto, J. P. & Oort, A. H. 1992 *Physics of Climate*. American Institute of Physics.

Pierrehumbert, R. T., Brogniez, H. & Roca, R. 2007 On the relative humidity of the atmosphere. In *The Global Circulation of the Atmosphere*, ed. Schneider T. and A. H. Sobel. Princeton University Press, pp. 143–185.

Rhines, P. 1975 Waves and turbulence on a beta-plane. *J. Fluid Mech.*, **69**, 417–443.

Rose, H. A. & Sulem, P.-L. 1978 Fully developed turbulence and statistical mechanics. *J. Phys. France*, 441–484.

Rossby, C.-G. 1945 On the propagation of frequencies and energy in certain types of oceanic and atmospheric waves. *J. Meteorol.*, **2**, 187–203.

Salmon, R. 1978 Two-layer quasigeostrophic turbulence in a simple special case. *Geophys. Astrophys. Fluid Dyn.*, **10**, 25–52.

Salmon, R. 1998 *Lectures on Geophysical Fluid Dynamics*. Oxford University Press.
Schneider, T. & Walker, C. C. 2006 Self-organization of atmospheric macroturbulence into critical states of weak nonlinear eddy–eddy interactions. *J. Atmos. Sci.*, **63**, 1569–1586.
Schneider, T. & Walker, C. C. 2008: Scaling laws and regime transitions of macroturbulence in dry atmospheres. *J. Atmos. Sci.*, **65**, 2153–2173.
Smith, K. S. 2004 Comments on the k^{-3} and $k^{-5/3}$ energy spectrum of atmospheric turbulence: quasigeostrophic two-level model simulation. *J. Atmos. Sci.*, **61**, 937–942.
Smith, L. M. & Waleffe, F. 1999 Transfer of energy to two-dimensional large scales in forced, rotating three dimensional turbulence. *Phys. Fluids*, **11**, 1608–1622.
Straus, D. M. & Ditlevsen, P. 1999 Two-dimensional turbulence properties of the ECMWF reanalysis. *Tellus*, **51A**, 749–772.
Tran, C. & Bowman, J. 2003 On the dual cascade in two dimensional turbulence. *Physica D*, **176**, 242–255.
Tran, C. & Bowman, J. 2004 Robustness of the inverse cascade in two-dimensional turbulence. *Phys. Rev. E*, **69**, 036303.
Tran, C. V. & Shepherd, T. G. 2002 Constraints on the spectral distribution of energy and enstrophy dissipation in forced two-dimensional turbulence. *Physica D*, **165**, 199–212.
Tung, K. K. & Orlando, W. 2003a On the differences between 2D and QG turbulence. *Discrete Contin. Dyn. Syst. B*, **3**, 145–162.
Tung, K. K. & Orlando, W. W. 2003b The k^{-3} and $k^{-5/3}$ energy spectrum of atmospheric turbulence: quasigeostrophic two-level model simulation. *J. Atmos. Sci.*, **60**, 824–835.
Vallis, G. K. 2006 *Atmospheric and Oceanic Fluid Dynamics*. Cambridge University Press.
Vallis, G. K. & Maltrud, M. E. 1993 Generation of mean flows and jets on a beta plane and over topography. *J. Phys. Oceanogr.*, **23**, 1346–1362.
Whitaker, J. S. & Sardeshmukh, P. D. 1998 A linear theory of extratropical synoptic eddy statistics. *J. Atmos. Sci.*, **55**, 237–258.
Zhang, Y. & Held, I. M. 1999 A linear stochastic model of a GCM's midlatitude storm tracks. *J. Atmos. Sci.*, **56**, 3416–3435.

9

Stochastic versus deterministic backscatter of potential enstrophy in geostrophic turbulence

BALASUBRAMANYA T. NADIGA

Downgradient mixing of potential vorticity and its variants are commonly employed to model the effects of unresolved geostrophic turbulence on resolved scales. This is motivated by the (inviscid and unforced) particle-wise conservation of potential vorticity and the mean forward or downscale cascade of potential enstrophy in geostrophic turbulence. By examining the statistical distribution of the transfer of potential enstrophy from mean or filtered motions to eddy or subfilter motions, we find that the mean forward cascade results from the forward-scatter being only slightly greater than the backscatter. Downgradient mixing ideas do not recognise such equitable mean-eddy or large-scale–small-scale interactions and consequently model only the mean effect of forward cascade; the importance of capturing the effects of backscatter – the forcing of resolved scales by unresolved scales – is only beginning to be recognised. While recent attempts to model the effects of backscatter on resolved scales have taken a stochastic approach, our analysis suggests that these effects are amenable to being modelled deterministically.

9.1 Introduction

The geostrophically balanced nature of flows at large scales in the atmosphere and oceans makes the quasi-geostrophic equations a good model for these scales (Pedlosky 1987; Vallis 2006); most studies of geostrophic turbulence use this framework. The conservation of quasi-geostrophic (QG) potential vorticity (PV) (in the absence of forcing and dissipation) along fluid particle paths in the QG system, places constraints on scale interactions. In particular, the dual conservation of energy and potential enstrophy in this system leads, away from boundaries and under certain other conditions, to a dual inertial range (Charney 1971) analogous to that found in two-dimensional turbulence (Kraichnan 1967): an inverse cascade of energy and a forward cascade of potential enstrophy.

We will consider the issue of transfer of potential enstrophy between the filtered or averaged motion and the residual or eddy motions in the range of scales of forward cascade of potential enstrophy.[1] An important observation about this transfer is that since it is due entirely to inviscid processes, the transfer is not sign-definite. That is to say, while we expect from homogeneous turbulence phenomenology that the *mean* (domain-integrated) transfer of potential enstrophy from mean or filtered motions to eddy or residual motions is positive (forward cascade), there can be regions where the transfer is negative. Such (localised) reversal of transfer in a regime of mean forward cascade is called backscatter [e.g. see (Pope 2000)]. In the following, we will show that backscatter of potential enstrophy in geostrophic turbulence is large. That is, the (mean) forward cascade is obtained as a small difference between a large forward-scatter and a similarly large backscatter.

The simplest eddy closure employed in the modelling of ocean circulation is that of an eddy viscosity, designed to represent the mean damping of the resolved scales by the unresolved scales. Such a viscosity may be based on mixing length ideas or the local magnitude of strain as in the classical Smagorinsky model (Pope 2000). However, the largeness of backscatter would imply that if we wanted to capture the effects of backscatter while using an eddy viscosity closure, a stochastic backscatter model would be required [e.g. as in Leith (1990); Jung *et al.* (2005)]. Alternatively, we find that the transfer (both forward-scatter and backscatter) can be recovered using a nonlinear combination of mean or filtered gradients to represent the eddy flux, suggesting that a large fraction of the effects of backscatter may be representable by a deterministic closure based on such a nonlinear combination.

The issue of backscatter has traditionally been considered only in the context of spatial-scale-based decomposition (filtering) as occurs in large eddy simulation formalism. However, a straightforward extension of the 'locally reversed transfer' ideas to the Reynolds-averaging formalism allows for a consideration of backsctter in the Reynolds-averaging formalism as well. Thus, in this chapter, we will consider eddy-resolving simulations of the classical wind-driven ocean circulation and analyse backscatter of potential enstrophy from both the Reynolds-averaging and filtering perspectives. First, the governing equations and the problem set-up that we work with are described. Next, backscatter in the Reynolds-averaging and scale-decomposition approaches are considered and the nonlinear gradient model is motivated. Finally, results are presented which suggest that a closure based

[1] Since the dissipation of potential enstrophy at large scales is minimal, this transfer appears as a sink in the filtered or averaged potential enstrophy evolution equation and as a production term in the residual or eddy potential enstrophy evolution equation.

on the nonlinear gradient model should be able to capture effects of backscatter deterministically, and the chapter ends with a discussion.

9.2 Governing equations and problem description

Consider the evolution of quasi-geostrophic potential vorticity in the layered form. Using standard notation,

$$\frac{\partial q_i}{\partial t} + \boldsymbol{u}_i \cdot \nabla q_i = F_i + D_i, \quad \nabla \cdot \boldsymbol{u}_i = 0, \tag{9.2.1}$$

where q is potential vorticity, $\boldsymbol{u}$ is velocity, F is forcing, D is dissipation and subscript i refers to layer number. The non-divergent nature of the two-dimensional advecting geostrophic velocity allows for the introduction of a stream function ψ such that $\boldsymbol{u} = k \times \nabla\psi$. In (9.2.1), potential vorticity, e.g. for a two layer system, is given as

$$q_i = \nabla^2\psi_i + (-1)^i \frac{f_0^2}{g' H_i}(\psi_1 - \psi_2) + \beta y,$$

with index $i = 1$ corresponding to the top layer and 2 to the bottom layer. For a multilayer system with more than two layers, the stretching term in the definition of potential vorticity above has contributions from both the top and bottom interfaces for layers other than the top and bottom layers. The effect of (shallow) topography is felt only by the bottom layer and enters the definition of PV of the bottom layer as another term $f_0/H_N Z_b$, where Z_b is the bottom elevation.

For convenience, 9.2.1 may be written as

$$\frac{\partial q}{\partial t} + \nabla \cdot (\boldsymbol{u}q) = F + D, \quad \nabla \cdot \boldsymbol{u} = 0. \tag{9.2.2}$$

Our interest being in the oceans, we consider the classic configuration used in Holland & Rhines (1980) to study eddy-induced ocean circulation, noting the generation of short Rossby waves by the process of reflection of long Rossby waves at western boundaries, unlike in the doubly periodic and zonally periodic channel configurations. The set-up consists of a two-layer ocean basin on a midlatitude beta plane. A characteristic non-uniform ocean stratification is considered in which the undisturbed top layer depth H_1 is 1 km and the bottom layer depth H_2 is 4 km. The lateral geometry consists of a square domain with a latitudinal and longitudinal extent of 2560 km. The midlatitude beta plane is such that the Rossby deformation radius is about 40 km and the grid spacing is 10 km. The forcing consists of a steady double gyre wind stress with a peak amplitude of 1 dyne cm^{-2}. The dissipation consists of a combination of bottom drag and small-scale mixing of

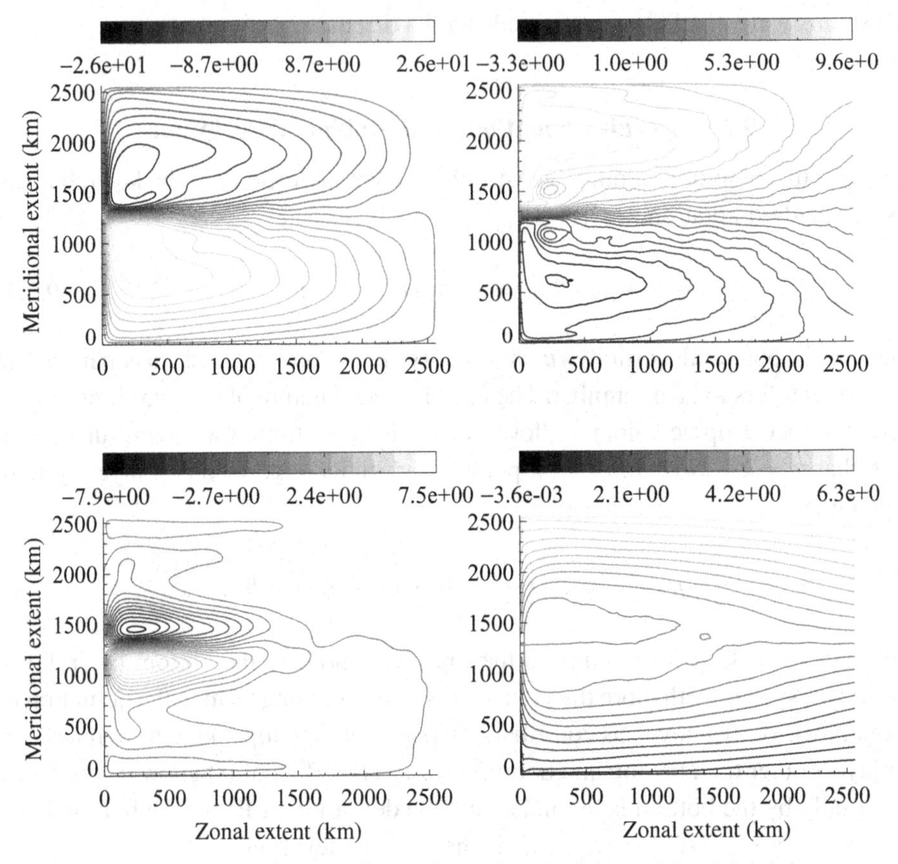

Figure 9.1 Time-mean circulation (left) and potential vorticity (right) in non-dimensional units. Top row is top layer. Bottom row is bottom layer.

relative vorticity that is either Laplacian or biharmonic but with a (Munk) scale defined as $(\nu_2/\beta)^{1/3}$ or $(\nu_4/\beta)^{1/5}$ of about 10 km.

An enstrophy and energy conserving form of spatial discretisation is used for the Jacobian (Arakawa 1966) and time stepping is carried out using an adaptive fifth order embedded Runge–Kutte Cash–Karp scheme (Press *et al.* 1992) to minimise time-truncation errors. An advantage of this choice of numerical discretisation is that the resultant code is nonlinearly stable and allows for simulations at any level of dissipation including no dissipation at all.

The long time-mean circulation and potential vorticity in a representative case are shown in Fig. 9.1 for the two layers. The phenomenology of this flow is classic and often referred to in connection with the potential-vorticity homogenisation theory. The reader is referred to Holland & Rhines (1980) for details. While the mean circulation in the lower layer is entirely eddy-driven, the mean circulation in

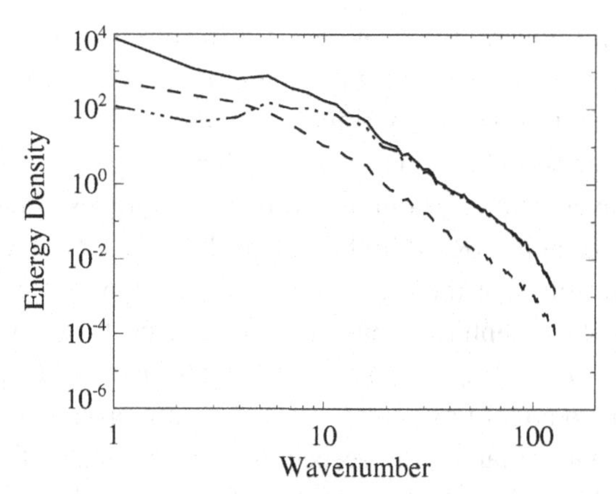

Figure 9.2 Spectra of total energy (solid line), barotropic energy (dashed line) and baroclinic kinetic energy (dot–dashed line). Note that (1) only the total energy is inviscidly conserved; and (2) all spectra fall off steeply at the scale of filtering.

the upper layer is modified at O(1) by the eddies. The ratio of the eddy to mean kinetic energy is about 3 in the upper layer and about 7 in the bottom layer. We have carried out similar computations in three-layer configurations in order to have a layer that is shielded from both direct wind-stress curl forcing and bottom friction, and the qualitative nature of the flows remains unchanged.

That the flow is resolved should be clear from Fig. 9.2 where the horizontal spectral distributions of the total energy (solid line), barotropic (dashed line) and baroclinic (dot–dashed line) kinetic energies are shown. The bulk of the total energy resides in the large-scale sloping of the isopycnals (layer interface) and constitutes the available potential energy of the system, as seen by the difference between the total energy and the sum of the baroclinic kinetic energy and the barotropic kinetic energy. And the energy flow in the system from available potential energy to baroclinic kinetic energy to barotropic energy follows the classic phenomenological picture of geostrophic turbulence [e.g. Salmon (1998)]. Considering the non-uniform stratification, the distribution of kinetic energy in the top layer looks more like the distribution of kinetic energy in the baroclinic mode and likewise with the lower-layer–barotropic pair. This is consistent with the present interpretation of altimetric data on large scales (Smith & Vallis 2001).

9.3 Backscatter in the Reynolds-averaging approach

Ocean circulation is characterised by interactions over a vast range of spatial and temporal scales. Consequently direct numerical simulation (DNS) of ocean circulation on climate time scales is unlikely in the foreseeable future. The problem of modelling ocean circulation is then one of how best to abstract important physics represented in the full governing equations at a lower cost. Note that while the Navier–Stokes equations are the governing equations for ocean circulation, the previous statement holds for further approximations of the system such as the primitive equations, the quasi-geostrophic equations and others.

One choice is to average the system over time or ensembles. Such an averaging is called Reynolds-averaging (RA) in classical turbulence (Pope 2000). A model based on RA aims to solve for the mean aspects of the flow. Thus, while a model based on RA is capable of capturing many important aspects of a turbulent flow, it is unable to predict accurately spatio-temporal characteristics of the flow. One way to improve on this is to adopt unsteady RA. In this case, averages are considered over time intervals that are large with respect to the time scale of turbulence, but small compared to the variability time scales of interest. Clearly unsteady RA-based models can be successful for flows with a distinct separation of turbulent and variability time scales.

Reynolds averaging proceeds by assuming a decomposition of the form

$$\boldsymbol{u} = \overline{\boldsymbol{u}} + \boldsymbol{u}',$$

where $\overline{\boldsymbol{u}}$ is the ensemble mean, but often replaced by a time mean. With such a decomposition, $\overline{\overline{\boldsymbol{u}}} = \overline{\boldsymbol{u}}$ and $\overline{\boldsymbol{u}'} = 0$. The evolution of mean PV is then given by

$$\frac{\partial \overline{q}}{\partial t} + \nabla \cdot (\overline{\boldsymbol{u}}\, \overline{q}) = \overline{F} + \overline{D} - \nabla \cdot \Sigma, \tag{9.3.1}$$

where

$$\Sigma = \overline{\boldsymbol{u}q} - \overline{\boldsymbol{u}}\, \overline{q} = \overline{\boldsymbol{u}'q'}. \tag{9.3.2}$$

Evolution of mean potential enstrophy is obtained by multiplying the above equation with $\overline{q}$ as

$$\frac{\partial \overline{Z}}{\partial t} + \nabla \cdot (\overline{\boldsymbol{u}}\overline{Z} + \overline{q}\Sigma) = \overline{F}\overline{q} + \overline{D}\overline{q} - \mathcal{T}_{\overline{Z}}, \tag{9.3.3}$$

where $\overline{Z} = \overline{q}^2/2$ and

$$\mathcal{T}_{\overline{Z}} = -\Sigma \cdot \nabla \overline{q} = -\overline{\boldsymbol{u}'q'} \cdot \nabla \overline{q}. \tag{9.3.4}$$

Further, since PV is an advected scalar in an inviscid and unforced setting, the evolution equation of its variance (eddy potential enstrophy) is revealing and is given by

$$\frac{\partial \overline{z}}{\partial t} + \nabla \cdot (\overline{\boldsymbol{u}z}) = \overline{F'q'} + \overline{D'q'} + \mathcal{T}_{\overline{Z}}, \tag{9.3.5}$$

where $z = q'^2/2$, and $\overline{z}$ is eddy potential enstrophy.

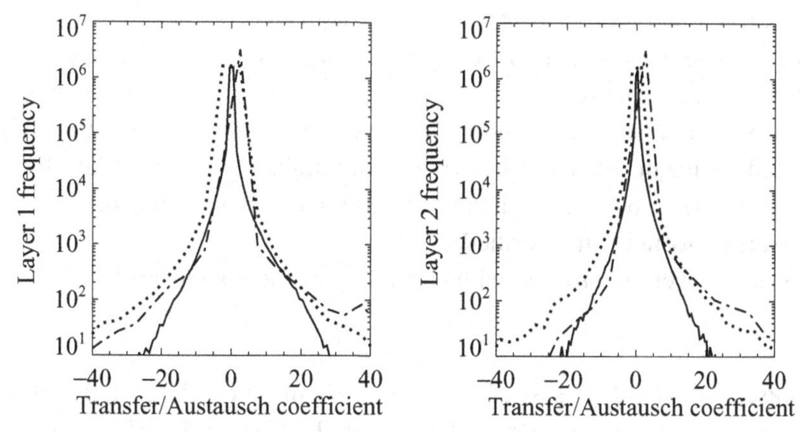

Figure 9.3 Distribution of the actual transfer of potential enstrophy between eddies and mean using Reynolds decomposition ($\mathcal{T}_{\overline{Z}}$) is shown in the solid line. The figure on the left is for the top layer; the figure on the right for the bottom layer; the transfers are normalised by their respective standard deviations. The mean values are positive (0.01 and 0.08 respectively in the two layers) indicating mean forward cascade of potential enstrophy. However, the smallness of the mean with respect to the standard deviation indicates large forward scatter and backscatter. The thick dotted line shows the statistical distribution of κ, the non-dimensional Austausch coefficient (again normalised by its standard deviation) in the two layers. The dot–dashed line shows the statistical distribution of the transfer using the nonlinear gradient model (9.4.10). The transfer resulting from the nonlinear gradient model in the Reynolds-averaging context compares well with the actual transfer, even though the nonlinear gradient model is motivated and derived in the filtering formalism.

From (9.3.3) and (9.3.5) it is clear that $\mathcal{T}_{\overline{Z}}$ is a transfer of potential enstrophy from the mean to the eddies. Considering (9.3.5), with steady wind stress as we presently have, variations of (upper) layer thickness do not change PV forcing in quasi-geostrophy ($F \propto 1/H_i^0$ where H_i^0 is the undisturbed depth and so $F' = 0$). Thus, in a statistically stationary turbulent flow driven by steady forcing, if dissipation is negative definite in (9.3.5), since the second term on the left only serves to redistribute, it is clear that $\mathcal{T}_{\overline{Z}}$ has to have a net (i.e. in the domain integrated sense) positive value.

Figure 9.3 shows the distribution of the transfer of potential enstrophy between eddies and mean using Reynolds decomposition ($\mathcal{T}_{\overline{Z}}$) in the two layers. In this case and all other cases the transfers are normalised by the standard deviation of the distribution. Distribution of the transfer is obtained by considering the transfer at all interior grid points at a few different times separated by the eddy-averaging time (e.g. $\overline{\boldsymbol{u}'q'}^{\tau}(\boldsymbol{x}, T) \cdot \nabla\overline{q}^{\tau}(\boldsymbol{x}, T)$ at various T, and where τ is the averaging time).

The mean values of the potential enstrophy transfers are positive indicating forward cascade in the mean, but the values are small with respect to the standard deviation (0.01 and 0.08 respectively in the two layers) indicating that the mean forward cascade of potential enstrophy is achieved as a small difference between large forward- and backscatters.

Figure 9.3 can also be interpreted in terms of alignment between the eddy flux of PV and its mean gradient: there is a mean alignment of the eddy flux down the mean gradient, but the alignment is extremely weak, indicating that a local downgradient closure is not verified.[2]

Thus, if one were to proceed with a local downgradient closure[3]

$$\overline{\boldsymbol{u}'q'} = -\kappa \nabla \overline{q}, \tag{9.3.6}$$

the backscatter would look stochastic and thus the modelling of the effects of backscatter would require a stochastic model. The heavy dotted line in Fig. 9.3 shows the statistical distribution of κ, the Austausch coefficient in the two layers defined in terms of the transfer as

$$\kappa = \frac{\mathcal{T}_{\overline{Z}}}{|\nabla \overline{q}|^2} = -\frac{\overline{\boldsymbol{u}'q'} \cdot \nabla \overline{q}}{|\nabla \overline{q}|^2}.$$

Again while the mean values are positive, they are extremely small compared to the standard deviations of the distributions. The spatial distribution of the Austausch coefficient in the two layers in Fig. 9.4 not only shows that it can be both positive and negative as has been previously noted by many authors [e.g. see Holland & Rhines (1980) or Marshall and Shutts (1981)], but that it can be almost randomly distributed.

9.4 Backscatter in the scale decomposition approach

In classical turbulence, the computational cost of DNS of large Reynolds number turbulent flows increases as the cube of the Reynolds number (Pope 2000) and is

[2] Time-averaging of the eddy flux is performed over a period of about 3.4 gyre turnaround times where the lateral scale is the domain size and velocity scale is the Sverdrup velocity. Varying the averaging time from 0 (instantaneous) to 34 gyre turnaround times improves the mean inclination in the lower layer from close to $\pi/2$ to about 2.2, with almost no change to the mean inclination in the upper layer (1.6). The averaging over long times such as 34 eddy turnover times would be appropriate for an RA-based steady simulation; for that a local downgradient eddy-flux approximation is reasonable for the lower layer only. Averaging over about 3.4 eddy turnover times seems appropriate for an unsteady RA-based simulation, and for this reason, we fix averaging time at about 3.4 eddy turnover times for all RA-based analysis. For this case a local downgradient eddy-flux approximation is becoming inappropriate for either of the layers.

[3] Clearly for this approximation to be valid locally, inspection of (9.3.5) (assuming statistical stationarity and steady forcing again) advection of perturbation potential enstrophy would have to be negligible. As Rhines and Holland (1979) point out, this is the case when the lateral scale of the mean potential-vorticity field far exceeds the displacement of fluid particles over a few eddy periods. In effect this is a restatement, from a Lagrangian point of view, of the requirement of a scale separation between the turbulence that is being parameterised and the scales of the flow that are being studied. Unfortunately, this is not the case in eddy-permitting simulations of ocean circulation with their intense western boundary currents and strongly curved flows.

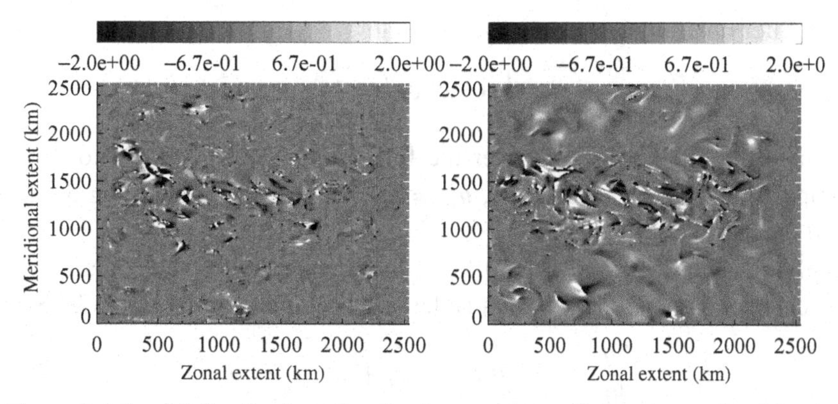

Figure 9.4 Spatial distribution of κ, the Austausch coefficient (normalised by the standard deviation of its distribution) in the top layer (left) and bottom layer (right).

therefore prohibitive. Further, RA-based approaches have failed to provide detailed local spatio-temporal flow characteristics in a predictive physics-based fashion. On the other hand, it is almost always the case that in fully resolved computations, a disproportionately high fraction of the computational effort is expended on the smaller scales whereas energy is predominantly contained in the larger scales (Pope 2000; Geurts 2004). This has led to the technique of large eddy simulation (LES) in which the large-scale unsteady motions that are driven by the specifics of the flow geometry and forcing and that are not universal are computed explicitly and the smaller, subgrid motions (that are presumably more universal) are modelled (Pope 2000; Geurts 2004). [Historically, however, LES had its origin with the use of the Smagorinski model to represent unresolved scales in simulations of geophysical flows (Pope 2000).] Large Eddy Simulation is a turbulence-modelling approach that is intermediate between DNS and RA-based approaches.

In LES, the resolution of energy-containing eddies that dominate flow physics is made computationally feasible by introducing a formal scale separation (Pope 2000). The scale separation is achieved by applying a low-pass filter to the original equations. To this end, consider filtering at scale l so that fields $\boldsymbol{u}$, q, etc.... can be split into large-scale and small-scale components as

$$\boldsymbol{u} = \boldsymbol{u}_{>l} + \boldsymbol{u}_{<l} = \boldsymbol{u}_l + \boldsymbol{u}_s,$$

where

$$\boldsymbol{u}_{>l}(\boldsymbol{x}) = \boldsymbol{u}_l(\boldsymbol{x}) = \int_D G(\boldsymbol{x} - \boldsymbol{x}')\boldsymbol{u}(\boldsymbol{x}')\mathrm{d}\boldsymbol{x}',$$

$$\boldsymbol{u}_{<l}(\boldsymbol{x}) = \boldsymbol{u}_s(\boldsymbol{x}) = \boldsymbol{u} - \boldsymbol{u}_l,$$

and the filter function G is normalised so that

$$\int_D G(\boldsymbol{x}')\mathrm{d}\boldsymbol{x}' = 1,$$

and where the integrations are over the full domain D. In contrast to Reynolds decomposition, however, generally, $\boldsymbol{u}_{ll} \neq \boldsymbol{u}_l$ and $\boldsymbol{u}_{sl} \neq 0$ on applying the filter a second time.

Applying such a filter to (9.2.2) leads to an equation for the evolution of the large-scale component of PV which is the primary object of interest in LES:

$$\frac{\partial q_l}{\partial t} + \nabla \cdot (\boldsymbol{u}_l q_l) = F_l + D_l - \nabla \cdot \boldsymbol{\sigma}, \tag{9.4.1}$$

where

$$\sigma = (\boldsymbol{u}q)_l - \boldsymbol{u}_l q_l \tag{9.4.2}$$

is the turbulent subfilter PV flux. This turbulent subgrid PV flux may in turn be written in terms of the Leonard stress, cross stress and Reynolds stress (Pope 2000) as

$$\sigma = \underbrace{(\boldsymbol{u}_l q_l)_l - \boldsymbol{u}_l q_l}_{\text{Leonard stress}} + \underbrace{(\boldsymbol{u}_l q_s)_l + (\boldsymbol{u}_s q_l)_l}_{\text{cross stress}} + \underbrace{(\boldsymbol{u}_s q_s)_l}_{\text{Reynolds stress}} . \tag{9.4.3}$$

However, while σ itself is Galilean-invariant, the above Leonard- and cross-stresses are not Galilean-invariant. Thus when these component stresses are considered individually, the following decomposition, originally due to Germano (1986), is preferable:

$$\sigma = \underbrace{(\boldsymbol{u}_l q_l)_l - \boldsymbol{u}_{ll} q_{ll}}_{\text{Leonard stress}} + \underbrace{(\boldsymbol{u}_l q_s)_l + (\boldsymbol{u}_s q_l)_l - \boldsymbol{u}_{ll} q_{sl} - \boldsymbol{u}_{sl} q_{ll}}_{\text{cross stress}} + \underbrace{(\boldsymbol{u}_s q_s)_l - \boldsymbol{u}_{sl} q_{sl}}_{\text{Reynolds stress}} . \tag{9.4.4}$$

The filtered equations, which are the object of simulation on a grid with a resolution commensurate with the filter in LES, are then closed by modelling subgrid-scale (SGS) stresses to account for the effect of the unresolved small-scale eddies. In this case (9.4.1) will be closed on modelling the turbulent subgrid PV flux σ.

We next consider the transfer of enstrophy from large scales to small scales at every location in physical space. To this end, multiplying (9.4.1) by q_l leads to an equation for the evolution of large-scale potential enstrophy $Z_{\text{large}} = q_l^2/2$:

$$\frac{\partial Z_{\text{large}}}{\partial t} + \nabla \cdot (\boldsymbol{u}_l Z_{\text{large}} + q_l \sigma) = F_l q_l + D_l q_l - \Pi_Z, \tag{9.4.5}$$

where

$$\Pi_Z = -\sigma \cdot \nabla q_l = -[(\boldsymbol{u}q)_l - \boldsymbol{u}_l q_l] \cdot \nabla q_l. \tag{9.4.6}$$

The flux on the left side can only spatially redistribute large-scale potential enstrophy, whereas the term Π_Z represents the flux of potential enstrophy out of large scales and into small scales. To wit, the term Π_Z appears with the opposite sign in the equation governing the evolution of the potential enstrophy involving small scales:

$$\frac{\partial Z_{\text{res}}}{\partial t} + \nabla \cdot (uZ - \boldsymbol{u}_l Z_{\text{large}} - q_l \sigma) = Fq - F_l q_l + Dq - D_l q_l + \Pi_Z, \quad (9.4.7)$$

where

$$Z_{\text{res}} = (q^2 - q_l^2)/2.$$

An irreversible forward cascade of potential enstrophy requires that the domain integral of Π_Z be positive: on integrating (9.4.5) over the domain, since the dissipation of enstrophy at large scales is small, in a statistically stationary state, enstrophy input by forcing is balanced by a flux of enstrophy out of the large scales and into the small scales. For this to happen, the subfilter PV flux would have to have a net alignment down the gradient of the large-scale PV in the domain integrated sense.

We choose a filter that corresponds to an inversion of the Helmholtz operator:

$$q_l = \left(1 - \frac{L_{\text{f}}^2}{4\pi^2}\nabla^2\right)^{-1} q, \quad (9.4.8)$$

where L_{f} is the filter width and $q_l = q$ on the boundaries and so also for any of the other variables. We choose a filter width of 63 km, about 1.5 times the Rossby deformation radius of 40 km. There are only minor differences on varying the filter width from 40 km to 80 km and these differences are not considered further. The nature of the spectrum in Fig. 9.2 suggests that our results are not likely to be sensitively dependent on the width of the filter over a wider range of filter widths.

Similar to Fig. 9.3, Fig. 9.5 shows the distribution of the transfer of potential enstrophy between the filtered and subfilter scales using the LES formalism in the two layers. The statistical distribution of the eddy diffusivity of PV, if one were to attempt a subfilter diffusion model, is shown in the heavy dotted line in Fig. 9.5. In close analogy with the explanation of the corresponding figures for Reynolds-averaging, it should be clear that the backscatter would appear stochastic if viewed from a subfilter PV diffusion perspective.

9.4.1 The nonlinear-gradient model

In eddy-permitting simulations, some of the range of scales of turbulence are explicitly resolved. Therefore, information about the structure of turbulence at these scales is readily available. In LES formalism, there is a class of models that attempt to model the smaller unresolved scales of turbulence based on the

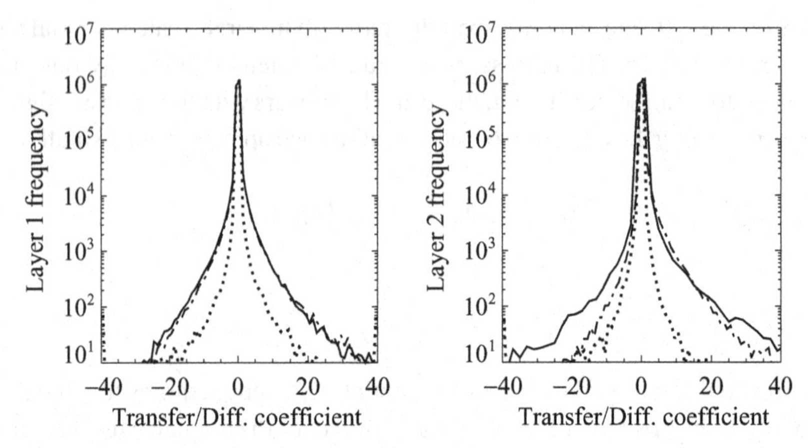

Figure 9.5 Statistical distribution of the actual transfer of potential enstrophy between filtered scales and subfilter scales is shown in the solid line. The figure on left (right) is for top (bottom) layer. The transfer is normalised by the standard deviations of the respective distributions. The statistical distribution of the coefficient of eddy diffusivity of PV in the two layers is shown in the heavy dotted line. The dot–dashed line shows the statistical distribution obtained using the nonlinear gradient model (9.4.10).

assumption that the structure of the turbulent velocity field at scales below the filter scale is the same as the structure of the turbulent velocity field at scales just above the filter scale (Meneveau & Katz 2000).

Further expansion of the velocity field in a Taylor series and performing filtering analytically results in

$$(u_i u_j)_l \propto \frac{\partial u_{li}}{\partial x_k}\frac{\partial u_{lj}}{\partial x_k}, \tag{9.4.9}$$

a quadratic nonlinear combination of resolved gradients for the subgrid model. The interested reader is referred to Meneveau & Katz (2000) for a comprehensive review of the nonlinear-gradient model.

Equivalently, expansion of $\boldsymbol{u}_l$ and q_l in the Galilean-invariant form of the Leonard-stress component of the subfilter eddy flux of PV (9.4.4) in a Taylor series:

$$\begin{aligned}(\boldsymbol{u}_l q_l)_l - \boldsymbol{u}_{ll} q_{ll} = &\int \mathrm{d}\boldsymbol{x}'\; G(\boldsymbol{x}-\boldsymbol{x}')\left(\boldsymbol{u}_l(\boldsymbol{x}) + (x'-x)_j \frac{\partial u_{li}}{\partial x_j}(\boldsymbol{x})\right)\\ &\times\left(q_l(\boldsymbol{x}) + (x'-x)_j \frac{\partial q_l}{\partial x_j}(\boldsymbol{x})\right) - \int \mathrm{d}\boldsymbol{x}'\; G(\boldsymbol{x}-\boldsymbol{x}')\\ &\times\left(\boldsymbol{u}_l(\boldsymbol{x}) + (x'-x)_j \frac{\partial u_{li}}{\partial x_j}(\boldsymbol{x})\right)\\ &* \int \mathrm{d}\boldsymbol{x}'\; G(\boldsymbol{x}-\boldsymbol{x}')\left(q_l(\boldsymbol{x}) + (x'-x)_j \frac{\partial q_l}{\partial x_j}(\boldsymbol{x})\right)\end{aligned}$$

produces at the first order

$$\sigma = C_{2l} \frac{\partial u_{li}}{\partial x_j} \frac{\partial q_l}{\partial x_i} = C_{2l} \nabla \boldsymbol{u}_l \cdot \nabla q_l, \tag{9.4.10}$$

again a quadratic nonlinear combination of resolved gradients and where C_{2l} is the second moment of the filter used. In the two-dimensional context, this model has been derived by (Eyink 2001) without the self-similarity assumption, but rather by assuming scale-locality of contributions to σ at scales smaller than the filter scale; its use has been investigated by Chen *et al.* (2003) and Bouchet (2003).

Figure 9.5 compares the statistical distributions of the actual transfer (solid line) and the transfer resulting from the nonlinear model (9.4.10). Evidently, the nonlinear-gradient model can deterministically capture both the forward scatter and the backscatter. This is much like the excellent agreement found in Chen *et al.* in the purely two-dimensional context.

Although the nonlinear model was motivated and derived using the filtering formalism of LES, we check its ability to model the two-way interactions (forward- and backscatter) between the mean and eddying motions in the Reynolds-averaging context in Fig. 9.3. The comparison of the actual transfer with the transfer resulting from the nonlinear gradient model is again good, although not as good as in the scale-decomposition approach (Fig. 9.5). This requires further investigation.

9.5 Discussion

Given its multiscale nature, direct numerical simulation of ocean circulation on time scales of interest are unlikely. Modelling of the effects of a substantial range of unresolved scales on larger-scale circulation is therefore necessary. While Reynolds-averaging-based modelling is appropriate for coarse-resolution simulations with resolutions of the order of a 100 km, an LES approach seems more appropriate for the eddy-permitting and eddy-resolving simulations that are presently possible for the interannual to decadal time scale.

Using eddy-resolving simulations, our research group has examined the nature of transfer of potential enstrophy between the filtered or averaged motion and the residual or eddy motions in the range of scales of forward cascade of potential enstrophy. We did this in both the Reynolds-averaging approach and the scale-decomposition approach in an inhomogeneous (basin) setting. In either of these approaches, a mean forward cascade of potential enstrophy is expected from the phenomenology of geostrophic turbulence and is verified. However, the mean forward cascade is found to be extremely weak in comparison to the magnitudes of the forward-scatter and backscatter in both approaches. The same analysis also shows the poor alignment of the eddy flux of PV in the downgradient direction. On the other hand, transfers obtained using the nonlinear-gradient model compare well with the actual transfer. Stated in another way, a large fraction of the effects of the unresolved scales on the resolved scales are correlated to the resolved structures and are representable by a nonlinear combination of resolved gradients. These findings are of significance both to deterministic and stochastic representation of subgrid processes in geostrophic turbulence and we will return to them after first considering the following.

While it is clear that it is only the divergent component of the eddy flux that directly drives mean circulation, since a decomposition of the eddy flux into rotational and divergent components is non-unique (arbitrary), we prefer to analyse the full flux. In spite of the non-uniquess of decomposing the eddy flux of PV, in light of the utility of defining a residual component of eddy flux of PV as in temporal Eulerian mean (TEM) and temporal residual mean (TRM) [e.g. see Eden *et al.* (2007)], we are exploring the issue further and will report on it elsewhere.

In Nadiga (2008), only the orientation of the eddy PV flux was considered with respect to the mean gradient of PV and the same nonlinear combination of gradients considered here. In that article, like here, the eddy PV flux aligned well with the nonlinear combination of gradients in the filtering approach, whereas in the Reynolds-averaging approach, the alignment was not as good. However, our present analysis of the potential-enstrophy transfer (see Fig. 9.3) suggests that the nonlinear combination of gradients may be a good basis to develop a parameterisation. This suggests the importance of further consideration of the distribution functions conditioned on appropriate physically motivated criteria. As an example, the orientation of the eddy fluxes may be important only when the fluxes themselves are significantly large, say compared to the external forcing (Marshall 2008, personal communication). We are in the process of studying such conditional distributions of orientation of eddy PV fluxes.

Back to the significance of our findings. First, the nearly symmetric nature of the distribution of transfer highlights the two-way nature of the communication between the eddy or subfilter scales and the filtered or mean motions. There is simultaneously (a) a drain of enstrophy from filtered/mean motion to subfilter/ eddy motions and (b) backscatter from subfilter/eddy scales to filtered/mean scales. The two processes are of comparable importance and it is only a slight difference between the two processes that results in the net forward cascade nature of the transfer. Eddy (subgrid) viscosity approaches, on the other hand, end up representing only the small net-downgradient nature of the subfilter resolved-scale interactions. With the situation being similar in the forward energy cascade regime of three-dimensional turbulence, improved subgrid models have been obtained by representing the two processes distinctly (Leith 1990; Chasnov 1991). This forms the basis of stochastic LES in the context of three-dimensional turbulence.

One may similarly anticipate better subgrid models of geostrophic turbulence when the drain and backscatter processes are both represented, rather than just the net effect. In fact, Jung *et al.* (2005) have successfully employed such a strategy in an operational atmospheric circulation model to reduce certain model biases and improve the simulation of certain weather regimes. In more idealised oceanic set-ups Berloff (2005), Nadiga *et al.* (2005), Duan & Nadiga (2007) and Nadiga & Livescu (2007) have preformed statistical analyses of subfilter terms and reported preliminary results from such stochastic subgrid closures.

Thus, if the subgrid modelling problem is approached from the point of view of (scalar) eddy viscosity, then, a stochastic representation of backscatter has shown promise. However, our finding that transfers obtained using the nonlinear-gradient model compare well with the actual transfer suggests that a parameterisation of the eddy flux based on the nonlinear-gradient model has the potential to model the forward- and backscatter rather naturally and deterministically. Thus we expect parameterisations based on the nonlinear model will be better than those based on downgradient closures.

In this article, we have conducted some a priori tests, wherein certain properties of models based only on the large-scale fields are tested against the corresponding properties of the actual subfilter turbulent fluxes in a resolved flow. While we find one such model to be good based on these tests, this is neither necessary nor sufficient for LES using this model to perform well. It only provides a good physical basis to design parameterisations; how well a model performs is best assessed by comparing results from simulations that use or don't use the model against available data, some of which may come from appropriately resolved simulations (a posteriori testing of LES).

Used directly, we find, like many others [e.g. Bouchet (2003)], that the nonlinear-gradient model (9.4.10) is generally unstable. On the other hand, we also find that (1) a related model based on a Pade approximation (Galdi & Layton, 2000) as opposed to the first-order Taylor series expansion in the derivation of the nonlinear model has better stability and that (2) in certain limits, such an approximation can be viewed in terms of a regularisation procedure [e.g. see Geurts (2004) or Holm & Nadiga (2003)]. Using such approaches, we have successfully modelled the eddy-driven mean flow aspects of the example circulation considered in this chapter at much lower resolutions. However, a discussion of those techniques and results are beyond the scope of this chapter, and we will present them elsewhere.

Finally, we expect that, although the findings presented here are not based on simulations of extensive ranges of parameters, the qualitative nature of these results will hold up in flow situations where eddies play an important role in shaping the large-scale and/or mean circulation.

Acknowledgements

It is my pleasure to thank Greg Eyink and John Marshall for discussions and Andy Majda for encouraging me to pursue and write up some early work that I presented at a workshop in Banff. BTN was supported in part by the Climate Change Prediction Program of DOE and the LDRD program at LANL (20030038DR).

References

Arakawa, A. 1966 Computational design for long-term numerical integration of the equations of fluid motion: two-dimensional incompressible flow, Part 1. *J. Comput. Phys.*, **1**, 119–143.

Berloff, P. S. 2005 Random-forcing model of the mesoscale oceanic eddies. *J. Fluid Mech.*, **529**, 71–95.

Bouchet, F. 2003 Parameterization of two-dimensional turbulence using an anisotropic maximum entropy production principle. http://arXiv:cond-mat/0305205.

Charney, J. G. 1971 Geostrophic turbulence, *J. Atmos. Sci.*, **28**, 1087–1095.

Chasnov, J. R. 1991 Simulation of the Kolmogorov inertial subrange using an improved subgrid model, *Phys. Fluids A*, **3**, 188–200.

Chen, S. Y., Ecke, R. E., Eyink, G. L., Wang, X. & Xiao, Z. L. 2003 Physical mechanisms of the two-dimensional enstrophy cascade. *Phys. Rev. Lett.*, **91**, 214501.

Duan, J. & Nadiga, B. T. 2007 Stochastic parameterization for large eddy simulation of geophysical flow. *Proc. Amer. Math. Soc.*, **135**, 1187–1196.

Eden, C., Greatbatch, R. J. & Olbers, D. 2007 Interpreting eddy fluxes. *J. Phys. Oceanogr.*, **37**, 1282–1296.

Eyink, G. L. 2001 Dissipation in turbulent solutions of 2D Euler equations. *Nonlinearity*, **14**, 787–802.

Galdi, G. P. & Layton, W. J. 2000 Approximation of the larger eddies in fluid motion II: a model for space filtered flow. *Math. Models Methods Appl. Sci.*, **10**, 343–350.

Germano, M. 1986 A proposal for a redefinition of the turbulent stresses in the filtered Navier–Stokes equations. *Phys. Fluids*, **29**, 2323–2324.

Geurts, B. J. 2004 *Elements of Direct and Large-Eddy Simulation*. Edwards.

Holland, W. R. & Rhines, P. B. 1980 An example of eddy-induced ocean circulation. *J. Phys. Oceanogr.*, **10**, 1010–1031.

Holm, D. D. & Nadiga, B. T. 2003 Modeling subgrid scales in the turbulent barotropic double gyre circulation. *J. Phys. Oceanogr.*, **33**, 2355–2365.

Jung, T., Palmer, T. N. & Shutts, G. J. 2005 Influence of a stochastic parameterization on the frequency of occurrence of North Pacific weather regimes in the ECMWF model. *Geophys. Res. Lett.*, **32**, 1–4.

Kraichnan, R. H. 1967 Inertial ranges in two-dimensional turbulence. *Phys. Fluids*, **10**, 1417–1423.

Leith, C. E. 1990 Stochastic backscatter in a subgrid-scale model: plane shear mixing layer, *Phys. Fluids A*, **2**, 297–299.

Marshall, J. C. & Shutts, G. 1981 A note on rotational and divergent eddy fluxes, *J. Phys. Oceanogr.*, **11**, 1677–1680.

Meneveau, C. & Katz, J. 2000 Scale-invariance and turbulence models for large-eddy simulation. *Annu. Rev. Fluid Mech.*, **32**, 1–32.

Nadiga, B. T. 2008 Orientation of eddy fluxes in geostrophic turbulence. *Phil. Trans. R. Soc. A*, **366**, 2491–2510.

Nadiga, B. T., Livescu, D. & McKay, C. Q. 2005 Stochastic Large Eddy Simulation of geostrophic turbulence. *Eos. Trans. AGU*, **86**(18); *Geophysical Research Abstracts*, vol. 7, 05488.
Nadiga, B. T. & Livescu, D. 2007 Instability of the perfect subgrid model in implicit-filtering large eddy simulation of geostrophic turbulence, *Phys. Rev. E*, **75**, 046303.
Pedlosky, J. 1987 *Geophysical Fluid Dynamics*, 2nd edition. Springer-Verlag.
Pope, S. B. 2000 *Turbulent Flows*. Cambridge University Press.
Press, W. H., Flannery, B. P., Teukolsky, S. A. & Vetterling, W. T. 1992 *Numerical Recipes in Fortran 77*, Cambridge University Press, pp. 708–716.
Rhines, P. B. & Holland, W. R. 1979 A theoretical discussion of eddy-driven mean flows. *Dyn. Atmos. Oceans*, **3**, 289–325.
Salmon, R. 1998 *Lectures on Geophysical Fluid Dynamics*, ch. 6. Oxford University Press.
Smith, K. S. & Vallis, G. K. 2001 The scales and equilibration of midocean eddies: freely evolving flow. *J. Phys. Oceanogr.*, **31**, 554–571.
Vallis, G. K. 2006 *Atmospheric and Oceanic Fluid Dynamics*, Cambridge University Press, Cambridge, 2006.

10

Stochastic theories for the irregularity of ENSO

RICHARD KLEEMAN

The El Niño/Southern Oscillation (ENSO) phenomenon is the dominant climatic fluctuation on interannual time scales. It is an irregular oscillation with a distinctive broadband spectrum. In this article, we discuss recent theories that seek to explain this irregularity. Particular attention is paid to explanations that involve the stochastic forcing of the slow ocean modes by fast atmospheric transients. We present a theoretical framework for analysing this picture of the irregularity and also discuss the results from a number of coupled ocean–atmosphere models. Finally, we briefly review the implications of the various explanations of ENSO irregularity for attempts to predict this economically significant phenomenon.

10.1 Observational background

On time scales ranging from the annual to the decadal, the dominant form of variability within the climate system is the El Niño/Southern Oscillation (ENSO) phenomenon whose centre of action is in the equatorial central and eastern Pacific. The variability is highly coherent spatially in both ocean and atmosphere and tends to be dominated by a relatively small number of large-scale patterns in both media.

Observations of ENSO have increased significantly in the past century as global shipping has become more prevalent and specialised oceanographic platforms have been put in place to monitor this important climatic effect. Prior to *c.* 1940, observations of both atmosphere and ocean in the equatorial Pacific were generally quite sparse and historical reconstructions of variables such as sea-surface temperature (SST) for that period rely heavily on observed historical global teleconnections in order to infer the equatorial patterns of variation. The standard empirical orthogonal function (EOF) analysis reveals that, for ENSO, only a small number of large-scale patterns are required to explain much of the variance for many

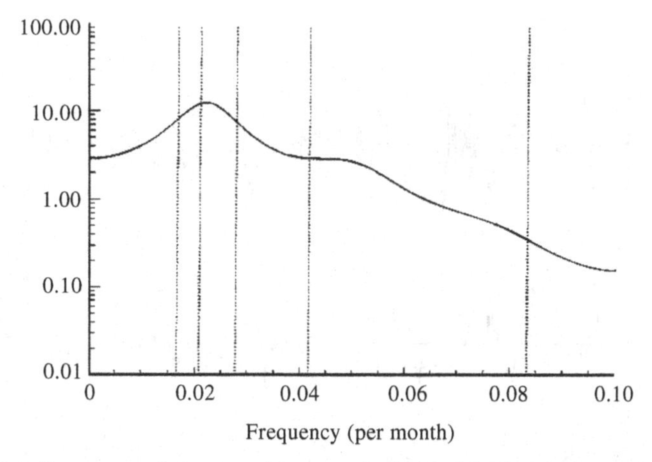

Figure 10.1 Spectrum of the monthly mean NINO3 SST index for ENSO for the period 1950–1993. Adapted from the work of Blanke *et al.* (1997).

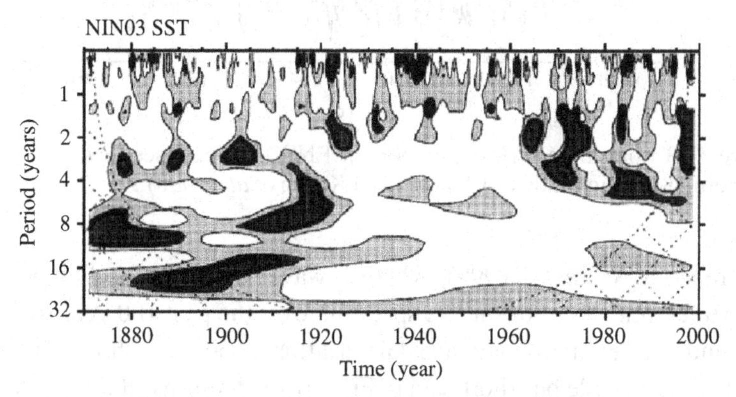

Figure 10.2 Wavelet analysis of the time-varying spectrum of the NINO3 SST index. Adapted from the work of Torrence & Webster (1999).

climatically important variables. Furthermore, the fact that such patterns are highly coherent in time between different climatic variables can assist considerably in the historical reconstruction of the dominant aspects of ENSO.

Based on the data from the more recent and reliable era, it is clear that ENSO is a broadband phenomenon with a spectral peak around four years. Figure 10.1 displays a spectral analysis by Blanke *et al.* (1997) of a widely used SST index (NINO3) of the eastern Pacific for the period 1950–1993.

The SST data can be extended back in time to the mid-nineteenth century using global EOFs to fill data-sparse regions that are prominent in the deep tropics before 1950 (e.g. Rayner *et al.* 2003). While the basic qualitative structure of the spectrum does not change much (see below), it is interesting to note, however, that there are significant decadal variations in the details. This is illustrated in Fig. 10.2, adapted

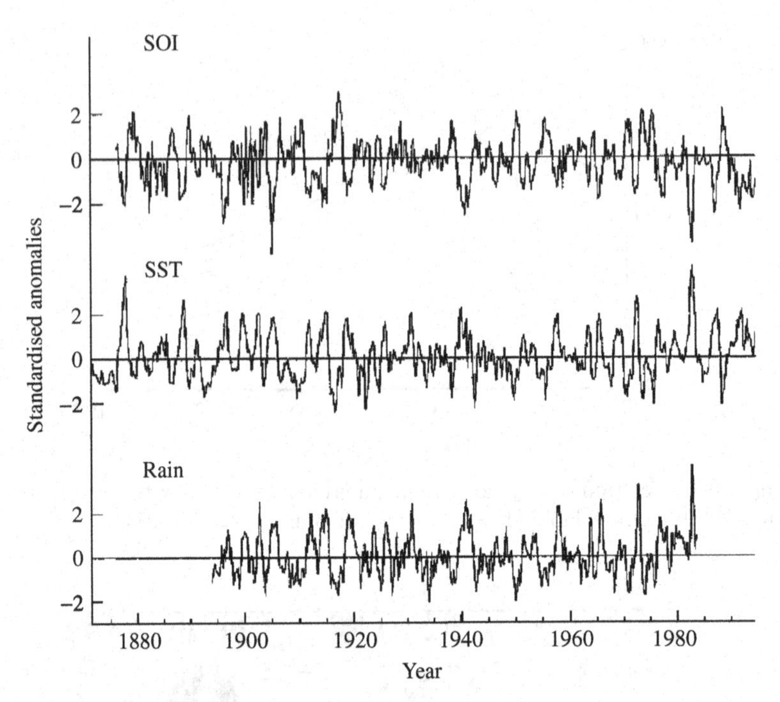

Figure 10.3 Long time series of important ENSO indices (see text). SOI is the Southern Oscillation Index. Adapted from Kestin *et al.* (1998).

from Torrence & Webster (1999), where a wavelet analysis technique has been used to extract time variations in the spectrum over the past 100 years or so. Given the apparently rather significant decadal variations, one must show a little caution in assigning the reliable but short data in Fig. 10.1 a definitive status. Note also that well-defined, rather narrow spectral peaks can occur in the data from time to time but are not persistent over the entire record.

A particularly robust method for extracting the dominant spectral character of ENSO has been performed by Kestin *et al.* (1998). Here, three completely independent data sources have been carefully compiled back to the late nineteenth century. These are derived from reconstructed SST data, station measurements of atmospheric pressure (the Southern Oscillation Index) as well as rain-gauge data from selected equatorial islands. These indices exhibit very high temporal coherence with a 21-year moving window correlation coefficient fluctuating between 0.6 and 0.8 during the length of the time series. Interestingly, the low-correlation periods correspond to quiescent periods for ENSO. The three time series and their spectra[1] are displayed in Figs. 10.3 and 10.4, respectively, and, while there are some minor

[1] The latter is unpublished but provided to the author by T. Kestin (2000, private communication).

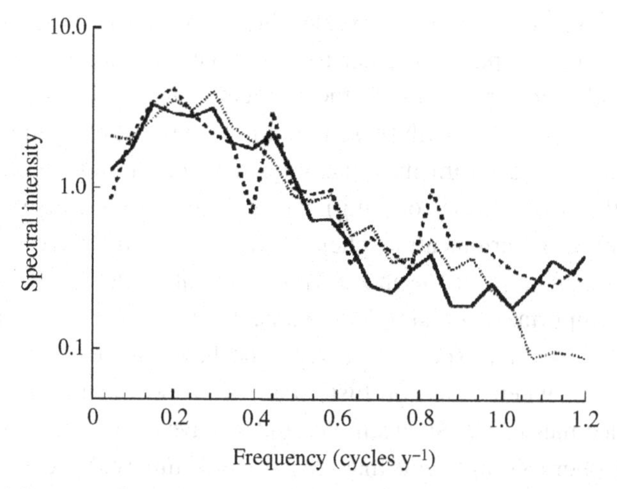

Figure 10.4 Spectra of the time series in Fig. 10.3 (T. Kestin 2000, private communication). Solid line, SOI; dotted line, NINO3; dashed line, rain.

differences, the overall common spectral shape is quite clear and qualitatively close to the spectra calculated using 45 years of recent reliable data.

Another feature of ENSO, which is very well established from many observational studies (e.g. Rasmusson & Carpenter 1982), is the phase locking to the seasonal cycle of warm events. These tend, in general, to occur towards the end of the calendar year. While there has been much discussion in the literature concerning the details of this phase locking (including the individual character of events and potential decadal variability), this basic aspect of the ENSO cycle appears beyond doubt.

In summary, both the spectrum and phase locking of the dominant mode of ENSO variability appear reasonably well established from the existing data. A theory of the irregularity of ENSO should be able to explain these important features. One would also hope that it should explain some of the decadally varying features of the phenomenon.

10.2 Theories for ENSO irregularity

The present theoretical understanding of ENSO is that it derives fundamentally from a coupling of the tropical ocean–atmosphere system. A comprehensive (but somewhat dated) review of the dynamical understanding of ENSO and coupled variability may be found in the excellent review by Neelin *et al.* (1998). The tropical atmospheric variability on interannual time scales in the Pacific can, in this view, be considered to be in approximate equilibrium with the underlying ocean temperature (SST).[2] On shorter time scales, however, moist convective instabilities are widely believed to play a crucial role in generating atmospheric variability. It seems natural therefore to divide the tropical dynamical system into fast and slow time scales associated with these two atmospheric forcing mechanisms. The slow time scale derives from the relaxation time of the equatorial ocean basin (modified by the atmospheric coupling), while the fast time scale has its genesis in the lifecycle times of various atmospheric convective disturbances.[3]

An interesting feature of the model development in the ENSO area has been the historical importance of simple models. In general, they have given good simulations of important aspects of ENSO and have also exhibited useful levels of prediction skill when tested on historical data (e.g. Cane *et al.* 1986; Barnett *et al.* 1993; Kleeman 1993). Such simple models have been characterised for many years by atmospheric components that are in equilibrium with the SST produced by their ocean component. Thus, in terms of the terminology introduced above, such models are entirely 'slow'. They have traditionally taken two forms: 'intermediate' and 'hybrid'; the difference being that intermediate models typically have an ocean with linear shallow water equations, while hybrid models have a fully nonlinear primitive equation ocean (an OGCM). Both types of model use highly simplified atmospheric components that use either shallow water equations or linear statistical models of the wind/SST relationship.

This conceptual timescale division introduced above facilitates the explanation of the two leading theories for ENSO irregularity. The first theory posits that the irregularity derives primarily from a chaotic interaction of various slow modes and therefore assumes implicitly that the fast modes are of secondary importance. The second theory contends instead that the slow timescale modes are not chaotic but are instead 'disrupted' in a stochastic fashion by the fast modes.

In this article, we shall concentrate heavily on detailing the second explanation, but it should be borne in mind that the first explanation has many adherents among climate scientists and so the area of ENSO irregularity is still not a settled question. We review in turn the basic ideas behind each of the two theories.

10.2.1 Slow-mode chaos

This theory was advanced by a number of investigators in the early 1990s (e.g. Chang *et al.* 1994; Jin *et al.* 1994; Tziperman *et al.* 1994). The principal idea is that the slow manifold of the tropical ocean–atmosphere system has a number of

[2] It forces the ocean in turn by the generation of various baroclinic (internal) waves that affect SST and hence the coupled paradigm.

[3] These can vary from hours to weeks depending on the nature and scale of the disturbances.

important time scales and the ocean is sufficiently nonlinear that modes with such time scales are able to interact and generate chaotic behaviour. Most focus has been on the interaction between the annual cycle and the most unstable coupled mode, which in most realistic models has a time scale of approximately three to five years. Less attention has been focused on the interaction between coupled (unstable) modes of differing time scales [see, however, Tziperman *et al.* (1995), who analyses the model of Zebiak & Cane (1987)].

The annual cycle has been found in many coupled models to show a significant interaction with the leading coupled mode. This often, but not always, manifests itself in the form of locking of the interannual variability to a fixed integral multiple of the annual cycle. A detailed and careful analysis of this effect may be found in Jin *et al.* (1996). The important physical parameters of this dynamical system problem are the ratio of the time scales of the forced and unstable modes as well as the growth rate of the latter mode. The authors mentioned above studied large parameter regions of a simplified (but realistic) coupled model in which these two factors showed variation over all realistically possible values. Overall, they found that the locking of the variability to various multiples of annual period was the most common behaviour. When the model was tuned to be between such multiples of the annual cycle, then chaos appeared. The route taken in the parameter space in approaching these chaotic regimes was one often seen in the transition to turbulence in simple dynamical models. These chaotic regions of the parameter space were interpreted as overlapping resonances of the system to the annual cycle forcing. Such resonances were the most commonly seen model behaviour, i.e. oscillations with periods of some integral multiple of the annual period. The power spectra associated with this form of chaos tend to show the overall pattern mentioned in Section 10.1; however, they usually also often displayed peaks at periods that were a (low) rational multiple of the annual period. This preference for low rational multiples of the forcing period is a common feature of these kinds of dynamical systems. The particular peaks emphasised in the chaotic spectrum varied according to the chosen model parameters.

10.2.2 Stochastic forcing of slow modes by fast modes

Many intermediate or hybrid coupled models have been developed in the past 25 years and have been consistent with the parameter space analysis of the particular simple coupled models discussed in Section 10.2.1, which usually exhibit a regular oscillation.[4] Such an oscillation may be either self-sustained or damped depending

[4] This statement is based on extensive consultations by the author with a large range of ENSO modellers.

on how the particular model is tuned. Motivated partially by this empirical observation, a different paradigm for irregularity to that discussed above was suggested in the late 1980s and 1990s (e.g. Battisti 1988; Kleeman & Power 1994; Penland & Matrosova 1994; Penland & Sardeshmukh 1995; Kleeman & Moore 1997; Moore & Kleeman 1999).

This new paradigm explicitly separated the fast and slow modes in the atmosphere by adding the former to simple models as a stochastic forcing term. The additional fast-mode random forcing was found to robustly disrupt the slow scales and convert the original regular oscillation to an irregular one. It was also observed that damped regular oscillations could be transformed by this mechanism to self-sustained irregular oscillations, i.e. made to much better resemble the observed ENSO behaviour. This latter effect occurs through the flow of energy from the fast to slow modes. It is interesting to observe how the nature of early ENSO modelling with slaved atmospheric components and unrealistic regular oscillations facilitated this theoretical development.

As we shall see in Section 10.3, an important feature of the fast-mode stochastic forcing is its required spatial coherency. The dominant ENSO modes are typically rather large scale and require for their excitation forcing with a similarly large scale. It was realised therefore that the efficient stochastic forcing of ENSO could arise only from large-scale synoptic atmospheric transients such as the Madden–Julian Oscillation (MJO) and associated westerly wind bursts (WWB) or inter-tropical convergence zone easterly waves. Whether or not any particular coherent convective phenomenon is responsible for ENSO irregularity still remains somewhat uncertain since the form of the most efficient stochastic forcing can vary from model to model. However, it is worth noting that for a range of particular coupled models, this issue has been directly addressed (see Zavala-Garay *et al.* 2008) and the conclusion is that the coherent aspects associated with the MJO, at least, can indeed be very important.

When simple models are stochastically forced in the manner described above, it is usually rather easy to produce an irregular oscillation with a spectrum qualitatively matching that shown in Fig. 10.4. It is also possible to produce a phase locking of warm events to the annual cycle similar to observations. The physical mechanism behind this latter effect is easy to discern and appears physically plausible. Much more detail on all these points may be found in Sections 10.3 and 10.4.

10.3 Analysis of the stochastic forcing of complex dynamical systems

As we noted previously, a central question in investigating the plausibility of stochastic forcing as a mechanism for ENSO irregularity is in determining exactly what kinds of random forcing are most efficient in disrupting the slow manifold of the dynamical system. After all, if the plausible candidates for random forcing are very inefficient in such disruption, then the theory becomes less attractive as an explanation of ENSO irregularity.

This question has been analysed comprehensively in the context of systems in which the growth of variance of important dynamical variables can be explained by linearised dynamics. This appears to be a reasonable assumption for ENSO where linearised analysis has historically proved very useful (e.g. Hirst 1986; Battisti & Hirst 1989). It also appears appropriate on empirical modelling grounds for the analysis of variance growth over short intervals (three to six months). Finally, a posteriori forcing of nonlinear models with the results of the linearised analysis confirms the validity of the approximation used (e.g. Moore & Kleeman 1999).

The stochastic forcing of high-dimensional linear dynamical systems has been exhaustively investigated in Farrell & Ioannou (1995), Kleeman & Moore (1997) and Gardiner (2004). We present the second approach here, as it is adapted to suit the needs of the problem presently under consideration.

Consider a multidimensional stochastic differential equation. If we time discretise the solution, we obtain

$$\boldsymbol{u}_{\mu+1} = R(\mu+1,\ \mu)\boldsymbol{u}_\mu + \Delta t f_\mu, \tag{10.3.1}$$

where $\boldsymbol{u}$ is a vector in the sense that it may represent spatial variation and also many physical variables. Time indices are denoted by Greek subscripts. The operator R is the so-called propagator that shifts a state vector forward in time and, finally, f is a stochastic forcing term whose statistics are assumed to satisfy

$$\left.\begin{aligned} \langle f_{j\lambda}\rangle &= 0, \\ \langle f_{i\mu} f_{j\nu}\rangle &= C_{ij}^{\mu\nu}, \end{aligned}\right\} \tag{10.3.2}$$

where we are using Latin subscripts to denote the vector indices. If we iterate equation (10.3.1) from some initial time $\mu = 0$, then we obtain

$$\boldsymbol{u}_\mu = R(\mu,\ 0)\boldsymbol{u}_0 + \Delta t \sum_{\lambda=0}^{\mu-1} R(\mu,\ \lambda+1) f_\lambda. \tag{10.3.3}$$

If we now further assume for simplicity that the noise is white in time (a reasonable assumption for our purposes since the atmospheric synoptic time scale is very short for climate problems), then we may write

$$\langle f_{i\mu} f_{j\nu}\rangle = \frac{1}{\Delta t}\delta_{\mu\nu} C_{ij}.$$

From equation (10.3.3), we may now easily write down expressions for the first and second moments (mean and covariance) of $\boldsymbol{u}_\mu$ as

$$\langle \boldsymbol{u}_\mu \rangle = R(\boldsymbol{\mu}, 0)\boldsymbol{u}_0,$$
$$\langle \boldsymbol{u}_\mu, \boldsymbol{u}_\mu \rangle = \Delta t \sum_{\lambda=0}^{\mu-1} R(\mu, \lambda + 1) C R^*(\mu, \lambda + 1),$$

where the asterisk denotes the transpose or adjoint operator. Taking the continuous limit, we obtain

$$\langle \boldsymbol{u}(t) \rangle = R(t, 0)\boldsymbol{u}(0),$$
$$\langle \boldsymbol{u}(t), \boldsymbol{u}(t) \rangle = \int_0^t R(t, t')C R^*(t, t')\mathrm{d}t'.$$

Let us now consider the variance with respect to some index (e.g. average eastern equatorial SST)

$$\mathrm{var}(t) = X_{ij} \langle \boldsymbol{u}^i(t), \boldsymbol{u}^j(t) \rangle,$$

where we are assuming the summation convention for repeated Latin indices and X_{ij} can be considered the 'metric' matrix for our index of interest. It is now easy to show that

$$\mathrm{var}(t) = \mathrm{trace}\{ZC\},$$
$$Z \equiv \int_0^t R^*(t, t')X R(t, t')\mathrm{d}t'.$$

Note that we have completely separated the stochastic forcing represented by the covariance matrix C from the dynamics represented by the operator Z. It is easy to show that both these operators are positive (and hence Hermitian) and therefore have positive eigenvalues. We can therefore write

$$\mathrm{trace}\{ZC\} = \sum_{n,m} p_n q_m (\boldsymbol{P}_n, \boldsymbol{Q}_m)^2,$$

where the lower-case p and q are the eigenvalues of, respectively, Z and C, while the upper case $\boldsymbol{P}$ and $\boldsymbol{Q}$ are the corresponding eigenvectors. The inner product squared here can be interpreted as vector projection. Thus, if the eigenvectors of the noise forcing covariance matrix project onto the dynamical eigenvectors (we call these stochastic optimals), then there will be significant variance (or uncertainty) growth in our index of interest. Obviously, if the noise eigenvectors (often called EOFs) resemble the stochastic optimals with largest eigenvalues, then maximal variance growth will occur. We have therefore a very convenient framework for analysing the susceptibility of dynamical systems to disruption by noise.

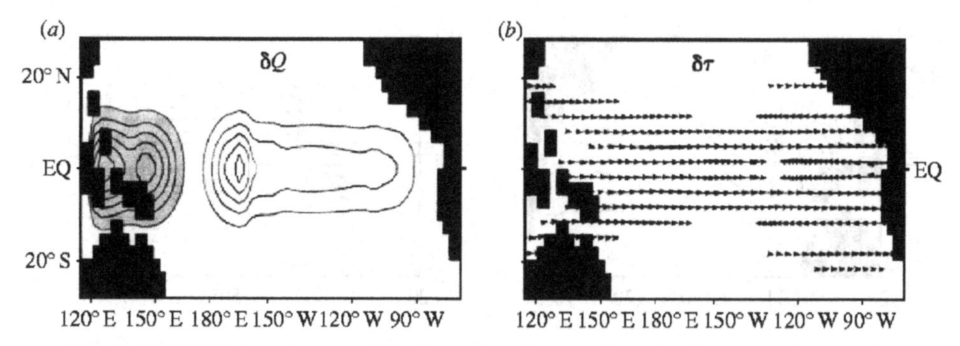

Figure 10.5 The dominant stochastic optimal for an intermediate coupled model. (*a*) Heat flux and (*b*) wind stress. Shading in (*a*) indicates negative values. Adapted from Moore & Kleeman (1999).

10.4 Stochastically forced model results

The spectrum (and eigenvectors) of the operator Z discussed in Section 10.3 have been evaluated for a range of linearised ENSO intermediate and hybrid models (see Kleeman & Moore 1997; Tang *et al.* 2005; Moore *et al.* 2006), and, in the models studied, it is highly peaked with most of the variance growth being caused by the first two eigenvectors (i.e. the p_1 and p_2 values are much greater than the others). These stochastic optimals are therefore crucial for whether large variance growth can occur. Figure 10.5 shows the spatial patterns of heat and momentum flux associated with these optimals for a particular intermediate coupled model.

Patterns of forcing such as this within the coupled model quickly grow into SST and wind-stress disturbances resembling the so-called observed westerly wind burst, as shown in Fig. 10.6.

This signature of a disturbance often associated with the MJO suggests that this large-scale pattern of internal atmospheric variability may be favourably configured to disrupt the ENSO dynamical system. It also states that only noise with large-scale spatial coherency will be effective at disruption. These results are from one coupled model only but the general qualitative conclusions hold for many models (for other results with different models, see Moore & Kleeman 2001). The detailed nature of the stochastic optimal can show some variation from model to model and also depends on the linearisation background state used.

If the ENSO intermediate model above is forced by white noise with the spatial coherency of the stochastic optimals, then an irregular oscillation is induced. Figure 10.7 shows the result of such forcing on the above intermediate coupled model with such noise. Without the noise, a perfectly regular decaying oscillation is observed.

The filled circles show December of each year, showing that the observed seasonal synchronisation of large warm events is also achieved. The reason for

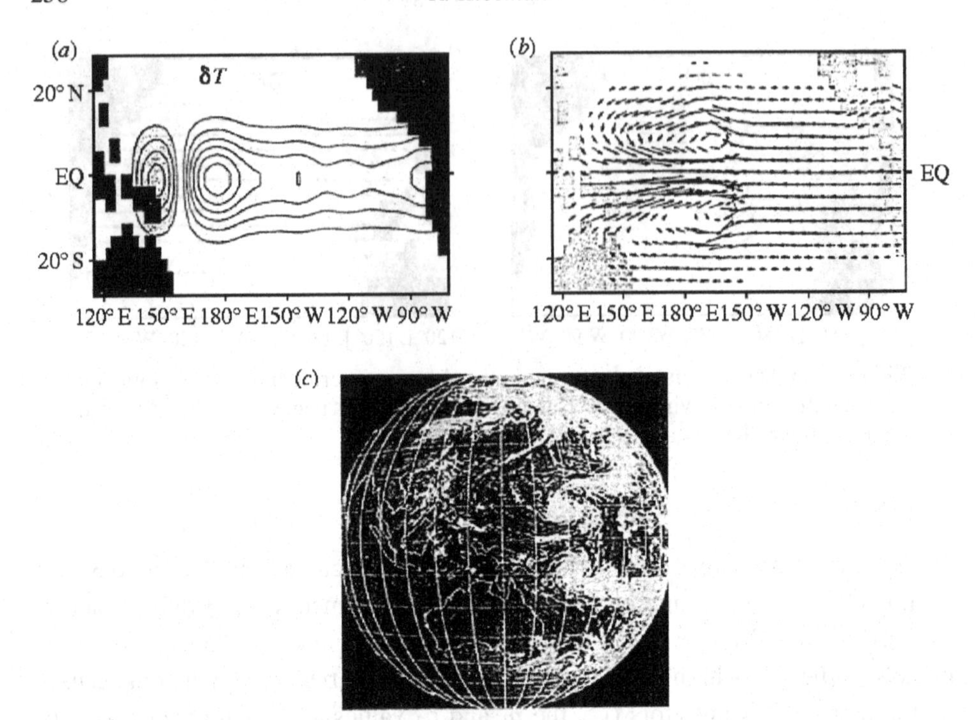

Figure 10.6 The short time-scale response of an intermediate coupled model to the forcing shown in Fig. 10.5. (*a*) The SST anomaly (shading indicates negative values) and (*b*) the wind stress. For reference, the cloud patterns associated with a westerly wind burst are shown in (*c*). Note the double cyclone structure that is consistent with the winds shown in (*b*). Adapted from Moore & Kleeman (1999).

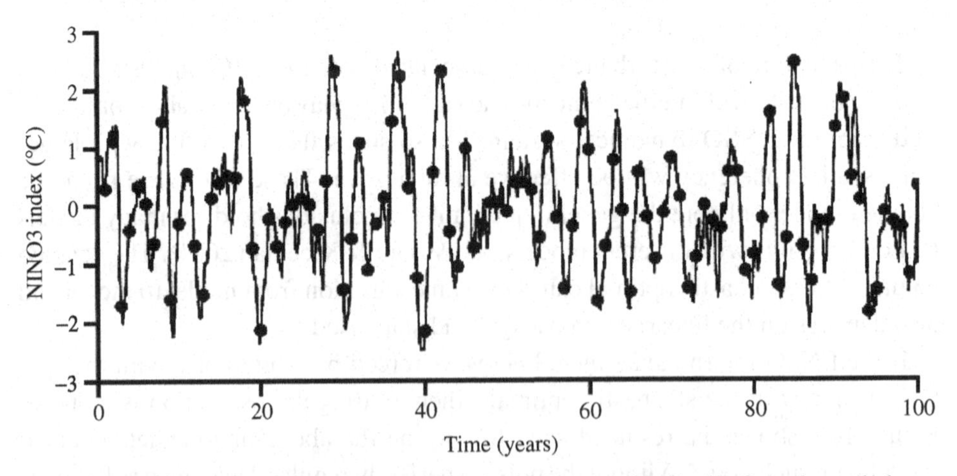

Figure 10.7 A plot of the NINO3 SST index from a stochastically forced coupled model. The filled circles indicate December. Note the frequent locking of warm event peaks to this time of year. Adapted from Moore & Kleeman (1999).

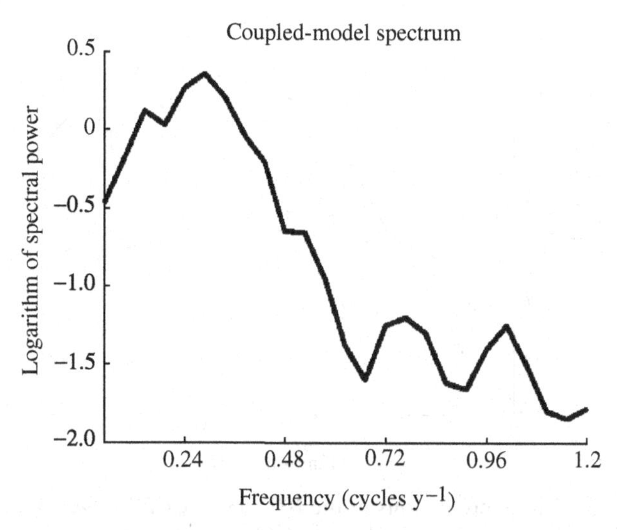

Figure 10.8 Spectrum corresponding to the time series in Fig. 10.7.

the synchronisation is related to the seasonal and ENSO cycle variation of the instability of the system to small perturbations. In the model displayed, this tends to be greatest in the northern spring and in the lead up to warm events. At that time, convective anomalies in the atmosphere are able to develop with the greatest zonal extent since the mean SST is high at the mentioned phases of both the seasonal and ENSO cycles. It is also worth noting that this synchronisation scenario has recently been confirmed in the careful observational study of Hendon *et al.* (2007). The irregular oscillation in Fig. 10.7 shows a clear resemblance to the observed pattern in Fig. 10.3, and the spectrum displayed in Fig. 10.8 is qualitatively the same as the observed spectrum in Fig. 10.4.

The irregular behaviour noted is particularly robust as one can vary the amplitude of the forcing by some orders of magnitude without qualitative effect. It has now been observed by many other investigators using a range of different models (examples include, but are not restricted to, Blanke *et al.* 1997, Eckert & Latif 1997 and Zhang *et al.* 2003). In addition, long integrations of this stochastic model show that there is a decadal variation of its spectrum qualitatively similar to the observations seen in Fig. 10.2. Such agreement is not surprising as it may also be observed in very simple stochastically forced oscillators as demonstrated in Fig. 10.9 of Kestin *et al.* (1998) for the case of an AR(3) process.

The variance growth in time predicted by this theory of ENSO indices has the form of rather rapid growth for the first six to nine months with considerably slower growth subsequently. It is interesting that this pattern is also observed in

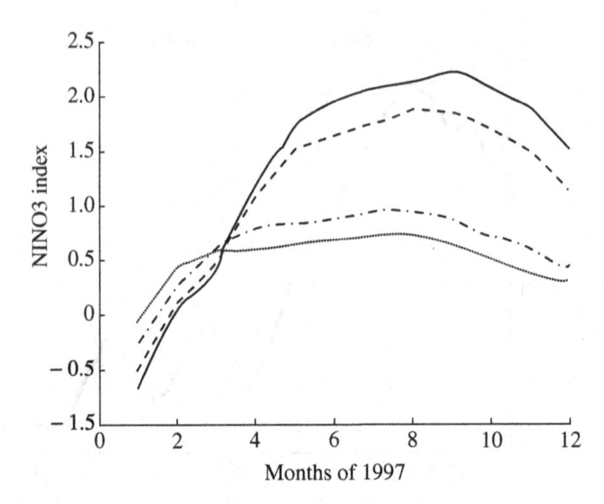

Figure 10.9 Real-time predictions of the NINO3 index from the operational coupled model prediction system of Kleeman (1994) for the first four months of 1997 (dotted line, January start; dashed–dotted line, February start; dashed line, March start; solid line, April start). Values before a particular prediction are from the data assimilation phase of the system. The observed warm event peaked in December with the NINO3 index = 3.68, therefore all predictions here were underestimates of the true event.

physically complete coupled models (coupled general circulation models, CGCMs; see Stockdale *et al.* 1998) that, of course, have inbuilt atmospheric transients as part of their atmospheric components.

A further detailed comparison of the simple-model results discussed above with those from CGCMs is difficult owing to the complexity of the latter and the fact that their overall simulation of the observed seasonal cycle, interannual variability and moist convection is often still suboptimal. Nevertheless, such models are improving and a recent model of Lengaigne *et al.* (2004) shows reasonable simulations of all these features. This particular model shows strong sensitivity of predictions to disruption by wind anomalies with the structure of WWB and, in that respect, resembles the simple-model behaviour discussed above.

This latter model is also able to capture with surprising accuracy the qualitative high-frequency evolution of strong ENSO events. The detailed evolution of zonal wind, SST and zonal currents is very well depicted when compared with the good observational record of the 1997 warm event. Ensemble experiments with very closely aligned initial conditions reveal the strong sensitivity of predictions of warm events. Additionally, they show that such ensembles may be shifted markedly towards strong warm events by the insertion of a short-lived westerly wind anomaly with a spatial structure resembling that detailed earlier in this section.

10.5 Predictability implications

The accurate prediction of the ENSO phenomenon is of great societal value. The ENSO warm and cold events are often responsible for huge economic dislocation due to the induced global large-scale precipitation and temperature anomalies (Ropelewski & Halpert 1987; Halpert & Ropelewski 1992). For this reason, considerable energy has been devoted to accurately forecasting ENSO (Kirtman *et al.* 2002) and many operational models with historical skill are currently used to make routine forecasts.

Evidently, the nature of the irregularity of ENSO is central to the issue of potential limits to the predictability of this climatic variation. Different physical mechanisms for irregularity have probably very different implications for a fundamental predictability limitation. At present, it remains unclear the degree to which present forecast errors are due to such limiting factors or to the more prosaic model error. Certainly, clear limitations in practical prediction were seen in the recent large warm event of late 1997, as documented by Barnston *et al.* (1999).

Predictability implications of the two mechanisms outlined in Section 10.2 have been addressed using idealised twin experiments to investigate the potential error growth due to uncertain initial conditions (e.g. Goswami & Shukla 1991; Kleeman & Moore 1997). In general, the error growth patterns in the two cases have very different characteristics with chaotic (and unstochastically forced) models exhibiting much longer time scales than those seen in stochastic models (years in the first case versus months in the second). In addition, an important sensitivity has been found in the predictability results from stochastic models by Thompson & Battisti (2001). The limitation time scale was found to be a strong function of the instability time scale of the coupled mode within the system. More unstable coupled modes lengthened the predictability because it made such modes (which are evidently central to any prediction) less susceptible to disruption by random forcing. The greater the coupled-mode instability, the more pronounced and persistent the resulting oscillation. Such oscillations are by their nature less able to have their phase shifted by random perturbations (a nice illustration of this idea may be found in simple climate models studied in Kleeman 2002).

The results from Section 10.3 suggest that large-scale convective fluctuations such as the MJO (and associated westerly wind events) may be important factors in limiting predictability. By their very nature, the phase and amplitude of such episodes are not predictable beyond a month at most. Thus, if they are efficient in forcing the coupled system, then they are clear candidates for the disruption of predictions. There is some evidence from an operational prediction model (documented in Kleeman 1994) that this may have indeed occurred in the case of the 1997 warm event. Figure 10.9 displays real-time operational predictions of the

NINO3 index made one month apart from January through April of that year. All predictions were suggesting warming later in the year; however, the amplitude of this shows a sudden jump between February and March with the later predictions being significantly more accurate. This behaviour suggests that something in the model initialisation during March 1997 caused the predictions to respond rather dramatically and beneficially. The examination of the wind data for that month (the main source of coupled model initialisation) shows that, indeed, there was a quite strong equatorial westerly wind event near the international date line in the first half of that month.

The case for the influence of the MJO on the ENSO development has also been made strongly from an observational perspective by Zhang & Gottschalck (2002) and Batstone & Hendon (2005). These authors show that a large fraction of equatorial oceanic Kelvin wave activity preceding ENSO events can be traced back to the MJO activity.

Finally, it is worth noting that several authors (Fedorov 2002; Flügel *et al.* 2004) have proposed, for differing reasons, that the stochastic influence on ENSO predictability may exhibit significant decadal variation.

10.6 Outlook

Perhaps the least defined aspect of the paradigm outlined here concerns the detailed nature of the stochastic forcing operating within the coupled system. In particular, while it is clear that the forcing most efficient at disrupting the system is large scale in character, its precise nature is clearly dependent on the detailed way in which the coupled dynamics are represented within a model. This dependency manifests itself in the differing stochastic optimals seen in different coupled models.

In a physically complete CGCM, such uncertainty derives from two important and only moderately well-understood physical parameterisations: atmospheric tropical convection and oceanic vertical mixing. Both these physical processes strongly influence the basic nature of the coupled system. It is to be expected that further deeper understanding of the stochastic forcing of ENSO will occur when understanding of these two physical processes increases.

Recent theoretical and modelling works (e.g. Perez *et al.* 2005; Jin *et al.* 2007) have suggested that the stochastic forcing operating within the coupled system should be dependent on the phase of ENSO. This amounts to an assumption of multiplicative stochastic forcing as opposed to the simple additive forcing discussed in Section 10.4. Convective disturbances are confined generally to high SST regions. Such warm-pool regions are a strong function of the ENSO phase and, furthermore, there is evidence from atmospheric studies that the zonal extent of the warm pool influences the development of large-scale convective disturbances.

These factors are all physical grounds for expecting multiplicative effects to be important. Clearly, this area deserves more investigation.

References

Barnett, T. P., Graham, N., Pazan, S. *et al.* 1993 ENSO and ENSO-related predictability. Part I: prediction of Equatorial Pacific sea surface temperature with a hybrid coupled ocean-atmosphere model. *J. Climate*, **6**, 1545–1566 (doi:10.1175/1520-0442(1993)006<1545:EAERPP>2.0.CO;2).

Barnston, A. G., Glantz, M. H. & He, Y. 1999 Predictive skill of statistical and dynamical climate models in forecasts of SST during the 1997–98 El Niño episode and the 1998 La Niña onset. *Bull. Am. Meteorol. Soc.*, **80**, 217–243 (doi:10.1175/1520-0477(1999)080<0217:PSOSAD>2.0.CO;2).

Batstone, C. & Hendon, H. H. 2005 Characteristics of stochastic variability associated with ENSO and the role of the MJO. *J. Climate*, **18**, 1773–1789 (doi:10.1175/JCLI3374.1).

Battisti, D. S. 1988 Dynamics and thermodynamics of interannual variability in the tropical atmosphere/ocean system. PhD thesis, University of Washington.

Battisti, D. S. & Hirst, A. C. 1989 Interannual variability in a tropical atmosphere–ocean model: influence of the basic state, ocean geometry and nonlinearity. *J. Atmos. Sci.*, **46**, 1687–1712 (doi:10.1175/1520-0469(1989)046<1687:IVIATA>2.0.CO;2).

Blanke, B., Neelin, J. D. & Gutzler, D. 1997 Estimating the effects of stochastic wind stress forcing on ENSO irregularity. *J. Climate*, **10**, 1473–1486 (doi:10.1175/1520-0442(1997)010<1473:ETEOSW>2.0.CO;2).

Cane, M. A., Zebiak, S. E. & Dolan, S. C. 1986 Experimental forecasts of El Niño. *Nature*, **321**, 827–832 (doi:10.1038/321827a0).

Chang, P., Wang, B., Li, T. & Ji, L. 1994 Interactions between the seasonal cycle and the Southern Oscillation – frequency entrainment and chaos in a coupled ocean-atmosphere model. *Geophys. Res. Lett.*, **21**, 2817–2820 (doi:10.1029/94GL02759).

Eckert, C. & Latif, M. 1997 Predictability of a stochastically forced hybrid coupled model of El Niño. *J. Climate*, **10**, 1488–1504 (doi:10.1175/1520-0442(1997)010<1488:POASFH>2.0.CO;2).

Farrell, B. F. & Ioannou, P. J. 1995 Stochastic dynamics of the mid-latitude atmospheric jet. *J. Atmos. Sci.*, **52**, 1642–1656 (doi:10.1175/1520-0469(1995)052<1642:SDOTMA>2.0.CO;2).

Fedorov, A. V. 2002 The response of the coupled tropical ocean–atmosphere to westerly wind bursts. *Q. J. R. Meteorol. Soc.*, **128**, 1–23 (doi:10.1002/qj.200212857901).

Flügel, M., Chang, P. & Penland, C. 2004 Identification of dynamical regimes in an intermediate coupled ocean–atmosphere model. *J. Climate*, **13**, 2105–2115.

Gardiner, C. W. 2004 *Handbook of Stochastic Methods for Physics, Chemistry and the Natural Sciences*. Springer Series in Synergetics, vol. **13**. Springer-Verlag.

Goswami, B. N. & Shukla, J. 1991 Predictability of a coupled ocean-atmosphere model. *J. Climate*, **4**, 3–22 (doi:10.1175/1520-0442(1991)004<0003:POACOA>2.0.CO;2).

Halpert, M. S. & Ropelewski, C. F. 1992 Surface temperature patterns associated with the Southern Oscillation. *J. Climate*, **5**, 577–593 (doi:10.1175/1520-0442(1992)005<0577:STPAWT>2.0.CO;2).

Hendon, H. H., Wheeler, M. C. & Zhang, C. 2007 Seasonal dependence of the MJO–ENSO relationship. *J. Climate*, **20**, 531–543 (doi:10.1175/JCLI4003.1).

Hirst, A. C. 1986 Unstable and damped equatorial modes in simple coupled ocean–atmosphere models. *J. Atmos. Sci.*, **43**, 606–632 (doi:10.1175/1520-0469(1986)043<0606:UADEMI>2.0.CO;2).

Jin, F.-F., Neelin, J. D. & Ghil, M. 1994 El Niño on the devil's staircase: annual subharmonic steps to chaos. *Science*, **264**, 70–72 (doi:10.1126/science.264.5155.70).

Jin, F.-F., Neelin, J. D. & Ghil, M. 1996 El Niño/Southern Oscillation and the annual cycle: subharmonic frequency-locking and aperiodicity. *Physica D*, **98**, 442–465 (doi:10.1016/0167-2789(96)00111-X).

Jin, F.-F., Lin, L., Timmermann, A. & Zhao, J. 2007 Ensemble-mean dynamics of the ENSO recharge oscillator under state-dependent stochastic forcing. *Geophys. Res. Lett.*, **34**, L03807 (doi:10.1029/2006GL027372).

Kestin, T. S., Karoly, D. J., Yano, J. & Rayner, N. A. 1998 Time-frequency variability of ENSO and stochastic simulations. *J. Climate*, **11**, 2258–2272 (doi:10.1175/1520-0442(1998)011<2258:TFVOEA>2.0.CO;2).

Kirtman, B. P., Shukla, J., Balmaseda, M. *et al.* 2002 Current status of ENSO forecast skill. A report to the climate variability and predictability (CLIVAR) numerical experimentation group (NEG), CLIVAR working group on seasonal to interannual prediction. See http://www.clivar.org/publications/wg_reports/wgsip/nino3/report.htm

Kleeman, R. 1993 On the dependence of hindcast skill on ocean thermodynamics in a coupled ocean–atmosphere model. *J. Climate*, **6**, 2012–2033 (doi:10.1175/1520-0442(1993)006<2012:OTDOHS>2.0.CO;2).

Kleeman, R. 1994 Forecasts of tropical Pacific SST using a low order coupled ocean–atmosphere dynamical model. *NOAA Experimental Long-Lead Forecast Bulletin*. NOAA.

Kleeman, R. 2002 Measuring dynamical prediction utility using relative entropy. *J. Atmos. Sci.*, **59**, 2057–2072 (doi:10.1175/1520-0469(2002)059<2057:MDPUUR>2.0.CO;2).

Kleeman, R. & Moore, A. M. 1997 A theory for the limitation of ENSO predictability due to stochastic atmospheric transients. *J. Atmos. Sci.*, **54**, 753–767 (doi:10.1175/1520-0469(1997)054<0753:ATFTLO>2.0.CO;2).

Kleeman, R. & Power, S. B. 1994 Limits to predictability in a coupled ocean–atmosphere model due to atmospheric noise. *Tellus A*, **46**, 529–540 (doi:10.1034/j.1600-0870.1994.00014.x).

Lengaigne, M., Guilyardi, E., Boulanger, J. P. *et al.* 2004 Triggering of El Niño by westerly wind events in a coupled general circulation model. *Clim. Dyn.*, **23**, 601–620 (doi:10.1007/s00382-004-0457-2).

Moore, A. M. & Kleeman, R. 1999 Stochastic forcing of ENSO by the intraseasonal oscillation. *J. Climate*, **12**, 1199–1220 (doi:10.1175/1520-0442(1999)012<1199:SFOEBT>2.0.CO;2).

Moore, A. M. & Kleeman, R. 2001 The differences between the optimal perturbations of coupled models of ENSO. *J. Climate*, **14**, 138–163 (doi:10.1175/1520-0442(2001)014<0138:TDBTOP>2.0.CO;2).

Moore, A. M., Zavala-Garay, J., Tang, Y. *et al.* 2006 Optimal forcing patterns for coupled models of ENSO. *J. Climate*, **19**, 4683–4699 (doi:10.1175/JCLI3870.1).

Neelin, J. D., Battisti, D. S., Hirst, A. C. *et al.* 1998 ENSO theory. *J. Geophys. Res. C Oceans*, **103**, 14.

Penland, C. & Matrosova, L. 1994 A balance condition for stochastic numerical models with applications to the El Niño–Southern Oscillation. *J. Climate*, **7**, 1352–1372 (doi:10.1175/1520-0442(1994)007<1352:ABCFSN>2.0.CO;2).

Penland, C. & Sardeshmukh, P. D. 1995 The optimal growth of tropical sea surface temperature anomalies. *J. Climate*, **8**, 1999–2024 (doi:10.1175/1520-0442(1995)008<1999:TOGOTS>2.0.CO;2).

Perez, C. L., Moore, A. M., Zavala-Garay, J. & Kleeman, R. 2005 A comparison of the influence of additive and multiplicative stochastic forcing on a coupled model of ENSO. *J. Climate*, **18**, 5066–5085 (doi:10.1175/JCLI3596.1).

Rasmusson, E. M. & Carpenter, T. H. 1982 Variations in tropical sea surface temperature and surface wind fields associated with the Southern Oscillation/El Niño. *Mon. Weather Rev.*, **110**, 354–384 (doi:10.1175/1520-0493(1982)110<0354:VITSST>2.0.CO;2).

Rayner, N. A., Parker, D. E., Horton, E. B. *et al.* 2003 Global analyses of sea surface temperature, sea ice, and night marine air temperature since the late nineteenth century. *J. Geophys. Res.*, **108**, 4407 (doi:10.1029/2002JD002670,2003).

Ropelewski, C. F. & Halpert, M. S. 1987 Global and regional scale precipitation patterns associated with the El Niño/Southern Oscillation. *Mon. Weather Rev.*, **115**, 1606–1626 (doi:10.1175/1520-0493(1987)115<1606:GARSPP>2.0.CO;2).

Stockdale, T. N., Anderson, D. L.T., Alves, J. O.S. & Balmaseda, M. A. 1998 Global seasonal rainfall forecasts using a coupled ocean–atmosphere model. *Nature*, **392**, 370–373 (doi:10.1038/32861).

Tang, Y., Kleeman, R. & Moore, A. M. 2005 On the reliability of ENSO dynamical predictions. *J. Atmos Sci.*, **62**, 1770–1791 (doi:10.1175/JAS3445.1).

Thompson, C. J. & Battisti, D. S. 2001 A linear stochastic dynamical model of ENSO. Part II: analysis. *J. Climate*, **14**, 445–466 (doi:10.1175/1520-0442(2001)014<0445:ALSDMO>2.0.CO;2).

Torrence, C. & Webster, P. J. 1999 Interdecadal changes in the ENSO-monsoon system. *J. Climate*, **12**, 2679–2690. (doi:10.1175/1520-0442(1999)012<2679:ICITEM>2.0.CO;2).

Tziperman, E., Stone, L., Cane, M. A. & Jarosh, H. 1994 El Niño chaos: overlapping of resonances between the seasonal cycle and the Pacific ocean–atmosphere oscillator. *Science*, **264**, 72–74 (doi:10.1126/science.264.5155.72).

Tziperman, E., Cane, M. A. & Zebiak, S. E. 1995 Irregularity and locking to the seasonal cycle in an ENSO prediction model as explained by the quasi-periodicity route to chaos. *J. Atmos. Sci.*, **52**, 293–306 (doi:10.1175/1520-0469(1995)052<0293:IALTTS>2.0.CO;2).

Zavala-Garay, J., Zhang, C., Moore, A. M. *et al.* 2008 Sensitivity of hybrid ENSO models to uncoupled atmospheric variability. *J. Climate*, **21**, 3704–3721 (doi:10.1175/2007JCLI1188.1).

Zebiak, S. E. & Cane, M. A. 1987 A Model El Niño–Southern Oscillation. *Mon. Weather Rev.*, **115**, 2262–2278 (doi:10.1175/1520-0493(1987)115<2262:AMENO>2.0.CO;2).

Zhang, C. & Gottschalck, J. 2002 SST anomalies of ENSO and the Madden–Julian Oscillation in the equatorial Pacific. *J. Climate*, **15**, 2429–2445 (doi:10.1175/1520-0442(2002)015<2429:SAOEAT>2.0.CO;2).

Zhang, L., Flügel, M. & Chang, P. 2003 Testing the stochastic mechanism for low-frequency variations in ENSO predictability. *Geophys. Res. Lett.*, **30**, 1630 (doi:10.1029/2003GL017505).

11

Stochastic models of the meridional overturning circulation: time scales and patterns of variability

ADAM H. MONAHAN, JULIE ALEXANDER AND ANDREW J. WEAVER

The global meridional overturning circulation (MOC) varies over a wide range of space and time scales in response to fluctuating 'weather' perturbations that may be modelled as stochastic forcing. This chapter reviews model studies of the effects of climate noise on decadal to centennial MOC variability, on transitions between the MOC regimes and on the dynamics of Dansgaard–Oeschger events characteristic of glacial periods.

11.1 Introduction

The meridional overturning circulation (MOC), which represents oceanic mass transport in the two-dimensional meridional/vertical plane, is a fundamental diagnostic for understanding the role of the ocean in past, present and future climates. In particular, it is believed that sudden changes in the Atlantic MOC (AMOC) are associated with abrupt climate changes prevalent in high-resolution proxy records over the last glacial cycle (e.g. Clark *et al.* 2002; Rahmstorf 2002). The importance of the AMOC to climate lies in its association with much of the total oceanic poleward heat transport in the present-day Atlantic, peaking at approximately 1.2 ± 0.3 PW at 24° N (e.g. Ganachaud & Wunsch 2000). Owing to the importance of the AMOC in the transport of heat to the North Atlantic, the variability and stability of the AMOC (particularly in response to anthropogenic forcing) is a subject of considerable scientific interest (e.g. Wood *et al.* 2003; Meehl *et al.* 2007).

This collection of currents has traditionally been referred to as the thermohaline circulation, a term which is problematic because it emphasises density gradients while implicitly downplaying the importance of mechanical forcing by the winds and interior turbulent mixing. In fact, ocean density gradients are largely set up by the winds via Ekman currents and surface buoyancy fluxes. Furthermore, mechanical forcing provides the energy source for driving these circulations (see Kuhlbrodt *et al.* 2007 for a review), although there is no simple or necessary link between mechanical energy supply and oceanic heat transport (e.g. Saenko & Weaver 2004). Ocean stratification determines the structure and strength of the response to this mechanical driving. Inherent in the use of the terms MOC and AMOC is the fact that wind and buoyancy forcing are inseparable, and that wind and tidal forcing play a fundamental role in providing the energy required for turbulent mixing within the ocean.

The existence and structure of the MOC is fundamentally connected with the locations of deep water formation in the ocean. The two main constituent water masses of the deep North Atlantic Ocean – North Atlantic Deep Water (NADW) at the bottom and Labrador Sea Water at an intermediate level – are currently formed in the Greenland–Iceland–Norwegian seas and the Labrador Sea, respectively. Deep convection also occurs at a number of locations around Antarctica, but the dense bottom water is susceptible to being trapped by topographic sills (as in the Bransfield Strait) or local circulation patterns (not excluding the Antarctic Circumpolar Current, ACC). In the Southern Ocean, around the southern tip of South America, an enhanced formation of low-salinity Antarctic Intermediate Water (AAIW) also occurs. The AAIW plays a critical role in linking the Pacific and Atlantic oceans and, in particular, in determining the stability of the AMOC (Saenko *et al.* 2003; Weaver *et al.* 2003). Today's climate has no sources of deep water in the North Pacific.

The various forcings that influence (and are in turn influenced by) the MOC (e.g. evaporative and heat fluxes, precipitation, sea-ice advection and melting, wind stresses, ocean eddy activity) display variability over a broad range of space and time scales, with a substantial concentration on time scales of atmospheric variability (subannual to interannual). When studying the dynamics of the MOC on time scales from decades to millennia, it is often convenient to represent these fluctuations as rapidly decorrelating stochastic processes (e.g. Hasselmann 1976). In this chapter, the influence of stochastic forcing on the dynamics of the MOC is considered. Section 11.2 examines the role of atmospheric fluctuations in driving MOC variability on decadal to centennial time scales. The influence of fluctuating forcing on transitions between MOC regimes is discussed in Section 11.3. Section 11.4 addresses the importance of these fluctuations in driving millennial-scale variability during glacial periods. Finally, conclusions are presented in Section 11.5.

A fundamental limitation in the study of MOC variability is the paucity of observational evidence. Palaeoclimate data provide evidence of changes in the NADW formation and the AMOC strength on millennial and longer time scales (e.g. Clark *et al.* 2002; McManus *et al.* 2004), while instrumental observations are providing a first characterisation of MOC variability on subannual (e.g. Cunningham *et al.*

2007) and interannual to interdecadal time scales (e.g. Bindoff *et al.* 2007). However, direct observations of the MOC of sufficiently high-temporal resolution over sufficiently long periods to address the questions of MOC variability considered in this review are not presently available. This chapter consequently focuses primarily on model studies of the MOC and its response to fluctuating forcing. As observational data sets improve, it is to be expected that a more complete comparison of data with model simulations will be possible.

11.2 Decadal- to centennial-scale variability

Climate variability on time scales ranging from decades to centuries has been identified in the instrumental and observational records of the North Atlantic, associated with a wide range of atmospheric, oceanic and biological processes (e.g. IPCC 2007). While the mechanisms responsible for the low-frequency variability of the climate system are not fully understood, the ocean circulation is believed to play a key role owing to its large thermal inertia. In particular, the variability of the AMOC has received a great deal of theoretical and modelling attention owing to its role as a major transporter of heat to high latitudes in the North Atlantic. In both simple and complex oceanic models, the MOC exhibits variability on decadal and centennial time scales (e.g. Weaver 1995; Bindoff *et al.* 2007), yet the instrumental–observational data sets are generally shorter than 100 years.

Three primary classes of mechanisms have been proposed to explain natural decadal- to centennial-scale variability of the MOC: (i) damped, uncoupled ocean modes excited by atmospheric variability; (ii) unstable, uncoupled ocean modes that express themselves spontaneously; and (iii) unstable or weakly damped, coupled modes of the ocean–atmosphere system. Numerous modelling studies have found self-sustained MOC oscillations associated with diffusive, advective and convective processes (e.g. Weaver & Sarachik 1991; Delworth *et al.* 1993; Weaver *et al.* 1993; Chen & Ghil 1995; Greatbatch & Zhang 1995; Rivin & Tziperman 1997; Fanning & Weaver 1998; Arzel *et al.* 2006). This section considers the major results of those studies, which have found that the maintenance of oscillatory variability of the MOC requires the input of energy from fluctuating atmospheric forcing.

Stochastic forcing has traditionally been used to represent high-frequency variability in surface fluxes. If the climate system has no internal oscillatory mode of variability, then the ocean integrates short-term atmospheric fluctuations, transforming the essentially white-noise atmospheric forcing into a red response ocean signal (e.g. Hasselmann 1976). However, if the climate system has preferred (if damped) modes of variability, then stochastic forcing results in peaks in the response spectrum at the characteristic time scales of that variability. In general, self-sustained MOC oscillations are due to internal nonlinearities in the climate system, while oscillations driven by stochastic atmospheric forcing can be largely accounted for by linear dynamics and would not exist without the energy from the external stochastic forcing.

In studies of simulated MOC variability on decadal to centennial time scales, stochastic forcing has been used to model fluctuating freshwater fluxes (e.g. Mikolajewicz & Maier-Reimer 1990; Spall 1993; Weaver *et al.* 1993; Weisse *et al.* 1994; Pierce *et al.* 1995; Aeberhardt *et al.* 2000), thermal fluxes (e.g. Griffies & Tziperman 1995; Saravanan & McWilliams 1997; Weaver & Valcke 1998; Kravtsov & Ghil 2004) and surface winds (e.g. Holland *et al.* 2000, 2001; Herbaut *et al.* 2002). A fundamental challenge in the parameterisation of these fluctuations by stochastic processes is a lack of knowledge of the forcing fluctuation amplitude, decorrelation time, length scale and distribution. The available observational evidence indicates that the dominant modes of atmospheric variability are essentially white in time but may not be white in space (e.g. Saravanan *et al.* 2000).

Identifying spatial patterns of stochastic forcings that effectively excite oscillations in the modelled AMOC was the focus of studies by Capotondi & Holland (1997), Saravanan & McWilliams (1997) and Tziperman & Ioannou (2002). Analysing the linearised dynamics of a three-box model (representing the polar ocean, the midlatitude surface ocean and the deep ocean) using generalised stability theory (e.g. Farrell & Ioannou 1996), Tziperman & Ioannou (2002) determined the optimal spatial structure of the noise that results in maximal variance of the AMOC variability. They found that the optimal forcing induces low-frequency variability by exciting salinity variability modes of the AMOC. While a three-box model can be useful for theoretical studies, it is too idealised to be quantitatively accurate. In particular, it is impossible to answer the question of whether observations project onto the predicted optimal modes: more complex models are needed to answer this question. Using a three-dimensional ocean model with idealised basin geometry, Capotondi & Holland (1997) analysed decadal variability by considering the spatial pattern of stochastic forcing as a variable of the problem. The period of oscillation of the simulated AMOC was found to be independent of the spatial pattern, leading to the conclusion that the variability at the decadal time scale is an internal mode of the system and not associated with some characteristics of the forcing (although the amplitude of the response was found to depend on the spatial structure of the forcing). Saravanan & McWilliams (1997) found that spatial resonance, defined as the forcing of a system with a spatial pattern that results in oscillations at a preferred frequency not present in the internal dynamics of the system, was responsible for exciting the oceanic decadal oscillation in a coupled atmosphere–ocean model. Eliminating the spatial correlations in the forcing was found to substantially reduce the variance associated with the interdecadal oscillation of the AMOC. The spatial pattern of the dominant mode of surface heat flux interacted with a single oceanic mode to induce the AMOC oscillations. Other studies have identified spatial patterns of buoyancy flux variations bearing a strong resemblance to the North Atlantic Oscillation (NAO) and which drive multidecadal to centennial AMOC fluctuations associated with damped internal oscillatory modes of the model ocean (Mikolajewicz & Maier-Reimer 1990; Delworth & Greatbatch 2000; Bentsen *et al.* 2004; Dong & Sutton 2005). On the other hand, the model of Spall (1993) produced sea-surface salinity variance as a direct response to the stochastic forcing and not an internal mode of variability. As hypothesised by Hasselmann (1976), Spall found that an undisturbed straightforward integration of the white-noise freshwater flux anomalies took place in the Labrador Sea, and it was the irregularly occurring salinity anomalies here that were responsible for the decadal variability in the North Atlantic.

Model studies have also considered the response of the AMOC to fluctuating surface wind stresses. The random wind-forcing field used by Holland *et al.* (2001) was random in time but had a spatial pattern similar to that of observations. The lack of specific time scales in the forcing indicates that the preferred time scales of the model's response were due to internal model physics and not to external forcing. This study showed that the AMOC variability was excited by the stochastic freshwater forcing provided by variable wind-driven Arctic ice export and responded linearly to this forcing at interdecadal time scales. Herbaut *et al.* (2002) described a damped mode of the ocean system requiring stochastic NAO-like wind-stress anomalies to maintain the oscillation. In this study, wind-stress forcing drove anomalous currents; the resulting advection of the mean temperature structure generates temperature anomalies that influenced the strength of the simulated AMOC.

Finally, modelling studies have suggested that the character of decadal- to centennial-scale variability in the MOC may be sensitive to the amplitude of the fluctuating forcing. The propagation of simulated salinity anomalies, which mediate the strength of the AMOC, has been shown to be facilitated by larger random freshwater forcing amplitude (e.g. Weaver *et al.* 1993; Skagseth & Mork 1998). Furthermore, changes in the strength of the random forcing have been found to cause transitions between different model oscillatory states (Aeberhardt *et al.* 2000).

Owing to model uncertainties and limited observations, it is not clear whether MOC variability on decadal through centennial time scales is self-sustained or driven by high-frequency variability. Direct comparisons of model results are complicated by variations in model complexity (ranging from box models through two-dimensional models to fully complex three-dimensional models); differences between ocean-only models and coupled ocean–atmosphere models; and the variety of methodologies used to identify the dominant 'modes' of decadal- to centennial-scale variability. Rivin & Tziperman (1997) suggested that the probability distribution function (PDF) of the MOC time series could be used to differentiate between linear noise-forced and nonlinear self-sustained oscillations. If the PDF is Gaussian when the stochastic forcing is Gaussian, then the oscillations result from the stochastic excitation of damped modes, while for nonlinear oscillations the PDF is strongly non-Gaussian. The refinement of tools such as this to distinguish between driven and self-sustained oscillatory variability will be an important component of determining the importance of fluctuating forcing in producing decadal- to centennial-scale variability in the MOC.

11.3 AMOC regimes

The present-day AMOC is characterised by a strong NADW formation in the Labrador and Nordic seas, but both palaeoclimate and modelling studies suggest that the AMOC can exist in other configurations (e.g. Rahmstorf 2002). There are two sets of feedbacks associated with these rearrangements of the AMOC, involving large-scale and local processes, respectively. In its present state, the AMOC transports warm salty water into the North Atlantic, where it is both cooled and freshened. The salt advected northward helps maintain the high densities of water in the North Atlantic and the vigorous formation of NADW. A reduction in deep water formation as a result of surface freshening reduces the poleward transport of salt and amplifies the initial perturbation; global circulation model (GCM) simulations demonstrate that if the initial perturbation is sufficiently strong, then this large-scale advective feedback can drive the AMOC to another stable steady state in which the NADW formation and the overturning circulation are essentially turned off (e.g. Kuhlbrodt *et al.* 2007). The second set of local feedbacks involves the formation of NADW through deep convection, which homogenises the water column into a convectively neutral state and transports relatively fresh water to depth. If convection is reduced due to surface freshening, then the reduced sinking flux of freshwater can amplify the initial surface freshening and (for a sufficiently strong perturbation) shut off convection all together (e.g. Rahmstorf 2001). It is a generic result that the stability properties of nonlinear systems are affected by the presence of noise, and environmental fluctuations that affect the AMOC are ubiquitous (e.g. Weaver *et al.* 1999). The effects of climate noise on transitions between the AMOC regimes will be considered in this section.

The simplest model that captures the AMOC bistability associated with large-scale advective feedbacks is that introduced by Stommel (1961), in which the circulation is associated with the exchange of heat and salt between two well-mixed boxes (representing the high- and low-latitude ocean in a single hemisphere) forced by specified freshwater fluxes and temperature relaxation to an externally specified difference. Denoting by ΔT and ΔS the interbox temperature and salinity differences, respectively, the respective tendencies can be expressed as

$$\frac{\mathrm{d}}{\mathrm{d}t}\Delta T = -q(\Delta\rho, t)\Delta T + \Gamma(\Delta T_0 - \Delta T), \quad (11.3.1)$$

$$\frac{\mathrm{d}}{\mathrm{d}t}\Delta S = -q(\Delta\rho, t)\Delta S + F(t), \quad (11.3.2)$$

where ΔT_0 is the externally imposed interbox temperature difference to which the system relaxes on a time scale of Γ^{-1} and F is the imposed freshwater forcing. The interbox exchange, q, is assumed to depend on the interbox density difference $\Delta\rho = \alpha_S \Delta S - \alpha_T \Delta T$ (α_S and α_T are, respectively, the haline and thermal expansivities of seawater) as $q(\Delta\rho, t) = \beta(t) + f(|\Delta\rho|)$, where $\beta(t)$ is a 'diffusive' exchange (including interbox fluxes mediated by gyre circulations; Monahan 2002c) and the function $f(|\Delta\rho|)$ models advective exchange by the overturning circulation (e.g. Stommel 1961; Cessi 1994). As in Monahan (2002a), we focus on the model $f(|\Delta\rho|) = c|\Delta\rho|$; the following results are not qualitatively sensitive to this parameterisation.

This already highly idealised system can be further simplified by taking the temperature relaxation time scale to be much faster than the interbox exchange time scale (a reasonable approximation), so to leading order $\Delta T \simeq \Delta T_0$ and we obtain a differential equation in ΔS alone. The influence of high-frequency fluctuations on the overturning will be modelled by taking $\beta(t)$ and $F(t)$ each to be the sum of a fixed mean and random fluctuations. Non-dimensionalising the resulting equation as in Monahan (2002a), we obtain the stochastic differential equation (SDE)

$$\frac{\mathrm{d}}{\mathrm{d}t}y = -(b_0 + |1 - y|)y - y \circ \eta + \mu + \xi, \quad (11.3.3)$$

where y is the non-dimensional salinity difference; b_0 and η are, respectively, the mean and fluctuations of the non-dimensional diffusive exchange parameter; and μ and ξ are, respectively, the mean and fluctuations of the freshwater flux. The open circle indicates that, for η modelled as white noise, the SDE is interpreted in the Stratonovich sense (i.e. as the white noise approximation to an autocorrelated process; Gardiner 1997; Penland 2003). Such highly idealised models of the effects of 'weather' forcing on the AMOC have been considered in a number of studies (e.g. Stommel & Young 1993; Cessi 1994; Bryan & Hansen 1995; Lohmann & Schneider 1999; Timmermann & Lohmann 2000; Vélez-Belichí *et al.* 2001; Monahan 2002a,b; Monahan *et al.* 2002; Kleinen *et al.* 2002).

A plot of steady-state solutions of equation (11.3.3) for $b_0 = 0$ in the absence of fluctuations ($\eta = \xi = 0$) is given in Fig. 11.1*a*. For a range of values of the freshwater forcing parameter $0 \le \mu \le 0.25$, the idealised model admits three

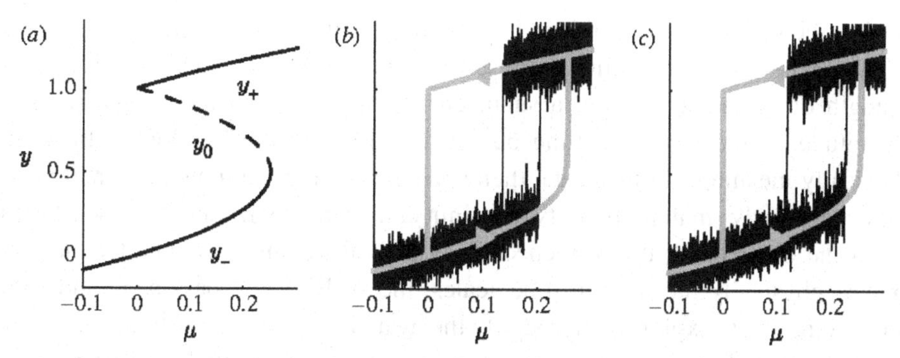

Figure 11.1 (*a*) Bifurcation structure of the AMOC box model equation (11.3.3) with $\eta = \xi = 0$ and $b_0 = 0$. The solid (dashed) lines are stable (unstable) solution branches. (*b*) Grey curve, hysteresis loop traced out by deterministic model as freshwater forcing μ is increased from -0.1 to 0.3 and then back to -0.1. Black curve, realisation of hysteresis curve for the stochastic model with $\eta = \sigma_1 \dot{W}_1$, $\xi = \sigma_2 \dot{W}_2$, where $\sigma_1 = \sigma_2 = 0.075$ and $\dot{W}_i$ are the independent white-noise processes. (*c*) As in (*b*), for a second realisation of the stochastic model. With a very high probability, the stochastic hysteresis loops are smaller than the deterministic ones.

steady states. Two of these steady states are stable, y_+ and y_-, with weak (strong) overturning circulations and strong (weak) interbox density differences, respectively. The third steady state y_0 is unstable. This interval of bistability is bounded by fold bifurcations, beyond which only a single steady-state solution exists. If the parameter μ is increased from below the lower bifurcation point to above the upper bifurcation point and then decreased again to below the lower bifurcation, the hysteresis loop displayed in Fig. 11.1b,c is traced out. Because the steady states are equilibrium solutions, deterministic transitions between solution branches for an evolving freshwater forcing $\mu(t)$ will generally occur somewhat beyond the bifurcation point; that is, there will be a slight overshoot (e.g. Berglund & Gentz 2006). The fact that three-dimensional coupled ocean–atmosphere models produce analogous hysteresis structures (e.g. Weaver & Hughes 1994; Rahmstorf *et al.* 2005; Kuhlbrodt *et al.* 2007) suggests that this AMOC model captures the essential physics of the advective feedback bistability (although its predictions cannot be expected to be quantitatively meaningful).

In the absence of climate noise, the AMOC regime transitions can only occur if μ is moved beyond one of the fold bifurcations. However, in the presence of noise, spontaneous transitions between regimes can occur within the region of bistability; if the noise is unbounded (as with Gaussian fluctuations), then transitions are possible everywhere that both states exist. For η and ξ white-noise processes, the mean transition time from y_- to y_+ can be computed analytically (e.g. Cessi 1994; Monahan 2002a) and takes the form $\tau(y_- \to y_+) \sim \exp\,[-V(y_-, y_0)/\Sigma^2]$, where Σ is a measure of the noise level and $V(y_-, y_0)$ is a measure of the 'potential barrier' the system must overcome to pass from y_- to y_+ (an analogous expression holds for the reverse transition). These transition rates are highly sensitive to the strength of the noise forcing; a small change in Σ can change τ by orders of magnitude. If the transition time out of an AMOC regime is longer than any physically meaningful time scale, then spontaneous regime transitions will occur with vanishingly small probability (the limit considered by Bryan & Hansen 1995). If, on the other hand, the residence times of both regimes are smaller than the longest physically meaningful time scales, the AMOC will pass back and forth between regimes, exploring thoroughly the available set of states. In this case, the signature of multiple regimes will be multimodality of the stationary (long-term equilibrium) PDF. For η and ξ white-noise processes, the stationary PDF of the process governed by equation (11.3.3) can be determined analytically (Cessi 1994; Timmermann & Lohmann 2000; Monahan 2002a). Finally, if the noise strength is so large that typical excursions of $y(t)$ are much larger than the separation between y_- and y_+, then the system will not feel the presence of the different deterministic equilibria and the two peaks of the stationary PDF will not be well separated (the limit considered by Stommel & Young 1993); this high-noise case is not relevant to the AMOC.

If the intensity of the climate noise is independent of the state of the AMOC (i.e. if the noise is additive), then the primary effect of stochastic fluctuations is inducing transitions between the AMOC regimes (or exciting oscillations, as discussed in Section 11.2). However, if the intensity of one or more of the noise processes is dependent on the state of the system (i.e. if the noise is multiplicative), then the noise itself can alter qualitative aspects of the dynamics (e.g. Penland 2003). In equation (11.3.3), the fluctuations in diffusive exchange η enter multiplicatively (when η is taken to be white noise), and their intensity has an effect on the multimodality of the stationary AMOC PDF. As the intensity of η increases, the range of freshwater forcings μ over which the PDF is bimodal shifts and contracts, eventually vanishing (Timmermann & Lohmann 2000; Monahan 2002a). The domain of bistability is also altered if η and ξ are correlated (as they might be expected to be physically; Monahan 2002b). In this way, the structure of the AMOC regimes (rather than just their occupation statistics) is influenced by the stochastic climate forcing.

Shifts in the domain of bimodality produced by multiplicative noise persist when η is allowed to have a non-zero autocorrelation time (i.e. to be red noise). In this situation, the dynamics of the vector (y, η) is governed by a two-dimensional SDE with a stationary PDF that is multimodal outside of the range of freshwater forcing values μ for which the deterministic part of the dynamics has multiple equilibria (Monahan 2002b; Monahan *et al.* 2002; Timmermann & Lohmann 2000). Bimodality can occur in the absence of multiple deterministic equilibria when there is a neighbourhood of the (y, η) state space without a fixed point but with a local minimum of the magnitude of the deterministic tendency. Passing through such a region, the system slows down but does not come to a halt. Random fluctuations can drive the system from the neighbourhood of the fixed point into this 'sluggish' region, where it slows down and lingers, building up probability mass; the resulting PDF can be bimodal (Monahan 2002b).

For moderate values of the noise intensity, the residence times of the two model regimes generally differ by orders of magnitude. It follows that, for such noise levels, the probability of being in one regime is orders of magnitude greater than that of being in the other, so while the PDF is technically bimodal it is effectively unimodal. This phenomenon and its consequences were discussed in Monahan (2002a,b), where it was referred to as regime stabilisation by noise. One particularly significant consequence is that if the freshwater forcing approaches the fold bifurcation where the current AMOC regime vanishes, then with high probability a transition to the other regime will occur before the bifurcation point is reached (Fig. 11.1b,c). The point at which the transition occurs is random, with a mean value that depends both on the noise level and on the rate at which μ is changing (e.g. Berglund & Gentz 2006). This effect is seen in both idealised models (Monahan 2002a) and more comprehensive coupled atmosphere–ocean models (Wang *et al.* 1999; Knutti & Stocker 2002). These transitions become less predictable as the bifurcation point is approached (Knutti & Stocker 2002). In the presence of climate noise, the AMOC regime shifts become less predictable as they become more likely.

The response of the modelled AMOC to weather noise does provide a potential 'early warning system' for regime transitions (Kleinen *et al.* 2002; Held & Kleinen 2004; Lenton *et al.* 2008). As the bifurcation point is approached, the real part of at least one eigenvalue of the dynamics linearised around the equilibrium state approaches zero (as at the bifurcation point, stable and unstable steady states coalesce). It follows that there is at least one state-space direction along which the feedbacks driving the system towards the equilibrium state weaken. As these feedbacks become weaker, perturbations excited along this weakly damped direction acquire larger amplitude and return more slowly to the equilibrium state. In consequence, the autocorrelation time scale and the variance of AMOC fluctuations both increase. Increasing trends in either of these statistics could herald the approach of a bifurcation point [as Carpenter & Brock (2006) have also noted in an ecological context], and these trends could potentially be measured using operational AMOC monitoring networks. The practical use of such an early warning system would depend on the ability to statistically distinguish anthropogenic trends from variability intrinsic to the climate system (Lenton *et al.* 2008).

The influence of fluctuating climate forcing on AMOC regime dynamics in more sophisticated ocean models indicates that the conclusions drawn from box models are qualitatively robust. In the zonally averaged two-dimensional ocean model used to study the response of centennial-scale AMOC variability to fluctuating freshwater forcing, Mysak *et al.* (1993) found noise-induced transitions between three distinct AMOC regimes in the limit of large horizontal diffusivity and small vertical diffusivity. Eyink (2005) obtained analytical solutions for the stationary PDF of the AMOC in a similar model. The deterministic equilibria and response to stochastic forcing in this model are broadly consistent with the box model results presented above, despite the considerably greater sophistication of the model. Furthermore, the AMOC transition early warning system originally characterised in a box model by Kleinen *et al.* (2002) was shown to be characteristic of a coupled ocean–atmosphere model of intermediate complexity in Held & Kleinen (2004).

The discussion of AMOC regimes has so far focused on those associated with large-scale advective feedbacks. In the absence of stochastic forcing, local convective feedbacks lead either to multimodal or oscillatory behaviour in models (e.g. Weaver *et al.* 1993; Cessi 1996; Rahmstorf 2001). For the former case, the effects of climate noise are broadly the same as those associated with the advective feedback multiple equilibria (Kuhlbrodt *et al.* 2001; Kuhlbrodt & Monahan 2003). For the latter case, stochastic forcing can modify the character of the oscillations and the parameter range over which they occur (Weaver *et al.* 1993; Cessi 1996), as well as driving the AMOC between different oscillating regimes (Aeberhardt *et al.* 2000). Furthermore, modelled oceanic diffusive processes have been shown to produce millennial-scale oscillatory responses with time scales strongly dependent on fluctuating freshwater forcing (e.g. Weaver & Hughes 1994).

11.4 Stochastic resonance and Dansgaard–Oeschger events

Evidence from palaeoclimate records, particularly high-latitude ice and ocean sediment cores, demonstrates that the climate of the last glacial period was characterised by a succession of abrupt shifts between relatively cold (stadial) and relatively warm (interstadial) states (e.g. Rahmstorf 2002). These transitions, known as Dansgaard–Oeschger (DO) events, are evident in the records of ice oxygen isotopic composition (δ^{18}O; a measure of high-latitude temperature) and calcium concentration (a measure of atmospheric dustiness) from the Greenland Ice Core Project (GRIP) presented in Fig. 11.2. In particular, the joint distribution of δ^{18}O with the logarithm of the Ca concentration is manifestly bimodal with clearly separated stadial and interstadial regimes (Fuhrer *et al.* 1999). It is evident from Fig. 11.2 that while DO events occur with millennial time scales, they do not simply reflect a regular sinusoidal oscillation of the climate system.

While feedbacks in many different components of the climate system are involved in DO events (as external periodic forcing might also be; see below),

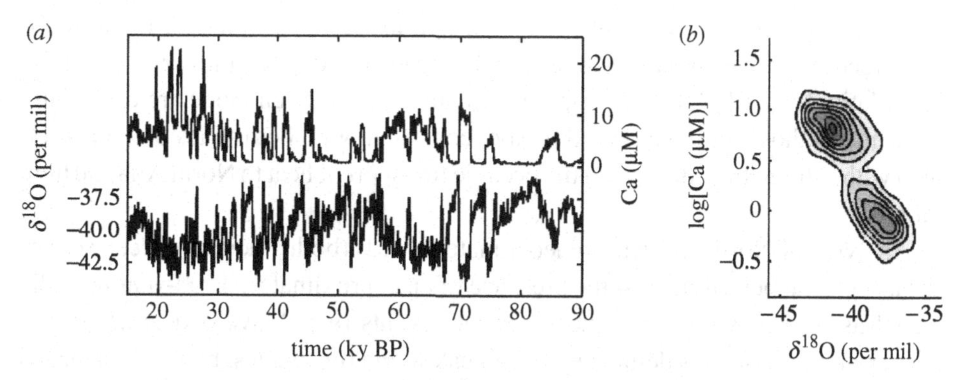

Figure 11.2 Oxygen isotope and Ca concentration data from the GRIP ice core. The δ^{18}O record measures the excess of O^{18} over O^{16} in the ice and is a measure of high-latitude temperature (higher δ^{18}O corresponding to warmer temperatures), while the Ca record is a measure of atmospheric dustiness (Fuhrer *et al.* 1999). (*a*) δ^{18}O and Ca concentration from 15 000 to 90 000 years BP, linearly interpolated to a nominally uniform temporal resolution of 100 years. (*b*) Joint PDF of δ^{18}O and log Ca.

the AMOC is believed to play a central role in stadial/interstadial transitions as a result of (i) the importance of the AMOC for the transport of heat to the high-latitude North Atlantic, and (ii) the potentially nonlinear response of the AMOC to buoyancy forcing (e.g. Clark *et al.* 2002; Rahmstorf 2002). Palaeoclimate data indicate that the climate system is much more variable during glacial periods than during interglacials, and provide evidence of changes in the location and rate of NADW formation associated with stadial/interstadial transitions (Clark *et al.* 2002). Modelling studies suggest that the stability properties of the AMOC are also considerably different between glacial and interglacial periods (e.g. Ganopolski & Rahmstorf 2001; Schmittner *et al.* 2002): during glacial periods, a stadial circulation state with NADW formation in the subpolar North Atlantic is found to be stable, while an interstadial state with NADW formation in the Nordic seas is unstable but easily excited from the stadial state by freshwater perturbations to the North Atlantic. While the interstadial state is not steady, the trajectory of the system is relatively slow in its immediate neighbourhood. This picture is consistent with the observed evolution of DO events: a rapid transition from stadial to interstadial is followed by a gradual relaxation of the AMOC towards the stadial state with a final rapid shift. Note that simple models of the AMOC discussed in Section 11.3 demonstrate that it is possible for the PDF of the stochastic system to be bimodal (as seen in the palaeoclimate records; Fig. 11.2) even when the deterministic component of the system has a single fixed point, if the deterministic tendency takes a local minimum in some neighbourhood into which the system is easily excited. Evidence for a third circulation state with essentially no NADW production (the so-called 'Heinrich mode') is also found in palaeoclimate records and in climate models (e.g. Rahmstorf 2002; Schmittner *et al.* 2002). High-latitude North Atlantic temperatures during a Heinrich mode are essentially the same as those during a stadial period, consistent with both states being associated with a dramatic reduction in the oceanic transport of heat to North Atlantic high latitudes.

Analyses of North Atlantic palaeoclimate records (both glacial and deep sea) of the last glacial period suggest the presence of an approximately 1500-year periodic signal associated with the sequence of DO events (e.g. Mayewski *et al.* 1997). The existence of a well-defined spectral peak in the time series, however, depends on both the data set (e.g. Ditlevsen *et al.* 2005) and the time period (e.g. Schulz 2002) under consideration, and has been suggested to be a spurious signal resulting from aliasing of the annual cycle (Wunsch 2000). Furthermore, there is evidence (still controversial, as discussed below) that DO events do not occur with strictly regular periodicity, but are separated by time intervals that are approximately integer multiples of 1470 years (Alley *et al.* 2001; Schulz 2002; Rahmstorf 2003). Such behaviour is the hallmark of the phenomenon of *stochastic resonance* (SR) in which the addition of noise to a periodic sub-threshold signal results in approximately periodic crossings of some particular threshold. On occasion, one or more crossings will be missed, so intervals between successive crossings will cluster together around integer multiples of the period of the forcing; the distribution of these crossing times will decay exponentially (Alley *et al.* 2001). Stochastic resonance was originally introduced as a model for glacial/interglacial transitions in response to Milankovitch forcing (Benzi *et al.* 1982), and, although it does not appear to be relevant in this original context, SR has since been found to be characteristic of a broad range of physical and biological systems (e.g. Gammaitoni *et al.* 1998).

Vélez-Belichí *et al.* (2001) used a stochastic box model (such as in Section 11.3) to demonstrate SR in the AMOC in response to periodic forcing; although this study focused on Milankovitch forcings rather than the millennial time scales characteristic of DO events, it demonstrated that SR occurs over a broad range of driving periods. Stochastic resonance on millennial time scales was demonstrated by Ganopolski & Rahmstorf (2002) in a more sophisticated coupled atmosphere–ocean model forced by boundary conditions appropriate for the last glacial period. In this study, a small sinusoidal freshwater perturbation with a period of 1470 years was applied to the North Atlantic, along with stochastic freshwater fluxes (with amplitudes comparable to present-day interannual variability). In response, the simulated AMOC displayed SR, alternating between stadial and interstadial circulation states with a distribution of transition times that compared favourably with those of DO events in the GRIP ice core. Stochastic resonance could be achieved with realistic noise levels because the modelled interstadial state is easily excited from the stadial state by freshwater perturbations to the North Atlantic (Ganopolski & Rahmstorf 2001; Schmittner *et al.* 2002). In Ganopolski & Rahmstorf (2002), the 1470-year periodic forcing was some external perturbation of unknown provenance; using the same model, Braun *et al.* (2005) suggested that the forcing may in fact arise through the superposition of 87- and 210-year Gleissberg and DeVries solar cycles. When forced by a linear combination of North Atlantic freshwater forcings with these periodicities (or a more realistic modulation of the 11-year solar cycle by the Gleissberg cycle), the model responds with a 1470-year almost-periodic alternation between stadial and interstadial states (for certain forcing parameter ranges). That the modelled AMOC should respond on millennial time scales to this shorter time-scale forcing was attributed to a combination of strongly nonlinear dynamics and the long intrinsic AMOC adjustment time scales.

Using an idealised coupled atmosphere–ocean–sea-ice model, Timmermann *et al.* (2003) suggested a variation on the SR mechanism for driving cycles of DO events. Instead of being driven by an external periodic signal, DO events in this model occur through a combination of noise and a periodic limit cycle internal to the climate system itself in a phenomenon known as *coherence resonance* (CR) or *autonomous* SR (also discussed in Ganopolski & Rahmstorf 2002). An essential (and testable) difference between SR and CR is that, in the former, the phase of the external driving signal and the resulting transitions is fixed, while, in the latter, the phase of the oscillation can drift as a result of internal climate system interactions.

Stochastic resonance is a meaningful candidate mechanism for driving DO event cycles only to the extent that the distribution of observed intertransition times clusters around integer multiples of a single (1470-year) time scale, with transitions that are phase-locked to this external periodic forcing. An exponential distribution of transition times without this clustering would be suggestive of stadial/interstadial transitions being driven by climate noise alone, without a periodic forcing (e.g. Ditlevsen *et al.* 2007), while a looser clustering of transition times without locking to a periodic signal of fixed phase would be suggestive of CR. Distinguishing between these alternatives is a statistical problem complicated by (i) the difficulty of defining precisely when a transition has occurred, (ii) problems with ice-core chronology, and (iii) the diversity of null hypotheses against which the data can be compared. Time-series analyses of high-latitude palaeoclimate records by Roe & Steig (2004) and Ditlevsen *et al.* (2005, 2007) suggested that a statistical model with stadial/interstadial transitions driven by climate noise without a preferred periodic forcing provides a better fit to the observations than a stochastically resonant model, particularly for the newly obtained North Greenland Ice Core Project (NGRIP) ice core (Ditlevsen *et al.* 2007). As well, the influence of solar variability on the onset of DO events has been questioned (Muscheler & Beer 2006). Owing to the numerous uncertainties involved in the reconstruction of past climates, SR remains a controversial mechanism for the pacing of DO cycles.

11.5 Conclusions

Variability of the oceanic MOC is an important component of variability in the climate system on time scales from decades through centuries to millennia. Modelling studies suggest that MOC variability on these 'climate' time scales may be strongly influenced by fluctuations in surface forcing on much shorter weather time scales. Such high-frequency forcing has typically entered models of the MOC through stochastic processes parameterising unresolved atmospheric processes, representing fluctuations in surface buoyancy and (less often) mechanical fluxes. A significant gap in our understanding of the importance of stochastic forcing comes not through its role in external forcing but rather through the means that the effects of this external forcing are parameterised in ocean and climate models. The development of physically consistent representations of fluctuations in unresolved scales and their dependence upon resolved scales – that is, of stochastic subgrid-scale parameterisations – remains a significant physical and mathematical challenge (e.g. Palmer 2001). In particular, mechanical forcing provides the energy necessary to drive the thermohaline circulation both via direct mixing (through both wind- and tidally generated internal wave breaking) or through wind-driven upwelling in the Southern Ocean (e.g. Kuhlbrodt *et al.* 2007). A natural question arises as to whether or not the circulation in large-scale ocean models is sensitive to random fluctuations in mixing associated with the internal wave field, which is patchy and episodic.

Recent observations of MOC transport at 26.5° N made from the rapid climate change (RAPID) mooring array measure variability in the maximum meridional overturning, with a standard deviation of 5.6 Sv around the mean value of 18.7 Sv on subannual time scales from 2004 to 2005 (Cunningham *et al.* 2007). The existence of such variability has important consequences for predictability of the MOC, both because it can make detection of trends difficult and because these fluctuations themselves may influence the timing of transitions between circulation regimes. Weather is not simply noise superimposed upon climate: interactions across time scales are essential to the dynamics of the climate system. Probability theory and stochastic dynamics provide natural tools for investigating the connection between weather and climate; this chapter has presented an overview of the insights into the variability of the MOC that have so far resulted from the application of these tools to this problem. The study of the interaction between weather and climate in the MOC is relatively young: as models and the observational record both improve, we are confident that this perspective will play an increasingly large role in understanding the dynamics of this fundamentally important component of the climate system.

Acknowledgements

The ice-core data were generously provided by Eric Wolff (British Antarctic Survey), Sigfus Johnsen (University of Copenhagen) and Bernhard Stauffer (University of Bern). A.H.M. and A.J.W. acknowledge support from Natural Sciences and Engineering Research Council of Canada and from the Canadian Foundation for Climate and Atmospheric Sciences (CFCAS). J.A. acknowledges support from CFCAS. A.H.M. also acknowledges support from the Canadian Institute for Advanced Research Earth System Evolution Program.

References

Aeberhardt, M., Blatter, M. & Stocker, T. F. 2000 Variability on the century time scale and regime changes in a stochastically forced zonally averaged ocean–atmosphere model. *Geophys. Res. Lett.*, **27**, 1303–1306 (doi:10.1029/1999GL011103).

Alley, R. B., Anandakrishnan, S. & Jung, P. 2001 Stochastic resonance in the North Atlantic. *Paleoceanogr.*, **16**, 190–198 (doi:10.1029/2000PA000518).

Arzel, O., Huck, T. & Verdiere, A. C. 2006 The different nature of the interdecadal variability of the thermohaline circulation under mixed and flux boundary conditions. *J. Phys. Oceanogr.*, **36**, 1703–1718 (doi:10.1175/JPO2938.1).

Bentsen, M., Drange, H., Furevik, T. & Zhou, T. 2004 Simulated variability of the Atlantic meridional overturning circulation. *Clim. Dyn.*, **22**, 701–720 (doi:10.1007/s00382–004–0397-x).

Benzi, R., Parisi, G., Sutera, A. & Vulpiani, A. 1982 Stochastic resonance in climate change. *Tellus*, **34**, 10–16.

Berglund, N. & Gentz, B. 2006 *Noise-Induced Phenomena in Slow–Fast Dynamical Systems: a Sample-Paths Approach.* Springer-Verlag.

Bindoff, N. L. *et al.* 2007 Observations: oceanic climate change and sea level. *Climate Change 2007: the Physical Science Basis. Contribution of Working Group I to the Fourth Assessment Report of the Intergovernmental Panel on Climate Change*, ed. S. Solomon, D. Qin, M. Manning *et al.*, pp. 385–432. Cambridge University Press.

Braun, H., Christl, M., Rahmstorf, S. *et al.* 2005 Possible solar origin of the 1,470-year glacial climate cycle demonstrated in a coupled model. *Nature*, **438**, 208–211 (doi:10.1038/nature04121).

Bryan, K. & Hansen, F. C. 1995 A stochastic model of North Atlantic climate variability on decade-to-century time scales. In *The Natural Variability of the Climate System on 10–100 Year Time Scales*, ed. D. G. Martinon, K. Bryan, M. Ghil *et al.*, pp. 355–362. National Academy of Sciences.

Capotondi, A. & Holland, W. 1997 Decadal variability in an idealised ocean model and its sensitivity to surface boundary conditions. *J. Phys. Oceanogr.*, **27**, 1072–1093 (doi:10.1175/1520–0485(1997)027<1072:DVIAIO>2.0.CO;2).

Carpenter, S. R. & Brock, W.A. 2006 Rising variance: a leading indicator of ecological transition. *Ecol. Lett.*, **9**, 311–318 (doi:10.111/j.1461–0248.2005.00877.x).

Cessi, P. 1994 A simple box model of stochastically forced thermohaline flow. *J. Phys. Oceanogr.*, **24**, 1911–1920 (doi:10.1175/1520–0485(1994)024<1911:ASBMOS>2.0.CO;2).

Cessi, P. 1996 Convective adjustment and thermohaline excitability. *J. Phys. Oceanogr.*, **26**, 481–491 (doi:10.1175/1520–0485(1996)026<0481:CAATE>2.0.CO;2).

Chen, F. & Ghil, M. 1995 Interdecadal variability of the thermohaline circulation and high-latitude surface fluxes. *J. Phys. Oceanogr.*, **25**, 2547–2567 (doi:10.1175/1520–0485(1995)025<2547:IVOTTC>2.0.CO;2).

Clark, P. U., Pisias, N. G., Stocker, T. F. & Weaver, A. J. 2002 The role of the thermohaline circulation in abrupt climate change. *Nature*, **415**, 863–869 (doi:10.1038/415863a).

Cunningham, S.A. *et al.* 2007 Temporal variability of the Atlantic meridional overturning circulation at 26° N. *Science*, **317**, 935–938 (doi:10.1126/science.1141304).

Delworth, T. & Greatbatch, R.J. 2000 Multidecadal thermohaline circulation variability driven by atmospheric surface flux forcing. *J. Climate*, **13**, 1481–1494 (doi:10.1175/1520–0442(2000)013<1481:MTCVDB>2.0.CO;2).

Delworth, T., Manabe, S. & Stouffer, R. 1993 Interdecadal variations of the thermohaline circulation in a coupled ocean-atmosphere model. *J. Climate*, **6**, 1993–2011 (doi:10.1175/1520–0442(1993)006<1993:IVOTTC>2.0.CO;2).

Ditlevsen, P. D., Kristensen, M. S. & Andersen, K. K. 2005 The recurrence time of Dansgaard–Oeschger events and limits on the possible periodic component. *J. Climate*, **18**, 2594–2603 (doi:10.1175/JCLI3437.1).

Ditlevsen, P. D., Andersen, K. K. & Svensson, A. 2007 The DO-climate events are probably noise induced: statistical investigation of the claimed 1470 years cycle. *Clim. Past*, **3**, 129–134.

Dong, B. & Sutton, R. T. 2005 Mechanism of interdecadal thermohaline circulation variability in a coupled ocean–atmosphere GCM. *J. Climate*, **18**, 1117–1135 (doi:10.1175/JCLI3328.1).

Eyink, G. 2005 Statistical hydrodynamics of the thermohaline circulation in a two-dimensional model. *Tellus*, **A57**, 100–115 (doi:10.1111/j.1600–0870.2005.00082.x).

Fanning, A. & Weaver, A. J. 1998 Thermohaline variability: the effects of horizontal resolution and diffusion. *J. Climate*, **11**, 709–715 (doi:10.1175/1520–0442(1998)011<0709:TVTEOH>2.0.CO;2).

Farrell, B. F. & Ioannou, P. J. 1996 Generalized stability theory. Part I: autonomous operators. *J. Atmos. Sci.*, **53**, 2025–2040 (doi:10.1175/1520–0469(1996)053<2025:GSTPIA>2.0.CO;2).

Fuhrer, K., Wolff, E. W. & Johnsen, S. J. 1999 Timescales for dust variability in the Greenland Ice Core Project (GRIP) ice core in the last 100,000 years. *J. Geophys. Res.*, **104**, 31043–31052 (doi:10.1029/1999JD900929).

Gammaitoni, L., Hänggi, P., Jung, P. & Marchesoni, F. 1998 Stochastic resonance. *Rev. Mod. Phys.*, **70**, 223–287 (doi:10.1103/RevModPhys.70.223).

Ganachaud, A. & Wunsch, C. 2000 Improved estimates of global ocean circulation, heat transport and mixing from hydrographic data. *Nature*, **408**, 453–456 (doi:10.1038/35044048).

Ganopolski, A. & Rahmstorf, S. 2001 Rapid changes of glacial climate simulated in a coupled climate model. *Nature*, **499**, 153–158 (doi:10.1038/35051500).

Ganopolski, A. & Rahmstorf, S. 2002 Abrupt glacial climate changes due to stochastic resonance. *Phys. Rev. Lett.*, **88**, 038501 (doi:10.1103/PhysRevLett.88.038501).

Gardiner, C. W. 1997 *Handbook of Stochastic Methods for Physics, Chemistry, and the Natural Sciences*, 2nd edn. Springer-Verlag.

Greatbatch, R. J. & Zhang, S. 1995 An interdecadal oscillation in an idealised ocean basin forced by constant heat flux. *J. Climate*, **8**, 81–91 (doi:10.1175/1520–0442(1995)008<0081:AIOIAI>2.0.CO;2).

Griffies, S. M. & Tziperman, E. 1995 A linear thermohaline oscillator driven by stochastic atmospheric forcing. *J. Climate*, **8**, 2440–2453 (doi:10.1175/1520–0442(1995)008<2440:ALTODB>2.0.CO;2).

Hasselmann, K. 1976 Stochastic climate models. Part I: theory. *Tellus*, **28**, 473–484.

Held, H. & Kleinen, T. 2004 Detection of climate system bifurcations by degenerate fingerprinting. *Geophys. Res. Lett.*, **31**, L23207 (doi:10.1029/2004GL020972).

Herbaut, C., Sirven, J. & Fevrier, S. 2002 Response of a simplified oceanic general circulation model to idealised NAO-like stochastic forcing. *J. Phys. Oceanogr.*, **32**, 3182–3192 (doi:10.1175/1520–0485(2002)032<3182:ROASOG>2.0.CO;2).

Holland, M. M., Brasket, A. J. & Weaver, A. J. 2000 The impact of rising atmospheric CO_2 levels on low frequency North Atlantic climate variability. *Geophys. Res. Lett.*, **27**, 1519–1522 (doi:10.1029/1999GL010949).

Holland, M. M., Bitz, C. M., Eby, M. & Weaver, A. J. 2001 The role of ice–ocean interactions in the variability of the North Atlantic thermohaline circulation. *J. Climate*, **14**, 656–675 (doi:10.1175/1520–0442(2001)014<0656:TROIOI>2.0.CO;2).

IPCC 2007 *Climate Change 2007: the Physical Science Basis. Contribution of Working Group I to the Fourth Assessment Report of the Intergovernmental Panel on Climate Change*, ed. S. Solomon, D. Qin, M. Manning *et al.* Cambridge University Press.

Kleinen, T., Held, H. & Petschel-Held, G. 2002 The potential role of spectral properties in detecting thresholds in the Earth system: application to the thermohaline circulation. *Ocean Dyn.*, **53**, 53–63 (doi:10.1007/s10236–002–0023–6).

Knutti, R. & Stocker, T. F. 2002 Limited predictability of the future thermohaline circulation close to an instability threshold. *J. Climate*, **15**, 179–186 (doi:10.1175/1520–0442(2002)015<0179:LPOTFT>2.0.CO;2).

Kravtsov, S. & Ghil, M. 2004 Interdecadal variability in a hybrid coupled ocean–atmosphere–sea ice model. *J. Phys. Oceanogr.*, **34**, 1756–1775 (doi:10.1175/1520–0485(2004)034<1756:IVIAHC>2.0.CO;2).

Kuhlbrodt, T. & Monahan, A. H. 2003 Stochastic stability of open-ocean deep convection. *J. Phys. Oceanogr.*, **33**, 2764–2780 (doi:10.1175/1520–0485(2003)033<2764:SSOODC>2.0.CO;2).

Kuhlbrodt, T., Titz, S., Feudel, U. & Rahmstorf, S. 2001 A simple model of seasonal open ocean convection. Part II: Labrador Sea stability and stochastic forcing. *Ocean Dyn.*, **52**, 36–49 (doi:10.1007/s10236–001–8175–3).

Kuhlbrodt, T., Griesel, A., Montoya, M. *et al.* 2007 On the driving processes of the Atlantic meridional overturning circulation. *Rev. Geophys.*, **45**, RG2001 (doi:10.1029/2004RG000166).

Lenton, T. M., Held, H., Kriegler, E. *et al.* 2008 Tipping elements in the Earth's climate system. *Proc. Natl Acad. Sci. USA*, **105**, 1786–1793 (doi:10.1073/pnas.0705414105).

Lohmann, G. & Schneider, J. 1999 Dynamics and predictability of Stommel's box model. A phase-space perspective with implications for decadal climate variability. *Tellus*, **A51**, 326–336 (doi:10.1034/j.1600-0870.1999.t01-1-00012.x).

Mayewski, P. A., Meeker, L. D., Twickler, M. S. *et al.* 1997 Major features and forcing of high-latitude northern hemisphere atmospheric circulation using a

110,000-year-long glaciochemical series. *J. Geopys. Res.*, **102**, 26345–26366 (doi:10.1029/96JC03365).

McManus, J. F., Francois, R., Gherardi, J.-M., Keigwin, L. D. & Brown-Leger, S. 2004 Collapse and rapid resumption of Atlantic meridional circulation linked to deglacial climate changes. *Nature*, **428**, 834–837 (doi:10.1038/nature02494).

Meehl, G.A. *et al.* 2007 Global climate change projections. In *Climate Change 2007: The Physical Science Basis. Contribution of Working Group I to the Fourth Assessment Report of the Intergovernmental Panel on Climate Change*, ed. S. Solomon, D. Qin, M. Manning *et al.*, pp. 747–846. Cambridge University Press.

Mikolajewicz, U. & Maier-Reimer, E. 1990 Internal secular variability in an ocean general circulation model. *Clim. Dyn.*, **4**, 145–156 (doi:10.1007/BF00209518).

Monahan, A. H. 2002a Stabilization of climate regimes by noise in a simple model of the thermohaline circulation. *J. Phys. Oceanogr.*, **32**, 2072–2085 (doi:10.1175/1520–0485(2002)032<2072:SOCRBN>2.0.CO;2).

Monahan, A. H. 2002b Correlation effects in a simple stochastic model of the thermohaline circulation. *Stochastics Dyn.*, **2**, 437–462 (doi:10.1142/S0219493702000510).

Monahan, A. H. 2002c Lyapunov exponents of a simple stochastic model of the thermally and wind-driven ocean circulation. *Dyn. Atmos. Ocean*, **35**, 363–388 (doi:10.1016/S0377–0265(02)00049–0).

Monahan, A. H., Timmermann, A. & Lohmann, G. 2002 Comments on "Noise-induced transitions in a simplified model of the thermohaline circulation". *J. Phys. Oceanogr.*, **32**, 1112–1116 (doi:10.1175/1520–0485(2002)032<1112:CONITI>2.0.CO;2).

Muscheler, R. & Beer, J. 2006 Solar forced Dansgaard/Oeschger events? *Geophys. Res. Lett.*, **33**, L20706 (doi:10.1029/2006GL026779).

Mysak, L. A., Stocker, T. F. & Huang, F. 1993 Century-scale variability in a randomly-forced, two-dimensional thermohaline ocean circulation model. *Clim. Dyn.*, **8**, 103–116 (doi:10.1007/BF00208091).

Palmer, T. N. 2001 A nonlinear dynamical perspective on model error: a proposal for nonlocal stochastic–dynamic parameterisation in weather and climate prediction models. *Q. J. R. Meteorol. Soc.*, **127**, 279–304 (doi:10.1002/qj.49712757202).

Penland, C. 2003 A stochastic approach to nonlinear dynamics: a review. *Bull. Am. Met. Soc.*, **84**, ES43–ES52 (doi:10.1175/BAMS-84-7-Penland).

Pierce, D. W., Barnett, T. P. & Mikolajewicz, U. 1995 Competing roles of heat and freshwater flux in forcing thermohaline oscillations. *J. Phys. Oceanogr.*, **23**, 2046–2064 (doi:10.1175/1520–0485(1995)025<2046:CROHAF>2.0.CO;2).

Rahmstorf, S. 2001 A simple model of seasonal open ocean convection. Part I: theory. *Ocean Dyn.*, **52**, 26–35 (doi:10.1007/s10236-001-8174-4).

Rahmstorf, S. 2002 Ocean circulation and climate during the past 120,000 years. *Nature*, **419**, 207–214 (doi:10.1038/nature01090).

Rahmstorf, S. 2003 Timing of abrupt climate change: a precise clock. *Geophys. Res. Lett.*, **30**, 1510 (doi:10.1029/2003GL017115).

Rahmstorf, S. *et al.* 2005 Thermohaline circulation hysteresis: a model intercomparison. *Geophys. Res. Lett.*, **32**, L23605 (doi:10.1029/2005GL023655).

Rivin, I. & Tziperman, E. 1997 Linear versus self-sustained interdecadal thermohaline variability in a coupled box model. *J. Phys. Oceanogr.*, **27**, 1216–1232 (doi:10.1175/1520–0485(1997)027<1216:LVSSIT>2.0.CO;2).

Roe, E. G. & Steig, E. J. 2004 Characterization of millennial-scale climate variability. *J. Climate*, **17**, 1929–1944 (doi:10.1175/1520–0442(2004)017<1929:COMCV>2.0.CO;2).

Saenko, O. A. & Weaver, A. J. 2004 What drives heat transport in the Atlantic: sensitivity to mechanical energy supply and buoyancy forcing in the Southern Ocean. *Geophys. Res. Lett.*, **31**, L20305 (doi:10.1029/2004GL020671).

Saenko, O. A., Weaver, A. J. & Gregory, J. M. 2003 On the link between the two modes of the ocean thermohaline circulation and the formation of global-scale water masses. *J. Climate*, **16**, 2797–2801 (doi:10.1175/1520–0442(2003)016<2797:OTLBTT>2.0.CO;2).

Saravanan, R. & McWilliams, J. C. 1997 Stochasticity and spatial resonance in interdecadal climate fluctuations. *J. Climate*, **10**, 2299–2320 (doi:10.1175/1520–0442(1997)010<2299:SASRII>2.0.CO;2).

Saravanan, R., Danabasoglu, G., Doney, S. & McWilliams, J. C. 2000 Decadal variability and predictability in the midlatitude ocean–atmosphere system. *J. Climate*, **13**, 1073–1097 (doi:10.1175/1520–0442(2000)013<1073:DVAPIT>2.0.CO;2).

Schmittner, A., Yoshimori, M. & Weaver, A. J. 2002 Instability of glacial climate in a model of the ocean–atmosphere–cryosphere system. *Science*, **295**, 1489–1493 (doi:10.1126/science.1066174).

Schulz, M. 2002 On the 1470-year pacing of Dansgaard–Oeschger warm events. *Paleoceanogr.*, **17**, 1014 (doi:10.1029/2000PA000571).

Skagseth, Ø. & Mork, K. A. 1998 Stability of the thermohaline circulation to noisy surface buoyancy forcing for the present and a warm climate in an ocean general circulation model. *J. Phys. Oceanogr.*, **28**, 842–857 (doi:10.1175/1520–0485(1998)028<0842:SOTTCT>2.0.CO;2).

Spall, M.A. 1993 Variability of sea surface salinity in stochastically forced systems. *Clim. Dyn.*, **8**, 151–160 (doi:10.1007/BF00208094).

Stommel, H. M. 1961 Thermohaline convection with two stable regimes of flow. *Tellus*, **13**, 224–230.

Stommel, H. M. & Young, W. R. 1993 The average T-S relation of a stochastically forced box model. *J. Phys. Oceanogr.*, **23**, 151–158 (doi:10.1175/1520–0485(1993)023<0151:TAROAS>2.0.CO;2).

Timmermann, A. & Lohmann, G. 2000 Noise-induced transitions in a simplified model of the thermohaline circulation. *J. Phys. Oceanogr.*, **30**, 1891–1900 (doi:10.1175/1520–0485(2000)030<1891:NITIAS>2.0.CO;2).

Timmermann, A., Gildor, H., Schulz, M. & Tziperman, E. 2003 Coherent resonant millennial-scale climate oscillations triggered by massive meltwater pulses. *J. Climate*, **16**, 2569–2585 (doi:10.1175/1520–0442(2003)016<2569:CRMCOT>2.0.CO;2).

Tziperman, E. & Ioannou, P. J. 2002 Transient growth and optimal excitation of thermohaline variability. *J. Phys. Oceanogr.*, **32**, 3427–3435 (doi:10.1175/1520–0485(2002)032<3427:TGAOEO>2.0.CO;2).

Vélez-Belichí, P., Alvarez, A., Colet, P., Tintoré, J. & Haney, R. L. 2001 Stochastic resonance in the thermohaline circulation. *Geophys. Res. Lett.*, **28**, 2053–2056 (doi:10.1029/2000GL012091).

Wang, X., Stone, P. H. & Marotzke, J. 1999 Global thermohaline circulation. Part I: sensitivity to atmospheric moisture transport. *J. Climate*, **12**, 71–82 (doi:10.1175/1520–0442(1999)012<0071:GTCPIS>2.0.CO;2).

Weaver, A. J. 1995 Decadal-to-millennial internal oceanic variability in coarse-resolution ocean general-circulation models. In *The Natural Variability of the Climate System on 10–100 Year Time Scales*. National Academy of Sciences.

Weaver, A. J. & Hughes, T. M. C. 1994 Rapid interglacial climate fluctuations driven by North Atlantic ocean circulation. *Nature*, **367**, 447–450 (doi:10.1038/367447a0).

Weaver, A. J. & Sarachik, E. 1991 Evidence for decadal variability in an ocean general circulation model: an advective mechanism. *Atmos.-Ocean*, **29**, 197–231.

Weaver, A. J. & Valcke, S. 1998 On the variability of the thermohaline circulation in the GFDL coupled model. *J. Climate*, **11**, 759–767 (doi:10.1175/1520–0442(1998)011<0759:OTVOTT>2.0.CO;2).

Weaver, A. J., Marotzke, J., Cummins, P. F. & Sarachik, E. S. 1993 Stability and variability of the thermohaline circulation. *J. Phys. Oceanogr.*, **23**, 39–60 (doi:10.1175/1520–0485(1993)023<0039:SAVOTT>2.0.CO;2).

Weaver, A. J., Bitz, C. M., Fanning, A. F. & Holland, M. M. 1999 Thermohaline circulation: high-latitude phenomena and the difference between the Pacific and the Atlantic. *Annu. Rev. Earth Planet. Sci.*, **27**, 231–285 (doi:10.1146/annurev.earth.27.1.231).

Weaver, A. J., Saenko, O. A., Clark, P. U. & Mitrovica, J. X. 2003 Meltwater pulse 1A from Antarctica as a trigger of the Bølling-Allerød warm interval. *Science*, **299**, 1709–1713 (doi:10.1126/science.1081002).

Weisse, R., Mikolajewicz, U. & Maier-Reimer, E. 1994 Decadal variability of the North Atlantic in an ocean general circulation model. *J. Geophys. Res.*, **99**, 12411–12421 (doi:10.1029/94JC00524).

Wood, R. A., Vellinga, M. & Thorpe, R. 2003 Global warming and thermohaline circulation stability. *Phil. Trans. R. Soc. A*, **361**, 1961–1975 (doi:10.1098/rsta.2003.1245).

Wunsch, C. 2000 On sharp spectral lines in the climate record and the millennial peak. *Paleoceanogr.*, **15**, 417–424 (doi:10.1029/1999PA000468).

12

The Atlantic Multidecadal Oscillation: a stochastic dynamical systems view

LEELA M. FRANKCOMBE, HENK A. DIJKSTRA
AND ANNA S. VON DER HEYDT

We provide a dynamical systems framework to understand the Atlantic Multidecadal Oscillation (AMO) and show that this framework is in many ways similar to that of the El Niño/Southern Oscillation. A so-called minimal primitive equation model is used to represent the Atlantic Ocean circulation. Within this minimal model, we identify a normal mode of multidecadal variability that can destabilise the background climate state through a Hopf bifurcation. Next, we argue that noise is setting the amplitude of the sea-surface temperature variability associated with this normal mode. The results provide support that a stochastic Hopf bifurcation is involved in the multidecadal variability as observed in the North Atlantic.

12.1 Introduction

Several pronounced large-scale patterns of variability are known in the present climate system. On the interannual time scale, the El Niño/Southern Oscillation (ENSO) phenomenon provides a dominant pattern in sea-surface temperature (SST) variability, which is localised in the equatorial Pacific (Philander 1990). The Atlantic Multidecadal Oscillation (AMO), which is associated with a basin-wide temperature anomaly in the North Atlantic, is the clearest phenomenon on the multidecadal time scale (Enfield *et al.* 2001). Of course, many ENSO cycles have been measured over the past decades, while the AMO is recorded over one cycle at Mont. Basic theory of both the ENSO and the AMO should explain (i) the physics of the SST pattern and its propagation (if any), (ii) the physics of the dominant time scale of variability and (iii) the processes controlling the amplitude of the SST pattern.

For ENSO, such a basic theory exists (Neelin *et al.* 1998). An important aspect in the development of the theory was the availability of a so-called minimal model of the coupled equatorial ocean–atmosphere system, the Zebiak–Cane (ZC) model (Zebiak & Cane 1987). The solutions of the ZC model have been extensively analysed with the ocean–atmosphere coupling strength μ (the amount of wind stress per SST anomaly) as an important control parameter (Neelin *et al.* 1998). In the ZC model, the tropical Pacific annual mean climate state can become unstable when μ crosses a critical value μ_c. When $\mu > \mu_c$, specific time-dependent perturbations grow in time leading to oscillatory behaviour on an interannual time scale.

In terms of dynamical systems theory (Dijkstra 2005), a Hopf bifurcation occurs at $\mu = \mu_c$ in the ZC model. The simplest dynamical system (Guckenheimer & Holmes 1990) exhibiting such a bifurcation is the system

$$\frac{dx}{dt} = (\mu - \mu_c)x - \omega y - x(x^2 + y^2), \tag{12.1.1a}$$

$$\frac{dy}{dt} = (\mu - \mu_c)y + \omega x - y(x^2 + y^2), \tag{12.1.1b}$$

having two degrees of freedom (x, y). In polar coordinates (r, θ) with $x = r\cos\theta$ and $y = r\sin\theta$, equations (12.1.1a) and (12.1.1b) can be written as

$$\frac{dr}{dt} = (\mu - \mu_c)r - r^3, \tag{12.1.2a}$$

$$\frac{d\theta}{dt} = \omega. \tag{12.1.2b}$$

For $\mu < \mu_c$, there is only one steady state, $r = 0$ (or $x = y = 0$). For $\mu > \mu_c$, however, there are two solutions of the steady equation (12.1.2a), i.e. $r = 0$ and $r = \sqrt{\mu - \mu_c}$. The latter corresponds through (12.1.2b) to a periodic orbit with angular frequency ω and period $2\pi/\omega$. Hence, at the Hopf bifurcation ($\mu = \mu_c$), periodic behaviour with a frequency ω is spontaneously generated through an instability of the trivial solution $x = y = 0$.

In the ENSO theory, when the ZC model is discretised under annual mean forcing, a large-dimensional dynamical system of the form

$$\frac{d\boldsymbol{X}}{dt} = \boldsymbol{f}(\boldsymbol{X}, \mu) \tag{12.1.3}$$

appears, where the state vector $\boldsymbol{X}$ consists of the dependent quantities in the model (e.g. SST, oceanic and atmospheric velocities) at each grid point and $\boldsymbol{f}$ contains the tendencies of all these state variables. In the same way as for the simple system (12.1.1a) and (12.1.1b), Hopf bifurcations are found by considering the stability of the annual mean Pacific climate state $\overline{\boldsymbol{X}}$ in the ZC model. Putting $\boldsymbol{X} = \overline{\boldsymbol{X}} + \tilde{\boldsymbol{X}}$ and $\tilde{\boldsymbol{X}} = e^{\sigma t}\hat{\boldsymbol{X}}$, we find by linearisation around $\overline{\boldsymbol{X}}$ that $\hat{\boldsymbol{X}}$ (the spatial pattern of the eigenmode) is determined by

$$\sigma\hat{\boldsymbol{X}} = \mathcal{J}(\overline{\boldsymbol{X}})\hat{\boldsymbol{X}}, \tag{12.1.4}$$

where $\mathcal{J}(\overline{X})$ is the Jacobian matrix of f at $\overline{X}$. In this case, a Hopf bifurcation occurs when a complex conjugate pair of eigenvalues $\sigma = \sigma_r \pm i\sigma_i$ crosses the imaginary axis as μ crosses μ_c. If the associated eigenvector is indicated by $\hat{X} = \hat{X}_r \pm i\hat{X}_i$, then the periodic orbit at the Hopf bifurcation has the form

$$\mathbf{\Phi}(t) = \cos(\sigma_i t)\hat{X}_r - \sin(\sigma_i t)\hat{X}_i, \tag{12.1.5}$$

and $\mathbf{\Phi}(t)$ defines a propagating pattern with a period $2\pi/\sigma_i$. Note that the period is internally determined by processes in the system and not externally imposed.

As $\hat{X}$ is a solution to (12.1.4), it is called a normal mode and, in the case of the discretised ZC model, it is usually referred to as the ENSO mode. By following this mode to smaller values of μ, it is found (Jin & Neelin 1993) that it splits up into two modes (in a so-called mode merger or mode splitter). One of these modes (an SST mode) is related to tendencies in the SST equation in the ZC model and the other mode (an equatorial ocean basin mode) is related to equatorial wave dynamics. The pattern of SST in the ENSO mode at μ_c is inherited from the SST mode while the interannual time scale is inherited from the equatorial basin mode (Jin & Neelin 1993).

While, in the ZC model, sustained ENSO-type oscillations are found when $\mu > \mu_c$, there is no interannual time-scale oscillatory behaviour when $\mu < \mu_c$ as the annual mean Pacific climate state is stable. However, when noise (e.g. representing atmospheric weather) is applied in the model, interannual oscillations are found for $\mu < \mu_c$. The simplest system exhibiting qualitatively the same behaviour is the stochastic Itô extension of the simple system (12.1.1), i.e. the dynamical system

$$dX_t = \left[(\mu - \mu_c)X_t - \omega Y_t - X_t\left(X_t^2 + Y_t^2\right)\right]dt + \lambda\, dW_t, \tag{12.1.6a}$$
$$dY_t = \left[(\mu - \mu_c)Y_t + \omega X_t - Y_t\left(X_t^2 + Y_t^2\right)\right]dt + \lambda\, dW_t \tag{12.1.6b}$$

where λ is the amplitude of the additive noise and W_t is a Wiener process with increment dW_t. The expectation value $E[R_t]$, where $R_t^2 = X_t^2 + Y_t^2$, resulting from the stochastic integration of the system (12.1.6) is shown in Fig. 12.1 for several values of λ; the deterministic case is shown for $\lambda = 0$. Clearly, there is a response for values $\mu < \mu_c$, which increases with increasing noise level λ.

The effect of noise on the variability in the ZC model has been systematically studied by Roulston & Neelin (2000) and the results are qualitatively similar to those in Fig. 12.1. For values $\mu < \mu_c$, white noise in the wind stress over the equatorial Pacific is able to excite the ENSO mode to substantial amplitude, while for values $\mu > \mu_c$ there are sustained oscillations that are not much affected by the noise. In both the cases, the spatial pattern and the time scale of propagation associated with the interannual variability do not depend on the precise noise characteristics as both are coupled to the ENSO mode. Stochastic noise in the wind stress mainly affects the amplitude of the pattern (Roulston & Neelin 2000).

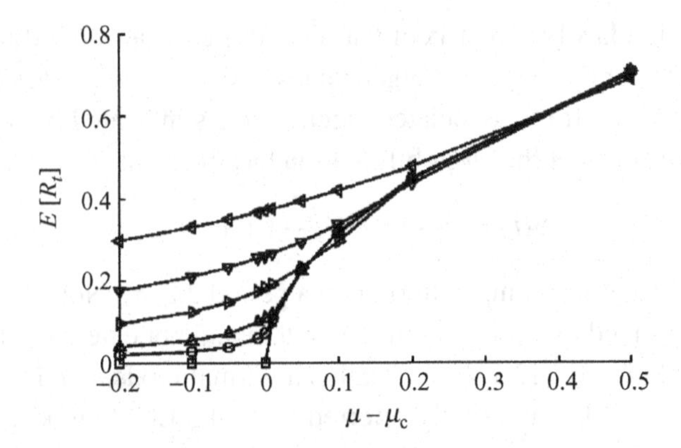

Figure 12.1 Response near a stochastic Hopf bifurcation at $\mu = \mu_c$ monitored through solutions of (12.1.6a) and (12.1.6b). In the deterministic case ($\lambda = 0$), $r = 0$ for $\mu < \mu_c$. When noise is included, the expectation value $E[R_t]$, where $R_t^2 = X_t^2 + Y_t^2$, increases with increasing λ for any value of $\mu - \mu_c$. In the cases with stochastic forcing, $E[R_t]$ is determined over a long time-interval integration. Square, $\lambda = 0$; circle, $\lambda = 0.01$; up triangle, $\lambda = 0.02$; right triangle, $\lambda = 0.05$; down triangle, $\lambda = 0.1$; left triangle, $\lambda = 0.2$.

Having introduced this stochastic dynamical systems view of ENSO, we now turn to the issue of whether a similar framework applies to the AMO. In fact, the main aim of this chapter is to show that such a framework can be developed and that the results are very promising. In Section 12.2, we shortly present the minimal model of the AMO and, in Section 12.3, we show that a Hopf bifurcation occurs in this model leading to multidecadal periodic oscillations. The origin of the AMO mode giving rise to this variability and the physics of its time scale and pattern are also provided in this section. The effect of additive noise is presented in Section 12.4 and the deformation of the mode due to continents is shown in Section 12.5. We conclude with a summary and discussion of the dynamical systems framework of the AMO in Section 12.6.

12.2 The minimal primitive equation model

A minimal model of the AMO was formulated by Greatbatch & Zhang (1995) and Chen & Ghil (1996) and consists of flow in an idealised three-dimensional Northern Hemispheric sector model forced only by a prescribed heat flux. We therefore consider ocean flows in a model domain on the sphere bounded by the longitudes $\phi_w = 286°$ (74° W) and $\phi_e = 350°$ (10° W) and by the latitudes $\theta_s =$ 10° N and $\theta_n = 74$ °N; the ocean basin has a constant depth H. The flows in this domain are forced by a restoring heat flux Q_{rest} (in W m^{-2}) given by

$$Q_{rest} = -\lambda_T (T^* - T_S), \qquad (12.2.1)$$

where λ_T (in W m^{-2} K^{-1}) is a constant surface heat exchange coefficient. The heat flux $Q_{\rm rest}$ is proportional to the temperature difference between the ocean temperature T^* taken at the surface and a prescribed atmospheric temperature T_S, chosen as

$$T_S(\theta) = T_0 + \frac{\Delta T}{2} \cos\left(\pi \frac{\theta - \theta_s}{\theta_n - \theta_s}\right), \tag{12.2.2}$$

where $T_0 = 15$ °C is a reference temperature and ΔT is the temperature difference between the southern and northern latitude of the domain. The forcing is distributed as a body forcing over the first (upper) layer of the ocean having a depth H_m.

Temperature differences in the ocean cause density differences according to

$$\rho = \rho_0[1 - \alpha_T(T^* - T_0)], \tag{12.2.3}$$

where α_T is the volumetric expansion coefficient and ρ_0 is a reference density. Inertia is neglected in the momentum equations because of the small Rossby number; we use the Boussinesq and hydrostatic approximations and represent the horizontal and vertical mixing of momentum and heat by constant eddy coefficients. With r_0 and Ω being the radius and angular velocity of the Earth, respectively, the governing equations for the zonal, meridional and vertical velocity u, v and w, respectively, the dynamic pressure p (the hydrostatic part has been subtracted) and the temperature $T = T^* - T_0$ become

$$-2\Omega v \sin\theta + \frac{1}{\rho_0 r_0 \cos\theta} \frac{\partial p}{\partial \phi} = A_V \frac{\partial^2 u}{\partial z^2} + A_H L_u(u, v), \tag{12.2.4a}$$

$$2\Omega u \sin\theta + \frac{1}{\rho_0 r_0} \frac{\partial p}{\partial \theta} = A_V \frac{\partial^2 v}{\partial z^2} + A_H L_v(u, v), \tag{12.2.4b}$$

$$\frac{\partial p}{\partial z} = \rho_0 g \alpha_T T, \tag{12.2.4c}$$

$$\frac{1}{r_0 \cos\theta}\left(\frac{\partial u}{\partial \phi} + \frac{\partial (v \cos\theta)}{\partial \theta}\right) + \frac{\partial w}{\partial z} = 0, \tag{12.2.4d}$$

$$\frac{DT}{dt} - \nabla_H \cdot (K_H \nabla_H T) - \frac{\partial}{\partial z}\left(K_V \frac{\partial T}{\partial z}\right) = \frac{(T_S - T^*)}{\tau_T} \mathcal{H}\left(\frac{z}{H_m} + 1\right), \tag{12.2.4e}$$

where $\mathcal{H}$ is a continuous approximation of the Heaviside function, C_p is the constant heat capacity, g is the gravitational acceleration and $\tau_T = \rho_0 C_p H_m / \lambda_T$ is the surface adjustment time scale of heat. In these equations, A_H and A_V are the horizontal and vertical momentum (eddy) viscosity, and K_H and K_V are the horizontal and vertical (eddy) diffusivity of heat, respectively. In addition, the operators in the above

Table 12.1 *Standard values of parameters used in the minimal primitive equation model*

$2\Omega = 1.4 \times 10^{-4}$ (s^{-1})	$r_0 = 6.4 \times 10^6$ (m)
$H = 4.0 \times 10^3$ (m)	$\tau_T = 3.0 \times 10$ (d)
$\alpha_T = 1.0 \times 10^{-4}$ (K^{-1})	$g = 9.8$ (m s^{-2})
$A_H = 1.6 \times 10^5$ ($m^2\,s^{-1}$)	$T_0 = 15.0$ (°C)
$\rho_0 = 1.0 \times 10^3$ (kg m^{-3})	$A_V = 1.0 \times 10^{-3}$ ($m^2\,s^{-1}$)
$K_H = 1.0 \times 10^3$ ($m^2\,s^{-1}$)	$K_V = 1.0 \times 10^{-4}$ ($m^2\,s^{-1}$)
$C_p = 4.2 \times 10^3$ (J kg^{-1} K^{-1})	$\Delta T = 20.0$ (° C)

equations are defined as

$$\frac{D}{dt} = \frac{\partial}{\partial t} + \frac{u}{r_0 \cos\theta}\frac{\partial}{\partial \phi} + \frac{v}{r_0}\frac{\partial}{\partial \theta} + w\frac{\partial}{\partial z},$$

$$\nabla_H \cdot (K_H \nabla_H) = \frac{1}{r_0^2 \cos\theta}\left[\frac{\partial}{\partial \phi}\left(\frac{K_H}{\cos\theta}\frac{\partial}{\partial \phi}\right) + \frac{\partial}{\partial \theta}\left(K_H \cos\theta \frac{\partial}{\partial \theta}\right)\right],$$

$$L_u(u, v) = \nabla_H^2 u + \frac{u}{r_0^2 \cos^2\theta} - \frac{2\sin\theta}{r_0^2\cos^2\theta}\frac{\partial v}{\partial \phi},$$

$$L_v(u, v) = \nabla_H^2 v + \frac{v}{r_0^2 \cos^2\theta} + \frac{2\sin\theta}{r_0^2\cos^2\theta}\frac{\partial u}{\partial \phi}.$$

Slip conditions and zero heat flux are assumed at the bottom boundary, while at all lateral boundaries no-slip and zero heat flux conditions are applied. As the forcing is represented as a body force over the first layer, slip and zero heat flux conditions apply at the ocean surface. Hence, the boundary conditions are

$$z = -H,\ 0: \quad \frac{\partial u}{\partial z} = \frac{\partial v}{\partial z} = w = \frac{\partial T}{\partial z} = 0, \tag{12.2.5a}$$

$$\phi = \phi_w,\ \phi_e: \quad u = v = w = \frac{\partial T}{\partial \phi} = 0, \tag{12.2.5b}$$

$$\theta = \theta_s,\ \theta_n: \quad u = v = w = \frac{\partial T}{\partial \theta} = 0. \tag{12.2.5c}$$

The parameters for the standard case are the same as in typical large-scale low-resolution ocean general circulation models and their values are listed in Table 12.1.

12.3 The AMO mode

To determine whether Hopf bifurcations occur within the minimal primitive equation model of Section 12.2, we use methods from numerical bifurcation theory

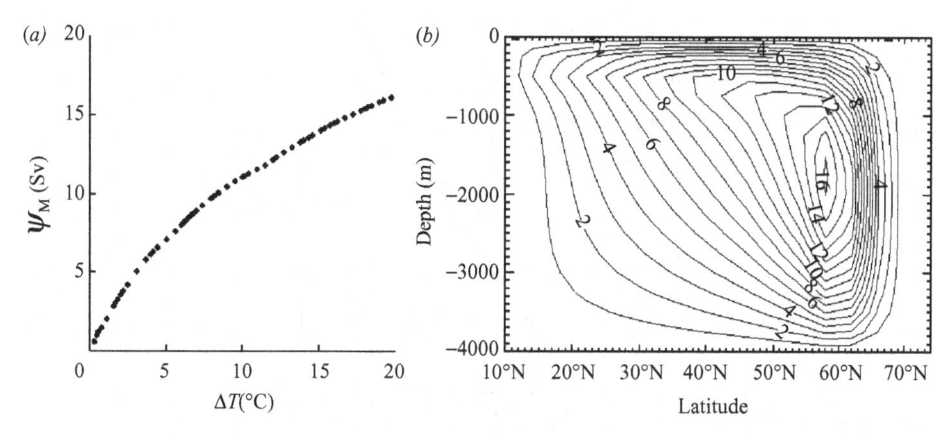

Figure 12.2 (*a*) Plot of the maximum meridional overturning (ψ_M) of the steady solution (in Sv) versus the equator-to-pole temperature difference ΔT (°C) under restoring conditions. (*b*) Plot of the meridional overturning streamfunction (contour values in Sv) for $\Delta T = 20$ °C.

(Dijkstra 2005). First, the governing equations (12.2.4a)–(12.2.4e) and boundary conditions (12.2.5a)–(12.2.5c) are discretised on an Arakawa B-grid using central spatial differences. In the results of this section we use a horizontal resolution of 4°. An equidistant grid with 16 levels is used in the vertical so that the first layer thickness is $H_m = 250$ m. The discretised system of equations can be written in the form (12.1.3) and a $16 \times 16 \times 16$ grid with five unknowns per point (u, v, w, p and T) leads to a dynamical system of dimension (the number of degrees of freedom) 20 480.

The steady equations of the form (12.1.3) are solved using a pseudo-arclength continuation method (Keller 1977). As the primary control parameter μ, we choose the equator-to-pole temperature difference ΔT. For every value of ΔT we calculate a steady solution of the minimal model under the restoring flux Q_{rest} in (12.2.1). For each steady flow pattern the maximum of the meridional overturning stream function (ψ_M) is calculated and plotted against ΔT in Fig. 12.2*a*. The meridional overturning stream function for $\Delta T = 20$ °C is plotted in Fig. 12.2*b*. The maximum of ψ occurs at approximately 55° N and the amplitude is approximately 16 Sv.

Next we diagnose the ocean–atmosphere heat flux Q_{pres} of each of the steady solutions and compute the linear stability of the steady solution under the heat flux $Q_{pres.}$ (where the subscript refers to 'prescribed'). To determine the linear stability we solve for the 'most dangerous' modes of the problem (12.1.4), i.e. those with the real part closest to the imaginary axis, and order the eigenvalues, $\sigma = \sigma_r + i\sigma_i$, according to the magnitude of their real part σ_r (the growth factor). The growth rate and period of the mode with the largest growth factor are plotted against ΔT in Fig. 12.3.

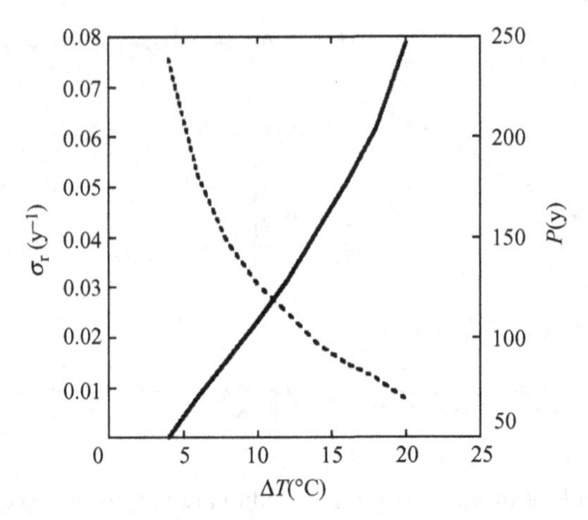

Figure 12.3 Growth factor σ_r (in y^{-1}, solid line) and period $P = 2\pi/\sigma_i$ (in y, dashed line) versus ΔT (in °C) of the AMO mode in the minimal primitive equation model under prescribed flux conditions.

For $\Delta T = 20$ °C the AMO mode has a positive growth factor and hence the background state, of which the meridional overturning stream function was shown in Fig. 12.2*b*, is unstable to the AMO mode. The period of the AMO mode is about 67 years at $\Delta T = 20$ °C, and it decreases with increasing ΔT (Fig. 12.3). From Fig. 12.3, we also see that the growth factor of the AMO mode decreases strongly with decreasing ΔT and becomes negative near $\Delta T_c \approx 4$ °C, where the Hopf bifurcation occurs. For $\Delta T < \Delta T_c$ the steady states are therefore linearly stable under the prescribed flux Q_{pres}.

It was shown in Dijkstra (2006) that, for small ΔT, the angular frequency of the AMO mode becomes zero and the complex conjugate pair of eigenvalues splits up into two real eigenvalues. The paths of the two different modes can be followed to the $\Delta T = 0$ limit, where the eigensolutions connect to those of the diffusion operator of the temperature equation, called SST modes in Dijkstra (2006). These SST modes can be ordered according to their zonal, meridional and vertical wavenumber, and it was found that the AMO mode connects to the (0, 0, 1) SST mode and the (1, 0, 0) SST mode at $\Delta T = 0$.

For each eigenvalue σ associated with the AMO mode, there is a corresponding eigenvector $\boldsymbol{X} = \boldsymbol{X}_r + i\boldsymbol{X}_i$ according to (12.1.4). In Fig. 12.4, the SST field of the real part of the eigenvector ($\boldsymbol{X}_r$) of the AMO mode is plotted for $\Delta T = 4$ °C (near a Hopf bifurcation) and $\Delta T = 20$ °C. A comparison of the pattern in Fig. 12.4*b* and the one in Fig. 4*d* of te Raa & Dijkstra (2002) demonstrates that the AMO mode here is the multidecadal mode as described in detail in te Raa & Dijkstra (2002). With increasing ΔT, the pattern becomes more localised in the northwestern part of the basin.

The physical mechanism of propagation of the AMO mode was presented in te Raa & Dijkstra (2002). This mechanism holds for every ΔT for which an

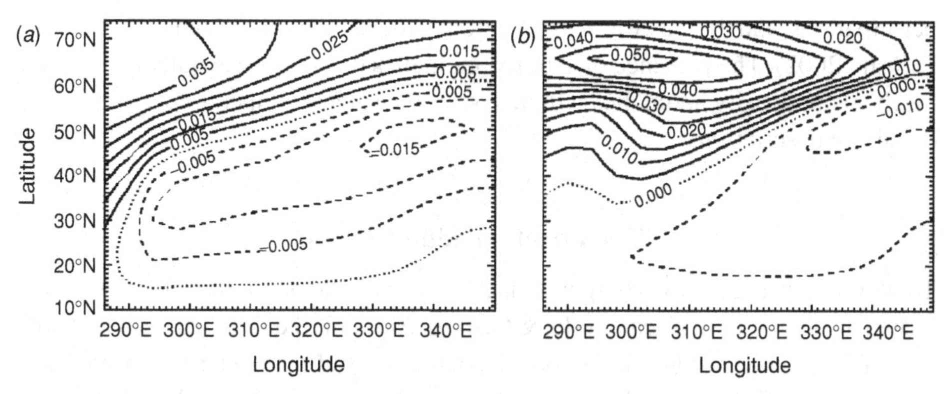

Figure 12.4 The SST pattern of the real part of the AMO mode for (*a*) $\Delta T = 4$ °C (near a Hopf bifurcation) and (*b*) $\Delta T = 20$ °C. Note that the amplitudes are arbitrary.

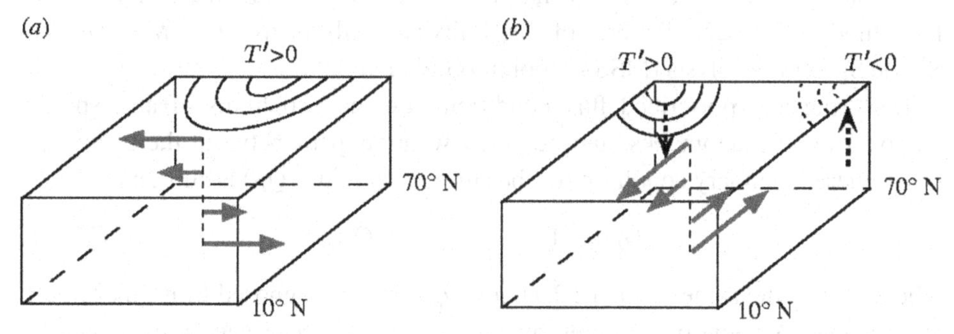

Figure 12.5 Schematic of the oscillation mechanism associated with the multidecadal mode caused by the westward propagation of the temperature anomalies T'. The phase difference between (*a*) and (*b*) is approximately a quarter period. For a further explanation, see text and te Raa & Dijkstra (2002).

oscillatory AMO mode is present (cf. Fig. 12.4). A slight generalisation (compared with that in te Raa & Dijkstra 2002) of this mechanism is provided with the help of Fig. 12.5. A warm anomaly in the north-central part of the basin causes a positive meridional perturbation temperature gradient, which induces (via the thermal wind balance) a negative zonal surface flow (Fig. 12.5*a*). The anomalous anticyclonic circulation around the warm anomaly causes southward (northward) advection of cold (warm) water to the east (west) of the anomaly, resulting in the westward phase propagation of the warm anomaly. Owing to this westward propagation, the zonal perturbation temperature gradient becomes negative, inducing a negative surface meridional flow (Fig. 12.5*b*). The resulting upwelling (downwelling) perturbations along the northern (southern) boundary cause a negative meridional perturbation temperature gradient, inducing a positive zonal surface flow, and the second half of the oscillation starts. The crucial elements in this oscillation mechanism are the phase difference between the zonal and meridional surface flow perturbations, and the westward propagation of the temperature anomalies (te Raa & Dijkstra 2002). The presence of salinity anomalies does not essentially change this mechanism; density anomalies will take over the role of temperature anomalies in the above description.

12.4 Effects of additive noise

To perform transient flow computations, we use version 3.1 of the GFDL Modular Ocean Model (MOM; Pacanowski & Griffies 2000) on the same domain and with the same forcing, resolution, boundary conditions and parameters as provided in Section 12.2. The only difference with the results in Section 12.3 is that here a stretched grid with 16 layers is used in the vertical so that the first four layers have a thickness of $H_m = 50$ m, with the thickness then increasing to 583 m in the lowest level. The internal mode in MOM has a time step of 1 day and the external mode has a time step of 225 s. Patterns of variability are analysed using the Multichannel Single Spectral Analysis (MSSA) toolkit (Ghil *et al.* 2002).

Restoring and prescribed flux conditions are the two limits of atmospheric damping of SST anomalies, and, to study what happens between these limits, a new general boundary condition for the surface heat flux (Q_D) is chosen as

$$Q_D = (1-\gamma)Q_{\text{rest}} + \gamma Q_{\text{pres}}, \tag{12.4.1}$$

where Q_{rest} is the same as in (12.2.1) and Q_{pres} is the diagnosed heat flux of each steady state. A value of γ representative of the real ocean can be estimated by examining the damping time scale of SST in the upper layer ocean. Under the forcing (12.4.1), the damping time scale τ_T is defined as

$$\tau_T = (1-\gamma)\frac{C_p H_m \rho_0}{\lambda_T}. \tag{12.4.2}$$

Using variables from the model ($\rho = 1 \times 10^3\,\text{kg}\,\text{m}^{-3}$, $C_p = 4.2 \times 10^3\,\text{J}\,\text{kg}^{-1}\,\text{K}^{-1}$ and $H_m = 50\,\text{m}$) and $\lambda_T = 20\,\text{W}\,\text{m}^{-2}\,\text{K}^{-1}$ and $\tau_T = 30$ d as a representative mid-latitude value (Barsugli & Battisti 1998) gives a value of $\gamma \approx 0.75$.

When γ decreases from $\gamma = 1$ (prescribed flux conditions), we expect that the growth factor of the AMO mode (as in Fig. 12.4*a*) will decrease as the atmospheric damping becomes larger. As for $\gamma = 0$ (restoring conditions), the growth rate of the AMO mode is negative (te Raa & Dijkstra 2003), and there must also be the same Hopf bifurcation (as in Section 12.3) somewhere between $\gamma = 0$ and 1. Under the heat flux (12.4.1), we therefore compute equilibrium states using time integration starting from the $\gamma = 0$ steady solution. For each solution obtained, the standard deviation of the SST over the box *B*: [46° N–62° N] × [74° W–50° W] is plotted in Fig. 12.6*a*. For $\gamma < 0.85$, there are no oscillations and, near $\gamma_c = 0.85$, the

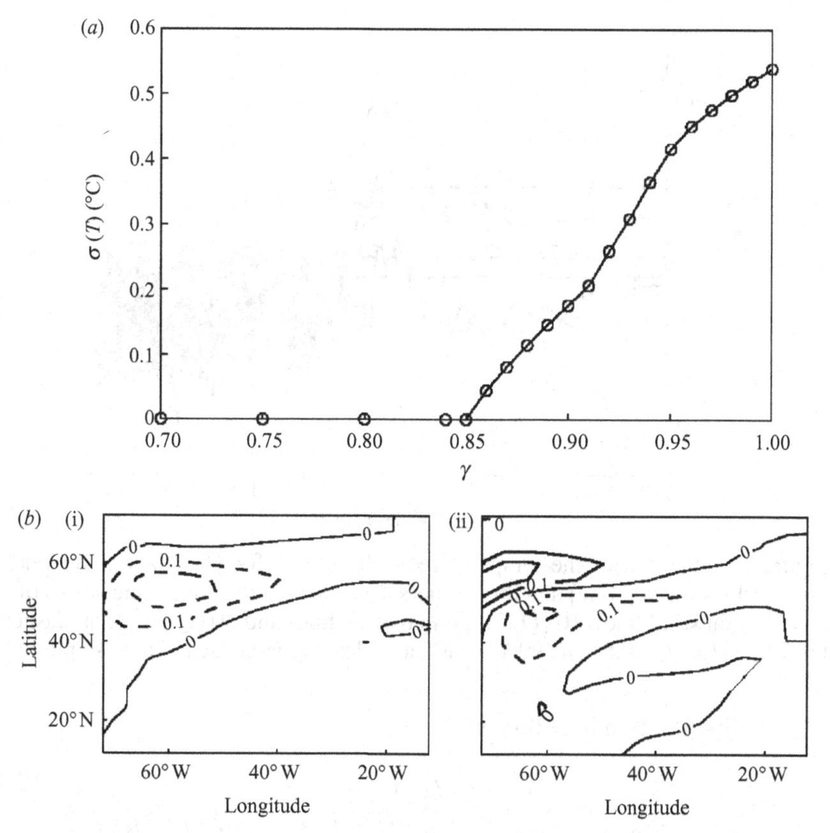

Figure 12.6 Deterministic case, i.e. no noise is applied. (*a*) Standard deviation of the SST over the box B [46° N–62° N] × [74° W–50° W] as a function of γ. Below $\gamma = 0.85$, the standard deviation is zero, so the mode is damped. (*b*) First two EOFs [(i) EOF 1, 49.2%; (ii) EOF 2, 48.4%] of the SST (92% of the variance), with $\gamma = 0.9$.

system undergoes the Hopf bifurcation and the multidecadal oscillations appear for $\gamma > \gamma_c$. The amplitude of the oscillations is measured by calculating the standard deviation of the box-averaged SST rather than the peak-to-peak amplitude, so that comparisons can be made with later simulations where the variability is not regular. The oscillations have periods decreasing from 53 y at $\gamma = 0.85$ to 45 y at $\gamma = 1$, and so the change in period with γ is much smaller than that with ΔT. The first two empirical orthogonal functions (EOFs) of the SST field, which together explain 92.0% of the variance, are shown for $\gamma = 0.9$ in Fig. 12.6*b*. We can see that the variance in temperature at the sea surface is concentrated in the northwest region of the basin similar to that of the AMO mode (Fig. 12.4*b*).

Next, we consider the effect of spatial and temporal coherence in the noise forcing under the conditions $\gamma < \gamma_c$ by comparing the responses of the model

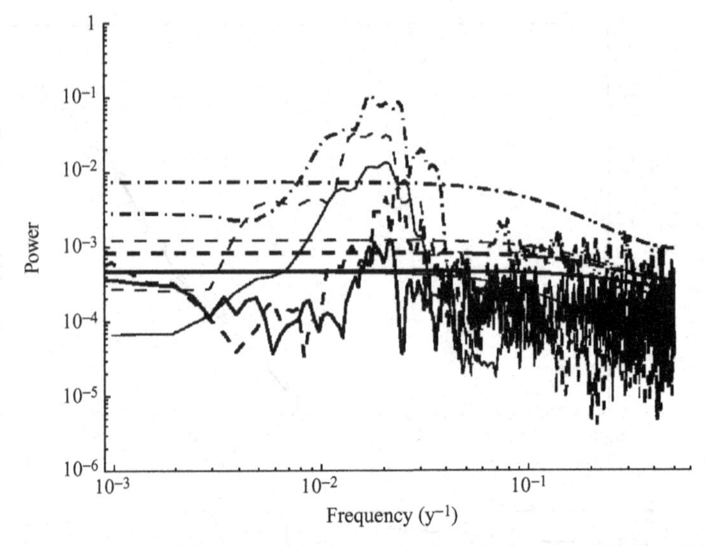

Figure 12.7 Spectra of the temperature in the box B for Q_W (thick solid line), $Q_{S,Wm}$ (thick dashed line) and three others using white noise with time scales of 1 (Q_{S,W_1}, thin solid line), 10 ($Q_{S,W_{10}}$, thin dashed line) and 30 ($Q_{S,W_{30}}$, dot–dashed line) days, for $\gamma = 0.8$. The 99% significance levels for each case are also plotted.

under the following two heat fluxes:

$$Q_W = Q_D + \lambda Z_{ij}(t), \tag{12.4.3a}$$

$$Q_{S,W_m} = Q_D + \lambda Z_m(t) \sin\left(\frac{\pi(i - i_e)}{i_w - i_e}\right) \sin\left(\frac{\pi(j - j_s)}{j_n - j_s}\right). \tag{12.4.3b}$$

In Q_W, λ is the amplitude of the noise and Z_{ij} is a normally distributed random variable that takes on a different value at each grid point (i, j) in space at each time step t. The noise in Q_W is thus uncorrelated in both space and time. In Q_{S,W_m}, $i_e \le i \le i_w$ and $j_s \le j \le j_n$ are the grid variables in the x and y directions. Furthermore, $Z_m(t)$ is a normally distributed random variable, where m indicates the number of days that this variable is persistent. The spatial pattern in (12.4.3b) is chosen as a rough approximation to variations in atmospheric heat fluxes seen over the North Atlantic (Cayan 1992), such as those associated with the North Atlantic Oscillation. In both Q_W and each Q_{S,W_m}, the amplitude (λ) of the noise was taken to be 10% of the difference between the minimum and maximum over the basin of the prescribed heat flux Q_{pres}, which is approximately 20 W m^{-2}.

In Fig. 12.7, the spectra for the case of $\gamma = 0.8$ with the addition of five different types of noise are shown. Although the noise added to the system has no preferred frequency, the spectrum shows a large peak at multidecadal frequencies. This is in contrast to the case of $\gamma = 0.8$ in the absence of noise, where neither the temperature nor the overturning strength vary at all. It can also be clearly seen that both the

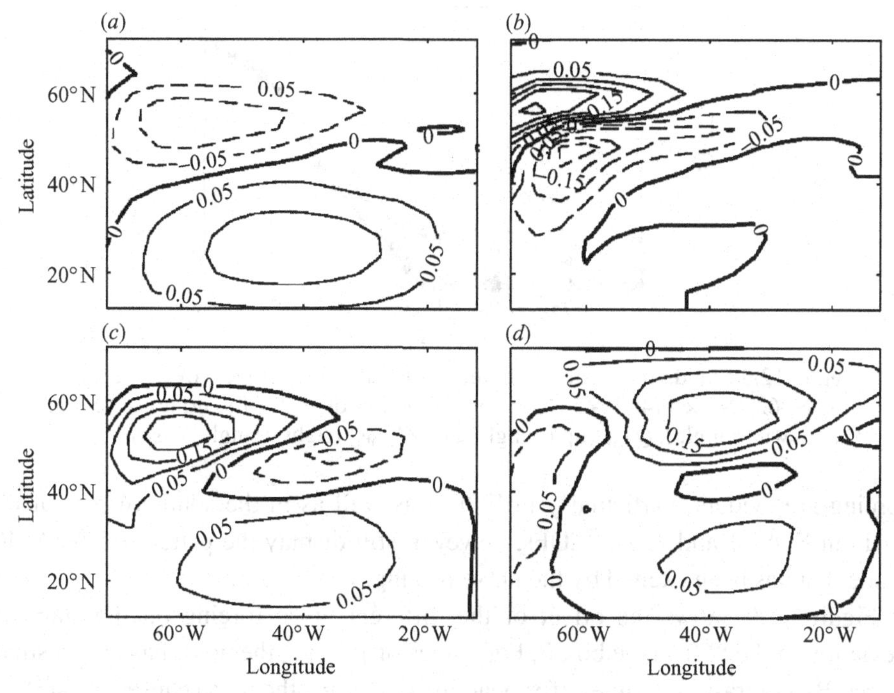

Figure 12.8 First four EOFs [(*a*) EOF 1, 23.8%; (*b*) EOF 2, 23.2%; (*c*) EOF 3, 8.2%; (*d*) EOF 4, 8.0%] of the SST for the $Q_{S,W_{30}}$ case, which together explain over 50% of the variance. The data were low-pass filtered to allow periods of 30–100 years.

spatial and the temporal correlations of the noise increase the height and breadth of the multidecadal peak. The multidecadal peak increases as the time scale of the persistence of the forcing increases. When the spatial coherence is removed (so that the noise added to each grid point is independent), the temporal coherence still causes large variations in temperature, but the power at multidecadal frequencies is greatly reduced (not shown).

When γ is decreased below the critical value γ_c, noise dominates the patterns seen in the MSSA. However, if the data are low-pass filtered to allow periods from 30 to 100 years, then the patterns of multidecadal variability can be seen. Figure 12.8 shows the first four EOFs of SST for the $Q_{S,W_{30}}$ case, with the EOFs accounting for 23.8, 23.2, 8.2 and 8.0% of the variance, respectively. Since the eigenvalues of these EOFs are so closely paired, we must take into account North's rule of thumb (North *et al.* 1982), which shows that the smaller the difference between the eigenvalues of two EOFs, the larger the error. In this case, the model integrations were long enough (2000 y) that the approximations for the typical error found using North's rule of thumb are small, giving some confidence in the interpretation of the EOFs. Signals from the sinusoidal spatial pattern of the noise

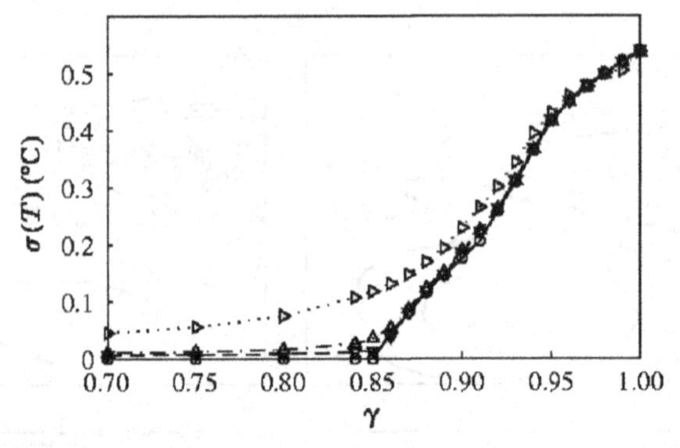

Figure 12.9 Standard deviation of the SST (in °C, averaged over the box B [46 °N–62° N] × [74° W–50° W]) as a function of γ, for the no noise (circle), Q_W (down triangle), $Q_{S,W}$ (up triangle) and $Q_{S,W_{30}}$ (right triangle) cases.

forcing are evident, particularly in EOF 4 as well as in the southern part of the basin in EOFs 1 and 3. The EOFs, however, still display the pattern of the AMO mode that has been excited by the noise forcing.

Figure 12.9 shows the effect of the different noise forcing on the standard deviation of the SST in the box B. For values of $\gamma > \gamma_c$, the noise has only a small effect. By contrast, for values of γ near and below γ_c, the noise causes the surface temperature to increase substantially. With the flux $Q_{S,W_{30}}$, the largest amplitude of the variability is achieved, showing that the spatial and temporal coherence in the noise are important to set the amplitude of the multidecadal variability. In addition, the amplitude of the variability versus γ, as shown in Fig. 12.9, is qualitatively very similar to that of the stochastic Hopf bifurcation in Fig. 12.1, showing that the AMO mode is excited by the atmospheric noise. The mechanism of this excitation is outside the scope of this chapter and is examined in detail in Frankcombe *et al.* (2009).

12.5 Deformation of the spatial pattern by continents

In order to compare the AMO mode of variability in the simple model to multidecadal variability found in GCMs as well at that observed in the real ocean we need to know how the spatial pattern of the AMO mode is modified by the presence of continents. For this we use the same model set-up as in Section 12.4 with the resolution increased to 2° × 2° and with 24 levels in the vertical. The model still has a uniform depth of 4000 m but continental boundaries are now included. Under prescribed flux boundary conditions the oscillatory mode has a period of about 42 y and a spatial pattern which is shown in the first two EOFs of SST in Fig. 12.10. The variability is still concentrated in the northwestern part of the basin but the pattern has been deformed by the continents.

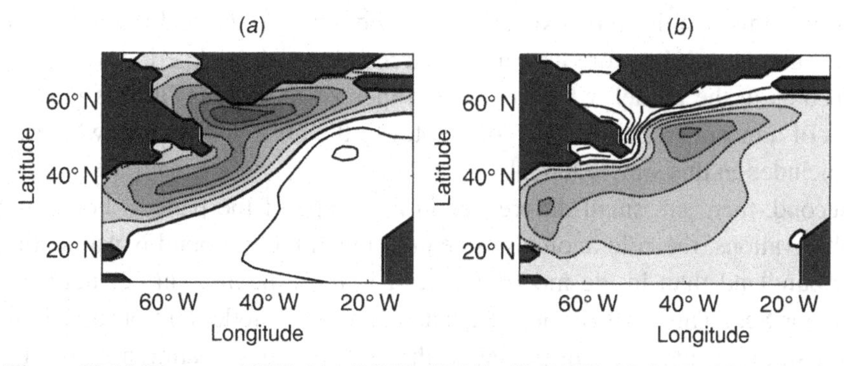

Figure 12.10 First two EOFs [(*a*) EOF1, 55.1%; (*b*) EOF2, 33.3%] of SST for the model configuration with continents, which together explain almost 90% of the variance. Contour interval is 0.5 °C, negative values are shaded.

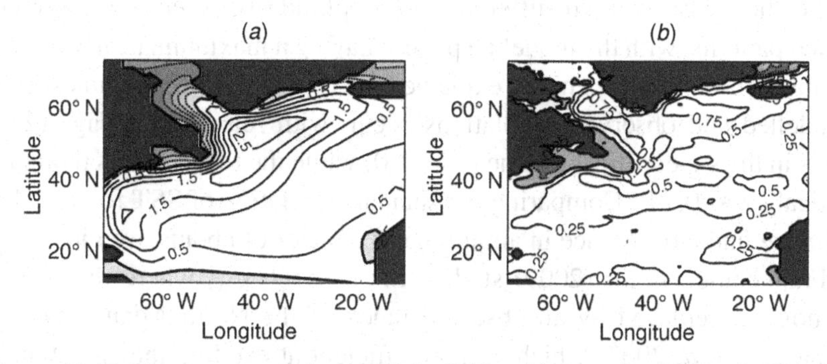

Figure 12.11 The pattern of the AMO in SST, with negative SSTs shaded. (*a*) Difference between the pattern of SST from years with maximum and minimum AMO indices in the model, contour interval 0.5 °C. (b) Sea-surface temperature averaged over the years 1970–1984 (negative AMO index) subtracted from the years 1950–1964 (positive AMO index), contour interval 0.25 °C.

Following Kushnir (1994) we take the difference in SSTs between the warm and cool phases of the AMO in both the model and observations. In the model, the SSTs from the year with the lowest AMO index were subtracted from those from the year with the highest AMO index (where the AMO index is the average SST over the North Atlantic, as defined by Enfield *et al.*, 2001). The resulting pattern is shown in Fig. 12.11(*a*). Observed SSTs were taken from the HadISST data set (Rayner *et al.* 2003) and averaged over 1950–1964 for the warm period and 1970–1984 for the cool period. The pattern that results when subtracting the cool from the warm period is shown in Fig. 12.11(*b*). There are notable similarities between the two patterns, with negative anomalies around Newfoundland and north of Iceland and the largest positive anomalies just south of Greenland. Several differences can also be observed. First, the pattern is not as coherent in observations as in the model.

However, this is only to be expected since the mode in the real ocean is thought to be damped and therefore its spatial pattern, amplitude and period will not be as defined as in the model, which is in the supercritical regime for this mode. The effect of atmospheric noise also contributes to the observed pattern, whereas it is not included in this simple model.

Second, there are small differences in the shape of the pattern. For example, in observations, the cold anomaly is centred much more around Nova Scotia and Newfoundland than in the model, where it is more intense further north in the Labrador Sea. This is also not unexpected since the model and observed climatologies are not identical. In particular the Gulf Stream separation point in such a coarse-resolution model is too far to the north compared to observations. This difference in mean states leads to differences in the spatial patterns of variability as well.

Third, there is a marked difference in amplitude between the observed and modelled patterns, with the modelled pattern having a maximum nearly three times larger than observations. This is explained by considering the regimes in which the modelled and observed oscillations occur, with reference to Fig. 12.9. The model is in the supercritical regime ($\gamma = 1.0$) while the real ocean is thought to be subcritical ($\gamma \approx 0.74$). Comparing the standard deviation of SST for $\gamma = 1.0$ and $\gamma = 0.8$ we find a difference in amplitude of a factor of about seven in the $Q_{S,W_{30}}$ case. Frankcombe *et al.* (2009) studied the case $Q_{\mathrm{S,NAO}}$ where the temporally white noise is replaced by an observed index of the North Atlantic Oscillation (Luterbacher *et al.* 2002) which is more efficient at exciting the AMO mode. In that case the difference in amplitudes between sub- and supercritical decreases to a factor of about five.

Therefore, despite some differences, the overall effect of including continents in the simple model is to deform the spatial pattern of the AMO mode into a form that resembles observations, giving us confidence that the multidecadal mode in the simple model is indeed the same phenomenon which gives rise to multidecadal variability in the real ocean.

12.6 Summary and discussion

Our aim with this chapter was to show that a similar dynamical systems framework can be formulated for the AMO as for ENSO. The minimal primitive equation model as presented in Section 12.2 takes the same role in the AMO theory as the Zebiak–Cane model in ENSO theory. In the ZC model, the ocean–atmosphere coupling strength μ serves as the main control parameter, while in the AMO theory the atmospheric damping time scale of SST anomalies, here mimicked by the parameter γ, has that role.

In both minimal models of ENSO and the AMO there is a normal mode, the 'most dangerous' mode (having the largest growth factor) which is able to destabilise the background state. The nature of the ENSO mode which destabilises the Pacific mean state at sufficiently large coupling strength is a merger between an ocean basin mode and a stationary SST mode. The mechanism of the ENSO mode propagation and time scale is known to be related to the dominant feedback mechanism (thermocline, upwelling, zonal advection) and the equatorial wave propagation (Neelin *et al.* 1998). In the AMO model, the 'most dangerous' normal mode is called the AMO mode and it results from a merger of two SST modes at small ΔT (Dijkstra, 2006). The propagation of the AMO mode is determined by a thermal wind response to a propagating temperature (or density) anomaly and the multidecadal time scale is set by the east–west propagation time of the temperature (or density) anomalies. The oscillation can be characterised by an out-of-phase response of the meridional and zonal overturning anomalies as shown in Fig. 12.5.

It is not known whether the tropical Pacific climate state is near a Hopf bifurcation (Fedorov and Philander 2000). Because of slow variations of this background state, it is likely that one ENSO event could occur in the supercritical regime (with less impact of noise) and the next in the subcritical regime (with noise controlling its amplitude). Similarly, it is not known whether the atmospheric damping time scale of temperature anomalies in the North Atlantic (in this paper mimicked by γ) induces a positive or negative growth factor of the AMO mode in the deterministic case as this also depends on the background state.

When additive noise is added, the ENSO mode can be excited below critical conditions (Roulston and Neelin 2000) and the ENSO variability results from a stochastic Hopf bifurcation as introduced in Section 12.1. Also for the AMO, the results here make a case that a stochastic Hopf bifurcation occurs. Oscillatory variability with an amplitude depending on the noise arises for values of $\gamma \leq \gamma_c$ while for $\gamma > \gamma_c$ the variability does not differ much from the deterministic case. The presence of noisy forcing continuously excites the variability, and a spectrum shows that the variability has the same multidecadal period as in the cases where $\gamma > \gamma_c$. Both spatial and temporal coherence in the random part of the heat flux forcing are important to excite the multidecadal variability to reasonable amplitude.

When the minimal model is run at a small value of γ such that the background state is very stable, each of the noise forcings is only able to cause very small variability (not shown). Hence, the presence of the AMO mode and the occurrence of the Hopf bifurcation certainly play a central role in the amplitude of the multidecadal variability. Other possible mechanisms such as a passive response of the ocean on the atmospheric noise (Hasselmann 1976) or the extension by Saravanan and McWilliams (1998), where the effect of horizontal advection leads to a preferred time scale, are therefore less likely. It is also interesting that in the noise-forced cases one sees normal mode patterns in the variability, instead of non-normal mode patterns (Farrell and Ioannou 1996). It is likely that the time scale on which non-normal modes grow is much faster than the typical time scale of the variability. Without knowing the non-normal modes for the minimal primitive equations model, it is difficult to assess their role in the multidecadal variability.

These results are for the minimal model, but how about more realistic models, i.e. extensions of the minimal model? Until now, only the AMO mode and the periodic oscillations under prescribed flux conditions have been studied in extensions of the minimal model. While continental shape is irrelevant for ENSO, the shape of the continents is essential for the deformation of the AMO mode into a pattern which resembles patterns obtained in coupled climate models (Delworth and Greatbatch 2000) and observations (Dijkstra *et al.* 2006). Results in idealised models with two basins showed that the AMO mode is localised in the sinking regions of the global thermohaline flow (von der Heydt and Dijkstra 2007). This indicates that the AMO mode is unique to the North Atlantic.

In summary, the results provided here indicate that a basic dynamical systems framework of the AMO can be formulated and that this framework is similar to that in ENSO theory. The central element is that the excitation of the AMO mode by atmospheric noise is causing the variability associated with the AMO. Although it is not easy to falsify this theory by the instrumental record, as the time scale is rather long in relation to the available data, we hope that the basic ideas will stimulate further analysis of model results and observations.

Acknowledgements

This work was funded by the Dutch Science Foundation (Earth and Life Sciences) through project ALW854.00.037 (L.M.F. and H.A.D.) and a VENI-grant (A.S.vdH.).

References

Barsugli, J. J. & Battisti, D. S. 1998 The basic effects of atmosphere–ocean thermal coupling on midlatitude variability. *J. Atmos. Sci.*, **55**, 477–493 (doi:10.1175/1520–0469(1998)055<0477:TBEOAO>2.0.CO;2).

Cayan, D. R. 1992 Latent and sensible heat flux anomalies over the northern oceans: the connection to monthly atmospheric circulation. *J. Climate*, **5**, 354–369 (doi:10.1175/1520-0442(1992)005<0354:LASHFA>2.0.CO;2).

Chen, F. & Ghil, M. 1996 Interdecadal variability in a hybrid coupled ocean–atmosphere model. *J. Phys. Oceanogr.*, **26**, 1561–1578 (doi:10.1175/1520-0485(1996)026<1561:IVIAHC>2.0.CO;2).

Delworth, T. L. & Greatbatch, R. J. 2000 Multidecadal thermohaline circulation variability driven by atmospheric surface flux forcing. *J. Climate*, **13**, 1481–1495 (doi:10.1175/1520-0442(2000)013<1481:MTCVDB>2.0.CO;2).

Dijkstra, H. A. 2005 *Nonlinear Physical Oceanography: a Dynamical Systems Approach to the Large Scale Ocean Circulation and El Niño*, 2nd edn., p. 532. Springer.

Dijkstra, H. A. 2006 Interaction of SST modes in the North Atlantic Ocean. *J. Phys. Oceanogr.*, **36**, 286–299 (doi:10.1175/JPO2851.1).

Dijkstra, H. A., te Raa, L. A., Schmeits, M. & Gerrits, J. 2006 On the physics of the Atlantic Multidecadal Oscillation. *Ocean Dyn.*, **56**, 36–50 (doi:10.1007/s10236-005-0043–0).

Enfield, D. B., Mestas-Nunes, A. M. & Trimble, P. 2001 The Atlantic Multidecadal Oscillation and its relation to rainfall and river flows in the continental US. *Geophys. Res. Lett.*, **28**, 2077–2080 (doi:10.1029/2000GL012745).

Farrell, B. F. & Ioannou, P. J. 1996 Generalized stability theory. I: autonomous operators. *J. Atmos. Sci.*, **53**, 2025–2040 (doi:10.1175/1520–0469(1996)053<2025:GSTPIA>2.0.CO;2).

Fedorov, A. & Philander, S. 2000 Is El Niño changing? *Science*, **288**, 1997–2002 (doi:10.1126/science.288.5473.1997).

Frankcombe, L. M., Dijkstra, H. A. & von der Heydt, A. 2009 Noise induced multidecadal variability in the North Atlantic: excitation of normal modes. *J. Phys. Oceanogr.*, **39**, 220–233.

Ghil, M., Allen, R. M., Dettinger, K. *et al.* 2002 Advanced spectral methods for climatic time series. *Rev. Geophys.*, **40**, 1003 (doi:10.1029/2000RG000092).

Greatbatch, R. J. & Zhang, S. 1995 An interdecadal oscillation in an idealized ocean basin forced by constant heat flux. *J. Climate*, **8**, 81–91 (doi:10.1175/1520-0442(1995)008<0081:AIOIAI>2.0.CO;2).

Guckenheimer, J. & Holmes, P. 1990 *Nonlinear Oscillations, Dynamical Systems and Bifurcations of Vector Fields*, 2nd edn. Springer-Verlag.

Hasselmann, K. 1976 Stochastic climate models. I: Theory. *Tellus*, **28**, 473–485.

Jin, F.-F. & Neelin, J. D. 1993 Modes of interannual tropical ocean–atmosphere interaction – a unified view. I: numerical results. *J. Atmos. Sci.*, **50**, 3477–3503 (doi:10.1175/1520-0469(1993)050<3477:MOITOI>2.0.CO;2).

Keller, H. B. 1977 Numerical solution of bifurcation and nonlinear eigenvalue problems. *Applications of Bifurcation Theory*, ed. P. H. Rabinowitz, pp. 359–384. Academic Press.

Kushnir, Y. 1994 Interdecadal variations in North Atlantic sea surface temperature and associated atmospheric conditions. *J. Phys. Oceanogr.*, **7**, 141–157.

Luterbacher, J., Xoplaki, E., Dietrich, D. *et al.* 2002 Exterding North Atlantic Oscillation reconstructions back to 1500. *Atmos. Sci. Lett.*, **2**, 114–124.

Neelin, J. D., Battisti, D. S., Hirst, A. C. *et al.* 1998 ENSO theory. *J. Geophys. Res.*, **103**, 14261–14290 (doi:10.1029/97JC03424).

North, G., Bell, T., Cahalan, R. & Moeng, F. 1982 Sampling errors in the estimation of empirical orthogonal functions. *Mon. Weather Rev.*, **110**, 699–706. (doi:10.1175/1520-0493(1982)110<0699:SEITEO>2.0.CO;2).

Pacanowski, R. C. & Griffies, S. M. 2000 *MOM 3.0 Manual*. See http://www.gfdl.gov/~smg/MOM/web/guide_parent/guide_parent.html.

Philander, S. G. H. 1990 *El Niño and the Southern Oscillation*. Academic Press.

Rayner, N. A., Parker, D. E., Horton, E. B. *et al.* 2003 Global analyses of sea surface temperature, sea ice, and night marine air temperature since the late nineteenth century. *J. Geophys. Res.*, **108**, doi:10.1029/2002JD002670.

Roulston, M. & Neelin, J. D. 2000 The response of an ENSO model to climate noise, weather noise and intraseasonal forcing. *Geophys. Res. Lett.*, **27**, 3723–3726 (doi:10.1029/2000GL011941).
Saravanan, R. & McWilliams, J. 1998 Advective ocean–atmosphere interaction: an analytical stochastic model with implications for decadal variability. *J. Climate*, **11**, 165–188 (doi:10.1175/1520-0442(1998)011<0165:AOAIAA>2.0.CO;2).
te Raa, L. A. & Dijkstra, H. A. 2002 Instability of the thermohaline ocean circulation on interdecadal time scales. *J. Phys. Oceanogr.*, **32**, 138–160 (doi:10.1175/1520-0485(2002)032<0138:IOTTOC>2.0.CO;2).
te Raa, L. A. & Dijkstra, H. A. 2003 Sensitivity of North Atlantic multidecadal variability to freshwater flux forcing. *J. Climate*, **16**, 2586–2601 (doi:10.1175/1520-0442(2003)016<2586:SONAMV>2.0.CO;2).
von der Heydt, A. & Dijkstra, H. A. 2007 Localization of multidecadal variability: I. Cross equatorial transport and interbasin exchange. *J. Phys. Oceanogr.*, **37**, 2401–2414 (doi:10.1175/JPO3133.1).
Zebiak, S. E. & Cane, M. A. 1987 A model El Niño-Southern oscillation. *Mon. Weather Rev.*, **115**, 2262–2278 (doi:10.1175/1520-0493(1987)115<2262:AMENO>2.0.CO;2).

13

Centennial-to-millennial-scale Holocene climate variability in the North Atlantic region induced by noise

MATTHIAS PRANGE, JOCHEM I. JONGMA AND MICHAEL SCHULZ

Under pre-industrial Holocene boundary conditions, a three-dimensional global climate model of intermediate complexity exhibits centennial-to-millennial-scale climate variability in the North Atlantic. The climate variability is associated with noise-induced 'on' and 'off' switches in Labrador Sea convection. On a multicentennial time scale these stochastic mode-transitions can be phase-locked to a small periodic freshwater forcing. These results suggest a stochastic resonance mechanism that can operate under Holocene conditions, involving changes in North Atlantic Deep Water formation as an important amplifying mechanism of relatively weak climate perturbations. We introduce a conceptual nonlinear stochastic model that reproduces the noise-induced transitions and highlights the importance of polar water flow towards the Labrador Sea in setting the stochastic time scale. Moreover, we present a new hypothesis in an attempt to explain an observed mid-Holocene mode shift in the power spectrum of North Atlantic climate variability from ~1500 to 600–1000 years. This hypothesis involves a mid-Holocene increase of polar water flow from the Greenland Sea into the Labrador Sea. The resulting decrease of the stochastic time scale favours the phase-locking to a smaller period in an applied multimodal external forcing, in accordance with stochastic resonance theory.

13.1 Introduction

Even though the climate of the Holocene (i.e. the past ~10 000 years) is generally regarded to be stable compared to the strongly fluctuating climate of the last ice age, a number of studies have revealed substantial Holocene climate variations in the North Atlantic at time scales ranging from centuries to a few millennia (e.g. O'Brien *et al.* 1995; Bond *et al.* 1997, 2001; Bianchi & McCave 1999; Schulz &

Paul 2002; Hall *et al.* 2004). The causes of these climate variations remain a source of debate. Hypotheses regarding their origin include internal oscillations of the climate system (e.g. Schulz & Paul 2002), external forces like variations in the Sun's radiative output (e.g. Bond *et al.* 2001) and/or a combination of the two. Based on analogies to larger-amplitude climate variations and proxy evidence, many authors have concluded that the Holocene climate fluctuations at centennial-to-millennial time scales, specifically those reconstructed for the North Atlantic region, are linked to variations in the rate of North Atlantic Deep Water (NADW) formation and associated changes in the Atlantic meridional overturning circulation (AMOC) and the oceanic northward heat transport (e.g. Bond *et al.* 2001; Oppo *et al.* 2003).

Recently, Berner *et al.* (2008) produced a new subpolar sea-surface temperature (SST) record with very high temporal resolution that provides new insights into Holocene climate variability in the North Atlantic. The record reveals Holocene multicentennial-to-millennial-scale SST variability on the order of 1–3° C southwest of Iceland. Climatic oscillations (note that we use the term 'oscillation' without implying strict periodicity) with 600- to 1000-, ~1500- and 2500-year periods are documented, with a time-dependent dominance of different periods through the Holocene. In the mid Holocene (7000–5000 years ago) a mode shift in the variability from higher (~1500 years) to lower (600–1000 years) periods is observed. Another important feature of the subpolar SST record is its correlation with reconstructed solar irradiance variability. So far, a theory for the centennial-to-millennial-scale North Atlantic climate variability, which accounts for both the relation to solar irradiance and the mid-Holocene shift in the power spectrum towards smaller periods, is lacking.

Based on results obtained with a global atmosphere–ocean model of intermediate complexity (Schulz *et al.* 2007; Jongma *et al.* 2007) we shall develop a conceptual model that may help to understand Holocene climate variability in the North Atlantic. The conceptual model involves noise-induced transitions between different states of the AMOC and a stochastic resonance mechanism to explain the subpolar SST record.

13.2 North Atlantic climate variability in an intermediate-complexity atmosphere–ocean model

To examine the potential of the climate system to generate centennial-to-millennial-scale oscillations, Schulz *et al.* (2007) used the global three-dimensional atmosphere–ocean model ECBilt-CLIO, version 3. This coupled model of intermediate complexity derives from the atmosphere model ECBilt (Opsteegh *et al.* 1998) and the ocean/sea-ice model CLIO (Goosse & Fichefet 1999). The atmospheric component solves the quasi-geostrophic equations with T21 resolution (~5.6°) for

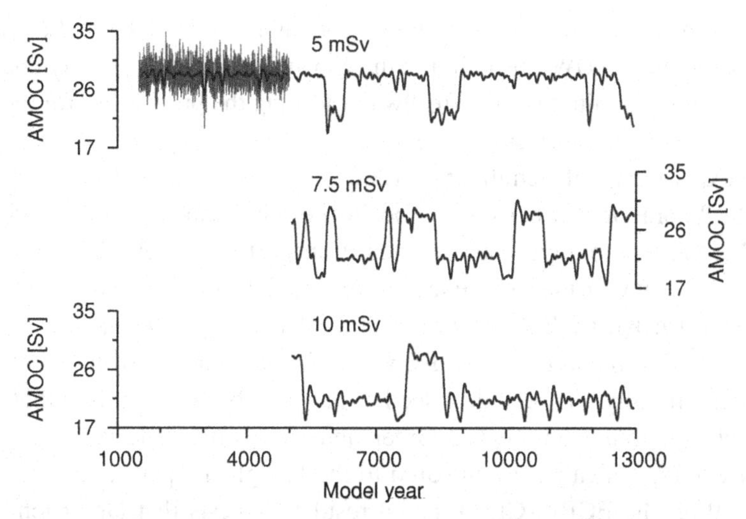

Figure 13.1 Time series of the maximum of the Atlantic meridional overturning circulation (AMOC), calculated north of 30° N and below 500 m water depth in ECBilt–CLIO. (Top) Unsmoothed annual values from the unperturbed control experiment (prior to model year 5000). The corresponding output from a 101-year wide Hanning filter is overlayed. After model year 5000 a 5 mSv freshwater perturbation is applied to the Labrador Sea. The resulting AMOC time series (smoothed) oscillates between approximately 22 and 28 Sv. (Centre) AMOC for a 7.5 mSv freshwater perturbation starting in model year 5000 (smoothed time series). (Bottom) As before, but for a 10 mSv forcing (Schulz *et al.* 2007).

three layers. Variability associated with large-scale weather patterns is explicitly computed. The primitive-equation, free-surface ocean component has a horizontal resolution of 3° and 20 levels in the vertical and uses a rotated subgrid in the North Atlantic Ocean to avoid the convergence of meridians near the north pole. It includes parameterisations for mixed-layer dynamics, downsloping currents and the Gent–McWilliams (1990) parameterisation of subgrid-scale processes. The ocean model is coupled to a thermodynamic–dynamic sea-ice model with viscous-plastic rheology.

For pre-industrial boundary conditions ECBilt-CLIO captures the two main deep convection sites in the modern North Atlantic. Consistent with observations, stable deep convection occurs in the Labrador Sea and in the Nordic seas (Schulz *et al.* 2007). On adding a weak, constant freshwater flux of 5–10 mSv (milli-Sverdrup; 1 Sv $= 10^6$ m^3s^{-1}) to the surface of the Labrador Sea, the character of the AMOC changes fundamentally (note that the applied freshwater forcings are at least one order of magnitude below the values typically used in 'freshwater-hosing experiments' to yield a complete shut-down of NADW formation). The forcing gives rise to a bimodal AMOC distribution with random timing of transitions between a strong and a weak overturning state (Fig. 13.1). Transitions between the strong state and the weak state are related to 'on' and 'off' switches in Labrador Sea convection, while NADW formation in the Nordic seas remains nearly unaffected. With increasing magnitude of the freshwater forcing the probability for the system to occupy the weak state rises.

In the experiments of Schulz *et al.* (2007) the cessation of deep-water formation in the Labrador Sea causes a drop in the temperature of the overlying air by ~3 °C (due to a reduced oceanic heat transport by the AMOC towards that region), consistent with the Holocene multicentennial-to-millennial-scale SST variability reconstructed by Berner *et al.* (2008). This temperature anomaly spreads over southern Greenland in the model, while air temperatures over central Greenland remain virtually unaffected. Indeed, it has long been recognised that temperature reconstructions from central Greenland show only little variability during the Holocene (specifically, if compared to the last glacial period; e.g. Grootes & Stuiver 1997). The ECBilt-CLIO model results suggests that larger temperature variations can be expected to be documented in southern Greenland, which is in agreement with palaeoclimatic evidence from borehole records (Dahl-Jensen *et al.* 1998).

The duration between subsequent 'on'/'off' switches in Labrador Sea convection and hence AMOC mode-transitions varies widely from 310 to 2660 years in the 7.5 mSv experiment (Fig. 13.1). The average duration is approximately 1420 years and the standard deviation amounts to 1020 years. Hence, the total range of values is almost completely covered by the interval of one standard deviation around the mean, supporting the notion that the durations are more or less uniformly distributed. The lack of a dominant time scale argues against a deterministic process controlling the timing of the AMOC oscillations and is easier to reconcile with a stochastic origin (note that we use the term 'stochastic' in a somewhat loose sense – in the coupled climate model any 'randomness' is generated by the interactions of deterministic processes which lead to high-frequency noise). Therefore, the simplest explanation for the AMOC oscillations involves a bistable system with noise-induced transitions from one mode of operation to the other.

Bistability of the system can be understood by means of a positive feedback between oceanic salt transport and Labrador Sea convection, according to the well-known 'Stommel feedback' (Stommel 1961). The NADW formation in the Labrador Sea drives a part of the AMOC which, in turn, transports salt from the subtropics towards the subpolar convective site. This salt supply (together with heat loss to the atmosphere) maintains a high density of surface waters and hence convection (this is the 'on' mode). A disruption of this 'conveyor belt' (e.g. by a randomly occuring negative density perturbation in the Labrador Sea) will result in a freshening of Labrador Sea surface water (due to excess precipitation and runoff) and may eventually lead to a halt of convection ('off' mode).

Given the noise-induced transitions between the states with and without convection in the Labrador Sea, the change in the ratio of the durations of the strong state to the weak state of the AMOC oscillations (cf. Fig. 13.1) can be explained. Starting from a strong state, a larger value of the continuous freshwater forcing brings the Labrador Sea closer to the point at which a random negative density anomaly can stop the deep mixing. Thus, the probability for a shut-down of convection in the Labrador Sea increases with increasing freshwater forcing. Once the system is in the weak mode, the likelihood for a large positive density anomaly determines how long this mode prevails. Since a larger freshwater forcing moves the Labrador Sea towards less dense surface waters, the probability for a weak-to-strong transition decreases with increasing freshwater forcing.

Even though the oscillations produced by ECBilt-CLIO might provide an explanation for Holocene centennial-to-millennial-scale variability in the North Atlantic, this explanation would not account for the phase-locking to the Sun's activity that has been suggested in previous studies (Bond *et al.* 1997, 2001; Berner *et al.* 2008). Jongma *et al.* (2007), however, have shown that the transitions between the two modes of the AMOC can be phase-locked to a small multicentennial periodic forcing.

13.3 Synchronisation with an external forcing

North Atlantic drift-ice and SST proxies correlate with reconstructed production rates of cosmogenic isotopes, which has been taken as evidence of a persistent solar influence on Holocene high-latitude climate (Bond *et al.* 1997, 2001; Berner *et al.* 2008). However, the direct effect of solar irradiance variations is considered too small to explain the observed climate variability in the Holocene (e.g. Rind 2002). Accordingly, an amplifying mechanism is required to explain centennial-to-millennial Holocene climate variability driven by a solar forcing. Jongma *et al.* (2007) demonstrated that changes in the NADW production rate can provide an amplifying mechanism of relatively weak climate perturbations during the Holocene – a mechanism that had previously been hypothesised by Bond *et al.* (2001).

To investigate whether ECBilt-CLIO's AMOC oscillations are susceptible to small external forcings, a periodically varying freshwater forcing was applied to the Labrador Sea instead of a constant one (Jongma *et al.* 2007). This freshwater flux varied sinusoidally between 5 and 10 mSv with a period of 500 years (the exact choice of the forcing period was not important; the authors just wanted to test whether the AMOC oscillations could be phase-locked to a *centennial-to-millennial-scale* forcing). It is important to realise that the amplitude of the forcing is sufficiently small to keep the system always in the bimodal regime (Fig. 13.1).

Accordingly, the periodic forcing was considered 'small' (or sub-threshold) since noise is required to trigger individual state switches.

In a 12 000-year long integration of ECBilt-CLIO, Jongma *et al.* (2007) found 10 AMOC state transitions to the weak mode (i.e. Labrador Sea convection 'off'-switches). A Rayleigh test showed that the timing of the mode-transitions was significantly ($p < 0.05$) phase-locked to the periodic forcing. Accordingly, the response of the system to the forcing occured approximately at integer multiples of the forcing period.

Intuitively, noise might be expected to weaken or obscure regular signals. However, in a bistable system the presence of noise can provide a crucial mechanism for the system to explore its possible states (Nicolis & Nicolis 1981). Consequently, the timing of the state switches in the experiment of Jongma *et al.* (2007) has a stochastic component. The response of the AMOC to the periodic forcing is both deterministic and stochastic, i.e. 'quasi-deterministic' (Freidlin & Wentzell 1998).

Noise-assisted amplification of a small periodic forcing is a stochastic resonance phenomenon (Benzi *et al.* 1981; Gammaitoni *et al.* 1998). It has been shown that on an 'interval of resonance' (a set of scale parameters for which chaotic or trivial behaviour of the system is excluded), there must exist a stochastic resonance point (Herrmann & Imkeller, 2005). The lower limit of this interval of resonance, below which the system remains practically in one of the states, coincides with a minimum exponential time scale for quasi-deterministic behaviour (Freidlin & Wentzell 1998; Freidlin 2000). The upper limit of this interval refers to a situation where the system switches chaotically between multiple states. By showing prolonged but finite noise-induced mode transitions that are significantly deterministic, Jongma *et al.* (2007) argued that the modelled climate system operates on this interval of resonance, implying that a stochastic resonance mechanism can be operational with respect to centennial-to-millennial North Atlantic climate variability in the Holocene. (Note that finding the stochastic resonance point, i.e. the time scale at which the system response is maximum, was not the authors' intention.)

13.4 A conceptual model of North Atlantic climate oscillations

Probably the simplest model that captures the *basic* mechanisms of the centennial-to-millennial-scale oscillations consists of two prognostic variables, θ and S, representing spatially averaged temperature and salinity of the upper (say, the topmost 200 m) Labrador Sea (Schulz *et al.* 2007). Hydrographic conditions in the upper Labrador Sea depend on the wind- and density-driven inflows from the subtropical Atlantic, the influx of polar water from the Greenland Sea (via the East Greenland Current through Denmark Strait) and surface fluxes. The deterministic governing equations for Labrador Sea temperature and salinity can therefore be formulated as

$$\frac{dS}{dt} = \frac{1}{V}[(q_1+\phi)S_1 + q_2S_2 - (q_1 + q_2 + \phi + P + P')S], \tag{13.1}$$

$$\frac{d\theta}{dt} = \frac{1}{V}[(q_1 + \phi)\theta_1 + q_2\theta_2 - (q_1 + q_2 + \phi)\theta] + H, \tag{13.2}$$

where V denotes the volume of the upper Labrador Sea (3×10^{14} m^3), q_1 is the wind-driven volume flux from the subtropical Atlantic, ϕ is the density-driven (overturning) volume flux from the subtropical Atlantic, q_2 is the volume flux from the Greenland Sea, P denotes a basic surface freshwater flux into the Labrador Sea (78 mSv), P' is a freshwater flux perturbation, H denotes the surface heat flux and t is time. The parameters S_1, θ_1 and S_2, θ_2 denote salinities and temperatures of the inflowing subtropical and polar water masses, respectively. To keep the model as simple as possible, the wind-driven volume fluxes are assumed to be constant ($q_1 =$ 2 Sv, $q_2 = 2$ Sv; after Fig. 10.50 in Dietrich *et al.* 1975). Likewise, any temperature and salinity changes in the water masses originating from the subtropical Atlantic and the Greenland Sea are neglected ($S_1 = 35.3$ psu, $\theta_1 = 8\,°$C, $S_2 = 34.6$ psu, $\theta_2 = 1.5\,°$C). The overturning ϕ couples the pair of differential equations through a linear dependence on the density difference between the Labrador Sea and the subtropical Atlantic, i.e. $\phi = \kappa[\alpha\ (\theta_1 - \theta) - \beta\ (S_1 - S)]$, where α and β denote the thermal and haline expansion coefficients of seawater ($\alpha = 0.1$ K^{-1}, $\beta =$ 0.8 psu^{-1}). The tuning parameter κ is set to 70 Sv. If ϕ becomes lower than zero, ϕ is set to zero (Labrador Sea convection 'off'). For the surface heat flux H, a simple restoring term is used, i.e. $H = (\theta_r - \theta)/\tau$, where $\theta_r = 0\,°$C is a relaxation temperature and $\tau = 1$ year is the relaxation time scale. This simple approach parameterises the damping of surface temperature anomalies by atmospheric heat advection and/or longwave radiation.

The stability behaviour of the simple deterministic model is illustrated by the calculation of an equilibrium hysteresis with respect to surface freshwater forcing. The 'Stommel feedback' along with the competition between thermal and saline forcings of the overturning results in multiple equilibria. Plotting the overturning ϕ against the anomalous freshwater forcing P' reveals a regime of bistability for intermediate forcing amplitudes (Fig. 13.2). With $P' = 0$ Sv, for instance, one stable equilibrium ('off' mode) has zero overturning and a Labrador Sea salinity of $S \cong 34.3$ psu, while the other stable equilibrium ('on' mode) yields an overturning of $\phi = 10$ Sv and a salinity of ~35.0 psu. For very large positive/negative values of P', only the 'off'/'on' mode of Labrador Sea overturning provides a stable solution. Since the inflow of polar water ($S_2 = 34.6$ psu) has a higher/lower salinity than the Labrador Sea in the 'off'/'on' mode, it will always counteract the current state

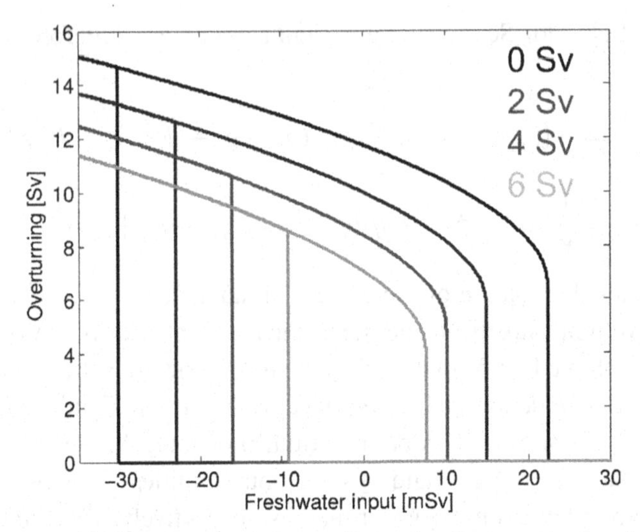

Figure 13.2 Equilibrium hysteresis loops (Labrador Sea overturning ϕ vs. freshwater input P') for the deterministic conceptual model with different values for the polar water inflow to the Labrador Sea ($q_2 = 0$ Sv, 2 Sv, 4 Sv, 6 Sv). The hysteresis loops have been obtained by slowly varying the freshwater input to the Labrador Sea (in the 'clockwise' direction).

of overturning in the Labrador Sea. To demonstrate this, we calculate hysteresis curves for different q_2 (Fig. 13.2). With increasing polar inflow, the hysteresis loop becomes narrower, i.e. smaller positive or negative freshwater perturbations P' are sufficient to induce a transition from one mode to the other. For $q_2 > 13$ Sv the hysteresis disappears and the system becomes monostable (not shown).

The simple deterministic model possesses no internal variability. In the bistable regime of ocean circulation, however, occasional state transitions can be introduced by a stochastic forcing component (cf. Cessi 1994). We add a term $\sigma\xi S_0/V$ to equation (13.1) to introduce a stochastic component in the salinity balance, where ξ represents Gaussian 'quasi white noise' with zero mean and unit variance, σ measures the standard deviation of the stochastic forcing and S_0 denotes a reference salinity (35 psu). We note that the stochastic forcing is *not* perfectly white in time due to a non-vanishing autocorrelation that is introduced by the three-day time step of the Euler scheme used to solve the differential equations. This rather simple approach is similar to numerical schemes used in previous studies on stochastically forced thermohaline flows (e.g. Cessi 1994; Monahan 2002). It is still important to realise that stochastic perturbations that are not white in time may yield significantly different results from those that are white (for an in-depth discussion, see e.g. Stastna & Peltier 2007). However, from a physical point of view it is not reasonable to expect the stochastic climate forcing to be a white-noise process and our simple approach can be interpreted as a particular form of fading memory, i.e. the noise

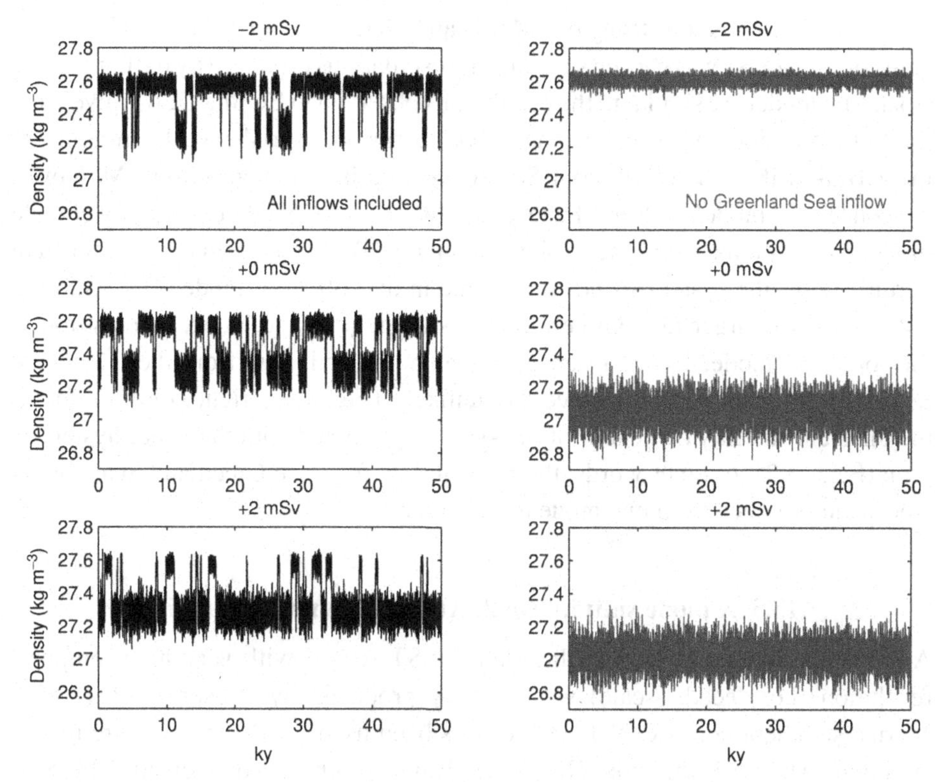

Figure 13.3 Typical time series of Labrador Sea density anomalies in the stochastically forced conceptual model with (left) all inflows included ($q_2 = 2$ Sv) and (right) polar water inflow set to zero ($q_2 = 0$). From top to bottom, the additional constant freshwater influx P' increases from –2 mSv to +2 mSv. Note that time-axis units are 10^3 years, i.e. kiloyears (Schulz *et al.* 2007).

has a perfect memory of three days and then loses all memory (we note that the three-day interval roughly matches the synoptic time scale of the climate system). Therefore, no attempts have been made to solve the stochastic differential equations with a more sophisticated Langevin approach (e.g. Kloeden & Platen 1992) in conjunction with perfectly white noise.

Setting $\sigma = 0.2$ Sv and $q_2 = 2$ Sv, and applying no additional freshwater flux ($P' = 0$ Sv), we obtain numerous transitions from one mode to the other during a 50 000 year integration, resulting in centennial-to-millennial variations of the overturning circulation (Fig. 13.3, left column). The average residence time in a circulation mode before switching back to the other state and, hence, the time scale of the oscillations, depends on the noise intensity (cf. Cessi 1994). The stronger the noise, the larger is the probability for a flip into the other state, and the shorter is the mean residence time in one mode. Note, however, that the underlying bistability of the deterministic system may be completely masked if the amplitude of the stochastic forcing is too strong (cf. Monahan 2002; Stommel & Young 1993). The value of $\sigma = 0.2$ Sv was chosen such that the oscillations produced by the conceptual stochastic model are similar to the oscillations observed in ECBilt-CLIO (given that the 'off'/'on' mode of the conceptual model corresponds to the weak/strong mode in ECBilt-CLIO, where Labrador Sea convection is switched off/on). Moreover, the conceptual model captures the response of the intermediate-complexity climate model to additional constant freshwater inputs. With positive/negative constant perturbations, the system spends more time in the 'off'/'on' mode (Fig. 13.3, left column), since larger random perturbations are required to induce a transition to the 'on'/'off' mode. In the absence of a polar water inflow from the Greenland Sea ($q_2 = 0$) state transitions become unlikely (Fig. 13.3, right column) due to the different stability behaviour of the system associated with the wider hysteresis loop (Fig. 13.2). In other words, the polar inflow from the Greenland Sea favours stochastic switches from one mode to the other.

13.5 A mode shift in North Atlantic climate variability

A Holocene diatom-based North Atlantic SST record with very high temporal resolution (i.e. decadal-scale) was recently produced by Berner *et al.* (2008). Marine sediment core LO09-14, which has been used for the temperature reconstruction, was retrieved from Reykjanes Ridge southwest of Iceland. The SST record documents climatic oscillations with 600- to 1000-, ~1500-, and 2500-year periodicities, with a time-dependent dominance of different periods through the Holocene. In particular, a striking mode shift is observed during the mid Holocene. Before this shift, the dominant period in SST variability is ~1500 years; after the mode shift, higher-frequency oscillations with a period of 600 to 1000 years prevail. Can such a mode shift be reproduced by our conceptual stochastic model?

To answer this question, we have to discover the processes that determine the time scale of the oscillations in the conceptual model. First and foremost, the average waiting time in one state before switching back to the other state is governed by the noise intensity, as already discussed in the previous section. However, since there is no evidence for a substantial shift in the noise structure over the North Atlantic, we assume a constant noise level throughout the Holocene. Another parameter that is of upmost importance for the time scale of the oscillations is q_2, the volume flux of polar water that flows from the Greenland Sea towards the Labrador Sea. As discussed above, the polar inflow always counteracts the current state of convection in the Labrador Sea and, hence, tends to reduce the width of the equilibrium hysteresis loops, i.e. the regime of bistability. Under a stochastic forcing with fixed intermediate noise level, this translates into a larger Kramers rate, i.e. a higher

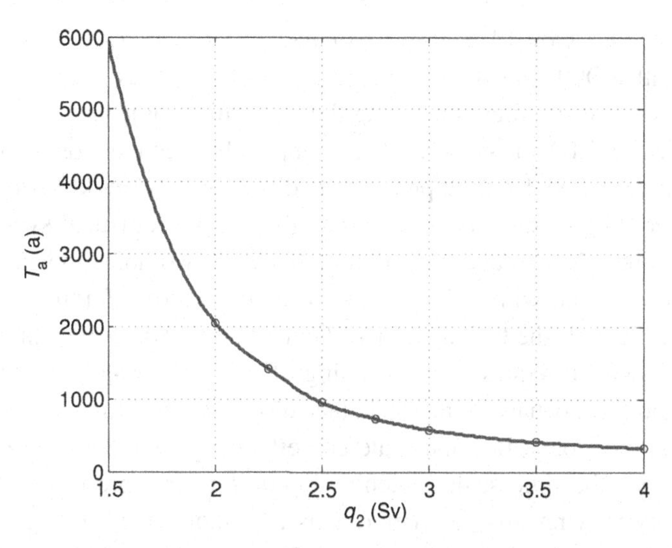

Figure 13.4 Average time interval T_a between two subsequent 'off'-to-'on'-switches in the conceptual stochastic model as a function of polar water inflow q_2. It is evident that a stronger polar water inflow to the Labrador Sea favours noise-induced transitions (i.e. reduction of the stochastic time scale). The values have been calculated from the Labrador Sea temperature time series (smoothed by a 70-year boxcar filter) of a 5×10^6-year integration.

frequency of switches between the two states of convection. Apart from the noise level, the polar water inflow q_2 is therefore the most efficient parameter to control the time scale of the oscillations in our conceptual model. Even relatively small changes in q_2 have a substantial effect on the stochastic time scale. For instance, the average time interval T_a between two subsequent 'off'-to-'on'-switches is about 2000 years for our standard set-up with $q_2 = 2$ Sv, and reduces to *c.* 1500 years for $q_2 = 2.25$ Sv and 750 years for $q_2 = 2.75$ Sv (Fig. 13.4). We therefore suggest that an increase in the flux of polar water from the Greenland Sea towards the Labrador Sea led to the mid-Holocene mode shift in North Atlantic climate variability towards a smaller period.

Indeed, there are indications from general circulation modelling that the supply of polar water from the Greenland Sea to the Labrador Sea was weaker in the early/mid Holocene than during the late Holocene. Climate simulations with the global coupled atmosphere–ocean model ECHO-G forced by varying orbital parameters show strong westerly wind anomalies over the North Atlantic between ~45° N and ~65° N during the early/mid Holocene (Lohmann *et al.* 2005). The stronger-than-present westerlies tended to block the westward flow of polar water from the East Greenland Current around the southern tip of Greenland and favoured a flow of polar water from the Greenland Sea towards the location of sediment core LO09-14 southwest of Iceland. The ECHO-G model result is consistent with LO09–14's diatom record which shows that the Greenland Current assemblage was most influential during the early/mid Holocene.

As in ECBilt-CLIO, the noise-induced hopping between the 'on' and 'off' states in the conceptual model can be phase-locked to a sub-threshold periodic external forcing. According to stochastic resonance theory, the statistical synchronisation is strongest when the average waiting time between two noise-induced state transitions is comparable with half the period of the external forcing, i.e. when T_a approximately equals the forcing period (Gammaitoni *et al.* 1998). In other words, stochastic resonance requires the matching of two time scales: the deterministic time scale and the stochastic time scale (see also Chapters 7 and 11).

Given that the stochastic time scale can efficiently be modulated by the polar water influx q_2, the time scale matching condition for a prescribed forcing can be fulfilled by varying this parameter. This is demonstrated in Fig. 13.5, where frequency distributions of the time interval T (defined as the time interval between two subsequent 'off'-to-'on'-switches) are plotted for $q_2 = 2.25$ Sv and $q_2 = 2.75$ Sv. Without any external periodic forcing, a wide range of transition times is likely (Fig. 13.5, upper panel). Both distributions have a positive skew with an exponential shape that is characteristic for stochastic transitions driven by noise alone. The skew (and hence the dispersion of T) is more pronounced in the distribution with smaller q_2, reflecting the longer stochastic time scale in the experiment with $q_2 = 2.25$ Sv ($T_a \cong 1500$ years) compared to the case with $q_2 = 2.75$ Sv ($T_a \cong 750$ years).

We now apply sub-threshold sinusoidal forcings to the system. Sub-threshold means that the amplitude of the forcing is smaller than half the width of the hysteresis loop of the deterministic model (Fig. 13.2). In other words, the sinusoidal forcing alone (i.e. without the help of stochastic perturbations) is unable to induce a state transition from the 'off' mode to the 'on' mode or vice versa. A simple deterministic response to the sinusoidal forcing (albeit obscured by the noise) is therefore excluded.

When a small (3 mSv amplitude) sinusoidal forcing with a 750-year period is applied to the experiment with $q_2 = 2.75$ Sv, the approximate matching of the deterministic and stochastic time scales leads to a strong resonance such that a sharp peak appears at $T = 750$ years in the frequency distribution (Fig. 13.5, middle panel). The resonance is weaker in the $q_2 = 2.25$ Sv case, where pronounced secondary peaks are visible at integer multiples of the forcing period due to frequent 'skipping' of the forcing beat. Such a distribution indicates that the system operates on the interval of resonance. However, the statistical synchronisation is not optimized.

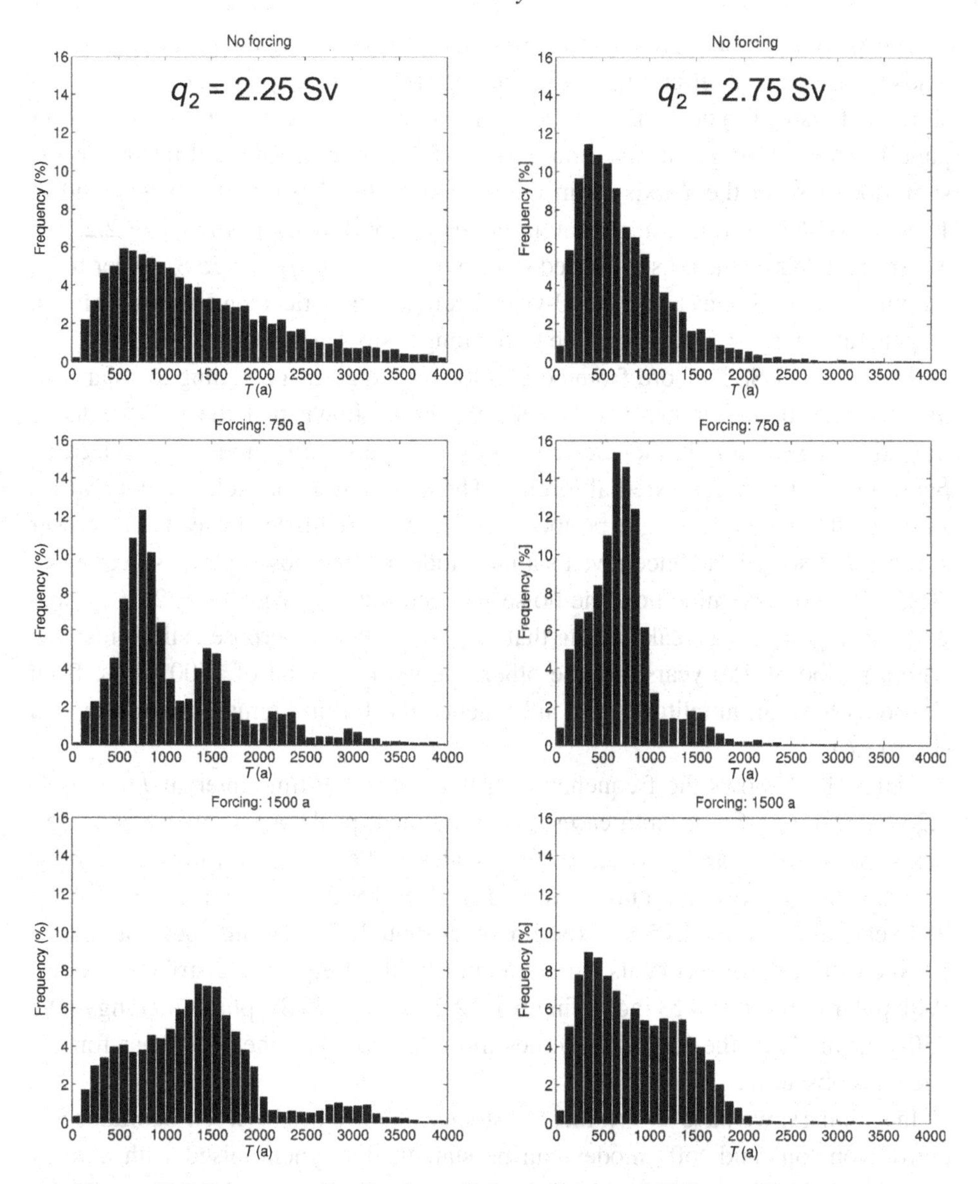

Figure 13.5 Frequency distributions of the time interval T (interval between two subsequent 'off'-to-'on'-switches) in the conceptual stochastic model for (left) $q_2 = 2.25$ Sv and (right) $q_2 = 2.75$ Sv. (Top) Without any external periodic forcing. (Centre) With sub-threshold (3 mSv amplitude) sinusoidal forcing of period 750 years. (Bottom) With sub-threshold (3 mSv amplitude) sinusoidal forcing of period 1500 years. The histograms are based on the Labrador Sea temperature time series (smoothed by a 70-year boxcar filter) of a 5×10^6-year integration.

The application of a forcing with 1500-year period (and again 3 mSv amplitude) leads to completely different results. The experiment with $q_2 = 2.75$ Sv does not show a statistical synchronisation with the sinusoidal forcing (Fig. 13.5, lower panel). Instead, the global maximum of the frequency distribution resides at the same location on the T-axis as in the unforced run. No maximum is found at 1500 years. By contrast, a maximum arises at 1500 years in the $q_2 = 2.25$ Sv experiment. While the noise-induced state switches in the $q_2 = 2.75$ Sv experiment are virtually 'immune' to the 1500-year forcing, a statistical synchronisation can be generated with $q_2 = 2.25$ Sv due to the longer stochastic time scale.

The subpolar SST record from core LO09-14 suggests a coupling to solar irradiance variations (Berner *et al.* 2008). We have shown that the noise-induced transitions between Labrador Sea convection 'on' and 'off' modes can indeed be phase-locked to a weak external forcing. The assumption that solar activity varies sinusoidally with one single period is however unrealistic. Instead, the power spectrum of solar irradiance reveals a multitude of superposed periods (Vonmoos 2005). We now examine how the noise-induced hopping with $q_2 = 2.25$ Sv and 2.75 Sv responds to a weak forcing that consists of two superposed sinusoids: one with a period of 750 years and the other one with a period of 1500 years. Both sinusoids have an amplitude of 3 mSv, hence the forcing remains sub-threshold (Fig. 13.6).

Figure 13.7 shows the frequency distributions of the time interval T for $q_2 =$ 2.25 Sv and 2.75 Sv. In both cases, two prominent peaks are visible: one at 750 years and one at 1500 years (i.e. the time scales of the forcing). However, for $q_2 =$ 2.25 Sv the global maximum is located at $T = 1500$ years, while it resides at 750 years in the $q_2 = 2.75$ Sv experiment. Although the forcing does not change between the two experiments, the structure of the frequency distribution does. With polar water inflow q_2 increasing from 2.25 Sv to 2.75 Sv, phase-locking to the 750-year mode of the forcing becomes more likely, while the 1500-year forcing becomes obscured.

In summary, we have shown that the stochastic switches between Labrador Sea convection 'on' and 'off' modes can be statistically synchronised with a weak external forcing that consists of different superposed sinusoids. In agreement with the theory of stochastic resonance, the strongest phase-locking occurs with that period in the forcing that is closest to the stochastic time scale of the noise-induced state transitions. We suggest that an increase in the polar water inflow during the mid Holocene led to a decrease of the stochastic time scale, thus favouring a synchronisation with a smaller period in the external (solar) forcing. This resulted in the 'observed' mid-Holocene mode shift of North Atlantic climate variability from ~1500 years in the early Holocene to 600–1000 years in the late Holocene.

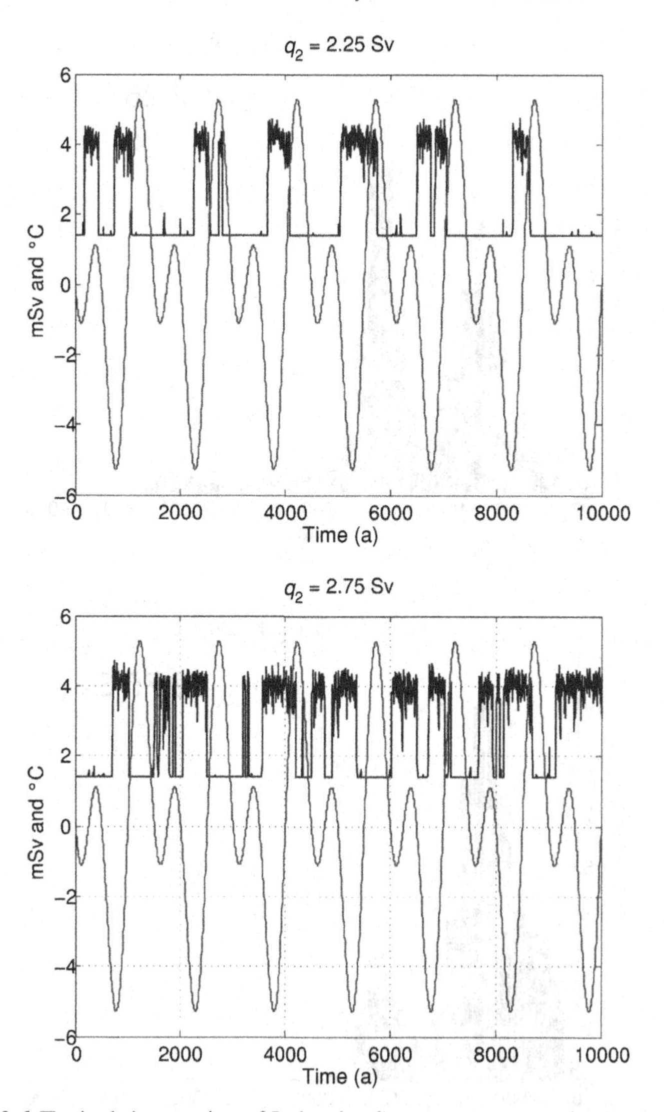

Figure 13.6 Typical time series of Labrador Sea temperature θ in the conceptual stochastic model with a weak (sub-threshold) forcing that consists of two superposed sinusoids (750 years and 1500 years; both sinusoids have an amplitude of 3 mSv) and a polar water inflow of (top) $q_2 = 2.25$ Sv and (bottom) $q_2 = 2.75$ Sv.

13.6 Discussion and conclusions

Centennial-to-millennial-scale North Atlantic climate variability in the coupled atmosphere–ocean model of intermediate complexity ECBilt–CLIO has been analysed (Schulz *et al.* 2007). The model exhibits internal oscillations of the AMOC that are characterised by a strong state with deep-water formation in both the Nordic seas

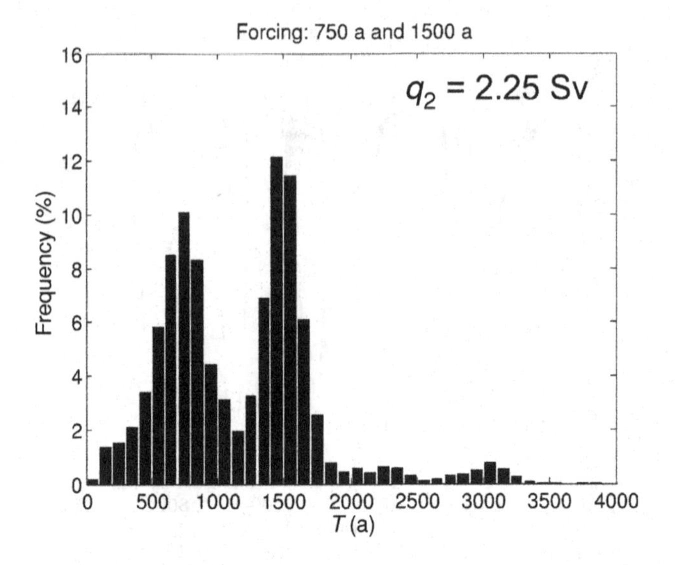

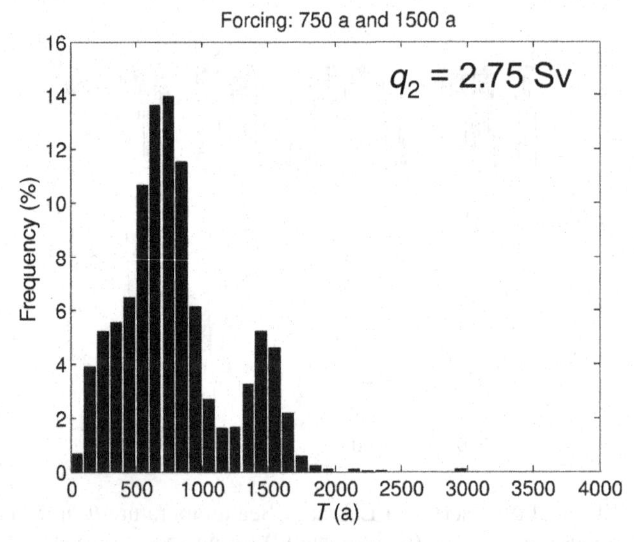

Figure 13.7 Frequency distributions of the time interval T (interval between two subsequent 'off'-to-'on'-switches) for (top) $q_2 = 2.25$ Sv and (bottom) $q_2 = 2.75$ Sv. A weak (sub-threshold) forcing that consists of two superposed sinusoids (periods of 750 and 1500 years) is applied (see Fig. 13.6).

and the Labrador Sea and a weak state in which deep water forms only in the Nordic seas. These oscillations can be attributed to an underlying bistability in Labrador Sea convection and state transitions induced by noise. In order to shift the system into the bistable regime, a small constant freshwater perturbation was applied to the Labrador Sea. The need for such a perturbation might be model-dependent. We note, however, that other modelling studies (cf. Kuhlbrodt *et al.* 2001; Wood *et al.* 1999) suggest that the Labrador Sea operates close to the border of the bistable regime.

Temperature fluctuations associated with the AMOC oscillations in the North Atlantic are consistent with reconstructions from the Holocene, while they are much smaller compared to variations associated with glacial Dansgaard–Oeschger events (e.g. Voelker *et al.* 2002). A palaeoceanographic reconstruction for the Labrador Sea suggests changes in deep convection and a state-switching behaviour on centennial-to-millennial time scales since at least ~8000–9000 years before present (Hillaire-Marcel *et al.* 2001). Even though the resolution of the record is insufficient to capture the variability in Labrador Sea deep-water formation found in the model, the data suggest nearly a dozen state-switches during the Holocene. The magnitudes of the corresponding surface salinity and density variations in the Labrador Sea are similar to the changes accompanying a state-shift in the climate model.

A simple nonlinear stochastic model reproduces the centennial-to-millennial-scale oscillations observed in ECBilt-CLIO and highlights the important role of polar water flow towards the Labrador Sea in setting the stochastic time scale. In both the conceptual model and in ECBilt-CLIO the noise-induced transitions can be synchronised with a weak (sub-threshold) centennial-to-millennial-scale external forcing. It has thus been demonstrated that phase-locked switches in Labrador Sea convection can provide an amplifying mechanism for relatively weak climate perturbations. Hence, the model results support the hypothesis (Bond *et al.* 2001) on the existence of a centennial-to-millennial-scale amplifying mechanism involving ocean–atmosphere interactions that can operate under Holocene boundary conditions. The amplification of the weak external forcing depends on the presence of noise. The results suggest a stochastic resonance mechanism that can operate under Holocene conditions and indicate that changes in the three-dimensional configuration of NADW formation can be an important component of centennial-to-millennial climate variability during interglacials. By contrast, stochastic resonance on the millennial Dansgaard–Oeschger time scale was demonstrated for glacial boundary conditions by Ganopolski & Rahmstorf (2002). However, the two-dimensional zonally averaged Atlantic Ocean in their climate model did not exhibit noise-induced transitions and stochastic resonance under Holocene conditions. We therefore conclude that a three-dimensional model set-up, which is capable of simulating NADW formation in both the Nordic seas and in the Labrador Sea, is crucial for modelling interglacial AMOC oscillations.

Finally, we have presented a new hypothesis in an attempt to explain an observed mid-Holocene mode shift in the power spectrum of North Atlantic SST variability from ~1500 to 600–1000 years (Berner *et al.* 2008). This hypothesis involves a mid-Holocene increase of polar water flow from the Greenland Sea into the Labrador

Sea. The resulting decrease of the stochastic time scale favoured the phase-locking to a smaller period in the multimodal external forcing, in accordance with stochastic resonance theory.

Even though the phase-locking to an external forcing in the ECBilt-CLIO and conceptual model experiments is an important step towards understanding the relation between solar forcing and climate variability, it is still unclear how variations in solar irradiance translate into climate forcing. Ganopolski and Rahmstorf (2002), Schulz *et al.* (2007), and Jongma *et al.* (2007) implemented the external forcing through freshwater perturbations. We note that there is no evidence for a strong effect of solar irradiance variability on the hydrologic cycle; subtle fluctuations, however, cannot be ruled out. Another open question concerns the periodicities involved in solar variability. The periods used here and in the study of Jongma *et al.* (2007), i.e. 500, 750 and 1500 years, were chosen rather arbitrarily with the only goal to investigate whether the noise-induced AMOC transitions are susceptible to multicentennial and millennial external forcings. There is, however, evidence from cosmogenic isotope records that the solar power spectrum includes multicentennial periods (Bond *et al.* 2001; Vonmoos, 2005). In addition, Braun *et al.* (2005) suggested that millennial-scale solar forcing may arise through the superposition of 87- and 210-year solar cycles.

In future studies, experiments with other three-dimensional climate models, specifically more comprehensive models, should be carried out, to test the robustness of the mechanism that gives rise to the low-frequency AMOC oscillations in the climate model of intermediate complexity. The approach, however, is not straightforward. Firstly, a fine-tuning of the comprehensive models might be necessary in order to shift Labrador Sea convection into a bistable regime. Secondly, an integration of 10 000 years or more with a comprehensive climate model requires enormous computational resources. The acceleration technique employed by Lohmann *et al.* (2005) to make the long-term integration feasible inhibits the simulation of low-frequency climate variability.

We finally note that the understanding of low-frequency oscillations of Holocene climate is not only of importance from a palaeoclimatic perspective. It is also essential to understand the origin and dynamics of these natural climate variations to predict their potential interference with the possible anthropogenic influence on climate.

Acknowledgements

This work was funded through the DFG Research Centre / Excellence Cluster 'The Ocean in the Earth System'.

References

Benzi, R., Sutera, A. & Vulpiani, A. 1981 The mechanism of stochastic resonance, *J. Phys. A Math. Gen.*, **14**, L453–L457.

Berner, K. S., Koç, N., Divine, D., Godtliebsen, F. & Moros, M. 2008 A decadal-scale Holocene sea surface temperature record from the subpolar North Atlantic constructed using diatoms and statistics and its relation to other climate parameters. *Paleoceanography*, **23**, PA2210 (doi:10.1029/2006PA001339).

Bianchi, G. G. & McCave, I. N. 1999 Holocene periodicity in North Atlantic climate and deep-ocean flow south of Iceland. *Nature*, **397**, 515–517.

Bond, G., Showers, W., Cheseby, M. *et al.* 1997 A pervasive millennial-scale cycle in North Atlantic Holocene and glacial climates. *Science*, **278**, 1257–1266.

Bond, G., Kromer, B., Beer, J. *et al.* 2001 Persistent solar influence on North Atlantic climate during the Holocene. *Science*, **294**, 2130–2136.

Braun, H., Christl, M., Rahmstorf, S. *et al.* 2005 Possible solar origin of the 1,470-year glacial climate cycle demonstrated in a coupled model. *Nature*, **438**, 208–211.

Cessi, P. 1994 A simple box model of stochastically forced thermohaline flow. *J. Phys. Oceanogr.*, **24**, 1911–1920.

Dahl-Jensen, D., Mosegaard, K., Gundestrup, N. *et al.* 1998 Past temperatures directly from the Greenland ice sheet. *Science*, **282**, 268–271.

Dietrich, G., Kalle, K., Krauss, W. & Siedler, G. 1975 *Allgemeine Meereskunde.* Gebr. Borntraeger.

Freidlin, M. I. 2000 Quasi-deterministic approximation, metastability and stochastic resonance. *Phys. D*, **137**, 333–352.

Freidlin, M. & Wentzell, A. D. 1998 *Random Perturbations of Dynamical Systems.* Springer-Verlag.

Gammaitoni, L., Haenggi, P., Jung, P. & Marchesoni, F. 1998 Stochastic resonance. *Rev. Mod. Phys.*, **70**, 223–287.

Ganopolski, A. & Rahmstorf, S. 2002 Abrupt glacial climate changes due to stochastic resonance. *Phys. Rev. Lett.*, **88**, 038501 (doi:10.1103/PhysRevLett.88.038501).

Gent, P. R. & McWilliams, J. C. 1990 Isopycnal mixing in ocean circulation models. *J. Phys. Oceanogr.*, **20**, 150–155.

Goosse, H. & Fichefet, T. 1999 Importance of ice–ocean interactions for the global ocean circulation: A model study. *J. Geophys. Res.*, **C104**, 23337–23355.

Grootes, P. M. & Stuiver, M. 1997 Oxygen 18/16 variability in Greenland snow and ice with 10^{-3}- to 10^{5}-year time resolution. *J. Geophys. Res.*, **C102**, 26455–26470.

Hall, I. R., Bianchi, G. G. & Evans, J. R. 2004 Centennial to millennial scale Holocene climate – deep water linkage in the North Atlantic. *Q. Sci. Rev.*, **23**, 1529–1536.

Herrmann, S. & Imkeller, P. 2005 The exit problem for diffusions with time – periodic drift and stochastic resonance. *Ann. Appl. Probab.*, **15**, 39–68.

Hillaire-Marcel, C., de Vernal, A., Bilodeau, G. & Weaver, A. J. 2001 Absence of deep-water formation in the Labrador Sea during the last interglacial period. *Nature*, **410**, 1073–1077.

Jongma, J. I., Prange, M., Renssen, H. & Schulz, M. 2007 Amplification of Holocene multicentennial climate forcing by mode transitions in North Atlantic overturning circulation. *Geophys. Res. Lett.*, **34**, L15706 (doi:10.1029/2007GL030642).

Kloeden, P. E. & Platen, E. 1992 *Numerical Solution of Stochastic Differential Equations.* Springer-Verlag.

Kuhlbrodt, T., Titz, S., Feudel, U. & Rahmstorf, S. 2001 A simple model of seasonal open ocean convection. Part II: Labrador Sea stability and stochastic forcing. *Ocean Dyn.*, **52**, 36–49.

Lohmann, G., Lorenz, S. & Prange, M. 2005 Northern high-latitude climate changes during the Holocene as simulated by circulation models. In *The Nordic Seas: An Integrated Perspective*, Geophysical Monograph 158, ed. H. Drange, T. Dokken, T. Furevik, R. Gerdes and W. Berger, pp. 273–288. American Geophysical Union.
Monahan, A. H. 2002 Stabilisation of climate regimes by noise in a simple model of the thermohaline circulation. *J. Phys. Oceanogr.*, **32**, 2072–2085.
Nicolis, C. & Nicolis, G. 1981 Stochastic aspects of climatic transitions: additive fluctuations. *Tellus*, **33**, 225–234.
O'Brien, S. R., Mayewski, P. A., Meeker, L. D. *et al.* 1995 Complexity of Holocene climate as reconstructed from a Greenland ice core. *Science*, **270**, 1962–1964.
Oppo, D. W., McManus, J. F. & Cullen, J. L. 2003 Deepwater variability in the Holocene epoch. *Nature*, **422**, 277–278.
Opsteegh, J. D., Haarsma, R. J., Selten, F. M. & Kattenberg, A. 1998 ECBilt: a dynamic alternative to mixed boundary conditions in ocean models. *Tellus*, **50A**, 348–367.
Rind, D. 2002 The Sun's role in climate variations. *Science*, **296**, 673–677.
Schulz, M. & Paul, A. 2002 Holocene climate variability on centennial-to-millennial time scales: 1. Climate records from the North-Atlantic realm. In *Climate Development and History of the North Atlantic Realm*, ed. G. Wefer, W. H. Berger, K.-E. Behre and E. Jansen, pp. 41–54. Springer-Verlag.
Schulz, M., Prange, M. & Klocker, A. 2007 Low-frequency oscillations of the Atlantic Ocean meridional overturning circulation in a coupled climate model. *Clim. Past*, **3**, 97–107.
Stastna, M. & Peltier, W. R. 2007 On box models of the North Atlantic thermohaline circulation: intrinsic and extrinsic millennial timescale variability in response to deterministic and stochastic forcing. *J. Geophys. Res.*, **112**, C10023 (doi:10.1029/2006JC003938).
Stommel, H. 1961 Thermohaline convection with two stable regimes of flow. *Tellus*, **13**, 224–228.
Stommel, H. & Young, W. 1993 The average T–S relation of a stochastically-forced box model. *J. Phys. Oceanogr.*, **23**, 151–158.
Voelker, A. H. L. and workshop participants 2002 Global distribution of centennial-scale records for Marine Isotope Stage (MIS) 3: a database. *Q. Sci. Rev.*, **21**, 1185–1212.
Vonmoos, M. 2005 Rekonstruktion der solaren Aktivität im Holozän mittels Beryllium-10 im GRIP Eisbohrkern. Ph.D. Thesis, ETH Zürich, Switzerland.
Wood, R., Keen, A., Mitchell, J. & Gregory, J. 1999 Changing spatial structure of the thermohaline circulation in response to atmospheric CO_2 forcing in a climate model. *Nature*, **399**, 572–575.

14

Cloud–radiation interactions and their uncertainty in climate models

ADRIAN M. TOMPKINS AND FRANCESCA DI GIUSEPPE

Clouds significantly modulate the radiation budget, but climate models are still integrated on grids too coarse to explicitly represent motions on the scale of clouds. This chapter illustrates how clouds interact with radiation and atmospheric motions. It also discusses how the physics of cloud–radiation interactions is portrayed in state-of-the-art climate models, and summarises future perspectives arising from new pertinent observations.

14.1 Introduction

Uncertainty in climate model integrations derives in part from their uncertainty in representing physical processes and their interactions. Despite ever-growing computing resources, climate models are still integrated on grids too coarse to explicitly represent motions on the scale of clouds, which thus have to be represented by simple models (termed 'parameterisations'). These define cloud quantities such as liquid water mass or cloud cover in terms of the climate model grid-resolved variables and often contain empiricisms that are difficult to justify from observations. Instead, clouds in climate models are mostly judged in terms of their bulk properties compared to global mean retrievals from satellites. Given that the first accurate satellite retrievals of a key bulk cloud quantity such as the cloud ice and its vertical distribution were only recently possible, it is hardly surprising that clouds were, and continue to be, assessed as one of the greatest uncertainties in climate modelling (Cess *et al.* 1990; Stephens 2005; Randall *et al.* 2007).

The task of representing clouds accurately in climate models is critical not only for their role in the hydrological cycle, but also due to their significant modulation of the radiation budget. Clouds absorb and emit radiation efficiently in the infrared spectrum. They thus act to warm the surface and the atmosphere below by absorbing radiation that would otherwise escape to higher levels, while emitting longwave radiation at a lower temperature both downwards and upwards. Clouds absorb and scatter short-wave radiation, and thus also act to cool lower-albedo surfaces. For low clouds, such as the stratocumulus decks, the reflective impact in the SW dominates and the net effect at the Earth's surface is cooling. High clouds have a larger effect in the long wave and can thus warm the Earth's surface.

Viewing clouds as static objects that act to warm or cool the surface is an over-simplification, however. The occurrence of clouds is directly linked to the dynamics of the atmosphere, with positive vertical motions leading to the requisite saturation of the atmosphere required to form clouds. Conversely, their impact on the vertical structure of radiative heating and consequently the atmospheric static stability implies that clouds themselves have the potential to affect atmospheric dynamics on spatial scales ranging from the cloud-scale to the meso-scale and beyond. Thus the two-way interaction between clouds and radiation through atmospheric dynamics is key to the simulation of the present-day climate in global climate models and to their sensitivity to future CO_2-enriched regimes. Indeed, in a recent critical review, Stephens (2005) emphasises how cloud–radiation–dynamical interactions are a key but oft-neglected component in our analysis of the role of clouds in future climate regimes.

The following text aims to illustrate how clouds interact with radiation and atmospheric motions and to portray the current state-of-the-art in cloud–radiation interactions as represented in climate models. After introducing the basic theory of cloud–radiation interaction, and the relevant assumptions made in current climate models, the role of such interactions in a number of atmospheric dynamic phenomena is discussed, ranging from cirrus and stratocumulus cloud systems to large-scale dynamic phenomena such as the Walker circulation and the Madden–Julian Oscillation. The current net assessment of clouds in general circulation models (GCMs) will be concluded with an outlook for future developments in cloud–radiation interactions in GCMs.

14.2 Physics of cloud–radiation interactions

14.2.1 Theory

In formally quantifying the interaction of clouds with atmospheric radiation one is faced with the twofold problem of characterising the cloud microphysical properties over a wide range of spatial scales in addition to solving a complex radiative transfer problem of a three-dimensional (3D) arbitrarily shaped element in a three-dimensional electromagnetic field. In general terms, the propagation of radiation through a medium is characterised by absorption, emission and scattering processes whose quantification is via the solution of the full radiative transfer equation which can be written in an abbreviated form as:

$$\mu\frac{\partial I}{\partial z} + (1-\mu^2)^{1/2}\left(\frac{\partial I}{\partial x}\cos\phi + \frac{\partial I}{\partial y}\sin\phi\right)$$
$$= -kI + k\omega_0 \int_0^{2\pi}\int_{-1}^{+1} \mathbb{P}(\mu,\phi:\mu',\phi')I(\mu',\phi')\mu' \mathrm{d}\mu' \mathrm{d}\phi'$$
$$+ S_0\mu_0\omega_0 k\mathbb{P}(\mu,\phi:\mu_0,\phi_0)\mathrm{e}^{-\int_{\vec{r}}^{\infty} k(\vec{r}')\mathrm{d}\vec{r}'}. \tag{14.1}$$

Here all parameters are functions of the frequency ν and the position $\vec{r}$. In a 3D formulation both the radiance field $I(\vec{r}, \mu, \phi)$ and the sources of radiation [last term in equation (14.1)] depend on angle and location. The parameter S_0 is the solar flux associated with a collimated beam incident on the cloud top having direction (μ_0, ϕ_0), μ is the cosine of the zenith angle, ϕ the azimuthal angle and $\vec{r} \equiv (x, y, z)$ is the vector that identifies the interaction point. The term $\tau = \int_{\vec{r}}^{\infty} k(\vec{r}')\mathrm{d}\vec{r}'$ represents the medium optical thickness ($k = \mathrm{e}^{-\tau}$ is the linear extinction coefficient) and includes the contribution through scattering (τ_s) and absorption (τ_a) by the cloud particles as well as the absorption due to the atmospheric gases (τ_g), and ω_0 is the single scattering albedo defined as the ratio of the scattering optical thickness and the total optical thickness (i.e. $\omega_0 = \frac{\tau_s}{\tau}$). Thus, $\omega_0 = 1$ for a non-absorbing cloud and $\omega_0 = 0$ when scattering is negligible, while $1 - \omega_0$ is the fraction of the incident radiation absorbed by a particle. The scattering phase function $\mathbb{P}(\mu, \phi: \mu', \phi')$ characterises the angular distribution of the scattered radiation field.

In the approximation of randomly oriented particles (generally true for water droplets but not for ice crystals), the phase function depends only on the scattering angle Θ that can be represented in terms of the direction of the incident radiation μ', ϕ' and the direction of the scattered radiation μ, ϕ as $\cos\Theta = \mu\mu' + \sqrt{1-\mu^2}\sqrt{1-\mu'^2}\cos(\phi - \phi')$. It is useful to define the asymmetry parameter $g = 2\pi\int_0^{\pi} \mathbb{P}(\vec{r}, \Theta)\sin\Theta\cos\Theta \mathrm{d}\Theta$ which ranges from -1 for complete backscattering, through 0 for isotropic scattering, to $+1$ for complete forward scattering. This parameter is frequently used in analytical formulae to approximate the phase functions $\mathbb{P}(\vec{r}, \Theta)$ (Henyey & Greenstein, 1941) when only the hemispheric integrated quantities are required (Hansen 1969).

Despite the complexity of equation (1.4.1), once a suitable description of the cloud optical properties and cloud geometry is available, a solution is analytically/ numerically possible. Most existing three-dimensional radiation models used to achieve this task have been based on Monte Carlo or spherical harmonic approaches. Monte Carlo methods explicitly model the passage of individual photons through the medium (Pincus, 2006).[1] The advantage is that energy conservation is guaranteed and that error estimates are derivable, whereas the drawback is that more complex statistics, such as radiances, three-dimensional fluxes in domain subregions, or fluxes integrated in any direction are cumbersome and computationally time-consuming to obtain. This is where the strength of the spherical harmonics approach (e.g. the SHDOM code of Evans 1998) lies, which explicitly solves the radiative transfer equation (14.1) and offers the possibility of simultaneous calculation of these complex diagnostics, but with an inferior achievable accuracy.

For such a calculation both *cloud geometry* and *cloud optical properties* must be known. The heart of any radiative transfer solution in clouds rests therefore with the suitable description of the cloud k (or the equivalent τ), ω_0 and $\mathbb{P}$ in terms of some appropriate microphysical variable such as the cloud droplet particle size distribution and the cloud water amount in addition to knowledge of the horizontal and vertical geometry of the cloud. The specification (parameterisation) method of these two aspects in climate models ultimately identifies the principal sources of biases in the final columnar energy budget calculation. In this section 'cloud geometry' refers to all aspects related to cloud geometry specification including the methods employed by GCMs to represent cloud unresolved subgrid-scale variability, while 'optical properties' include those techniques which relate microphysical cloud properties (such as water phase, particle size and shape distribution) to cloud bulk quantities. In the following two subsections these aspects will be analysed in turn.

14.2.2 Parameterising cloud geometry

The formal solution of (14.1) is too complex and time-consuming for operational use in any GCM. In any case, a full 3D solution is not necessary for modelling radiation transport from one GCM column to its neighbour, since this transport plays a negligible role with the horizontal grid resolutions currently employed by GCMs (e.g. Di Giuseppe & Tompkins, 2003b). Expending resources modelling 3D radiative interaction with a cloud system occurring on the *subgrid scale* is a superfluous exercise if the geometry (overlap, arrangement and so on) of that cloud system is uncertain. It is therefore common to reduce the complexity of the task; the two-stream approximation usually offers such a simplification and is employed in most GCMS (for example Geleyn & Hollingsworth 1979; Fouquart 1987; Chou *et al.* 1991; Fu & Liou 1992; Edwards & Slingo 1996; Mlawer *et al.* 1997, among others). The solution of equation (14.1) is performed for the hemispheric integrated

[1] The community Monte-Carlo Code has been developed in the framework of the 'Intercomparison 3D radiation codes' (I3RC) project and is freely available from http://i3rc.gsfc.nasa.gov/

upward ($F^{\uparrow}$) and downward ($F^{\downarrow}$) fluxes at plane-parallel boundaries of a stratified atmosphere. The 3D directionality of the solution is therefore lost and (14.1) simplifies to this system of integral–differential equations:

$$\frac{dF^{\uparrow}}{d\tau} = \alpha_1 F^{\uparrow} - \alpha_2 F^{\downarrow} - \begin{cases} (\alpha_1 - \alpha_2)\pi B & \text{infra-red} \\ \alpha_3 F^0 \mu_0 & \text{solar} \end{cases}, \tag{14.2}$$

$$\frac{dF^{\downarrow}}{d\tau} = \alpha_2 F^{\uparrow} - \alpha_1 F^{\downarrow} + \begin{cases} (\alpha_1 - \alpha_2)\pi B & \text{infra-red} \\ \alpha_4 F^0 \mu_0 & \text{solar} \end{cases}, \tag{14.3}$$

where the α_i coefficients depend on ω_0, on the form chosen for the phase function $\mathbb{P}$ and the solution methodology (e.g. Gauss-quadrature; Zdunkowski *et al.* 1982).

In the plane-parallel approach, not only is the atmosphere vertically stratified but clouds are also reduced to homogeneous slabs. Moreover, despite using varying vertical resolutions, most GCMs still assume clouds fill the model grid box entirely in the vertical direction (e.g. Brooks *et al.* 2005). The geometry of a cloudy atmosphere is therefore defined in terms of the horizontal cloud coverage at each height, and how these clouds overlap in the vertical.

When a one-dimensional (1D) radiative calculation is performed, besides the uncertainties which affect the estimation of optical parameters (addressed in the following subsection), several geometrical characteristics are thus neglected, such as the internal optical inhomogeneities of the clouds themselves, the interactions between separate clouds and the vertical cloud structure. Each of these effects can introduce a bias in the final heating rates.

To quantify the radiative biases produced by unresolved *spatial variability* of clouds it is common practice to perform a series of radiative transfer calculations through a given cloud field, for which subgrid scale inhomogeneities are progressively neglected (e.g. Cahalan *et al.* 1994a). Three methods are usually employed. The benchmark is a full 3D calculation that models multi-directional photon transport through a high-resolution rendition of the cloud. The second methodology still describes the cloud using a high-resolution 3D field, but the radiative calculation is performed column by column using a plane-parallel calculation. Thus cloud heterogeneity is included, but horizontal photon transport is ignored. Lastly, the plane-parallel (PP) method commonly employed in GCMs averages the cloud elements to horizontally homogeneous slabs over the spatial scales of a GCM grid box, and thus additionally neglects the impact of sub-cloud variability.[2] Figure 14.1 summarises the interaction between photons and a cloud field as it is seen using a 3D representation and the independent pixel approximation

[2] Note that GCMs can have implicit assumptions concerning subgrid-scale variability built into other parameterisations. Examples are cloud cover based on relative humidity (Sundqvist 1978) or microphysical thresholds for rainfall generation or evaporation (Kessler 1969).

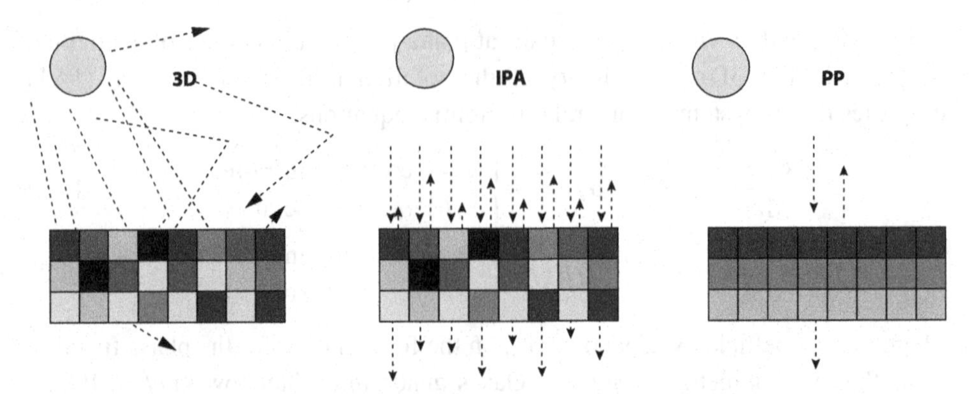

Figure 14.1 Interaction between solar radiation and a cloud field. In the three-dimensional solution photons can be transported across vertical columns. In the IPA and PP approximations this horizontal transport is inhibited.

(IPA) and PP approximations. Regions of different τ are indicated with different shadowing.

In summary, the difference between the IPA (sometimes also referred to as the Independent Column Approximation or ICA) and PP calculations describes the influence of horizontal in-cloud inhomogeneity (the PP bias) while the 3D/IPA contrast assesses the effect of horizontal photon transport due to the geometrical organisation of the cloud field (the IPA bias). From the analysis of the two different biases it is possible to quantify in which metrics the mean radiative properties of the atmosphere are affected by the cloud unresolved geometrical arrangement and/or their optical internal inhomogeneities. Figure 14.2 is an example of how the independent pixel concept is applied to clouds. In the plane-parallel approximation the reflection (R) is calculated for the mean cloud optical thickness ($\overline{\tau}$), while in the IPA the cloud reflection is obtained by averaging over several independent columns of different optical thickness. Since R is a convex function of the optical thickness, the inequality $R(\overline{\tau}) \geq \overline{R(\tau)}$ is always satisfied and the reflection of a horizontally inhomogeneous cloud is always smaller than the reflection of its homogeneous counterpart, provided that they have the same $\overline{\tau}$. Since the transmission (T) is a concave function of the optical thickness, the effect of the IPA is reversed with respect to the reflection and the inequality reads $T(\overline{\tau}) \leq \overline{T(\tau)}$. Since the absorption ($A$) is a linear combination of a convex and a concave function, its functional dependency on the optical thickness cannot be established a priori and the sign of the bias produced by the IPA calculation is not predictable.

There is general agreement that for overcast single-layer cloud systems such as stratocumulus, the IPA bias is much smaller than the PP bias (Cahalan *et al.* 1994a; Marshak *et al.* 1997; Di Giuseppe & Tompkins 2003a). Cahalan *et al.* (1994a) and Barker & Davies (1992b) claimed that the nonlinearity of albedo with increasing liquid water path results in substantial PP biases of around 15%. Cahalan *et al.*

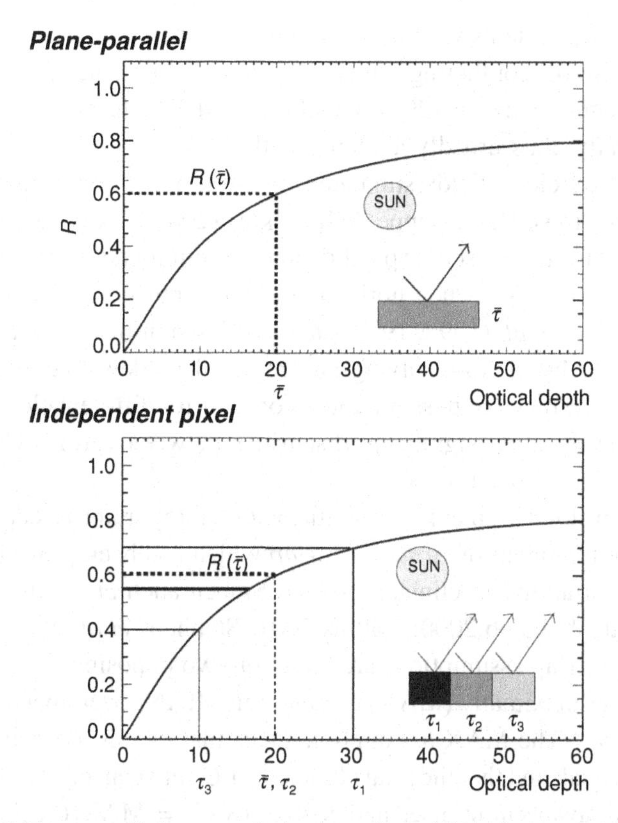

Figure 14.2 Plane-parallel approach (top panel) and IPA (bottom panel) for the reflection. The difference between $\overline{R(\tau)}$ and $R(\overline{\tau})$ constitutes the 'PP bias'. Original figure source: Zuidema (1998).

(1994a) even reported biases approaching 30% for very skewed liquid water distributions, while both Cahalan *et al.* (1994a) and Chambers *et al.* (1997b) found that in the same overcast conditions, IPA albedo biases are 1%, one order of magnitude smaller than the PP bias. Once such shallow clouds become broken this does not appear to be the case (McKee & Cox 1974; Aida 1977; Barker & Davies 1992a; O'Hirok & Gautier 1998) and the horizontal transport of radiation becomes an important source of error. Di Giuseppe & Tompkins 2003a) indicated that 3D effects in broken stratocumulus of 80% cover became relevant when the dominant horizontal spatial scale of the cloud fell below 5 km. This implies that GCMs that suffer from PP biases due to the neglected horizontal subgrid-scale variability of cloud water for such scenes could be making additional errors due to these 3D effects. Barker & Li (1997) also emphasised the role of 3D effects over shorter horizontal scales when discussing the purported cloud anomalous absorption effect (Cess *et al.* 1995; Byrne *et al.* 1996; Valero *et al.* 1997; Barker & Li 1997). Significant biases are found analysing cloud fields reconstructed from remote sensing techniques (Chambers *et al.* 1997a; Zuidema & Evans 1998).

For other cloud systems with more complex geometry such as deep convective regimes there is still conflicting opinion. Using satellite data to reconstruct such cloud fields, Vogelmann *et al.* (2001) concluded that 3D effects could be significant, especially locally, thus broadly agreeing with O'Hirok & Gautier (1998). Other studies have used cloud fields simulated by cloud resolving models (CRMs) in radiative calculations. Di Giuseppe & Tompkins (2003b) examined a single scene for a situation of unorganised tropical deep convection and documented biases in reflectance of up to 16% when horizontal photon transport is neglected. On the other hand, Barker *et al.* (1998, 1999) claimed less significant biases, using cloud scenes generated by cloud-resolving and regional models. Barker *et al.* (2003) portrayed the wide range of possible causes of error for different cloud regimes and emphasised the resulting significant disparities between various GCM radiation schemes applied to a given scene.

In addition to cloud horizontal variability another important aspect of 1D approximations is the treatment of *vertical overlap* which has been proved as important as horizontal variability in climate models (Stubenrauch *et al.* 1997; Chen *et al.* 2000; Morcrette & Jakob 2000; Collins 2001; Stephens *et al.* 2004; Di Giuseppe 2005). Cloud overlap assumptions are based on two opposing views, namely that clouds are either maximally (MAX) or randomly (RAN) overlapped (Morcrette & Fouquart, 1986).[3] The MAX assumption states that clouds are maximally correlated in the vertical, and that the total cloud cover C between any two layers (i and j, respectively) is always minimised and is given by $C_{i,j} = \mathrm{MAX}(C_i, C_j)$. The random overlap assumption on the other hand assumes no correlation exists between cloud layers, and C is given by $C_{i,j} = C_i + C_j - C_i C_j$. It has been frequently pointed out that the random overlap assumption suffers from an inherent and significant vertical resolution dependency (e.g. Bergman & Rasch 2002); the scheme essentially assumes a decorrelation length scale for clouds that is directly related to the vertical grid resolution.

Atmospheric models commonly employ parameterisations which use a mix of the MAX and RAN overlap approaches. Clouds in adjacent vertical levels are likely to form part of the same vertical cloud element, and therefore can be reasonably treated using maximum overlap, while clouds separated by a clear layer are less likely to be spatially correlated and may be better described by the random overlap approach. The combination of these two assumptions leads to the maximum–random (MAX–RAN) overlap scheme (e.g. Geleyn & Hollingsworth 1979). The

[3] There is also a third basic assumption described in Morcrette & Fouquart (1986) of minimal overlap, which would be relevant for describing overlap of regular cloud fields in which the elements are more evenly spaced than a random Poisson distribution. This invokes the concept of mutual exclusivity, and while there is some limited evidence from modelling (Cohen & Craig 2004) and observations (Zhu *et al.* 1992; Nair *et al.* 1998) that this can arise for certain cloud types over specific spatial scales this assumption has generally been discredited and is no longer in general use in models; it will not be discussed further here. Other simpler schemes that ignore subgrid cloud issues altogether, such as described by Stubenrauch *et al.* (1997), are also disregarded.

MAX–RAN assumption suffers much less from sensitivity to vertical resolution (Bergman & Rasch 2002) and it is commonly used in GCMs (e.g. Morcrette & Fouquart 1986; Stubenrauch *et al.* 1997; Chen *et al.* 2000; Collins 2001).

A modification to the MAX–RAN scheme was proposed by Hogan & Illingworth (2000) using retrieval data from the Chilbolton radar in the UK. They realised that vertically coherent cloud entities will decorrelate in height in the presence of wind shear (see also Lin & Mapes 2004) and that the maximum overlap assumption could thus underestimate cloud fraction in deep cloud systems. Hogan & Illingworth (2000) derived a length scale L_0 from radar data which described the rate at which a *continuous* cloud block exponentially decorrelates in the vertical (we therefore christen this approach the EXP–RAN scheme).

The EXP–RAN scheme was further generalised by Tompkins & Di Giuseppe (2006) to make the decorrelation additionally a function of the solar zenith angle (SZA), to also account for the changes in apparent cloud overlap that arise as the Sun descends (this scheme was termed the EXP–SZA–RAN overlap scheme). To illustrate the impact of different cloud overlap parameterisations on the top-of-atmosphere (TOA) energy balance, Figure 14.3 shows results from the full ECMWF global forecast model for a set of short three-hour forecasts. It is seen that locally the overlap parameterisation can impact TOA shortwave budgets in excess of 100 W m^{-2}, and that both the generalisations of Hogan & Illingworth (2000) and Tompkins & Di Giuseppe (2006) have similar order-of-magnitude impacts. These schemes were justified in terms of ground-based radar data from Europe, and the newly available Cloudsat satellite data (Stephens *et al.* 2002) will be instrumental in defining appropriate decorrelation length scales for the successful use of these approaches in global models.

14.2.3 Parameterising cloud optical properties

The formal expressions for the α_i parameters in equations (14.2)–(14.3) are given by the formulae:

$$\alpha_1 = D(1 - \omega_0\alpha), \tag{14.4}$$

$$\alpha_2 = D\omega_0\beta, \tag{14.5}$$

$$\alpha_3 = \omega_0\beta_0, \tag{14.6}$$

$$\alpha_4 = \omega_0(1 - \beta_0), \tag{14.7}$$

where D is the diffusivity factor which quantifies the multiple scattering contribution to the total radiation beam (D is given a value of 7/4 for solar wavelengths and 1.66 for longwave radiation) and α and β are the fractional forward and backward scattering coefficients for diffuse radiation. The coefficients α and β are usually parameterised in terms of the asymmetry parameter g of the phase function $\mathbb{P}$ and

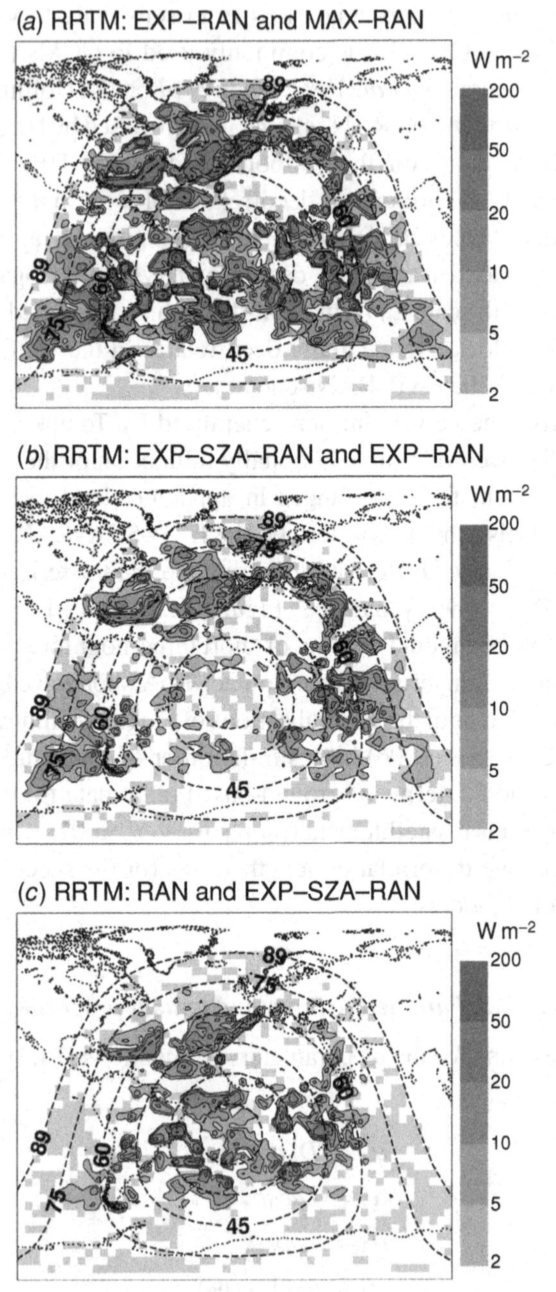

Figure 14.3 Difference in TOA net solar radiation (filled contours, units W m^{-2}) averaged over the first three hours of two forecasts using, respectively, the overlap assumptions of (*a*) MAX–RAN and EXP–RAN, (*b*) EXP–RAN and EXP–SZA–RAN and (*c*) EXP–SZA–RAN and RAN. The Mlawer & Clough (1997) Integrated Forecast System (IFS) shortwave parameterisation is used for the calculations. The dashed lines are contours of solar zenith angle and the light shaded squares indicate regions where cloud cover lies in the range of 25 to 75%; this is masked when the Sun is below the horizon.

the single scattering albedo ω_0. These parameters, together with the optical thickness τ, contain all the information related to the microphysical scattering/absorbing properties of cloud particles. Their parameterisation in terms of cloud microphysical quantities (such as the cloud water amount and the particle size distribution) is thus key for the calculation of heating rate profiles.

For spherical water cloud droplets, Mie theory applies. For single scattering events it predicts an intensely strong peak in the forward direction with special rainbow and glory effects in the backward direction (Liou 1980). In general for a polydispersion of water droplets both $\tau_s + \tau_a$ and g are relatively smooth functions of wavelength and thus can be specified over broad spectral intervals. The asymmetry parameter is usually within the range of 0.7 to 0.9, although more extreme values may be found for certain ice crystal habits (Macke *et al.* 1996).

On the contrary ω_0 varies strongly with wavelength, with $0.9 < \omega_0 < 1$ for solar radiation (0.2–3 μm) and $\omega_0 < 0.6$ in the longwave (>3 μm). Therefore in the shortwave scattering makes the greatest contribution to the extinction, while in the longwave the absorption is also relevant. In the shortwave, the scattering optical thickness for droplets is usually the most important parameter required. The difficulties in its evaluation lie in the specification of a cloud droplet size distribution $n(r)$ for the cloud in question. Analytical formulations have been sought for $n(r)$ and a number of 'standard' distributions have arisen over the years, each suited to describing a particular type of cloud (Diermendjian 1969; Hansen & Travis 1974), such as the standard gamma distribution given by

$$n(r) = \frac{(ab)^{-(1-2b)/b}}{\Gamma[(1-2b)/b]} r^{(1-3b)/b} e^{-r/ab}, \tag{14.8}$$

where $\int_r^{r+dr} n(r)dr$ is the number of particles per unit volume with between r and $r + dr$, Γ is the Gamma function, and a and b are adjustable parameters which control the mean and width of the distribution. Another distribution that is frequently used is based on the log-normal function (Hansen & Travis 1974)

$$n(r) = \frac{1}{\sqrt{2\pi}\sigma_g} \frac{1}{r} \exp\left[\frac{-(\ln r - \ln r_g)^2}{2\sigma_g^2}\right], \tag{14.9}$$

with size parameters r_g and σ_g. This distribution has a maximum just short of $r = r_g$. Each of these distributions is normalised such that

$$N = \int_0^\infty n(r)dr = 1. \tag{14.10}$$

Often it is advantageous to express the size parameters of the different distributions in terms of two common parameters. This is useful for intercomparisons between size distributions. The area-weighted mean, or effective radius r_{eff}, is used since larger particles tend to be more efficient scatterers, and similarly for the

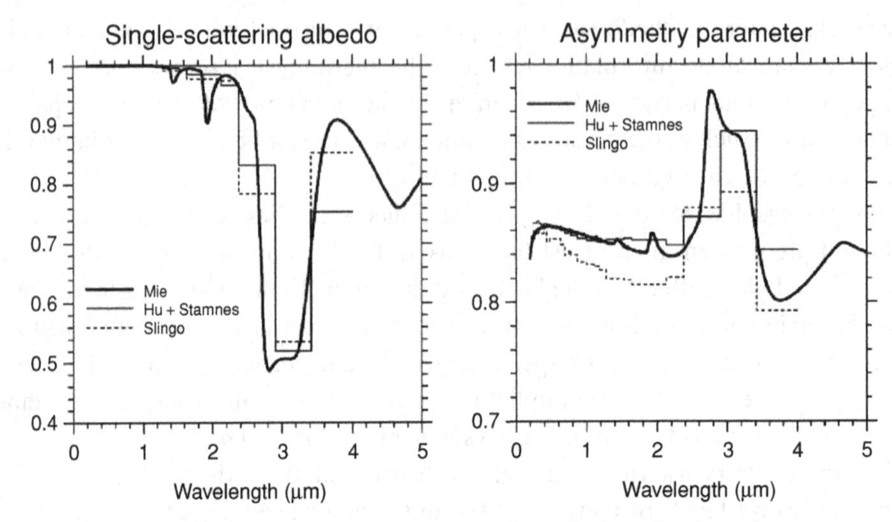

Figure 14.4 Spectral variation of single-scattering albedo and asymmetry parameter of water cloud with specified effective radius and liquid water content (source: Barker *et al.* 2003).

measure of the width of a distribution, the effective variance (v_{eff}) is used

$$r_{\text{eff}} = \frac{\int_{r_1}^{r_2} r^3 n(r) \text{d}r}{\int_{r_1}^{r_2} r^2 n(r) \text{d}r}, \tag{14.11}$$

$$v_{\text{eff}} = \frac{\int_{r_1}^{r_2} (r - r_{\text{eff}})^2 r^2 n(r) \text{d}r}{r_{\text{eff}}^2 \int_{r_1}^{r_2} r^2 n(r) \text{d}r}. \tag{14.12}$$

Examples of the spectral distributions of the single-scattering albedo and g are shown in Fig. 14.4 computed by Mie theory (Wiscombe 1980) using refractive indices from Segelstein (1981) for a gamma droplet size distribution with r_{eff} set to 10 μm and a value of 0.1 μm for v_{eff}. Also shown are estimates from the Slingo (1989) and Hu & Stamnes (1993) parameterisations at $r_{\text{eff}} = 10$ μm. Though differences are fairly minor for ω_0, the values adopted by Slingo (1989) for g are systematically lower by 0.02 across the entire visible and near IR, yielding clouds that are more reflective. A difference of 0.02 appears minimal, but a substantial reduction of r_{eff} to 6 μm is required to bring the estimate of g by Hu & Stamnes (1993) into line with that of Slingo (1989). This reduction of 4 μm in r_{eff} exceeds the differences estimated to have occurred in the most polluted regions over the industrial era due to increased sulphate aerosol concentrations (e.g. Lohmann & Feichter 1997).

The sensitivity of shortwave radiation calculations to *water* cloud optical parameters is shown in Fig. 14.5, where the reflection and absorption at a zenith angle of 60° as a function of the cloud liquid water path are reported. Two parameterisations are compared: the dotted curve is from Stephens (1984) while the solid lines are

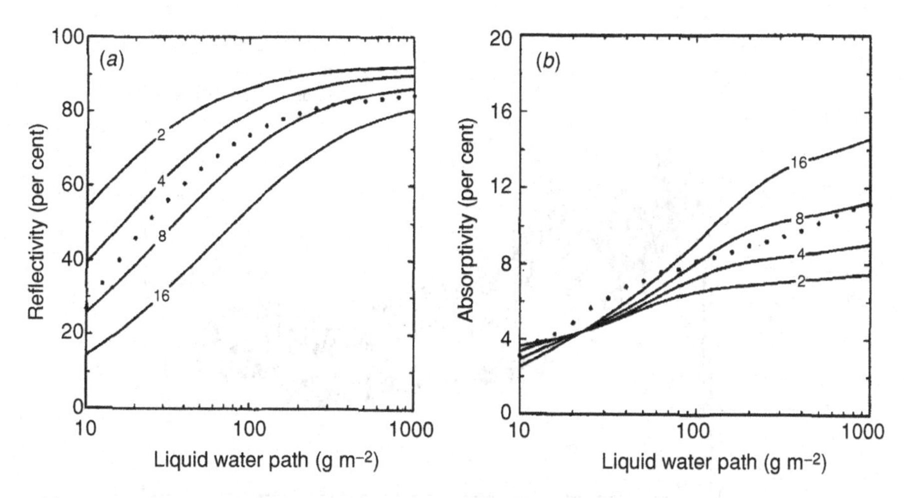

Figure 14.5 (*a*) Reflection and (*b*) absorption of water cloud at a zenith angle of 60° as a function of the cloud liquid water path. The solid lines are from Slingo (1989) for various values of the effective radius and the dotted line is from Stephens (1984). (Adapted from Slingo 1989).

from Slingo (1989) for various values of the effective radius. Reflection is very sensitive to r_{eff}, since smaller particles are more efficient scatterers, particularly in semi-opaque clouds. Unfortunately, aeroplane observations and satellite retrievals have shown that r_{eff} and the liquid water path can be quite uncertain and have substantial horizontal variability. Figure 14.6 shows measurements from observations on 10 July 1987 conducted with two aeroplanes within and above an extensive stratocumulus deck, taken during the First International Satellite Cloud Climatology Project (ISCCP) Regional Experiment (FIRE) campaign (Nakajima *et al.* 1991). It should be noted that even if the absorption of solar radiation by water droplets is often considered negligible in comparison to water vapour absorption, it can be as much as 10% of the latter. Stephens (1984) showed that overlap of the droplet and molecular absorptions in a consistent fashion is advisable.

Modelling radiative interaction with ice clouds involves a number of further complications related to their poorly understood microphysics. First, the ice nucleation process can occur by a variety of heterogeneous ice nucleation mechanisms (Pruppacher & Klett 1997) and thus the ice crystal number density is determined by the profligacy of suitable aerosols that can act as ice nuclei, which is a strongly nonlinear inverse function of temperature (e.g. Meyers *et al.* 1992). The matter is complicated further by the fact that it is not necessarily the current local temperature that determines ice crystal number, but the temperature history of the cloud volume during the period that the ice nucleation process was occurring. This implies that a diagnostic ice crystal number–temperature relationship may not always be appropriate.

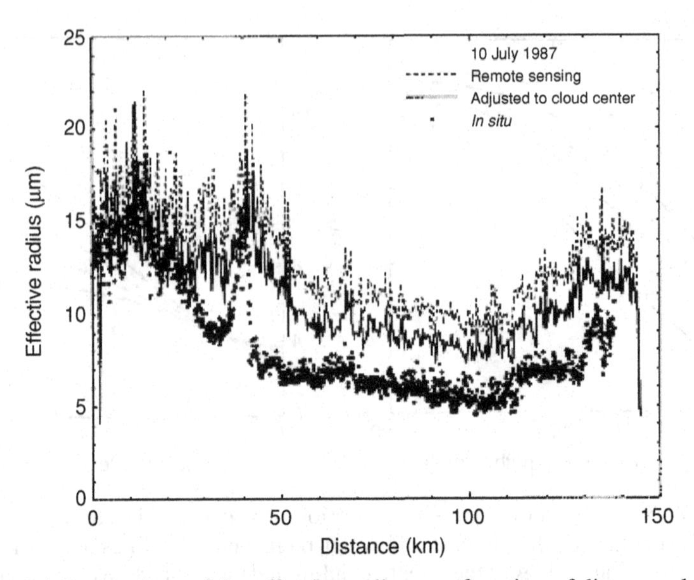

Figure 14.6 Comparison of the effective radius as a function of distance along the nadir track of the ER-2 aircraft as derived from the remote sensing (dashed line) and *in situ* measurements (solid circles). The solid line represents the expected values of effective radius at the geometric centre of the cloud layer. (Source: Nakajima *et al.* 1991.)

At very cold temperature below −38 °C, homogeneous ice nucleation can occur. This generally produces ice crystal number density orders of magnitude larger than the heterogeneous process (Demott *et al.* 2003), but which sensitively depends on time (measured in seconds and therefore much shorter than a typical GCM time step) that the relative humidity exceeds the critical threshold for nucleation to occur. Thus knowledge of the updraught velocity on the cloud scale is crucial (e.g. Gierens 2003; Ren & MacKenzie 2005). Lohmann and Kärcher (Lohmann & Kärcher 2002; Kärcher & Lohmann 2002) attempt to approximate the nucleation event analytically, and represent the cloud-scale velocity based on an integration of the turbulence kinetic energy equation; however, the determination of the ice crystal number density remains a key uncertainty in climate models and an active area of research. The importance is demonstrated by the inclusion of temperature-dependent ice crystal number density relationships in GCMs. Kristjansson *et al.* (2000) found that including a temperature-dependent effective particle size parameterisation resulted in significant radiative warming in the upper tropical troposphere and at high latitudes compared to the case where a globally uniform single-size ice particle size distribution was used. In addition to the direct impact on cloud radiative properties, the assumed ice crystal size affects the ice cloud sedimentation rates, and consequentially the ice cloud height and optical depths and thus ultimately radiative properties. This microphysical effect was also highlighted as significant by Jakob (2002).

Another complication of ice clouds is that ice crystals tend to be non-spherical, taking on a range of complex habit types, again according to temperature and local supersaturation. The general parameterisation approach is to use a gamma size distribution where r_{eff} refers to the radii of spheres that are equivalent in volume to a non-spherical ice crystal. As such, r_{eff} does not directly relate to a measurable crystal dimension. A complication is that there may not exist a representative phase function for all habit types. Baran *et al.* (2001) claim that a smooth phase function associated with randomly oriented aggregates could be more generally appropriate. That said, observations have shown that ice can often be vertically stratified into homogeneous layers of a single particle shape. Scattering by ellipsoids and cylinders is discussed by van de Hulst (1980) and ellipsoids and spheroids by Holt (1982) and more recently by Macke *et al.* (1996). A collection of work on non-spherical non-homogeneous scatterers can be found in Mischenko *et al.* (2000). The lower values of the scattering function at angles normal to the incident radiation in the Mie theory with respect to random aggregates (Baran *et al.* 2001) could result in an underestimation of 3D radiative effects. Both Kristjansson *et al.* (2000) and Jakob (2002) may underestimate the impact of ice crystal size uncertainty since neither study included the temperature and supersaturation impacts on ice crystal habit, and the consequential impact on the scattering phase function.

14.2.4 Assessment of uncertainties

In the previous subsections three cloud parameters have been identified as major sources of potential uncertainty in the specification of the radiative effect of clouds: overlap of fractional cloud, horizontal variability of condensate and optical properties. If we identify the radiative sensitivity as the derivative of TOA fluxes with respect to changes in one of these cloud aspects, then the radiative uncertainty can be defined as this product of radiative sensitivity and the estimated uncertainty in that parameter. As in the use of this approach for calculating climate feedback parameters (e.g. Cess *et al.* 1990; Arking 1991), it is assumed that a linear approximation is valid and that the cloud parameters are non-interacting. Nevertheless, this approach can be used to equivalently compare the importance of various sources of uncertainty, and prioritise the consequential parameterisation efforts for climate models and the directions of future observational missions.

Using this approach Barker & Räisänen (2005) showed that uncertainties in TOA shortwave fluxes for cloud fraction overlap and horizontal variability are of similar magnitude averaged over the globe but display distinct spatial differences. For this exploratory study, they used a single day's worth of data produced by a

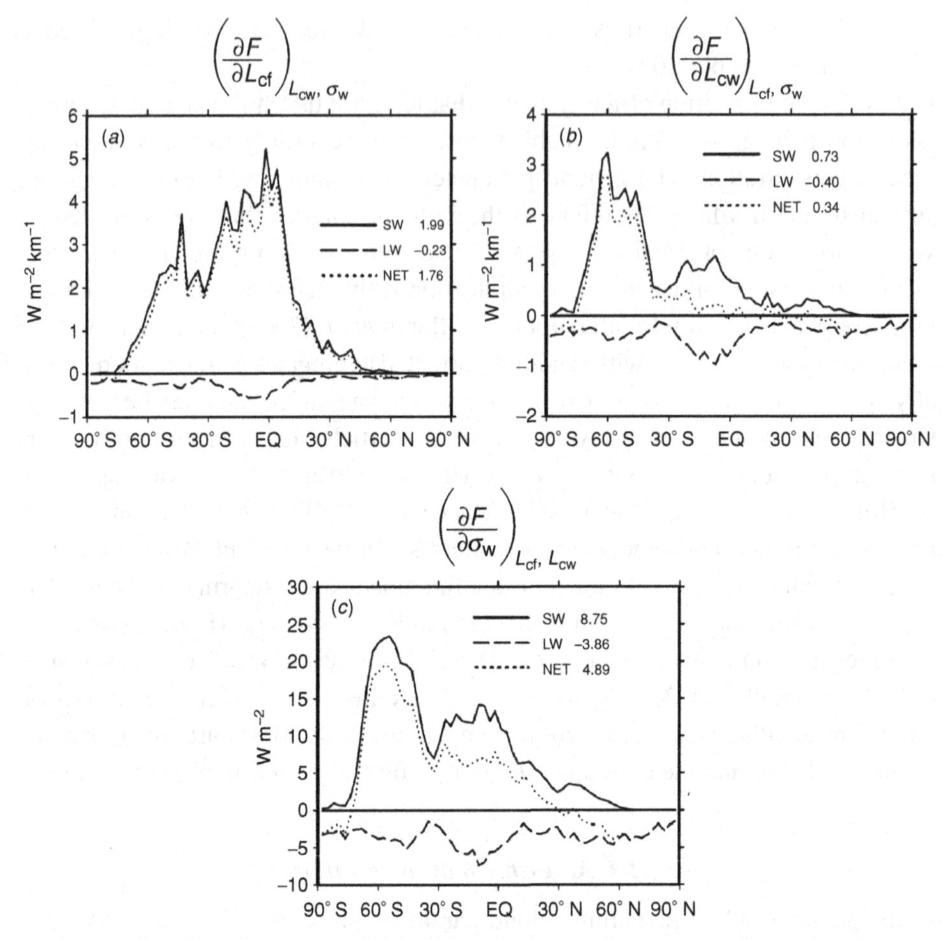

Figure 14.7 Zonal-mean sensitivities for shortwave, longwave and net radiative flux at TOA with respect to (*a*) vertical overlap of cloud fraction L_{cf}, (*b*) vertical overlap of condensate amount L_{cw} and (*c*) horizontal in-cloud inhomogeneities w. The partial derivatives are evaluated by systematically altering L_{cf}, L_{cw} and w around their actual values in the CSRM data set (see text for details). Corresponding global means are listed on each plot. (source: Barker & Räisänen, 2005.)

global array of cloud system resolving models (CSRM; Khairoutdinov & Randall 2001). Figure 14.7 shows zonal-mean sensitivities for shortwave, longwave and net radiative flux at TOA with respect to vertical overlap of cloud fraction L_{cf}, vertical overlap of condensate amount L_{cw} and horizontal in-cloud inhomogeneities σ_w.

For most regions, radiative sensitivities at solar wavelengths are several times larger than those at terrestrial wavelengths. Therefore when constructing parameterisations that account for unresolved interactions between clouds and radiation, attention should be placed on how well the parameterisation captures effects in the solar regime. Uncertainties associated with overlap of fractional cloud and horizontal variability are approximately as large as those associated with effective size of cloud particles (Barker & Räisänen 2005) at least for water clouds. Ice clouds are still very poorly known in terms of their microphysics and consequently the way they are represented in climate models. New observing missions such as CloudSat & CALIPSO (Stephens *et al.* 2002), which fly in formation with the National Aeronautics and Space Administration (NASA) Earth Observing System (EOS) PM (the Aqua Mission), are likely to provide valuable profiles of cloud ice and liquid content, optical depth, cloud type and aerosol properties. These observations, combined with wider swath radiometric data from EOS PM sensors, constitute a rich new source of information concerning the properties of clouds to improve parameterisation in global models and are discussed further in Section 14.4.

14.3 Cloud–radiation interactions in climate models

14.3.1 Validating cloud–radiation interactions

One methodology widely employed to validate cloud schemes in global climate models has been the examination of top-of-atmosphere net shortwave and outgoing longwave radiation budgets. Apart from errors in the basic thermodynamic state, errors in TOA radiative fluxes can derive from shortcomings in both the radiative assumptions employed and ascribed properties of clouds themselves. The latter can consist of inaccurate cloud liquid and ice water contents and misrepresented vertical distribution, cloud cover, cloud top heights, vertical overlap of cloud and incorrect treatment of sub-cloud variability; in other words a large variety of cloud macrophysical and microphysical properties.

Wielicki *et al.* (1995) pointed out that cloud and radiation observations from ground-based, *in-situ* or remote sensing platforms are inadequate to constrain the number of uncertainties in the parameterisation of cloud processes and their interaction with radiation. Nevertheless, the increasing availability of new remotely sensed cloud observations implies that basic cloud properties, such as liquid water content or total cloud cover, can be progressively constrained. This is illustrated by demonstrating the improvements in the representation of cloud–radiation parameterisations in the atmospheric model component of the European Centre for Medium Range Forecasts (ECMWF) forecast system.

Figure 14.8 shows the error in TOA net shortwave fluxes from two major releases of the model, compared to observations from CERES (Wielicki *et al.* 1996). The first version, titled cycle 23r4, was used operationally during 2000 and was the model version employed to generate the widely used ERA-40 reanalysis data set

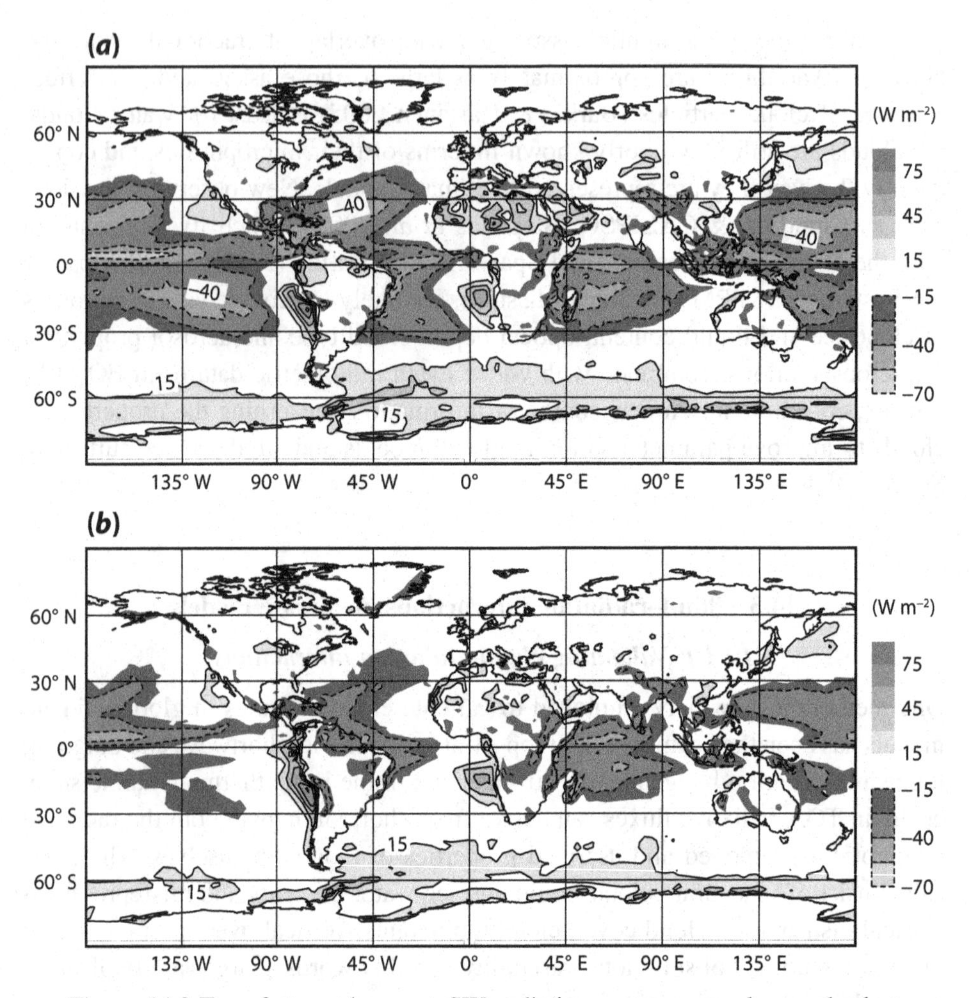

Figure 14.8 Top-of-atmosphere net SW radiation error averaged over the last 12 months of a three-member ensemble integration of (*a*) ECMWF model cycle 23r4 used in ERA-40 and system 2 of the seasonal forecasting suite and (*b*) 31r1, used in the system 3 of the seasonal forecast suite and EC-EARTH. The errors are with respect to observations from CERES. The mean and root-mean-square (RMS) errors between 50° N and 50° S are −12.6 and 24 in (*a*), and −6.7 and 17 in (*b*).

(Uppala *et al.* 2005), and was also used operationally for many years in the system 2 coupled seasonal forecast system (Palmer *et al.* 2004). The fields depict the mean fluxes averaged over three years of integrations forced by observed SSTs using an intermediate model resolution. This is contrasted to a second more recent model release, called cycle 31r1, which was operational during 2006 and 2007 and is used in the system 3 coupled seasonal forecast system, in addition to forming the atmospheric component of the new global climate model EC-EARTH.

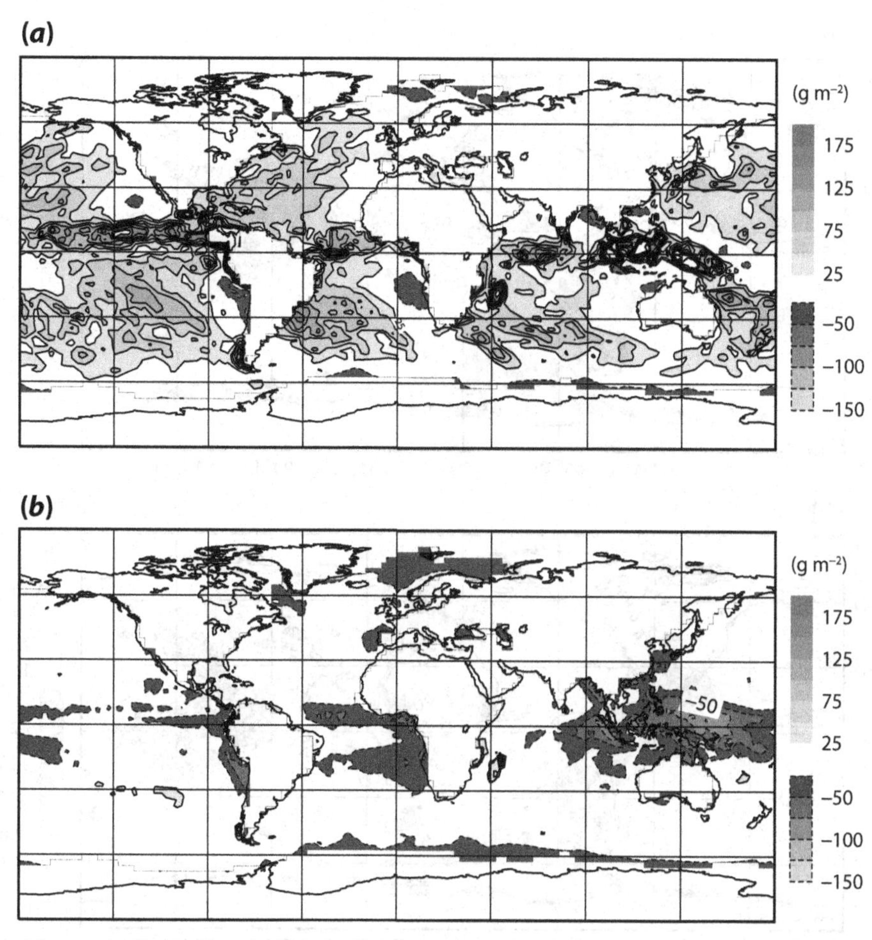

Figure 14.9 As Fig. 14.8 but showing errors in total column integrated cloud liquid water over oceans, compared to SSMI retrievals. Note that the severe over-prediction of cloud liquid water in the ERA-40 cycle 23r4 (*a*) has been (possibly over-) corrected by cycle 31r1 (*b*). The global mean and RMS errors are 38.5 and 65 in (*a*) and –13.3 and 25.3 (*b*).

Figure 14.8 reveals that both models are over-reflective in the tropical trade cumulus regimes, and under-reflective in the stratocumulus regions off the west coasts of the major continents of Africa and America. It is clear that the problem of the over-reflective trades is reduced with the newer model version. This improvement is directly attributable to the reduction in the total column liquid water (or liquid water path) that brought the model in much closer agreement to retrievals of this quantity derived from the Special Sensor Microwave/Imager (SSMI) satellite data (Wentz 1997, Fig. 14.9). This was confirmed using alternative retrievals from the Tropical Rainfall Measuring Mission (TRMM, not shown).

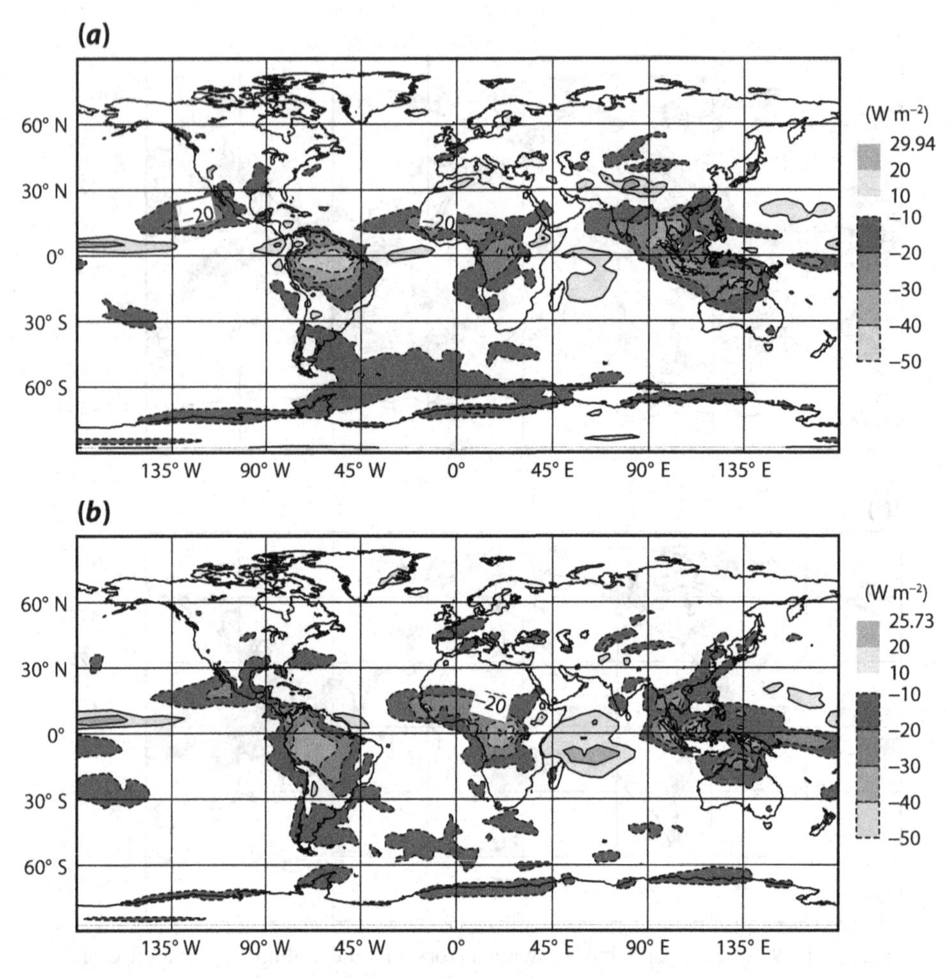

Figure 14.10 As Fig. 14.8 but for the TOA outgoing longwave radiation errors compared to CERES. The mean and RMS errors are –5.21 and 11 in (*a*) and −6.05 and 11 in (*b*).

The consistent net SW error in the stratocumulus regimes derives from a underestimation of cloud occurrence in these regions, which is evidenced by the comparison to the total column liquid, and confirmed by the underprediction of total cloud cover compared to satellite-derived products (not shown). While it is true that incorrect treatment of radiative properties of clouds and the vertical overlap are likely to contribute to the net shortwave error, the zero-order shortwave cloud radiative forcing in the stratocumulus regimes is determined by the total cloud cover and liquid water path, which can be reasonably constrained by remote observations.

The TOA infrared radiative forcing maps indicate that the maximum errors occur over the continents in the tropics (Fig 14.10), and thus are indicative of

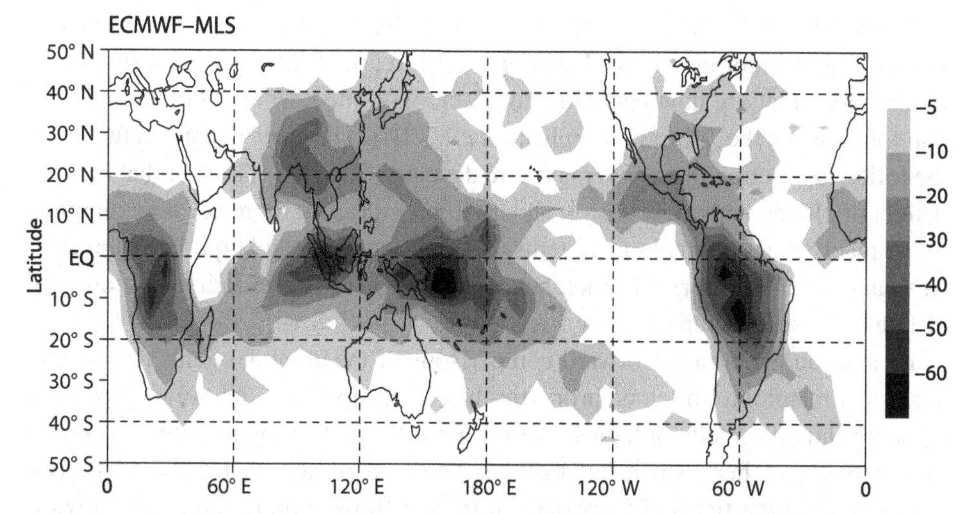

Figure 14.11 Per cent difference in 215 hPa ice water content of ECMWF analyses compared to MLS ice retrievals. The day 10 forecast ice was similar to the analysis for this pressure level. (Source: Li *et al.* 2007).

problems in the representation of land-based deep convection. Little evidence of an improvement is visible between the two model versions. In this instance, the uncertainty is larger, and the cause of the radiative biases more difficult to constrain since these cloud systems are tropospheric-deep, thus increasing the uncertainty due to cloud overlap treatment and also involving both the liquid, mixed and ice phase microphysics. Errors in cloud top height may also play a role. As pointed out in the previous section, the uncertainties due to the radiative properties of ice crystals are more substantial than those of liquid clouds. Moreover, until recently, even basic information such as the total ice water content of such clouds was highly uncertain. Recently progress has been with new microwave limb sounder (MLS) retrievals of upper tropospheric ice amounts (Li *et al.* 2005, 2007).

Figure 14.11 shows the mean percentage error in the analyses from 2004 and 2005 (a model version lying between seasonal forecasting system 2 and system 3) at 215 hPa, and reveals a strong underestimation in ice amounts in precisely the tropical continental regions that suffer from the strongest TOA infrared biases. Again, it should be emphasised that this does not discount other sources of radiative cloud parameterisation uncertainties as contributing to radiative budget errors. Indeed, Li *et al.* (2007) demonstrated that the model ice showed a significant decreasing trend during the forecast evolution at 147 hPa, indicating that the model also suffers an error in too-low cloud tops. Lacking the full vertical profile of ice mixing ratios from the MLS retrievals implies that it is not possible to discern what proportion of the infrared TOA budget error is due to this underprediction of ice.

The validation of cloud microphysical and cloud radiative property parameterisations is now entering a new phase for several reasons. The first consists of new observations that are now becoming available with the new cloud radar and lidar that form part of the 'A' train (Stephens *et al.* 2004). These are discussed further in the following section that concerns future developments, but it is clear that the new insight of clouds with these instruments will provide a far more accurate mapping of the macrostructure of cloud systems and their ice water and liquid distributions. Residual radiative errors will then be attributable to assumed radiative properties of liquid and ice particles.

In addition, the straightforward comparisons of mean model fields to satellite-derived climatologies of cloud properties has been supplemented over recent years by more intricate methodologies. We do not attempt to provide a full review of these techniques here, since this task has been performed admirably by the two extensive review articles of Stephens (2005) and Bony *et al.* (2006). These review articles highlight two methodologies in particular. The first involves passing model fields through a radiative transfer code to simulate direct satellite radiance or brightness temperature observations (known as forward modelling), rather than contrasting raw model fields with their equivalents that have been derived using retrieval techniques that contain numerous (and often hidden) assumptions. Early examples of this technique are given by Morcrette (1989, 1991) which has culminated in developments such as the widely used ISCCP cloud generator (which simulates the ISCCP cloud properties as per the retrievals of Rossow & Schiffer 1991), with which the cloud simulations of a variety of models were examined by Zhang *et al.* (2005). The concept is routinely extended to new cloud observing platforms such as Cloud Satellite (CloudSat; Haynes *et al.* 2007) and Cloud-Aerosol Lidar and Infrared Pathfinder Satellite Observation (CALIPSO; Chepfer *et al.* 2008).

The second methodology attempts to address one of the drawbacks of simple comparisons to annual or seasonal mean satellite-derived climatologies, namely that, while the spatial map of errors can be associated with certain geographically 'static' cloud regimes (i.e. lack of stratocumulus off the west coast of the major continents, or errors in the properties or occurrence of deep convection over the tropical land masses), the temporal averaging nevertheless results in the amalgamation of statistics for different dynamical regimes or different phases in the life cycle of regimes. The new methodology therefore invokes various compositing techniques to separate model error as a function of cloud and/or dynamical regimes, thus reducing the possibility of error cancellation between different cloud regimes.

One of the seminal studies to examine cloud radiative impact as a function of dynamical regime was conducted by Bony *et al.* (1997). Cloud radiative forcing

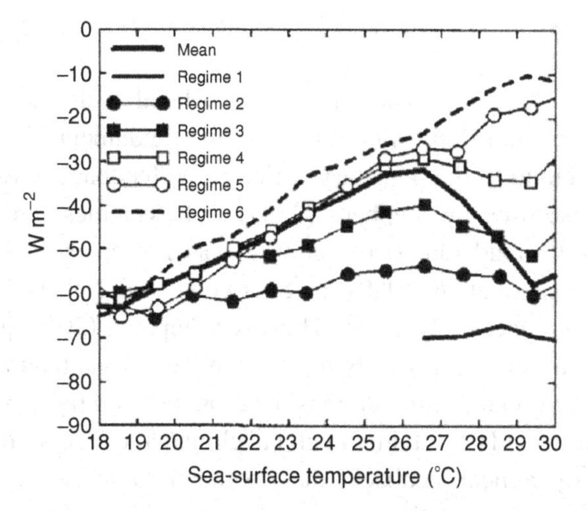

Figure 14.12 Outgoing longwave and net shortwave cloud forcing as a function of dynamical regime in the tropics, adapted from Bony *et al.* (1997). The regimes range from regime 1 of strong upward motion with $\omega < -40$ hPa d^{-1}, in 10 hPa divisions to the strong subsidence regime 6, for which $\omega > 30$ hPa d^{-1}. See Bony *et al.* (1997) or further details of the calculations.

was stratified in the tropics using the mid-tropospheric 500 hPa vertical velocity derived from analyses as a indication of the large-scale dynamical motion. Here we reproduce the shortwave cloud forcing from Bony *et al.* (1997) in Fig. 14.12, which clearly shows that shortwave forcing is a weak function of SST in dynamical ascending regimes. For descending regimes, which are often associated with stratocumulus forming in association with the resulting temperature inversion, there is a much stronger relationship between the shortwave cloud forcing and local SST.

This comparison does not identify the causality of these relationships, (warmer seas leading to thicker stratocumulus clouds, or thicker clouds shielding the surface and reducing SST), as pointed out by Bony *et al.* (1997), but the technique successfully separates the mean cloud properties into thermodynamic and dynamical effects. For example, it identifies that the increase in cloud shortwave forcing as a function of SST on average is a result of the increasing incidence of convergence and deep convection, and that changes in convective properties as a function of local thermodynamic conditions are negligible in relation. This was later confirmed in cloud-resolving model studies by Tompkins & Craig (1999), Wu & Moncrieff (1999) and Larson & Hartmann (2003). This compositing methodology has subsequently been widely adopted to investigate global model clouds (Klein & Jakob 1999; Tselioudis & Jakob 2002; Bony *et al.* 2004; Ringer & Allan 2004; Wyant *et al.* 2006) and was underlined as a key strategy for validating model clouds by Jakob (2003).

Work by Bony *et al.* (1997) and subsequent related studies led to the emphasis by Stephens (2005) and Bony *et al.* (2006) on the fundamental role that dynamics plays, from the cloud to the planetary scale, in the feedback between clouds and radiation. The assertion by Stephens (2005) that dynamics cannot be neglected when considering cloud radiative feedback was not a new concept; Slingo & Slingo (1988); Randall *et al.* (1989) and Sherwood *et al.* (1994), for example, also highlight the importance of dynamics. However, Stephens (2005) justifiably argued that the dynamical role has possibly received less attention than it deserves in the climate modelling literature. In our consideration of cloud radiative feedbacks and their representation and uncertainty in global climate models, we therefore identify cloud radiative dynamical feedbacks that operate across a range of spatial scales, and consider their representation in climate models.

14.3.2 Cloud radiative dynamical feedbacks

During the formative years of tropical dynamical research, the potential for radiative heating gradients to impact cloud dynamics was discounted on the grounds that the net radiative cooling rate was dominated locally by the heating rates associated with latent heat release in cloud systems. Gray & Jacobson Jr (1977) challenged this presiding view, proposing that horizontal gradients in infrared heating rates could drive mesoscale circulations. It is now accepted that, since clouds have the capacity to alter atmospheric radiative heating rates, the consequential changes to stability can lead to modification of dynamical circulations. These can be on the cloud scale, such as cloud-top cooling driven entrainment in stratocumulus layers, or can act over much larger scales, particularly in the tropics.

Dynamical effects resulting from cloud–radiation interactions acting on the cloud scale are not resolved by climate models, and thus if they are to be treated, must be parameterised within the cloud or subgrid-scale turbulence schemes. In contrast, dynamical feedback acting over larger scales of hundreds or thousands of kilometres may be represented explicitly in global climate models.

The purpose of this subsection is to discuss some examples of the roles of radiative impact of clouds on atmospheric dynamics, and to summarise their representation in climate models. As emphasised in the review of Stephens (2005), when global cloud radiative feedbacks are discussed in the context of present and future climates it is important to recall that the clouds are not static objects but rather proactive players in the atmospheric system that both respond to, and in turn partially determine, the atmospheric dynamical motions.

Dynamical radiative feedback on the cloud scale

One of the classic examples of cloud radiative dynamical feedback operating on the cloud scale is the case of radiatively driven turbulence playing a critical role in the evolution, both in terms of the growth and possible breakup, of stratocumulus cloud decks (Lilly 1968; Randall 1980; Hanson 1987). Stratocumulus cloud regimes are important to simulate correctly in climate models (e.g. Klein & Hartmann 1993; Miller 1997), having a significant impact on the TOA net shortwave radiative budget which is not offset by their infrared effect.

Despite their importance, the simulation of such low-level tropical cloud regimes poses long-standing difficulties for global models, evidenced in the ECMWF modelling system in Fig. 14.8, and confirmed when examining the cloud cover errors compared to MODIS or CloudSat. The ECMWF model is not alone, and this error is common in climate [and numerical weather prediction (NWP)] models. The radiatively driven cloud-top entrainment process is usually treated specifically, using the radiative flux divergence across the identified low cloud top to enhance the parameterised entrainment velocity (Moeng *et al.* 1999; van Zanten *et al.* 1999), and this was used in the Tiedtke (1993) cloud scheme in the operational ECMWF forecast model until the introduction of the moist-thermodynamic mass-flux/K-diffusion mixing scheme, which adopted a similar term for cloud-top entrainment in boundary-layer clouds (M. Köhler, personal communication). Such an approach depends critically on vertically resolving the inversion and the depth of the radiatively cooled cloud-top layer well (van Zanten *et al.* 1999). A full review of stratocumulus turbulent entrainment is beyond the scope of this chapter, and we refer the reader to Stevens (2002) for a more complete summary.

This specific treatment of the radiatively driven cloud-top entrainment in stratocumulus cloud layers forming under a strong temperature inversion implies that the same process is neglected in many other cloud regimes in which it can also play an important role. One common example is the role of radiatively driven turbulence in cirrus clouds. While simple parcel models did not reveal the requirement for strong turbulent entrainment to reproduce observed crystal size spectra (Lin *et al.* 1998), subsequent cloud-resolving modelling integrations such as those of Köhler (1999) and Gu & Liou (2001) demonstrated that interactive radiation can play a strong role in the growth and maintenance of cirrus clouds. Fusina *et al.* (2007) confirmed this using a cloud-resolving model with interactive radiation to investigate the formation and subsequent evolution of cirrus clouds in ice supersaturated regions. Unless the cloud boundary radiative heating is sufficient to alter the stability profile and promote turbulent mixing on the grid scale, this effect will be absent in climate models.

Further experiments by Krüger & Zulauf (2005) using both 2D and 3D high-resolution cloud-resolving models showed that the interaction between radiation and clouds is not just important for driving small-scale turbulence on scales much smaller than the cloud. In addition it was revealed that the differential heating between the cloud and its environment could drive mesoscale circulations that significantly enhance the cirrus anvil spreading rates, highlighting another potential important role for cloud radiative dynamical feedback on the near cloud scale.

Dynamical radiative feedback on the mesoscale

Cloud impacts on radiation may also modify the atmospheric circulations on the mesoscale. This is particularly true in the tropics where small temperature anomalies can alter the thermally direct circulations existing in weak horizontal temperature gradients. Outside of the polar regions, where the high-albedo snow-covered surface complicates matters, clouds mostly cool the surface and act to warm the atmosphere below them. Downwelling shortwave radiation is reduced at the surface, while the atmosphere below the cloud is warmed by the impact of longwave atmosphere. Gray & Jacobson Jr (1977) thus suggested that the differential atmospheric heating between cloudy and clear regions in the infrared could drive secondary mesoscale circulations, with the heating perturbation driving convergent motions towards the cloudy regions, reinforcing the existing circulation.

The importance of cloud–radiation interactions in determining the dynamics in the tropics can be investigated in idealised experiments using cloud resolving models (CRMs): non-hydrostatic models with adequately fine resolution to represent the motions within convective clouds. The numerical expense of running these models originally restricted their use to short-term integrations of the development of individual cloud systems, but by the 1990s resources were adequate to conduct longer integrations for many days to allow convective organisation to occur.

Some of these studies documented the impact of interactive radiation on high-resolution integrations of mesoscale organised systems such as polar lows (e.g. Craig 1995), tropical cyclones (Craig 1996) or tropical squall lines (Dharssi *et al.* 1997). These studies highlighted the impact that radiative forcing has on increasing convective activity. A second line of experiments integrated CRMs to a radiative–convective equilibrium state (e.g. Sui *et al.* 1994; Tao *et al.* 1996; Wu *et al.* 1998). Xu & Randall (1995) examined cloud radiative feedbacks in a 2D cloud-resolving model and concluded the Gray–Jacobson mechanism was minor compared to the diurnal modulation of the convection. However this study concentrated on the temporal modulation of the convection rather than the spatial organisation.

The first study to integrate a CRM to equilibrium in three dimensions using an interactive radiative scheme was conducted by Tompkins & Craig (1998) who reported that the convection within the domain strongly organised into a band

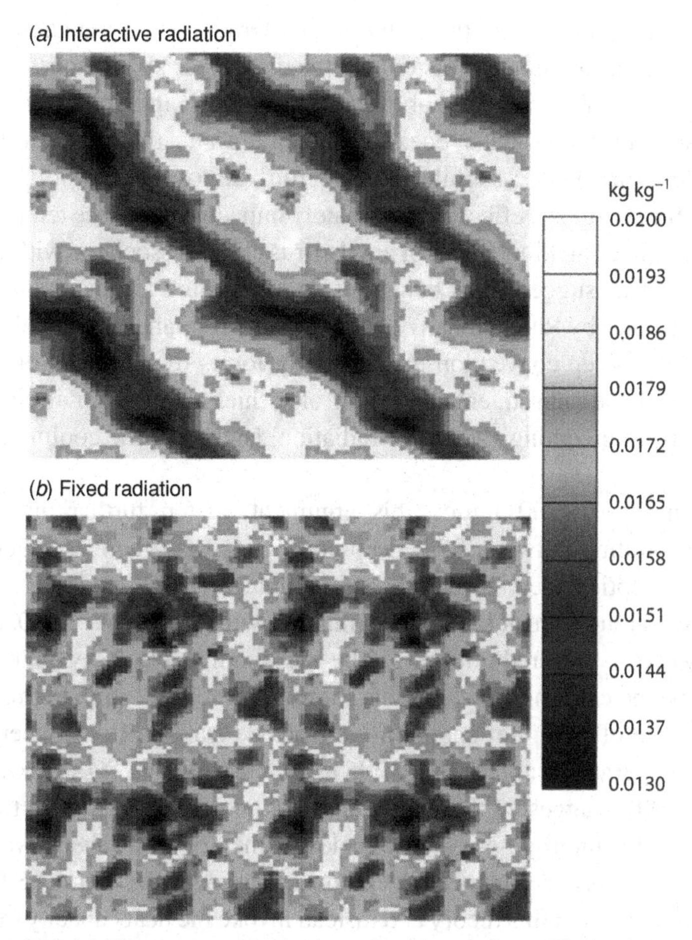

Figure 14.13 (*a*) Water vapour in the boundary layer after a 70-day integration of a 3D cloud resolving model using a 100×100 km domain size and an interactive shortwave and infrared radiation scheme that responds to the water vapour and cloud water content. The convective activity has strongly organised into bands that are associated with high water-vapour content, both in the boundary layer and above. The domain is repeated twice in each direction to emphasise the organisation. Starting from the scene in panel (*a*), the interactive radiation scheme was replaced by the domain and equilibrium-mean cooling profiles and the model integrated for a further four days. This resulted in the situation depicted in (*b*), where the strong organisation is lost, and convection occurs throughout the domain. (Adapted from Tompkins & Craig 1998.)

structure (Fig. 14.13). As a first indication of the potential for cloud radiative feedbacks to cause dynamical organisation of convection Tompkins & Craig (1998) demonstrated that replacing the interactively derived radiative cooling rates with horizontally homogeneous fixed radiative forcing of an equivalent magnitude broke up the organisation of convection within a few days. Note that these experiments did not separate the effect of the water-vapour structures on radiation, where the convecting regions are much moister than the drier clear-sky regions.

One drawback of many of the above CRM studies is that they were conducted using a fixed surface (usually ocean) temperature and therefore emphasise the atmospheric infrared effect. While clouds perturb atmospheric heating rates in the shortwave, the shortwave effect predominately impacts the surface temperatures. In fact the shortwave shielding effect of anvil cirrus clouds associated with convection in the tropics was suggested as being key to controlling the maximum observed SSTs in the tropics by Ramanathan *et al.* (1989) and Ramanathan & Collins (1991). The idea is that deep convection preferentially forms over regions of warm SST anomalies, and the subsequent formation of a thick tropical anvil cloud shields the surface from incoming shortwave radiation, leading to the cooling of the SST anomaly.

Hartmann *et al.* (2001) took this argument a step further and suggested that, if the coupling between SST, deep convection, and then subsequently convective cloud optical depth were strong enough, then the feedback would lead to convective cloudy areas having a 'similar' (within 10 Wm^{-2}) TOA effect on net (shortwave + infrared) fluxes as the surrounding clear-sky regions, which appears to be the case in tropical observations. One complication to this argument is the existence of SST 'hot-spots' in the tropics (Waliser, 1996) where the SST can reach temperatures approaching 305 K with significant TOA flux anomalies. Tompkins (2001) suggested these SST hot-spot anomalies were able to form and be maintained for multiple-week time scales by the dry atmospheric water-vapour anomalies that delay the onset of deep convection over the warm SST. Sobel & Gildor (2003) discount this theory and instead invoke the heat capacity of the ocean surface layer as the principle delaying mechanism that permits hot-spot formation.

Can climate models represent these dynamical radiative cloud feedbacks on the mesoscale? This will rather depend on the scale over which these interactions act, which is an aspect that is not clear from existing cloud-resolving studies which have not revealed the factors that set the spatial scale over which cloud–radiation dynamical feedback occurs. The CRM studies of Tompkins & Craig (1998) used a limited domain size of 100 km by 100 km, on the same order of scale as a typical global climate model grid box, and thus the infrared-driven organisation would be entirely a subgrid-scale phenomenon. However, it is obvious from the diagonal orientation of the convective band in Fig. 14.13 that the domain is artificially constraining the scale of organisation. More recent investigations using domains many times larger appear to confirm this. For example, convection in larger 3D integrations conducted by Bretherton *et al.* (2005) still collapsed to a single organised convective system. Bretherton *et al.* (2005) went on to present a simple model that described the onset time scale of organisation but was unable to offer a horizontal scaling argument. If the convective region in question concerned mesoscale organised convection spanning hundreds of kilometres then the resulting radiatively driven secondary circulation may well be resolvable by global models. Indeed, experiments with a coupled atmosphere–ocean model conducted by Li *et al.* (2000) were able to show that shortwave anvil shield effect operates (and dominates over other mechanisms for SST control) in their coupled model, although the issue of convective organisation was not discussed in that paper.

Dynamical radiative feedback on the planetary scale

The above discussion all points towards a potentially strong dynamical feedback between clouds, radiation and dynamics, especially in the tropics, with the interaction in the infrared spectrum acting predominately in the troposphere to reinforce circulations between clear-sky and cloudy regions thus increase the organisation of deep convection. In contrast, the interaction in the shortwave acts through the surface and tends to lead to more homogeneous SSTs and thus decrease the organisation of convection. Experiments with fixed SSTs will thus emphasise the former mechanism.

It has been proposed that radiative cloud dynamical feedback can operate over planetary spatial scales, playing a role in determining the strength of the Hadley and Walker circulations. Using a simplified physics model of the tropics, Raymond (2000) suggested that cloud radiative dynamical feedbacks are in fact instrumental in the establishment of the tropical Hadley circulations. Such circulation scales are resolvable by even low-resolution climate models, but to our best knowledge, the role that cloud radiative dynamical feedback plays in establishing the strength of such tropic-wide circulations in current state-of-the-art global climate models is yet to be extensively documented. However, Sohn (1999) calculated that cloud radiative feedback can enhance the strength of both the Hadley and Walker circulations, and in a full-physics GCM Sherwood *et al.* (1994) had earlier also highlighted the weakening of the Walker circulation when cloud radiative feedbacks were isolated. In their analysis, Bergman & Hendon (2000) estimate that cloud–radiation interactions contribute about 20% to the magnitude of low-latitude circulations. Using a highly simplified model, Tian & Ramanathan (2003) also emphasised this central role of differential radiative heating rates in driving the Hadley and Walker circulations. Applying horizontally uniform radiative heating in their model caused the Walker circulation to collapse and the Hadley circulation to reverse sign!

There is also the possibility that cloud–radiation interactions act to reinforce circulations that are set by other mechanisms. A good example, again in the tropics, is the Madden–Julian Oscillation (MJO): a large-scale, eastward-propagating convectively coupled feature of wavenumber 1 to 3 approximately that circumnavigates the tropics in around 45 to 60 days, and represents the dominant mode of intraseasonal tropical convective variability (Madden & Julian 1971, 1994). A number of mechanisms have been suggested for this imperfectly understood yet important phenomenon, including the interaction between deep convection and the large-scale water-vapour field (Bladé & Hartmann 1993; Grabowski & Moncrieff 2004), interaction between clouds and SST anomalies associated with the MJO (Hendon 2000; Woolnough *et al.* 2001) and direct cloud–radiation interaction in the IR (Flatau *et al.* 1997). The latter two mechanisms revisit the shortwave and infrared feedbacks discussed above, but this time the shortwave–cloud shield feedback is proposed as a method for promoting convective propagation, while the IR mechanism again plays a convective self-aggregation role. On the other hand, investigations by Lee *et al.* (2001) found that cloud–radiation interactions in their GCM simulations acted to damp convective organisation associated with eastward propagation waves, although they were able to moderate this effect by tuning the cloud microphysics to reduce the cloud radiative dynamical feedback strength.

One question that arises in the consideration of the above mechanisms is why, considering that the MJO is a large-scale, low-wavenumber phenomenon that should be resolvable in coarse-scale climate models, do (even coupled) climate models continue to have difficulty in reproducing the MJO (Slingo *et al.* 1996; Lin *et al.* 2006)? If cloud radiative dynamic feedbacks are the major mechanism for the MJO then it implies that there is a strong artificial damping of low-wavenumber oscillations, possibly by the mass-flux CAPE-closure convective parameterisation paradigm in widespread use in GCMs. Tompkins & Jung (2003) noted in the ECMWF global model that increasing the incidence of grid-scale convection, which is constrained to occur in moist regions of mid-tropospheric ascent, led to greater Kelvin-wave and MJO-like activity and went on to suggest that mass-flux convection schemes that place convection in regions of high boundary-layer moist static energy and are insensitive to mid-tropospheric humidity (as most schemes are, see Derbyshire *et al.* 2004, for example) could act to damp convectively coupled waves.

Vitart *et al.* (2007) demonstrated that the MJO signal is strongly damped in the ECMWF forecasting system operational during 2006. This is contrasted with the situation in 2008 where the model is able to maintain convective anomalies and predict the MJO evolution out to week 3 and beyond (F. Vitart, personal communication). One of the main changes in the system that led to this improvement was a revised convection scheme that is more sensitive to mid-tropospheric humidity (Bechtold *et al.* 2008).

It thus appears that correct representation of cloud–radiation interactions could be a necessary, but likely not sufficient, condition for simulating the Madden–Julian

Oscillation in global climate models. This appears to confirm the conclusions of Raymond (2001), obtained using a simple model of the tropics that was able to reproduce an eastward-propagating convective signal that is reminiscent of the MJO. The simplicity of the model used meant that Raymond (2001) was able to determine the origin of the organisation precisely, and established that the cloud radiative feedback in the IR played a critical role; however, inclusion of the convection sensitivity to dry or moist mid tropospheres was also necessary to establish the eastward-propagating mode. This conclusion was also reached in the studies of Bony & Emanuel (2005) and Zurovac-Jevtić *et al.* (2006), which used a simple linear model and a 2D aqua-planet model, respectively, to find that cloud–radiation interactions in the longwave could play a role in the large-scale organisation of convection and also in the the phase speed of propagation of planetary-scale disturbances such as the MJO, but that convection–humidity interactions were also of importance. This dependence on a number of feedback mechanisms implies that the sensitivity of the simulation of the MJO to a specific feedback mechanism in a particular model will be model-dependent, relying on the representation of the other relevant process interactions in the model.

14.4 Future perspectives

During approximately four decades of development in climate models there has been a considerable and steady progress with respect to their representation of cloud processes and their radiative impact. Certainly there has been an increase in the complexity of the representation of cloud processes, with climate models entertaining several prognostic variables for ice and liquid water categories as well as cloud amount. These variables are used directly as input to the radiation schemes with more justifiable assumptions concerning the cloud particle radiative properties.

The scientific community has taken great steps to create frameworks that facilitate the intercomparison of climate models starting with AMIP (Gates 1992) and leading on to CMIP (Meehl *et al.* 2000) and the IPCC framework. Validation efforts have matured from simple comparisons of cloud impact on TOA radiative fluxes to detailed comparisons to enhanced networks of detailed ground-based observations [Atmospheric Radiation Measurement (ARM) Program and Cloudnet (cloud network)] and retrievals of many cloud parameters from recently launched instruments such as CloudSat and CALIPSO, which is discussed further below. Validation techniques have evolved to include the use of conditional sampling and compositing techniques to isolate cloud–radiation interactions, in particular according to dynamical regimes. The earlier work highlighting cloud parameterisation inadequacies as a cause of model uncertainty (e.g. Cess *et al.* 1990, 1996) has meant that clouds have remained a focal point as recent methods for assessing climate model prediction uncertainty have improved (e.g. Murphy *et al.* 2004).

If one were to summarise the continuing limitations of current cloud–radiation interaction modelling there are two areas that deserve emphasis. The first relates to the relative paucity of cloud observations to date, with satellites at best providing column integrated estimates of total column ice and liquid water paths, cloud amount at high and, when not masked by high cloud, low levels, or cloud radiative forcing at the TOA in the shortwave and infrared. Until the introduction of the ground-based ARM sites and the more recent Cloudnet network, which provide high vertical resolution cloud profile observations combined with detailed radiative measurements at point locations, coordinated long-term ground-based observations of clouds were restricted to the synoptic cloud amount.

The second area of concern pertains to the column-model paradigm for parameterisation schemes, whereby all communication between horizontal neighbouring model grid cells occurs in the climate models that are governing equations, and parameterisation schemes operate on the grid-cell column. Within this grid-column, variability of cloud variables is at best poorly represented, and then usually inconsistently between the various parameterisation schemes. As an example, convective activity in one grid cell cannot induce convective triggering in neighbouring cells, making the representation of mesoscale organised convective systems problematic even in relatively high-resolution climate models that should start to resolve such systems. Even as horizontal resolutions eventually approach the cloud-resolving scales, current radiation parameterisations will not allow clouds to shade neighbouring grid cells.

Turning our attention to the lack of self-consistency for the subgrid-scale variability representation in cloud and radiative processes, radiation schemes must make diagnostic assumptions concerning parameters such as cloud droplet numbers for transfer calculations if this information is not available as a predicted parameter. These are often diverse from the assumptions hidden in the microphysics of the cloud scheme, however. Another example is the microphysics representation contained in the convective parameterisation and the large-scale cloud scheme, which are rarely consistent with each other, implying no convergence of microphysical behaviour as climate models move ever towards cloud-resolving resolutions. This indicates that even if a climate model were equipped with a Utopian convection scheme that perfectly predicted subgrid convective mass fluxes of an equivalent cloud-resolving climate model integration, the two model solutions would not be equivalent due to the disparities in the cloud physics.

Despite these weaknesses in the current modelling of cloud–radiation interactions there are several reasons to believe that the current decade is likely to produce dramatic progress in our understanding and representation of these processes.

14.4.1 New observations

As pointed out earlier in this chapter, one of the impediments to progress in cloud modelling has been the lack of pertinent global observations. To date, much of our present observational understanding of cloud properties has been inferred from their impact on the TOA radiative budgets such as CERES (Wielicki *et al.* 1996) or ERBE (Barkstrom 1984). Direct cloud observations from satellites have been restricted to column integrated quantities such as total column liquid water [SSMI; Wentz (1997)] or total cloud amount (e.g. ISCCP; Rossow & Schiffer 1991) with little vertical structure information available. Most of our knowledge concerning cloud vertical structure has been derived from isolated ground stations such as the excellent long-term measurements available from the ARM sites (Stokes & Schwartz 1994) or more recently Cloudnet over Europe (Illingworth *et al.* 2007). Even a key basic quantity such as the amount of cloud ice has until recently been very poorly observed from space, with limited information available for thin cirrus from TOVS (Rädel *et al.* 2003) and at coarse horizontal resolutions in the upper troposphere from microwave limb sounding retrievals (Li *et al.* 2005).

With the implementation of the 'A' train series of satellites, including the recent launches of the cloud lidar CALIPSO and the CloudSat 94 Ghz radar (Stephens *et al.* 2002) a new era of unprecedented global cloud observations has been introduced (see Fig. 14.14 for example transects illustrating the capabilities of the new CloudSat instrument). Early statistics of zonal-mean cloud cover from the new instrument are given by Kahn *et al.* (2008). The CALIPSO instrument is capable of detecting thin cloud layers, and the high spacial resolution of 330 m permits an improved evaluation of boundary-layer clouds (Chepfer *et al.* 2008). The new CloudSat and CALIPSO sensor combination permits retrievals of cloud ice and (to a lower accuracy) liquid cloud amounts at high vertical and horizontal resolution globally for the first time, allowing existing cloud parameterisation schemes to be validated and improvements to be made based on this understanding.

An example of an initial comparison of CloudSat ice retrievals for small and large ice particles at the 215 hPa pressure level to the full ECMWF ice fields is shown in Fig. 14.15. The model cycle is CY31r1, which is used both for the system 3 operational seasonal forecasts as well as the atmospheric component of the EC-EARTH model. The geographical agreement between the model and the observations is striking, and the full analysis of the ice amount contained in Woods *et al.* (unpublished work) shows that this model cycle was a significant improvement on previous model versions. The model ice mass-mixing ratios also appear to be in the correct range for this pressure level compared to the small-ice field, while the

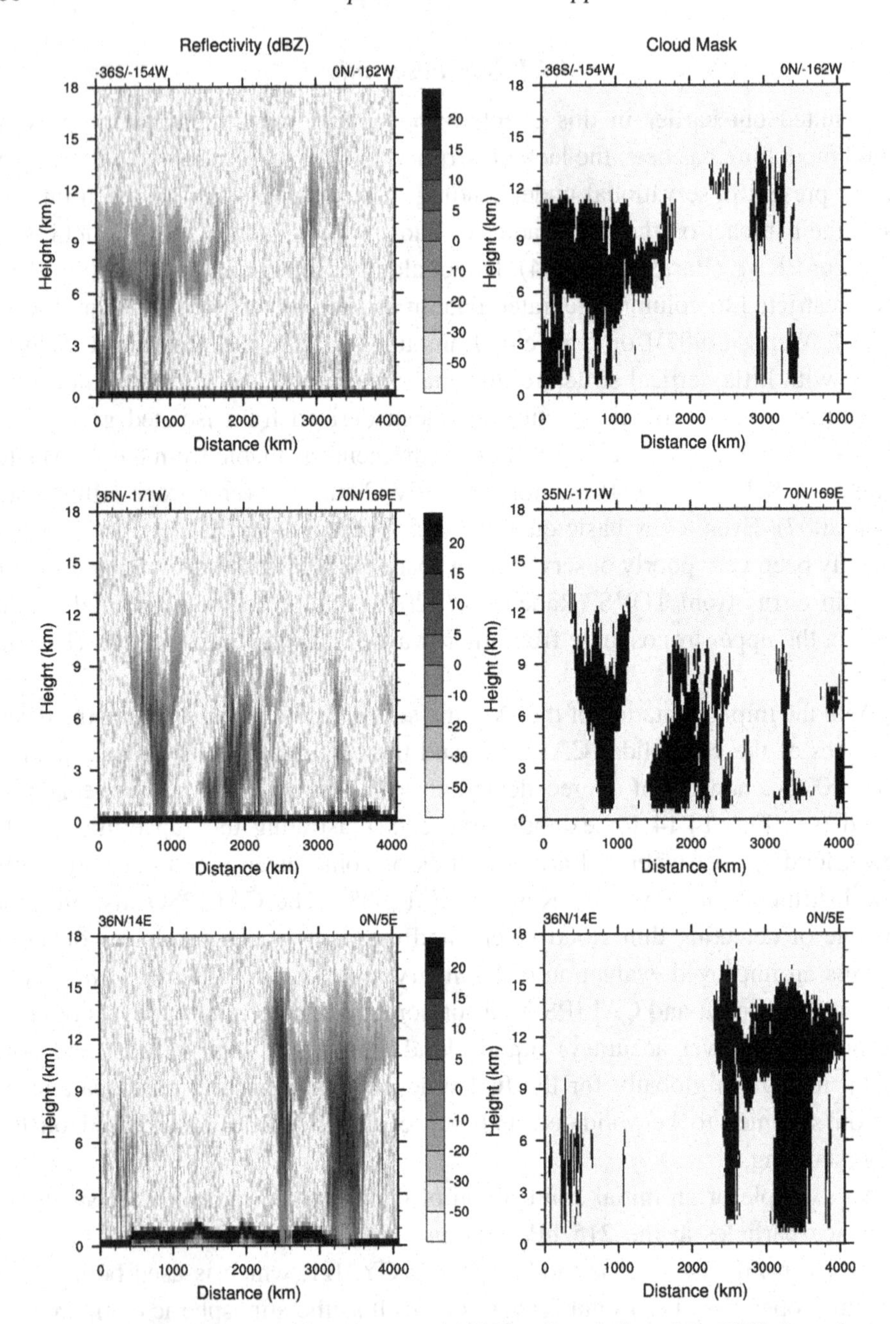

Figure 14.14 Three example transects of 4000 km length from July 2008 from the CloudSat instrument. The left panels show the raw radar reflectivity in dBZ, while the right panels show the equivalent scenes processed to give a cloud detection mark (standard level 2B product GEOPROF). The panel titles give the latitude/ longitude coordinates of the transect extremes.

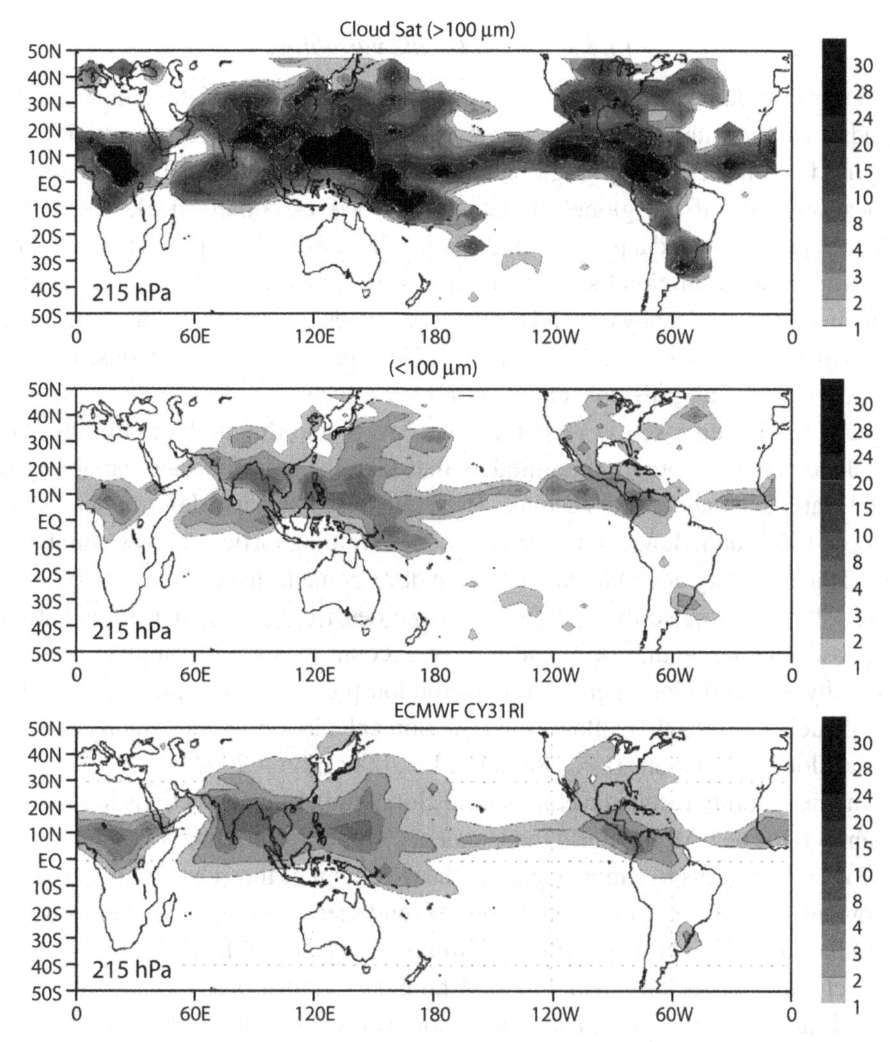

Figure 14.15 Retrievals of ice mass-mixing ratio (mg m^{-3}) with particle diameters greater/smaller (upper/middle panels) than 100 microns from CloudSat compared with (lower panel) ECMWF ice mixing ratio from cycle CY31R1. Adapted from Woods *et al.* (2008).

model underestimates cloud ice compared to the full CloudSat retrieval including large-ice particles. As well as permitting future model versions to have more accurate ice mass-mixing ratios, these new observations also allow the structure of clouds to be charted, so that the various vertical overlap schemes used in global climate models (Morcrette & Fouquart 1986; Hogan & Illingworth 2000; Tompkins & Di Giuseppe 2006), so crucial for the microphysical and radiative impact, can be validated globally.

14.4.2 Subgrid-scale variability

It is clear from the discussion in the preceding sections that the interaction between clouds, radiation and dynamics occurs across the whole range of spatial scales ranging from the cloud to the global. However, in the past, variability below the truncation scale of the global climate model was essentially neglected. Cloud schemes parameterised a partial cloud cover (diagnostically or prognostically) and the representation of cloud-scale variability was hidden in empirical assumptions contained in the microphysics and radiation schemes, such as the adjustment factor of Cahalan *et al.* (1994b) in the shortwave. This approach implies inconsistency in the treatment of subgrid-scale cloud variability effects.

Steps have been taken in recent years to overcome this problem by introducing cloud schemes into climate models that take the subgrid-scale variability of cloud water and vapour into account. These are based on early developments from Mellor (1977) and Bougeault (1982) which were diagnostic schemes suitable to model the planetary boundary layer. Recent developments have attempted to extend these schemes to make the distribution shape describing the subgrid fluctuations prognostic (Golaz *et al.* 2002) and to take account of a more complete range of microphysical and cloud generation/destruction processes (Tompkins 2002). The use of such schemes then allows the radiation calculation to consistently account for the cloud fluctuations (as in Barker & Fu 2000; Bäuml 2002).

Another parallel recent development in cloud–radiation interactions in radiation schemes is the new Monte Carlo Independent Column Approximation (Pincus *et al.* 2003). This attempts to simplify greatly the treatment of the cloud geometry in the following way. Instead of accounting for a cloud scene with varying cloud fraction as a function of height, with all the entailing complexity of handling flux transfer from cloud to clear and clear to cloud boundary transitions, the new approach instead samples the scene by taking a random vertical profile. Each random profile thus consists of a column of points which are either cloudy or clear, dispensing of the need to handle cloud overlap, *within the radiation scheme* (the overlap assumptions are instead involved in the the cloud sampling procedure, performed outside of the radiation calculation). Each band (or more accurately, each G-point or sub-band) in the radiative calculation uses a different random cloud sample to reduce the random sampling error, which Räisänen *et al.* (2005) prove to be mostly insignificant. The methodology holds much promise since it allows the simplification not just in terms of the cloud geometry through the handling of the cloud fractional mask, but also by accounting for sub-cloud-scale humidity and cloud water fluctuations through the overlap of the cloud water probability distribution functions (Pincus *et al.* 2005), a task that is extremely difficult to approach through conventional parameterisation methodologies.

14.4.3 High-resolution climate models and 'superparameterisation'

Computing resources are moving the modelling community inexorably towards cloud-resolving global integrations. While long-term climate integrations at cloud-resolving resolutions are unlikely to be achieved in the following decade, benchmark integrations have been performed for NWP integration time scales of several days (Tomita *et al.* 2005) and much can be learned from these. Often when global cloud-resolving integrations are discussed, their advantage is portrayed in terms of the fact that the parameterisation scheme for convection may be discarded. Their critique is that they do not represent a panacea since uncertain microphysical processes still require representation and that, even at resolutions on the order of a kilometre, boundary layer cloud processes are not resolved. However, an oft-overlooked aspect is that using cloud-resolving resolutions removes the requirement for a treatment of cloud amount and the vertical overlap of clouds that so critically impacts the calculation of cloud-affected radiative fluxes. The possibility of neglecting fractional cloud cover also renders the microphysical modelling problem far simpler, since the overlap assumptions also affect precipitation calculations in a strongly nonlinear way, and no separate memory of humidity evolution is required in the cloud-free and cloudy portions of the grid cell.

An intermediate promising step for representing cloud–radiation interactions in climate models with increased fidelity in the nearer-term future is represented by the so-called superparameterisation or 'cloud-resolving convective parameterisation' (CRCP; see Grabowski 2001; Randall *et al.* 2003; Khairoutdinov *et al.* 2005). This involves replacing the traditional convective parameterisation scheme with a limited-domain cloud-resolving model in each grid cell of the climate model (see schematic and further details in Fig. 14.16). While the approach is certainly more costly in computing resources than the traditional parameterisation schemes it can offer some of the potential advantages of a global CRM at a fraction of the cost. As with the global CRM, the vertical overlap of clouds must not be specified for the radiation or cloud microphysical processes since the arrangement of the clouds within the CRM domain is known. The influence of vertical wind shear on the cloud overlap is also accounted for explicitly with such an approach.

In the previous section, the development of the representation of subgrid-scale variability was discussed. One of the impediments to these developments has been the difficulty of obtaining information concerning thermodynamic and dynamic variability over small spatial scales from observations. Authors such as Tompkins (2002) attempt to supplement observations using information from limited-area cloud-resolving model integrations, but these are usually conducted for a small range of situations (e.g. Xu & Randall 1996) and do not adequately cover the

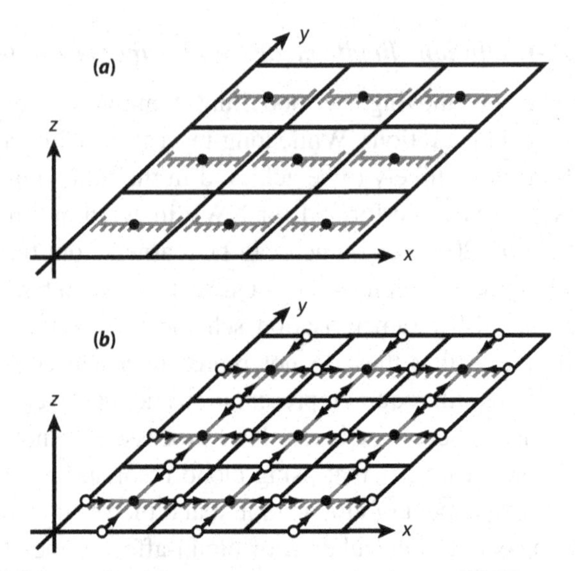

Figure 14.16 Schematic of the superparameterisation or CRCP approach adapted from Randall *et al.* (2003). (*a*) The first generation CRCP approach is illustrated, where a small domain 2D CRM is placed within each global model grid cell. Each CRM uses the global model resolved scale motions in the thermodynamic forcings (usually via a term $\omega\partial\phi/\partial p$ where ϕ represents both the potential temperature and water vapour, since the CRMs use cyclic rather than open boundary conditions). The CRM is then integrated for a period equivalent to the time step of the global model and the resulting tendencies from the CRM are then passed to the large-scale model in the same manner as a conventional convective parameterisation. Communication between grid-cells can only occur through the large-scale model equations, implying that some of the advantage of the CRM is lost, since resolved features such as squall-line may not propagate between adjacent cells. (*b*) An enhancement of the original approach, since CRM with open boundaries are used that may communicate between large-scale model cells. A version using this approach was tested by Jung & Arakawa (2005).

dynamical phase space for reliable parameterisation development. Global CRMs and global models using superparameterisations thus represent an extremely useful tool since they provide this information for a wide variety of dynamical regimes across the globe.

This superparameterisation approach is still in its infancy, and open issues still remain concerning the most effective manner to distribute available computing resources between the resolution of the host global model and the domain size, resolution and even dimensionality of the embedded CRM. Enhancing the methodology to allow communication between CRMs in neighbouring grid cells would also facilitate the representation of mesoscale organised convective systems.

Acknowledgements

The authors are very grateful for the detailed feedback, suggestions and corrections provided by Sandrine Bony (LMD/IPSL, France), Klaus Gierens (DLR, Germany) and Jean-Jacques Morcrette (ECMWF, UK). Anabel Bowen of ECMWF kindly reformatted a number of the figures at short notice.

References

Aida, M. A. 1977 Scattering of solar radiation as a function of cloud dimension and orientation. *Q. Spectrosc. Rad. Transfer*, **17**, 303–310.

Arking, A. 1991 The radiative effects of clouds and their impact on climate. *Bull. Amer. Meteorol. Soc.*, **72**, 795–813.

Baran, A. J., Francis, P. N., Labonnote, L. C. & Doutriaux-Boucher, M. 2001 A scattering phase function for ice cloud: tests of applicability using aircraft and satellite multi-angle multi-wavelength radiance measurements of cirrus. *Q. J. R. Meteorol. Soc.*, **127**, 2395–2416.

Barker, H. W. & Davies, J. A. 1992a Solar radiative fluxes for broken cloud fields above reflecting surfaces. *J. Atmos. Sci.*, **49**, 749–761.

Barker, H. W. & Davies, J. A. 1992b Solar radiative fluxes for stochastic scale invariant broken cloud fields. *J. Atmos. Sci.*, **49**, 1115–1126.

Barker, H. W. & Fu, Q. 2000 Assessment and optimization of the gamma weighted two-stream approximation. *J. Atmos. Sci.*, **57**, 1181–1188.

Barker, H. W. & Li, Z. Q. 1997 Interpreting shortwave albedo-transmittance plots: true or apparent anomalous absorption? *Geophys. Res. Lett.*, **24**, 2023–2026.

Barker, H. W., Morcrette, J.-J. & Alexander, G. D. 1998 Broadband solar fluxes and heating rates for atmospheres with 3D broken clouds. *Q. J. R. Meteorol. Soc.*, **124**, 1245–1271.

Barker, H. W. & Räisänen, P. 2005 Radiative sensitivities for cloud structural properties that are unresolved by conventional GCMs. *Q. J. R. Meteorol. Soc.*, **131**, 3103–3122.

Barker, H. W., Stephens, G. L., Partain, P. T. *et al.* 2003 Assessing 1D atmospheric solar radiative transfer models: interpretation and handling of unresolved clouds. *J. Climate*, **16**, 2676–2699.

Barker, H. W., Stephens, G. L. & Fu, Q. 1999 The sensitivity of domain averaged solar fluxes to assumptions about cloud geometry. *Q. J. R. Meteorol. Soc.*, **125**, 2127–2152.

Barkstrom, B. R. 1984 The Earth radiation budget experiment (ERBE). *Bull. Amer. Meteorol. Soc.*, **64**, 1170–1185.

Bäuml, G. 2002 Influence of sub-grid scale variability of clouds on the solar radiative transfer computations in the ECHAM5 climate model. Ph.D. thesis, University of Hamburg.

Bechtold, P., Kohler, M., Jung, T. *et al.* 2008 Advances in simulating atmospheric variability with the ECMWF model: From synoptic to decadal time-scales. *Q. J. R. Meteorol. Soc.*, **134**, 1337–1351.

Bergman, J. W. & Hendon, H. H. 2000 Cloud radiative forcing of the low-latitude tropospheric circulation: linear calculations. *J. Atmos. Sci.*, **57**, 2225–2245.

Bergman, J. W. & Rasch, P. J. 2002 Parameterizing vertically coherent cloud distributions. *J. Atmos. Sci.*, **59**, 2165–2182.

Bladé, I. & Hartmann, D. L. 1993 Tropical intraseasonal oscillations in a simple nonlinear model. *J. Atmos. Sci.*, **50**, 2922–2939.

Bony, S., Colman, R., Kattsov, V. M. *et al.* 2006 How well do we understand and evaluate climate change feedback processes? *J. Climate*, **19**, 3445–3482.

Bony, S., Dufresne, J. L., Le Treut, H., Morcrette, J.-J. & Senior, C. 2004 On dynamic and thermodynamic components of cloud changes. *Clim. Dyn.*, **22**, 71–86.

Bony, S. & Emanuel, K. A. 2005 On the role of moist processes in tropical intraseasonal variability: cloud–radiation and moisture–convection feedbacks. *J. Atmos. Sci.*, **62**, 2770–2789.

Bony, S., Lau, K.-M. & Sud, Y. C. 1997 Sea surface temperature and large-scale circulation influences on tropical greenhouse effect and cloud radiative forcing. *J. Climate*, **10**, 2055–2076.

Bougeault, P. 1982 Cloud ensemble relations based on the gamma probability distribution for the high-order models of the planetary boundary layer. *J. Atmos. Sci.*, **39**, 2691–2700.

Bretherton, C. S., Blossey, P. N. & Khairoutdinov, M. 2005 An energy-balance analysis of deep convective self-aggregation above uniform SST. *J. Atmos. Sci.*, **62**, 4273–4292.

Brooks, M. E., Hogan, R. J. & Illingworth, A. J. 2005 Parameterising the difference in cloud fraction defined by area and by volume as observed with radar and lidar. *J. Atmos. Sci.*, **62**, 2248–2260.

Byrne, R. N., Somerville, R. C. J. & Subasilar, B. 1996 Broken-cloud enhancement of solar radiation absorption. *J. Atmos. Sci.*, **53**, 878–886.

Cahalan, R. F., Ridgway, W. & Wiscombe, W. J. 1994a Independent pixel and Monte Carlo estimates of stratocumulus albedo. *J. Atmos. Sci.*, **51**, 3776–3790.

Cahalan, R. F., Ridgway, W., Wiscombe, W. J., Bell, T. L. & Snider, J. B. 1994b The albedo of fractal stratocumulus clouds. *J. Atmos. Sci.*, **51**, 2434–2455.

Cess, R. D., Zhang, M. H., Minnis, P. *et al.* 1995 Absorption of solar radiation by clouds: observations versus models. *Science*, **267**, 496–499.

Cess, R. D., Potter, G. L., Blanchet, J. P. *et al.* 1990 Intercomparison of climate feedback processes in 19 atmospheric general circulation models. *J. Geophys. Res.*, **95**, 16601–16615.

Cess, R. D., Zhang, M.-H., Ingram, W. H. *et al.* 1996 Intercomparison of climate feedback processes in 19 atmospheric general circulation models. *J. Geophys. Res.*, **101**, 12791–12794.

Chambers, L. H., Wielicki, B. A. & Evans, K. F. 1997a Accuracy of the independent pixel approximation for satellite estimates of oceanic boundary layer cloud optical depth. *J. Geophys. Res.*, **102**, 1779–1794.

Chambers, L. H., Wielicki & Evans, K. F. 1997b Independent pixel and two-dimensional estimates of Landsat-derived cloud field albedo. *J. Atmos. Sci.*, **54**, 1525–1532.

Chen, T., Zhang, Y. & Rossow, W. B. 2000 Sensitivity of atmospheric radiative heating rate profiles to variations of cloud layer overlap. *J. Climate*, **13**, 2941–2959.

Chepfer, H., Bony, S., Chiriaco, M. *et al.* 2008 Use of CALIPSO lidar observations to evaluate the cloudiness simulated by a climate model. *Geophys. Res. Lett.*, **35**, L15704.

Chou, M.-D., Krats, D. & Ridgway, W. 1991 Infrared radiation parameterization in numerical climate models. *J. Climate*, **4**, 424–437.

Cohen, B. G. & Craig, G. C. 2004 The response time of a convective cloud ensemble to a change in forcing. *Q. J. R. Meteorol. Soc.*, **130**, 933–944.

Collins, D. W. 2001 Parameterization of generalized cloud overlap for radiative calculations in general circulation models. *J. Atmos. Sci.*, **58**, 3224–3242.

Craig, G. C. 1995 Radiation and polar lows. *Q. J. R. Meteorol. Soc.*, **121**, 79–94.

Craig, G. C. 1996 Numerical experiments on radiation and tropical cyclones. *Q. J. R. Meteorol. Soc.*, **122**, 415–422.
Demott, P. J., Cziczo, D. J., Prenni, A. J. *et al.* 2003 Measurements of the concentrations and composition of nuclei for cirrus formation. *Proc. Nat. Acad. Sci.*, **100**, 14655–14660.
Derbyshire, S. H., Beau, I., Bechtold, P. *et al.* 2004 Sensitivity of moist convection to environmental humidity. *Q. J. R. Meteorol. Soc.*, **130**, 3055–3079.
Dharssi, I., Kershaw, R. & Tao, W.-K. 1997 Sensitivity of a simulated tropical squall line to long-wave radiation. *Q. J. R. Meteorol. Soc.*, **123**, 187–206.
Di Giuseppe, F. 2005 Sensitivity of 1D radiative biases to vertical cloud structure assumptions: validation with aircraft data. *Q. J. R. Meteorol. Soc.*, **131**, 1655–1676.
Di Giuseppe, F. & Tompkins, A. M. 2003a Effect of spatial organisation on solar radiative transfer in three-dimensional idealized stratocumulus cloud fields. *J. Atmos. Sci.*, **60**, 1774–1794.
Di Giuseppe, F. & Tompkins, A. M. 2003b Three dimensional radiative transfer in tropical deep convective clouds. *J. Geophys. Res.*, **108**, 4741, doi:10.1029/2003JD003392.
Diermendjian, D. 1969 *Electromagnetic Scattering on Spherical Polydisperions*. Elsevier.
Edwards, J. M. & Slingo, A. 1996 Studies with a flexible new radiation code. 1. Choosing a configuration for a large-scale model. *Q. J. R. Meteorol. Soc.*, **122**, 689–719.
Evans, K. F. 1998 The spherical harmonics discrete ordinate method for three-dimensional atmospheric radiative transfer. *J. Atmos. Sci.*, **55**, 429–446.
Flatau, M., Flatau, P. J., Phoebus, P. & Niller, P. P. 1997 The feedback between equatorial convection and local radiative and evaporative processes: the implications for intraseasonal oscillations. *J. Atmos. Sci.*, **54**, 2373–2386.
Fouquart, Y. 1987 Radiative transfer in climate modeling. *Physically-Based Modeling and Simulation of Climate and Climate Changes*, ed. M. E. Schlesinger, pp. 223–283. NATO Advanced Study Institute.
Fu, Q. & Liou, K. N. 1992 On the correlated k-distribution method for radiative transfer in nonhomogeneous atmospheres. *J. Atmos. Sci.*, **49**, 2139–2156.
Fusina, F., Spichtinger, P. & Lohmann, U. 2007 The impact of ice supersaturated regions and thin cirrus on radiation in the mid latitudes. *J. Geophys. Res.*, **112**, D24514, doi:10.1029/2007JD008449.
Gates,W. L. 1992 AMIP: the Atmospheric Model Intercomparison Project. *Bull. Amer. Meteorol. Soc.*, **73**, 1962–1970.
Geleyn, J. F. & Hollingsworth, A. 1979 An economical analytical method for the computation of the interaction between scattering and line absorption of radiation. *Contrib. Atmos. Phys.*, **52**, 1–16.
Gierens, K. 2003 On the transition between heterogeneous and homogeneous freezing. *Atmos. Chem. Phys.*, **3**, 437–446.
Golaz, J., Larson, V. E. & Cotton, W. R. 2002 A PDF-based parameterization for boundary layer clouds. Part I: method and model description. *J. Atmos. Sci.*, **59**, 3540–3551.
Grabowski, W. W. 2001 Coupling cloud processes with the large-scale dynamics using the cloud-resolving convection parameterization (CRCP). *J. Atmos. Sci.*, **58**, 978–997.
Grabowski, W. W. & Moncrieff, M. W. 2004 Moisture–convection feedback in the tropics. *Q. J. R. Meteorol. Soc.*, **130**, 3081–3104.
Gray, W. M. & Jacobson Jr, R. W. 1977 Diurnal-variation of deep cumulus convection. *Mon. Weather Rev.*, **105**, 1171–1188.
Gu, Y. & Liou, K. N. 2001 Radiation parameterization for three-dimensional inhomogeneous cirrus clouds: application to climate models. *J. Climate*, **14**, 2443–2457.

Hansen, J. E. 1969 Exact and approximate solutions for multiple scattering by cloudy and hazy planetary atmosphere. *J. Atmos. Sci.*, **26**, 478–487.
Hansen, J. E. & Travis, L. D. 1974 Light scattering in planetary atmospheres. *Space Sci. Rev.*, **16**, 527–610.
Hanson, H. 1987 Radiative/turbulent transfer interactions in layer clouds. *J. Atmos. Sci.*, **44**, 1287–1295.
Hartmann, D., Moy, L. & Fu, Q. 2001 Tropical convection and the energy balance at the top of the atmosphere. *J. Climate*, **14**, 4495–4511.
Haynes, J. M., Marchand, R. T., Luo, Z., Bodas-Salcedo, A. & Stephens, G. L. 2007 A multipurpose radar simulation package: QuickBeam. *Bull. Amer. Meteorol. Soc.*, **88**, 1723–1727.
Hendon, H. H. 2000 Impact of air–sea coupling on the Madden–Julian oscillation in a general circulation model. *J. Atmos. Sci.*, **57**, 3939–3952.
Henyey, L. G. & Greenstein, J. L. 1941 Diffuse radiation in the galaxy. *Astrophys. J.*, **93**, 70–83.
Hogan, R. J. & Illingworth, A. J. 2000 Deriving cloud overlap statistics from radar. *Q. J. R. Meteorol. Soc.*, **126**, 2903–2909.
Holt, A. R. 1982 Electromagnetic wave scattering by spheroids: a comparison of experimental and theoretical results. *IEEE Trans. Antennas Propag.*, **AP-30(4)**, 758–760.
Hu, Y. X. & Stamnes, K. 1993 An accurate parameterization of the radiative properties of water clouds suitable for use in climate models. *J. Climate*, **6**, 728–742.
Illingworth, A. J., Hogan, R. J., O'Connor, E. J. *et al.* 2007 Cloudnet – continuous evaluation of cloud profiles in seven operational models using groundbased observations. *Bull. Amer. Meteorol. Soc.*, **88**, 883–898.
Jakob, C. 2002 Ice clouds in numerical weather prediction models: progress, problems and prospects. In *Cirrus*, eds. D. K. Lynch, K. Sassen, D. Starr & G. Stephens, pp. 327–345. Oxford University Press.
Jakob, C. 2003 An improved strategy for the evaluation of cloud parameterizations in GCMs. *Bull. Amer. Meteorol. Soc.*, **84**, 1387–1401.
Jung, J. H. & Arakawa, A. 2005 Preliminary tests of multiscale modeling with a two-dimensional framework: sensitivity to coupling methods. *Mon. Weather Rev.*, **133**, 649–662.
Kahn, B. H., Chahine, M., Stephens, G. *et al.* 2008 Cloud type comparisons of AIRS, CloudSat, and CALIPSO cloud height and amount. *Atmos. Chem. Phys*, **8**, 1231–1248.
Kärcher, B. & Lohmann, U. 2002 A parameterization of cirrus cloud formation: homogeneous freezing of supercooled aerosols. *J. Geophys. Res.*, **107**, 4010, 10.1029/2001JD000470.
Kessler, E. 1969 *On the Distribution and Continuity of Water Substance in Atmospheric Circulation*, Meteorological Monography, 10. American Meteorological Society.
Khairoutdinov, M., Randall, D. & DeMott, C. 2005 Simulations of the atmospheric general circulation using a cloud-resolving model as a superparameterization of physical processes. *J. Atmos. Sci.*, **62**, 2136–2154.
Khairoutdinov, M. F. & Randall, D. A. 2001 A cloud resolving model as a cloud parameterization in the NCAR Community Climate System Model: preliminary results. *Geophys. Res. Lett.*, **28**, 3617–3620.
Klein, S. A. & Hartmann, D. L. 1993 The seasonal cycle of low stratiform cloud. *J. Climate*, **6**, 1587–1606.

Klein, S. A. & Jakob, C. 1999 Validation and sensitivities of frontal clouds simulated by the ECMWF model. *Mon. Weather Rev.*, **127**, 2514–2531.
Köhler, M. 1999 Explicit prediction of ice clouds in general circulation models. Ph.D. thesis, University of California Los Angeles.
Kristjansson, J. E., Edwards, J. M. & Mitchell, D. L. 2000 Impact of a new scheme for optical properties of ice crystals on climates of two GCMs. *J. Geophys. Res.*, **105**, 10063–10079.
Krüger, S. K. & Zulauf, M. A. 2005 Radiatively induced cirrus anvil spreading. Proceedings of the 15th Atmospheric Radiation Measurement (ARM) Science Team Meeting, ARM, US Department of Energy (available at http://www.arm.gov/publications).
Larson, K. & Hartmann, D. L. 2003 Interactions among cloud, water vapor, radiation, and large-scale circulation in the tropical climate. Part I: sensitivity to uniform sea surface temperature changes. *J. Climate*, **16**, 1425–1440.
Lee, M. I., Kang, I. S., Kim, J. K. & Mapes, B. E. 2001 Influence of cloud radiation interaction on simulating tropical intraseasonal oscillation with an atmospheric general circulation model. *J. Geophys. Res.*, **106**, 14219–14233.
Li, J.-L., Waliser, D. E., Jiang, J. H. *et al.* 2005 Comparisons of EOS MLS cloud ice measurements with ECMWF analyses and GCM simulations: initial results. *Geophys. Res. Lett.*, **32**, L18710 (doi:10.1029/2005GL023788).
Li, J. L., Jiang, J. H., Waliser, D. E. & Tompkins, A. M. 2007 Assessing consistency between EOS MLS and ECMWF analyzed and forecast estimates of cloud ice. *Geophys. Res. Lett.*, **34**, L08701, doi: 10.1029/2006GL029022.
Li, T., Hogan, T. F. & Chang, C. P. 2000 Dynamic and thermodynamic regulation of ocean warming. *J. Atmos. Sci.*, **57**, 3353–3365.
Lilly, D. K. 1968 Models of cloud-topped mixed layers under a strong inversion. *Q. J. R. Meteorol. Soc.*, **94**, 292–309.
Lin, H., Noone, K. J., Ström, J. & Heymsfield, A. J. 1998 Dynamical influences on cirrus cloud formation process. *J. Atmos. Sci.*, **55**, 1940–1949.
Lin, J. L. & Mapes, B. E. 2004 Wind shear effects on cloud–radiation feedback in the western Pacific warm pool. *Geophys. Res. Lett.*, **31**, L16118, (doi:10.1029/2004GL020199).
Lin, J. L., Kiladis, G. N. Mapes, B. E. *et al.* 2006 Tropical intraseasonal variability in 14 IPCC AR4 climate models. Part I: convective signals. *J. Climate*, **19**, 2665–2690.
Liou, K.-N. 1980 *An Introduction to Atmospheric Radiation*, International Geophysics Series. Accademic Press.
Lohmann, U. & Feichter, J. 1997 Impact of sulphate aerosols on albedo and lifetime of clouds. *J. Geophys. Res.*, **102**, 13685–13700.
Lohmann, U. & Kärcher, B. 2002 First interactive simulations of cirrus cloud formed by homogeneous freezing in the ECHAM general circulation model. *J. Geophys. Res.*, **107**, doi:10.1029/2001JD000767.
Macke, A., Mueller, J. & Raschke, E. 1996 Single scattering properties of atmospheric ice crystals. *J. Atmos. Sci.*, **53**, 2813–2825.
Madden, R. A. & Julian, P. R. 1971 Detection of a 40–50 day oscillation in the zonal wind in the tropical Pacific. *J. Atmos. Sci.*, **5**, 702–708.
Madden, R. A. & Julian, P. R. 1994 Observations of the 40–50-day tropical oscillations – a review. *Mon. Weather Rev.*, **122**, 814–837.
Marshak, A., Davis, A., Wiscombe, W. & Cahalan, R. F. 1997 Scale invariance in liquid water distributions in marine stratocumulus. 2 Multifractal properties and intermittency issues. *J. Atmos. Sci.*, **54**, 1423–1444.

McKee, T. B. & Cox, S. K. 1974 Scattering of visible radiation by finite clouds. *J. Atmos. Sci.*, **31**, 1885–1892.
Meehl, G. A., Boer, G. J., Covey, C., Latif, M. & Stouffer, R. J. 2000 The Coupled Model Intercomparison Project (CMIP). *Bull. Amer. Meteorol. Soc.*, **81**, 313–318.
Mellor, G. L. 1977 Gaussian cloud model relations. *J. Atmos. Sci.*, **34**, 356–358.
Meyers, M. P., DeMott, P. J. & Cotton, W. R. 1992 New primary ice nucleation parameterization in an explicit model. *J. Appl. Meteorol.*, **31**, 708–721.
Miller, R. 1997 Tropical thermostats and low cloud cover. *J. Climate*, **10**, 409–440.
Mischenko, M. I., Hovenier, J. W. & Travis, L. D. 2000 *Light Scattering by Non Spherical Particles: Theory Measurements and Applications*. Accademic Press.
Mlawer, E. J. & Clough, S. 1997 Shortwave and longwave enhancements in the Rapid Radiative Transfer Model. *Proceedings of the 7th Atmospheric Radiation Measurement (ARM) Science Team Meeting*. ARM, US Department of Energy (available at http://www.arm.gov/publications, CONF–9603149).
Mlawer, E. J., Taubman, S. J., Brown, P. D., Iacono, M. J. & Clough, S. A. 1997 Radiative transfer for inhomogeneous atmospheres: RRTM a validated correlated-k model for the longwave. *J. Geophys. Res.*, **102**, 16663–16682.
Moeng, C. H., Sullivan, P. P. & Stevens, B. 1999 Including radiative effects in an entrainment rate formula for buoyancy-driven PBLs. *J. Atmos. Sci.*, **56**, 1031–1049.
Morcrette, J.-J. 1989 Comparison of satellite-derived and model-generated diurnal cycles of cloudiness and brightness temperatures. *Adv. Space Res.*, **9**, 175–179.
Morcrette, J.-J. 1991 Evaluation of model-generated cloudiness: satellite-observed and model-generated diurnal variability of brightness temperature. *Mon. Weather Rev.*, **119**, 1205–1224.
Morcrette, J.-J. & Fouquart, Y. 1986 The overlapping of cloud layers in shortwave radiation parameterizations. *J. Atmos. Sci.*, **43**, 321–328.
Morcrette, J.-J. & Jakob, C. 2000 The response of the ECMWF model to changes in the cloud overlap assumption. *Mon. Weather Rev.*, **128**, 1707–1732.
Murphy, J. M., Sexton, D., Barnett, D. *et al.* 2004 Quantification of modelling uncertainties in a large ensemble of climate change simulations. *Nature*, **430**, 768–772.
Nair, U. S., Weger, R. C., Kuo, K. S. & Welch, R. M. 1998 Clustering randomness, and regularity in clouds fields 5. The nature of regular cumulus clouds fields. *J. Geophys. Res.*, **103**, 11363–11380.
Nakajima, T., King, M. D., Spinhirne, J. D. & Radke, L. F. 1991 Determination of the optical thickness and effective particle radius of clouds from reflected solar radiation measurements. Part II: marine stratocumulus observations. *J. Atmos. Sci.*, **48**, 728–751.
O'Hirok, W. & Gautier, C. 1998 A three-dimensional radiative transfer model to investigate the solar radiation within a cloudy atmosphere. Part I: spatial effects. *J. Atmos. Sci.*, **55**, 2162–2179.
Palmer, T., Alessandri, A., Andersen, U. *et al.* 2004 Development of a European multimodel ensemble system for seasonal-to-interannual prediction (DEMETER). *Bull. Amer. Meteorol. Soc.*, **85**, 853–872.
Pincus, R. 2006 Assessing tools for 3-dimensional radiative transfer. Proceedings of the 12th Conference on Atmospheric Radiation, American Meteorological Society, 10–14 July, Madison, WI.
Pincus, R., Barker, H. W. & Morcrette, J. J. 2003 A fast flexible, approximate technique for computing radiative transfer in inhomogeneous cloud fields. *J. Geophys. Res.*, **108**, 4376, doi:10.1029/2002JD003322.

Pincus, R., Hannay, C., Klein, S. A., Xu, K.-M. & Hemler, R. 2005 Overlap assumptions for assumed probability distribution function cloud schemes in largescale models. *J. Geophys. Res.*, **110**, D15S09, doi:10.1029/2004JD005100.
Pruppacher, H. R. & Klett, J. D. 1997 *The Microphysics of Clouds and Precipitation*. Kluwer Academic Publishers.
Rädel, G., Stubenrauch, C. J., Holz, R. & Mitchell, D. L. 2003 Retrieval of effective ice crystal size in the infrared: sensitivity study and global measurements from TIROS-N Operational Vertical Sounder. *J. Geophys. Res.*, **108**, 4281, doi:10.1029/2002JD002801.
Räisänen, P., Barker, H. W. & Cole, J. N. S. 2005 The Monte Carlo independent column approximation's conditional random noise: impact on simulated climate. *J. Climate*, **18**, 4715–4730.
Ramanathan, V. & Collins, W. 1991 Thermodynamic regulation of ocean warming by cirrus clouds deduced from observations of the 1987 El Nino. *Nature*, **351**, 27–32.
Ramanathan, V., Cess, R. D. Harrison, E. F. *et al.* 1989 Cloud-radiative forcing and climate: results from the Earth radiation budget experiment. *Science*, **243**, 57–63.
Randall, D. A. 1980 Conditional instability of the first kind upside-down. *J. Atmos. Sci.*, **37**, 125–130.
Randall, D. A., Wood, R. A., Bony, S. *et al.* 2007 Climate models and their evaluation. *Climate Change 2007: The Physical Science Basis. Contribution of Working Group I to the Fourth Assessment Report of the Intergovernmental Panel on Climate Change*, ed. S. Solomon, D. Din, M. Manning *et al.* Cambridge University Press.
Randall, D. A., Harshvardhan, D. A., Dazlich, D. A. & Corsetti, T. G. 1989 Interactions among radiation convection, and large-scale dynamics in a general circulation model. *J. Atmos. Sci.*, **46**, 1943–1970.
Randall, D. A., Khairoutdinov, M., Arakawa, A. & Grabowski, W. 2003 Breaking the cloud parameterization deadlock. *Bull. Amer. Meteorol. Soc.*, **84**, 1547–1564.
Raymond, D. J. 2000 The Hadley circulation as a radiative–convective instability. *J. Atmos. Sci.*, **57**, 1286–1297.
Raymond, D. J. 2001 A new model of the Madden–Julian Oscillation. *J. Atmos. Sci.*, **58**, 2807– 2819.
Ren, C. & MacKenzie, A. R. 2005 Cirrus parametrization and the role of ice nuclei. *Q. J. R. Meteorol. Soc.*, **131**, 1585–1605.
Ringer, M. A. & Allan, R. P. 2004 Evaluating climate model simulations of tropical cloud. *Tellus*, **A56**, 308–327.
Rossow, W. B. & Schiffer, R. A. 1991 ISCCP cloud data products. *Bull. Amer. Meteorol. Soc.*, **72**, 2–20.
Segelstein, D. J. 1981 The complex refractive index of water. M.Sc. thesis. University of Missouri-Kansas City.
Sherwood, S. C., Ramanathan, V., Barnett, T. P., Tyree, M. K. & Roeckner, E. 1994 Response of an atmospheric general-circulation model to radiative forcing of tropical clouds. *J. Geophys. Res.*, **99**, 20829–20845.
Slingo, A. 1989 A GCM Parameterization for the shortwave radiative properties of water clouds. *J. Atmos. Sci.*, **46**, 1419–1427.
Slingo, A. & Slingo, J. M. 1988 The response of a general-circulation model to cloud longwave radiative forcing. 1. Introduction and initial experiments. *Q. J. R. Meteorol. Soc.*, **114**, 1027–1062.
Slingo, J. M., Sperber, K. R., Boyle, J. S. *et al.* 1996 Intraseasonal oscillations in 15 atmospheric general circulation models: results from an AMIP diagnostic subproject. *Clim. Dyn.*, **12**, 325–357.

Sobel, A. H. & Gildor, H. 2003 A simple time-dependent model of SST hot spots. *J. Climate*, **16**, 3978–3992.
Sohn, B. J. 1999 Cloud-induced infrared radiative heating and its implications for large-scale tropical circulations. *J. Atmos. Sci.*, **56**, 2657–2672.
Stephens, G. 2005 Cloud feedbacks in the climate system: a critical review. *J. Climate*, **18**, 237–273.
Stephens, G. L. 1984 The parameterization of radiation for numerical weather prediction and climate models. *Mon. Weather Rev.*, **112**, 826–867.
Stephens, G. L., Vane, D. G., Boain, R. J. *et al.* 2002 The CloudSat mission and the A-Train. *Bull. Amer. Meteorol. Soc.*, **83**, 1771–1790.
Stephens, G. L., Wood, N. B. & Gabriel, P. M. 2004 An assessment of the parameterization of subgrid-scale cloud effects on radiative transfer. Part I: vertical overlap. *J. Atmos. Sci.*, **715**, 715–732.
Stevens, B. 2002 Entrainment in stratocumulus-topped mixed layers. *Q. J. R. Meteorol. Soc.*, **128**, 2663–2690.
Stokes, G. M. & Schwartz, S. E. 1994 The Atmospheric Radiation Measurement (ARM) Program: programmatic background and design of the cloud and radiation test bed. *Bull. Amer. Meteorol. Soc.*, **75**, 1201–1221.
Stubenrauch, C. J., Delgenio, A. D. & Rossow, W. B. 1997 Implementation of subgrid cloud vertical structure inside a GCM and its effect on the radiation budget. *J. Climate*, **10**, 273–287.
Sui, C. H., Lau, K.-M., Tao, W. K. & Simpson, J. 1994 The tropical water and energy cycles in a cumulus ensemble model. Part I: equilibrium climate. *J. Atmos. Sci.*, **51**, 711–728.
Sundqvist, H. 1978 A parameterization scheme for non-convective condensation including prediction of cloud water content. *Q. J. R. Meteorol. Soc.*, **104**, 677–690.
Tao, W. K., Lang, S., Simpson, J., Sui, C. H., Ferrier, B. & Chou, M. D. 1996 Mechanisms of cloud–radiation interaction in the tropics and midlatitudes. *J. Atmos. Sci.*, **53**, 2624–2651.
Tian, B. & Ramanathan, V. 2003 A simple moist tropical atmosphere model: the role of cloud radiative forcing. *J. Climate*, **16**, 2086–2092.
Tiedtke, M. 1993 Representation of clouds in large-scale models. *Mon. Weather Rev.*, **121**, 3040–3061.
Tomita, H., Miura, H., Iga, S., Nasuno, T. & Satoh, M. 2005 A global cloud resolving simulation: preliminary results from an aqua planet experiment. *Geophys. Res. Lett.*, **32**, L08805, doi:10.1029/2005GL022459.
Tompkins, A. M. 2001 On the relationship between tropical convection and sea surface temperature. *J. Climate*, **14**, 633–637.
Tompkins, A. M. 2002 A prognostic parameterization for the subgrid-scale variability of water vapor and clouds in large-scale models and its use to diagnose cloud cover. *J. Atmos. Sci.*, **59**, 1917–1942.
Tompkins, A. M. & Craig, G. C. 1998 Radiative–convective equilibrium in a three-dimensional cloud ensemble model. *Q. J. R. Meteorol. Soc.*, **124**, 2073–2097.
Tompkins, A. M. & Craig, G. C. 1999 Sensitivity of tropical convection to sea surface temperature in the absence of large-scale flow. *J. Climate*, **12**, 462–476.
Tompkins, A. M. & Di Giuseppe, F. 2006 Generalizing cloud overlap treatment to include solar zenith angle effects on cloud geometry. *J. Atmos. Sci.*, **64**, 2116–2125.

Tompkins, A. M. & Jung, T. 2003 Influence of process interactions on MJO-like convective structures in the IFS model. ECMWF/CLIVAR workshop on simulation and prediction of intra-seasonal variability with the emphasis on the MJO, ECMWF (available from http://www.ecmwf.int/publications, pp. 103–114).

Tselioudis, G. & Jakob, C. 2002 Evaluation of midlatitude cloud properties in a weather and a climate model: dependence on dynamic regime and spatial resolution. *J. Geophys. Res.*, **107**, 4781, doi:10.1029/2002JD002259.

Uppala, S. M., Kållberg, P. W., Simmons, A. J. *et al.* 2005 The ERA-40 re-analysis. *Q. J. R. Meteorol. Soc.*, **131**, 2961–3012.

Valero, F. P. J., Cess, R. D., Zhang, M. *et al.* 1997 Absorption of solar radiation by the cloudy atmosphere: interpretations of collocated aircraft measurements. *J Geophys. Res.*, **102**, 29917–29927.

van de Hulst, H. C. 1980 *Multiple Light Scattering*, vol. 1. Academic Press.

van Zanten, M., Duynkerke, P. & Cuijpers, J. 1999 Entrainment parameterization in convective boundary layers. *J. Atmos. Sci.*, **56**, 813–828.

Vitart, F., Woolnough, S., Balmaseda, M. A. & Tompkins, A. M. 2007 Monthly forecast of the Madden–Julian Oscillation using a coupled GCM. *Mon. Weather Rev.*, **135**, 2700–2715.

Vogelmann, A. M., Ramanathan, V. & Podgorny, I. A. 2001 Scale dependence of solar heating rates in convective cloud systems with implications to general circulation models. *J. Climate*, **14**, 1738–1752.

Waliser, D. E. 1996 Formation and limiting mechanisms for very high sea surface temperature: linking the dynamics and the thermodynamics. *J. Climate*, **9**, 161–188.

Wentz, F. J. 1997 A well-calibrated ocean algorithm for SSM/I. *J. Geophys. Res.*, **102**, 8703–8718.

Wielicki, B. A., Barkstrom, B. R., Harrison, E. F. *et al.* 1996 Clouds and the Earth's radiant energy system (CERES): an Earth observing system experiment. *Bull. Amer. Meteorol. Soc.*, **77**, 853–868.

Wielicki, B. A., Cess, R. D., King, M. D., Randall, A. D. & Harrison, E. F. 1995 Mission to planet Earth: role of clouds and radiation in climate. *Bull. Amer. Meteorol. Soc.*, **76**, 2125–2153.

Wiscombe, W. J. 1980 Improved Mie scattering algorithms. *Appl. Opt.*, **19**, 1505–1509.

Woolnough, S. J., Slingo, J. M. & Hoskins, B. J. 2001 The organisation of tropical convection by intraseasonal sea surface temperature anomalies. *Q. J. R. Meteorol. Soc.*, **127**, 887–907.

Wu, X., Grabowski, W. W. & Moncrieff, M. W. 1998 Long-term behavior of cloud systems in TOGACOARE and their interactions with radiative and surface processes. Part I: two-dimensional modeling study. *J. Atmos. Sci.*, **55**, 2693–2714.

Wu, X. & Moncrieff, M. W. 1999 Effects of sea surface temperature and largescale dynamics on the thermodynamic equilibrium state and convection over the tropical Western Pacific. *J. Geophys. Res.*, **104**, 6093–6100.

Wyant, M. C., Bretherton, C. S., Bacmeister, J. T. *et al.* 2006 A comparison of low-latitude cloud properties and their response to climate change in three AGCMs sorted into regimes using mid-tropospheric vertical velocity. *Clim. Dyn.*, **27**, 261–279.

Xu, K.-M. & Randall, D. A. 1995 Impact of interactive radiative-transfer on the macroscopic behavior of cumulus ensembles. 2. Mechanisms for cloud–radiation interactions. *J. Atmos. Sci.*, **52**, 800–817.

Xu, K.-M. & Randall, D. A. 1996 A semiempirical cloudiness parameterization for use in climate models. *J. Atmos. Sci.*, **53**, 3084–3102.

Zdunkowski, W. G., Panhans, W. G., Welch, R. M. & Korb, G. J. 1982 A radiation scheme for circulation and climate models. *Beitr. Phys. Atmos.*, **55**, 215–238.

Zhang, M. H., Lin, W. Y., Klein, S. *et al.* 2005 Comparing clouds and their seasonal variations in 10 atmospheric general circulation models with satellite measurements. *J. Geophys. Res.*, **110**, D15502 (doi:10.1029/2004JD005021).

Zhu, T., Lee, J., Weger, R. C. & Welch, R. M. 1992 Clustering randomness and regularity in cloud fields: 2. Cumulus cloud fields. *J. Geophys. Res.*, **97**, 20537–20558.

Zuidema, P. 1998 Radiative transfer through realistic clouds. Ph.D. Thesis, University of Colorado.

Zuidema, P. & Evans, K. F. 1998 On the validity of the independent pixel approximation for boundary layer clouds observed during ASTEX. *J. Geophys. Res.*, **103**, 6059–6074.

Zurovac-Jevtić, D., Bony, S. & Emanuel, K. 2006 On the role of clouds and moisture in tropical waves: a two-dimensional model study. *J. Atmos. Sci.*, **63**, 2140–2155.

15

Impact of a quasi-stochastic cellular automaton backscatter scheme on the systematic error and seasonal prediction skill of a global climate model

JUDITH BERNER, FRANCISCO J. DOBLAS-REYES, TIM N. PALMER, GLENN J. SHUTTS AND ANTJE WEISHEIMER

The impact of a nonlinear dynamic cellular automaton (CA) model, as a representation of the partially stochastic aspects of unresolved scales in global climate models, is studied in the European Centre for Medium-Range Weather Forecasts coupled ocean–atmosphere model. Two separate aspects are discussed: impact on the systematic error of the model, and impact on the skill of seasonal forecasts. Significant reductions of systematic error are found both in the tropics and in the extra-tropics. Such reductions can be understood in terms of the inherently nonlinear nature of climate, in particular how energy injected by the CA at the near-grid scale can backscatter nonlinearly to larger scales. In addition, significant improvements in the probabilistic skill of seasonal forecasts are found in terms of a number of different variables such as temperature, precipitation and sea-level pressure. Such increases in skill can be understood both in terms of the reduction of systematic error as mentioned above, and in terms of the impact on ensemble spread of the CA's representation of inherent model uncertainty.

15.1 Introduction

Ever since their introduction, numerical climate models have been formulated according to a rather precise prescription: represent the equations of motion as accurately as possible by projection onto a Galerkin basis down to some truncation scale, and represent the effect of unresolved scales on the resolved-scale motions through deterministic bulk-formula parameterisations. The tendencies associated with such parameterisations are determined by, and hence slaved to, the resolved-scale flow, typically at the truncation scale.

Such bulk-formula parameterisations are motivated by concepts in statistical mechanics. For example, the formulation of the effect of molecular viscosity is given by considering an ensemble of randomly moving molecules, the mean-free path of which is small compared with some macroscale of interest. Similarly, climate model parameterisations (e.g. of convection or gravity wave drag) are formulated by supposing that there exists an ensemble of subgrid processes (e.g. convective plumes or gravity waves) in quasi-equilibrium with the resolved-scale flow.

By using a simplified cloud-resolving model, Shutts & Palmer (2007) studied this statistical mechanical assumption quantitatively, in the case of convection in the tropics. In regions of strong convection it was found that the bulk-formula assumptions were in error by an order of magnitude; rather than determining the subgrid tendency, the truncation-scale motions provided only a partial constraint on some underlying probability distribution of subgrid tendencies.

Consistent with these findings, Palmer (2001) has suggested that subgrid motions should be represented by simplified nonlinear stochastic dynamic models, as an alternative to the deterministic bulk-formula approach. One consequence of such an approach is that a stochastic dynamic model can allow subgrid-scale energy to be backscattered onto the resolved-scale grid.

Stochastic kinetic energy backscatter has its origins in early three-dimensional turbulence closures that recognised the deficiencies of the eddy viscosity concept (Kraichnan 1976). Leith (1990) and Mason & Thomson (1992) demonstrated the benefits of using a stochastic backscatter term in large eddy simulations of turbulence and Frederiksen & Davies (1997) extended the technique to large-scale planetary motion. The idea is based on the notion that turbulent dissipation rate is the difference between upscale and downscale spectral transfer and that the upscale component is available to the resolved flow as a kinetic energy source. In a nonlinear system, such backscattered energy can in principle reduce the large-scale systematic error of the model. In addition, through multiple independent random samplings of the underlying stochastic processes, an ensemble of integrations is readily generated by such models, thus providing a representation of underlying model uncertainties.

One simple type of stochastic dynamic model considered by Palmer (2001) is based on the cellular automaton (CA). This approach has been further developed by Shutts (2005) and applied to medium-range forecast problems. In this chapter, the CA scheme of Shutts is applied to climate time scales. Specifically, the impact of the scheme, both on the systematic error of the European Centre for Medium-Range Weather Forecasts (ECMWF) coupled ocean–atmosphere model, and on the seasonal forecast skill of the same model, is studied.

A brief summary of the CA scheme is described in Section 15.2. The experimental design is given in Section 15.3, and results in terms of impact on systematic error and on seasonal forecast skill are given in Section 15.4. Conclusions are summarised in Section 15.5.

15.2 The CA backscatter scheme

In this section, we give a brief summary of Shutts' kinetic energy CA backscatter scheme (CASBS) and point out some minor differences from the original scheme. An extension of this scheme and its impact on medium-range ensemble forecasting and flow-dependent predictability and error growth is discussed in Berner *et al.* (2009).

Shutts (2005) argues that in numerical weather prediction (NWP) systematic kinetic energy is lost in both numerical integration schemes and parameterisations. For instance, errors in semi-Lagrangian departure-point interpolation cause a net energy sink, and kinetic energy released in deep convection does not find its way sufficiently into balanced flows and gravity wave generation. Consequently, the backscatter scheme aims at representing upscale error growth on synoptic and subsynoptic scales from convection, gravity and mountain/wave drag and numerical dissipation.

The backscatter scheme generates a flow-dependent stochastic kinetic energy source by introducing a stream function forcing

$$\psi(\varphi, \lambda, z, t) = \sqrt{D(\varphi, \lambda, z, t)}\phi(\varphi, \lambda, t), \tag{15.2.1}$$

where $D(\varphi, \lambda, z, t)$ is the instantaneous dissipation rate and $\phi(\varphi, \lambda, t)$ a realisation of the CA. The resulting stream-function forcing is then transformed to spectral space, converted into a vorticity increment and added to the dynamical equations. True to its nature, as a kinetic energy backscatter scheme, this will affect the kinetic energy spectrum of the model. In ensemble forecasts, CASBS is initialised with different choices of the CA pattern. Since these patterns can be chosen randomly, CASBS is effectively a stochastic subgrid parameterisation.

In the following paragraphs, we give some more details on the pattern generator and the computation of the total dissipation rate $D(\varphi, \lambda, z, t)$. The CA is a variant of the CA known as Conway's Game of Life (e.g. Gardner 1983). The CA domain covers the entire globe in a fine grid of cells that are regular in latitude/longitude coordinates; here 720×360 cells (Fig. 15.1*a*). Each cell can be alive or dead and living cells have up to 32 lives. Dead cells come to life if their nearest neighbours satisfy certain 'birth' conditions (e.g. when the number of them with 32 lives equals either 3, 4 or 5). When a dead cell comes to life it is given 32 lives and these are progressively lost in later steps when certain survival rules fail to be met. Only cells with the maximum number of lives (32) participate if the rule counts. In order to present a sufficiently smooth pattern to the forecast model, the cell values are averaged into 2° 'macro' cells comprising 4×4 CA cells and the subsequent array is smoothed again using passes of a 1–2–1 filter first in x (longitude) and then in y (latitude). The smoothed values are then normalised so that the domain average

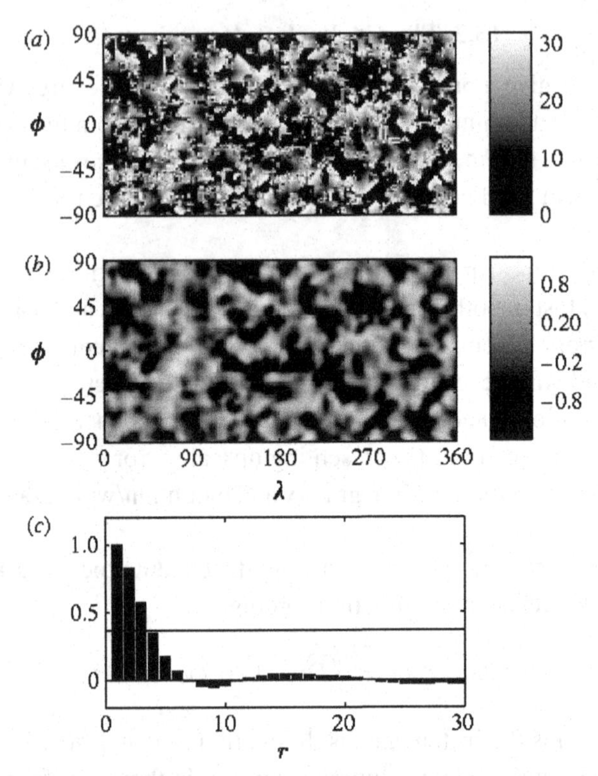

Figure 15.1 (*a*) Raw pattern of 720 × 360 cells generated by the fine-scale CA used in the CASBS. The shading depicts the number of lives. (*b*) Pattern after coarse-graining onto 180 × 90 macro cells, smoothing and normalisation (see text). The shading shows the pattern values after normalisation. (*c*) Spatial correlation as a function of the distance r between macro cells, in units of the width of a single macro cell. The horizontal line signifies the e-folding distance.

is unity (Fig. 15.1*b*). The resulting pattern looks very 'organic' and consists of clusters of densely populated cells with distinct fronts, reminiscent of mesoscale structures in the atmosphere. Further detail on the CASBS algorithm can be found in Shutts (2005).

The spatial structure of the smoothed coarse-grained CA pattern is analysed by calculating the spatial correlation as a function of the distance r between the macro cells, in units of the width of a single macro cell. The spatial correlation decreases with distance and oscillates at approximately zero for macro cells of seven or more units apart (Fig. 15.1*c*). The e-folding distance, i.e. the decorrelation scale, equals four units. Since the macro cell represents an area of $2 \times 2°$ the decorrelation scale is effectively approximately 8° or 800 km.

The total instantaneous dissipation rate $D(\varphi, \lambda, z, t)$ contains contributions from deep convection, numerical dissipation and gravity/mountain wave drag. We refer to Shutts (2005) for details on the computation of the numerical dissipation and contribution from gravity/mountain wave drag. Changes in the NWP model made it necessary to no longer base the dissipation from deep convection on convective updraught speeds; but instead a mass-flux formulation was used

$$D_{\text{conv}}(\varphi, \lambda, z, t) = \frac{\delta(\varphi, \lambda, z, t)}{\rho(\varphi, \lambda, z, t)^3} \frac{M(\varphi, \lambda, z, t)^2}{\beta^2},$$

where $M(\varphi, \lambda, z, t)$ is the updraft convective mass-flux rate in kg $(\text{m}^3\,\text{s})^{-1}$ from the convective parameterisations, $\delta(\varphi, \lambda, z, t)$ the updraft detrainment rate in kg $(\text{m}^3\,\text{s})^{-1}$, $\rho(\varphi, \lambda, z, t)$ the density and β an assumed detrainment cloud fraction for deep convection of $\beta = 2.6 \times 10^{-2}$. To focus on the large-scale structure, the total dissipation rate $D(\varphi, \lambda, z, t)$ is subsequently smoothed by applying a spectral filter that completely retains wavenumbers $n < 20$ and gradually reduces to zero for wavenumbers $20 < n < 30$.

The total dissipation rate per unit area typical of the ECMWF atmospheric model for the months December through February and its contributions from deep convection, numerical dissipation and gravity/mountain wave drag are shown in Fig. 15.2. For illustration only, we show density-weighted vertical averages of the three-dimensional dissipation fields. The dominating contributor to the total dissipation rate with a global mean of 0.99 $\text{W}\,\text{m}^{-2}$ is deep convection. Its maxima are in the deep convective regions of the tropics, especially over Indonesia, but also downstream of the Andes. The most active regions are seen in the summer hemisphere, and for the period June through August there are larger contributions from the Northern Hemisphere (not shown). With a global average of 0.56 $\text{W}\,\text{m}^{-2}$, the second largest contribution comes from the numerical dissipation. It is largest in the storm track regions and downstream of high orography such as Greenland, the Himalayas and the Andes. The global-mean dissipation from gravity and mountain wave drag is much smaller and occurs mainly in the lower tropospheric levels over orography. However locally, over the major mountain ranges, the dissipation rates may be very large indeed (e.g. over 100 $\text{W}\,\text{m}^{-2}$).

15.3 Experimental set-up

The experiments discussed in this chapter were performed using an ensemble of dynamical seasonal reforecasts. Two sets of experiments with and without the stochastic backscatter scheme were performed. In the following, the control model simulations without stochastic backscatter will be denoted as CTRL, whereas the experiments using the new stochastic physics scheme will be referred to as CASBS.

The model used to produce these two sets of reforecasts was integrated over seven months initialised twice a year on 1 May and 1 November at 00.00 hour

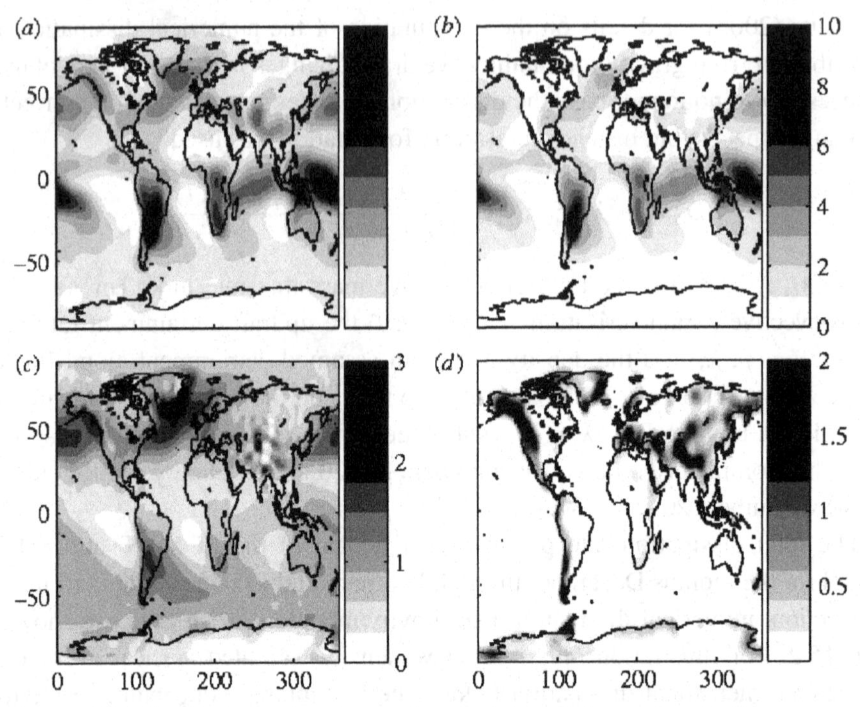

Figure 15.2 Vertically averaged annual mean (*a*) total dissipation (1.65) rate per unit area of the period December to February with contributions from (*b*) deep convection (0.99), (*c*) numerical dissipation (0.56), (*d*) gravity/mountain wave drag (0.09). Units are W m^{-2}. The global averages are indicated in brackets.

GMT over the 11-year period 1991–2001. Ensembles of nine members generated by different initial conditions were used for each start date, so that a total set of 22 nine-member seasonal reforecasts was available to analyse the results for both CTRL and CASBS.

As a forecast model, we used the ECMWF coupled atmosphere–ocean general circulation model (Anderson *et al.* 2007). The atmospheric component is the integrated forecasting system of ECMWF as used in operational medium-range weather forecasting in its version CY29R2. The model was run with a horizontal truncation of T_L95 and 40 vertical levels. The Hamburg Ocean Primitive Equation (HOPE) ocean model has a horizontal resolution of 1°, with an equatorial refinement of 0.3°, and 29 levels in the vertical. The coupler OASIS (version 2) is used to interpolate once per day between the oceanic and atmospheric grids.

The atmospheric initial conditions, including land surface conditions, come from ERA-40 (Uppala *et al.* 2005); the ocean initial conditions come from an ensemble of ocean analyses (Balmaseda *et al.* 2007). The nine initial-condition ensemble members were generated by adding small perturbations to the atmospheric and oceanic fields. For the initialisation of the atmosphere, perturbations based on

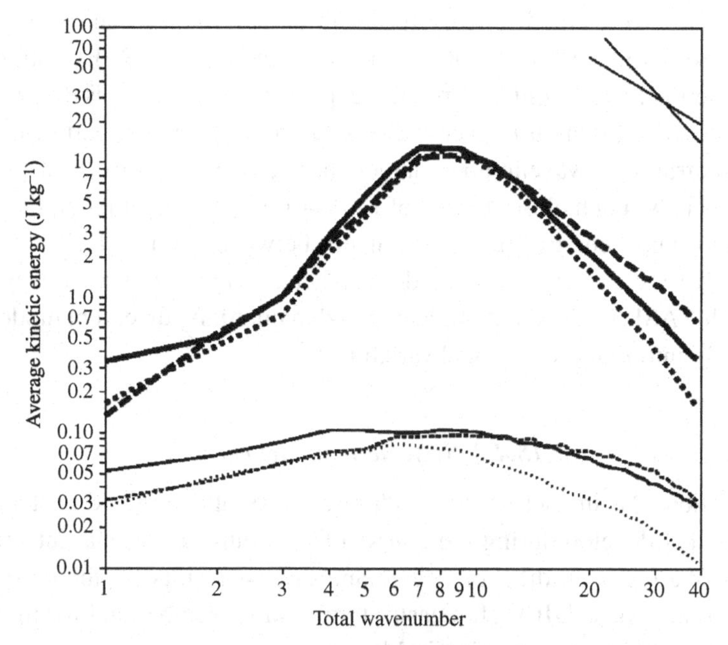

Figure 15.3 Mean kinetic energy spectra at 500 hPa of rotational (upper) and divergent (lower) component for simulations starting on 1 November over the period 1991–2001. The anomalies were computed as departures from the respective long-term mean. Solid line, ERA-40; dotted line, CTRL; dashed line, CASBS. Lines in upper right corner denote power-law behaviour with the slopes of -3 and $-5/3$.

singular vectors were applied in a similar way as in the operational medium-range ensemble forecasts (Buizza & Palmer 1995; Rodwell & Doblas-Reyes 2006). For the initialisation of the ocean, three different ocean analyses were created by adding daily wind-stress perturbations on the basis of the differences between two quasi-independent data sets. In addition, four sea-surface temperature (SST) perturbations, again based on the differences between two SST data sets, were added to or subtracted from the initial fields.

15.4 Results

15.4.1 Effect on kinetic energy spectrum

Figure 15.3 shows the rotational and divergent parts of the kinetic energy spectrum at 500 hPa for ERA-40 data, a control integration without stochastic backscatter (CTRL) and an integration with CASBS. The kinetic energy in CTRL is less than in ERA-40, suggesting that the model is underactive for all wavenumbers n, but especially so for $n > 10$. The stochastic backscatter scheme injects energy, so that the kinetic energy increases for all wavenumbers. This leads to an improvement in the rotational part of the kinetic energy spectrum in the synoptic band (wavenumbers 2–12) but to a slight overestimation for wavenumbers $n > 12$. By introducing a stream-function forcing, only the rotational part of the spectrum is directly forced. Nevertheless, a substantial impact is also seen in the divergent part of the kinetic energy spectrum for wavenumbers larger than $n = 5$. The divergent part of the spectrum is now much closer to that of ERA-40 (Fig 15.3), although there is now slightly too much energy for wavenumbers between 20 and 40. From this we conclude that the activity in the model has been changed in such a way that the model with CASBS is better at producing and maintaining divergent modes, which is especially important for tropical variability.

15.4.2 Systematic model error

In this section, the impact of the stochastic backscatter scheme on the systematic errors that develop during the course of the coupled seasonal integrations is described. We focus on the atmosphere and discuss results for the boreal winter (December–February, DJF) reforecasts starting in November and for the summer (June–August) reforecasts starting in May.

Figure 15.4 shows the systematic errors in simulating DJF mean total precipitation. The difference between CTRL and the Global Precipitation Climatology Project (GPCP) verification data set (Adler *et al.* 2003) is displayed in Fig. 15.4*a*. The control model generates excessive precipitation over the tropics. The inter-tropical convergence zone (ITCZ) over the tropical Pacific is particularly wet. The control reforecasts also generate too much precipitation in the Indonesian warm pool area. A dry bias can be seen over large parts of South America and Northern Australia. The errors in mid latitudes are in general smaller than in the tropics.

The systematic error of the CASBS runs is shown in Fig. 15.4*b*; Fig. 5.4*c* displays the difference CASBS minus CTRL. The impact of the stochastic backscatter scheme on precipitation errors can be summarised as reducing the wet bias over the tropical ITCZ, in particular over the Pacific north of the equator. This is due to both a southward shift of the convergence zone and a reduction of the overall rainfall amounts. A decrease of the excessive precipitation over the tropical Atlantic and the Indian Ocean can also be detected. The systematic errors in precipitation over land are mainly unaffected by the stochastic parameterisation scheme, although a positive impact can be found over Northern Australia.

Jung *et al.* (2005) showed that previous versions of the stochastic physics scheme induced significant changes in the extra-tropical tropospheric circulation in an uncoupled model. As a diagnostic of this impact, the Tibaldi & Molteni (1990) blocking detection index has been used to estimate the frequency of persistent blocking anticyclones in the simulations. The index computes the 500 hPa geopotential height meridional gradient for specific mid-latitude bands (of

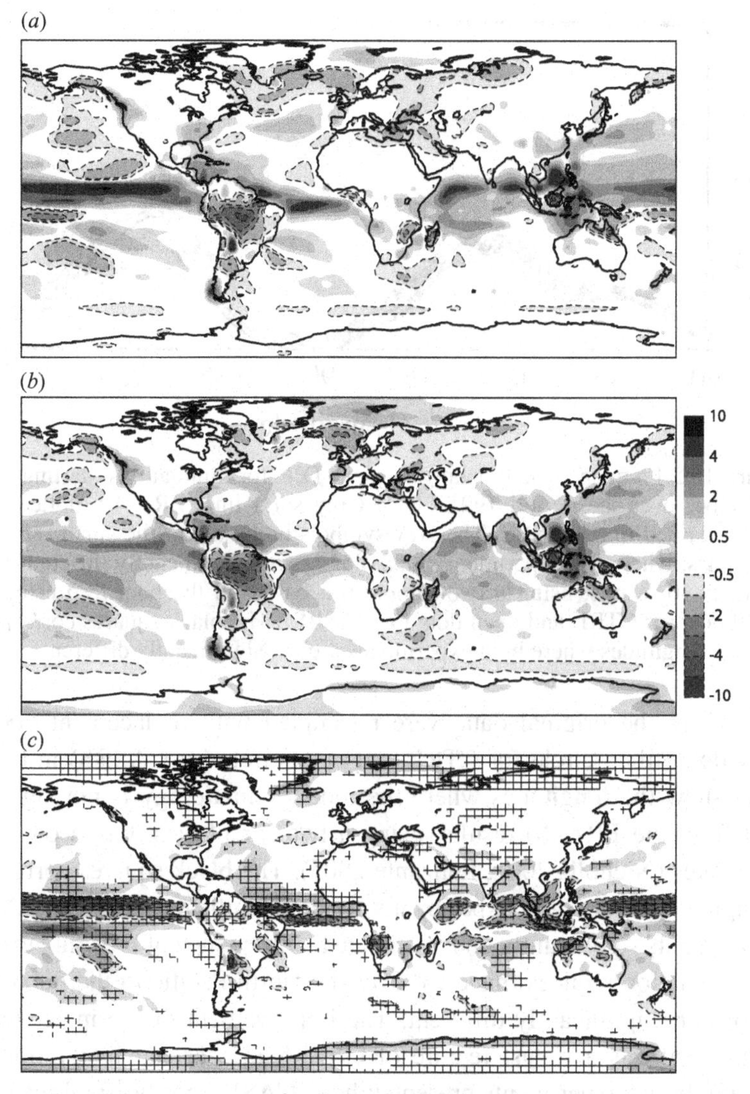

Figure 15.4 Systematic error of DJF total precipitation with respect to GPCP for simulations starting on 1 November over the period 1991–2001 for (*a*) CTRL and (*b*) CASBS. (*c*) Difference between CASBS and CTRL experiments; hatching indicates areas where the experiments are statistically significantly different with 95% confidence. Units are mm d^{-1}.

latitude between 40–60° N and 60–80° N) and classifies a longitude for a given time step as blocked, if in the northernmost latitude band a reverse of the westerly flow is detected. An event is classified as blocked if at least 10° of longitude are blocked over more than five consecutive days. Figure 15.5 shows the mean blocking frequency for DJF averaged over the years 1991–2001 and the nine ensemble members. Ninety per cent confidence intervals have been computed using a bootstrap

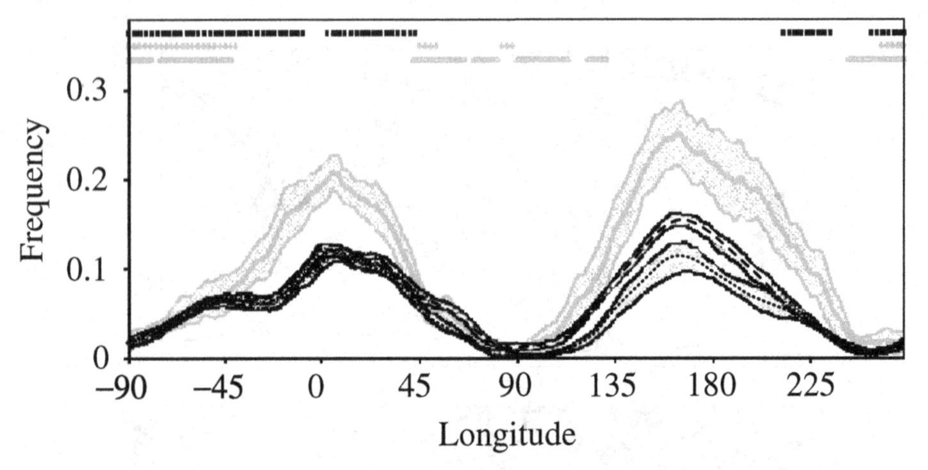

Figure 15.5 Frequency of blocking days in DJF for simulations starting on 1 November over the period 1991–2001. Grey solid line, ERA-40; dotted line, CTRL; dashed line, CASBS. The grey symbols in the top of the panel show the longitudes, where the experiment climatology is not significantly different from the ERA-40 results, using a two-sample test based on the bootstrap estimates (filled dots for CTRL and open dots for CASBS). The black squares correspond to those longitudes where both experiments are not significantly different.

method, where the original data were resampled with replacement 500 times. The grey dots (filled circle for CTRL and open circle for CASBS) in the top of the panel show the longitudes where the model climatology is not significantly different from the ERA-40 results, using a two-sample test based on the bootstrap estimates (Nicholls 2001; Lanzante 2005). The black squares correspond to those longitudes where both experiments are not significantly different. Although both experiments significantly underestimate the frequency of blocking events, the CASBS simulation reduces this error over the North Pacific to the point that the simulations are significantly different. The improvement of the mean state over the North Pacific is a robust feature of CASBS in seasonal integrations (Jung *et al.* 2005). In the experiments presented here, CASBS has no significant impact on Atlantic blocking.

In Fig. 15.6, we show the growth of systematic errors in SST over the Niño3.4 region (5° S–5° N, 170–120° W) in the eastern-to-central tropical Pacific, a key region for the development of the El Niño/Southern oscillation (ENSO) phenomenon, over the seven-month lead time of the reforecasts. The dotted curves show the SST bias for the CTRL simulation for the two start dates in May and November; the dashed ones that for the CASBS simulations. For both start dates and the two sets of experiments, the modelled SSTs are warmer than the climatological temperatures. However, for the May start dates, this warm bias is systematically reduced in the CASBS runs. For example, whereas in the CTRL experiment the

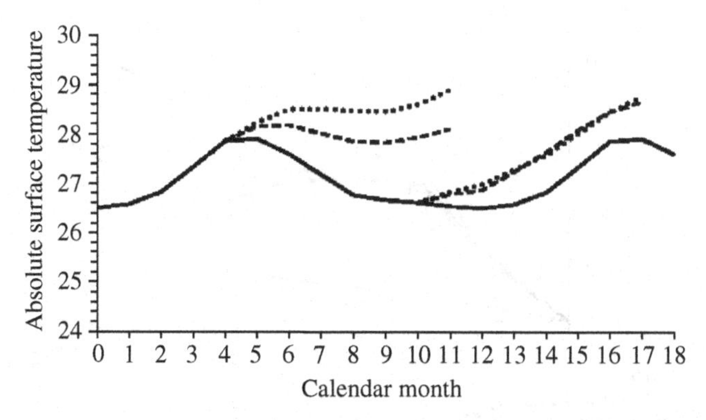

Figure 15.6 Sea-surface temperature (°C) over seven months in the Niño3.4 region (5° S–5° N, 170–120° W) for seasonal reforecasts starting on 1 May and 1 November over the period 1991–2001. Solid line, climatological annual cycle; dotted line, CTRL; dashed line, CASBS.

SST errors after three and seven months lead time come close to 1.5 and 2.5 K, respectively, they are reduced by nearly 30%, that is approximately 0.5 K after three months and almost 1 K after seven months, in the CASBS experiments. For the November start dates no difference between the two was found.

15.4.3 Forecast quality assessment

Various measures of forecast quality have been used to assess the relative merit of CASBS when compared with CTRL. The scores include the anomaly correlation of the ensemble mean coefficient (ACC) and, for dichotomous probability forecasts, the Brier skill score (BSS) with respect to climatology and the relative operating characteristic skill score (ROCSS; Jolliffe & Stephenson 2003). The BSS has also been decomposed into the sum of two components (Murphy 1986): the reliability (RELSS) term that measures the relative bias of conditional means, and the resolution (RESSS) term that measures the relative variance of the conditional means. The BSS decomposition used here includes two additional terms in the resolution component which account for the within-bin variance and covariance of the probability forecasts, as described in Stephenson *et al.* (2008). As the BSS depends strongly on the ensemble size for small ensembles, Ferro (2007) has developed an analytical expression to estimate the Brier score for different ensemble sizes using the Brier score value obtained from the sample. This expression that depends on the variance (also known as sharpness) of the probability forecasts, allows the computation of a theoretical BSS for infinite ensemble sizes (BSSI henceforth).

The seasonal forecast system performance has been traditionally tested on tropical Pacific SSTs. This is because the main source of seasonal predictability is

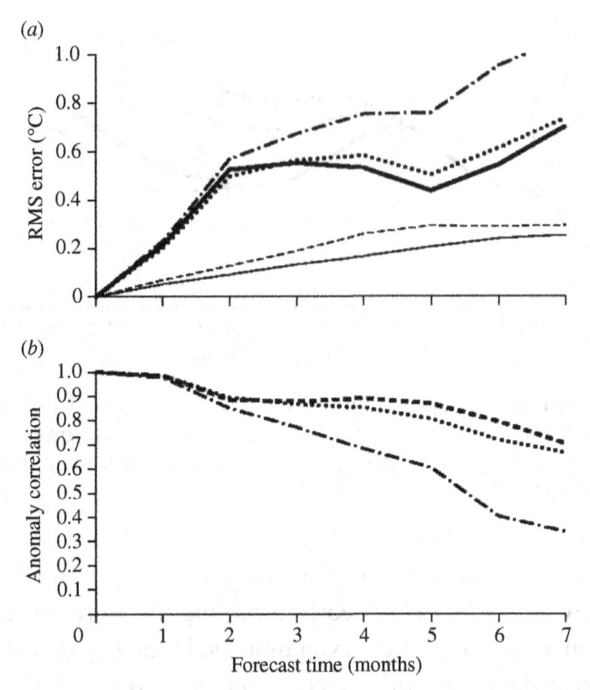

Figure 15.7 (*a*) Root-mean-square error of the ensemble mean (thick lines) versus lead time for anomalies of surface temperature over the Niño3.4 region (5° S–5° N, 170–120° W) of the CTRL (dotted line) and CASBS (dashed line) experiments for the May and November start dates in the period 1991–2001. The dot–dashed line corresponds to a simple statistical model based on persisting with the anomaly of the month previous to the start date. The thin lines show the spread estimated as the standard deviation of the ensemble members around the ensemble mean. (*b*) Anomaly correlation coefficient of the ensemble mean for CTRL (dotted line), CASBS (dashed line) and persistence (dot–dashed line).

assumed to come from the interannual variability related to ENSO. Figure 15.7 shows the root-mean-square (RMS) error and the anomaly correlation (ACC) of ensemble-mean surface temperature anomalies over the Niño3.4 region as a function of lead time, for the May and November start dates over the period 1991–2001. For comparison, the RMS error and forecast and ACC of a simple statistical model based on the persistence of the anomaly of the month previous to the start date are also shown. The accuracy of the reforecasts generally decreases with lead time, although both experiments have higher skill than this simple persistence. The smaller RMS error and higher ACC for CASBS are evident after the third month of the forecast. This improvement of CASBS over CTRL is mainly associated with the forecasts started in May; for these start dates CASBS also leads to a large reduction in the systematic error. Figure 15.7*a* also shows the spread of both experiments measured by the standard deviation of the ensemble members around the ensemble mean. For a reliable ensemble system, state-dependent uncertainty is expected to be

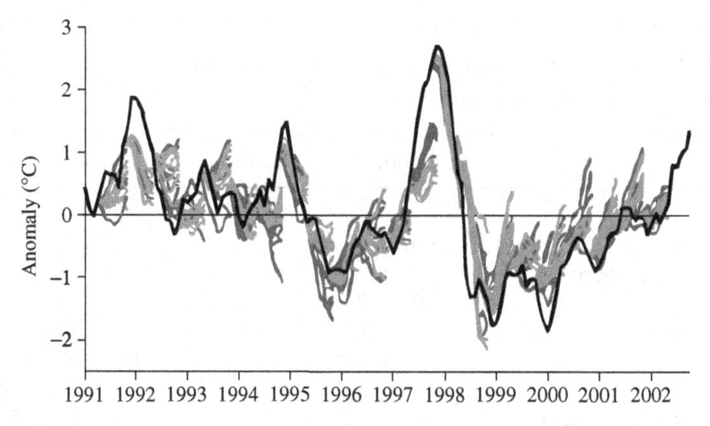

Figure 15.8 Ensemble forecasts of SST anomalies over the Niño3.4 region (5° S–5° N, 170–120° W) of the CTRL (light grey) and CASBS (dark grey) experiments for the May and November start dates in the period 1991–2001. The solid black line represents the observed monthly mean anomalies with respect to the mean seasonal cycle of the period 1971–2000.

modelled by the spread. In a well-calibrated system, spread and RMS error should have a similar magnitude, which is not the case for CTRL. However, both experiments underestimate the forecast uncertainty, although CASBS has an increased spread with respect to CTRL from the first month of the integration.

Figure 15.8 illustrates the different ensemble forecasts for each experiment in the Niño3.4 region. In general, CASBS has a larger spread than CTRL, and its ensemble members encompass the CTRL ensemble in most cases. Furthermore, there are cases where CASBS seems to yield a better forecast. For example, the onset of the extremely warm ENSO event in 1997/98 was better predicted in CASBS than in CTRL.

The full set of forecast quality measures has been computed for several variables (500 hPa geopotential height, 850 hPa temperature, precipitation, two-metre temperature and mean sea-level pressure) over a large number of regions (Table 15.1) and for three different events in the case of the probability forecasts: anomalies above the upper tercile, above the median and below the lower tercile. The terciles and the median have been computed separately for each experiment and for the reference data set to take into account the different systematic errors in the simulated distributions. Figure 15.9 shows the scatter plots of BSSI and ROCSS of CTRL versus CASBS for both start dates, the first four regions in Table 15.1, the three events mentioned above, and several forecast ranges (first month and forecast periods 1–3, 2–4, 3–5 and 4–6 months), making a total of 600 cases. For both skill scores there is a large range of values, from unskilful predictions (lower than zero) to values close to 0.5 and 1 for BSSI and ROCSS, respectively. Frequently, CASBS performs better than CTRL, as the larger number of points above the diagonal

Table 15.1 *Regions used in the computation of the forecast quality measures. (The first four regions include land and ocean grid points, while only land points have been considered in the rest of the regions.)*

	Latitude (south, north)	Longitude (west, east)
Europe	35° to 75°	−12.5° to 42.5°
North America	30° to 70°	−130° to −60°
Northern Hemisphere	30° to 87.5°	0° to 360°
Tropics	−20° to 20°	0° to 360°
Mediterranean	30° to 47.5°	−10° to 40°
Australia	−45° to −11°	110° to 155°
Amazon	−20° to 12°	−82.5° to −35°
Southern South America	−55° to −20°	−75° to −35°
Western North America	30° to 60°	−130° to −82.5°
Eastern North America	25° to 50°	−85° to −60°
Northern Europe	47.5° to 75°	−10° to 40°
West Africa	−12.5° to 17.5°	−20° to 22.5°
East Africa	−12.5° to 17.5°	22.5° to 52.5°
Southern Africa	−35° to −12.5°	−10° to 52.5°
Southeast Asia	−10° to 20°	95° to 155°
East Asia	20° to 50°	100° to 145°
Southern Asia	5° to 30°	65° to 100°
Central Asia	30° to 50°	40° to 75°
North Asia	50° to 70°	40° to 180°

suggests. The difference of the skill scores for a given variable, region, lead time, start date and event, between CASBS and CTRL, has been tested for significance using a two-sample test and a bootstrap method with 1000 samples. While CTRL is significantly better (with 95% confidence) than CASBS in three cases for BSSI and in six cases for ROCSS, CASBS is significantly better than CTRL in 251 and 118 cases, respectively.

This overwhelming superiority of CASBS over CTRL in terms of forecast quality is also evidenced in Table 15.2, which summarises the results for several skill scores and all the regions in Table 15.1. The large proportion of cases where RELSS is better for CASBS than for CTRL suggests that the increase in BSS and BSSI can be largely attributed to an improvement in reliability. This might be linked to the alleviated underdispersion of the ensemble when using CASBS. A less intuitive result is the large number of cases where the ROCSS in CASBS is significantly higher than in CTRL. This improvement in terms of forecast resolution (of which ROCSS is a measure) is especially important because the resolution of a forecast system can only be enhanced using additional sources of forecast information, while the reliability can be improved a posteriori using climatological information. Improved resolution could arise owing to the smaller systematic error in CASBS.

Table 15.2 *Number of cases in which one of the experiments is significantly better (with 95% confidence) than the other for different forecast quality measures. (The skill scores have been computed for both start dates, all the regions in Table 15.1, the variables 500 hPa geopotential height, 850 hPa temperature, precipitation, 2-m temperature and mean sea-level pressure, and several forecast times (first month and seasonal forecast periods 1–3, 2–4, 3–5 and 4–6 months). This gives 380 cases for the ACC and 1140 for the rest of the scores. The event anomalies above the upper tercile and the median and below the lower tercile have been considered for the probability forecasts.)*

	ACC	BSS	BSSI	RELSS	RESSS	ROCSS
CTRL	23	62	58	23	58	66
CASBS	114	644	729	735	149	303

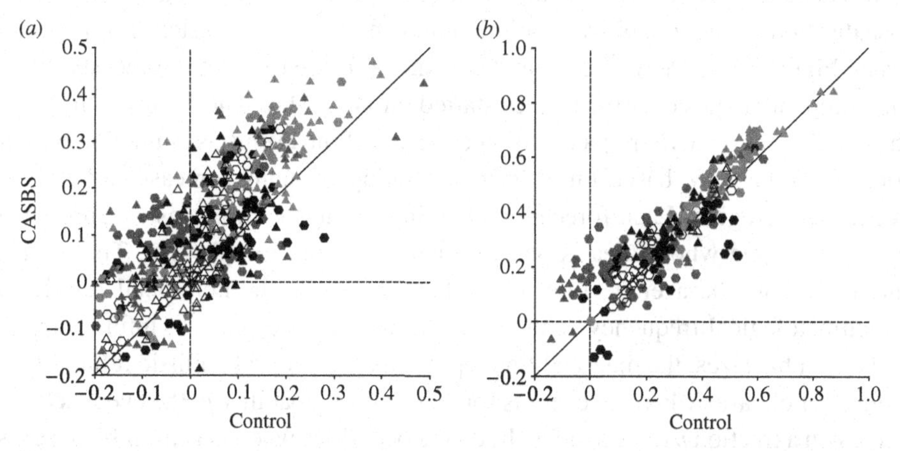

Figure 15.9 Brier skill score for (*a*) infinite ensemble size and (*b*) ROCSS for CASBS (ordinate) versus CTRL (abscissa). The scatter plots collect results for both start dates, the first four regions in Table 15.1 [tropics in black, Northern Hemisphere in dark grey, Europe in light grey and North America open symbols], the variables 500 hPa geopotential height, 850 hPa temperature, precipitation, 2-m temperature and mean sea-level pressure, several forecast times (first month and forecast periods 1–3, 2–4, 3–5 and 4–6 months) and the event anomalies above the upper tercile and the median and below the lower tercile. Circles (triangles) have been used for the results of the May (November) start date. The symbols below and to the left of the no-skill dotted lines have less skill than a climatological forecast.

The improvements due to CASBS can be highly relevant in specific cases. Figure 15.9 shows that most of the cases with negative skill scores for the European region for CTRL have positive skill for CASBS, especially for the May start date (note that the negative skill cases over Europe are displayed with dark-grey dots in Fig. 15.9). This improvement can be appreciated in more detail in Fig. 15.10, where the ROCSS for the predictions of 500 hPa geopotential height and 850 hPa temperature is shown as a function of the start date and the forecast range. While the forecast quality decreases with lead time (from left to right, with the November start date results starting after the first five sets of bars), all the sample values (i.e. those obtained with the reforecasts without resampling, which are displayed with black dots) for CASBS are positive, which is not always the case for CTRL. Confidence intervals (95%) obtained with the bootstrap method described above are displayed around the skill score sample value. They suggest that it is more likely to obtain a ROCSS significantly different from zero (colour bar clear of the zero-skill horizontal line) with CASBS than with the control.

Attribute diagrams (Hsu & Murphy 1986) offer a comprehensive illustration of the benefits of CASBS in terms of forecast quality. These diagrams allow the visualisation of the reliability, resolution and sharpness of a system for a specific event. Figure 15.11 shows the attribute diagrams for first-month reforecasts of precipitation anomalies over the tropics started on the 1 May. The diagrams illustrate the conditional relative frequency of occurrence of the events as a function of their forecast probability, based on a discrete binning of many forecast probabilities taken over a region. Each forecast probability bin in the diagrams is represented by a grey circle whose area is proportional to the bin sample size. The vertical line represents the average forecast probability, while the horizontal line is for the climatological frequency of the event. In the idealised case of infinite sample and ensemble sizes, the diagonal line represents perfect probabilistic reliability: if from a set of cases where an event is forecast with probability p, the event actually occurs on a fraction p of occasions. If, on the occasions where an ensemble forecast predicts some event with probability p, the event occurs in reality on a fraction q of times, then if p is sufficiently different from q, the ensemble forecast probabilities are not reliable. This case will appear in the diagram as a point away from the diagonal. If the corresponding curve is shallower than the diagonal, the forecast system is said to be overconfident, while if it is steeper the system will be underconfident. The sum of the horizontal squared distance of all the points to the diagonal (weighted by the sample size of each bin) is a measure of the lack of reliability of the system as measured by the Brier score. In the same way, the sum of the vertical distance of the points to the horizontal line corresponding to the climatological frequency of the event measures the forecast resolution, i.e. the ability of the system to produce reliable forecasts that differ from the naive probability. This means that

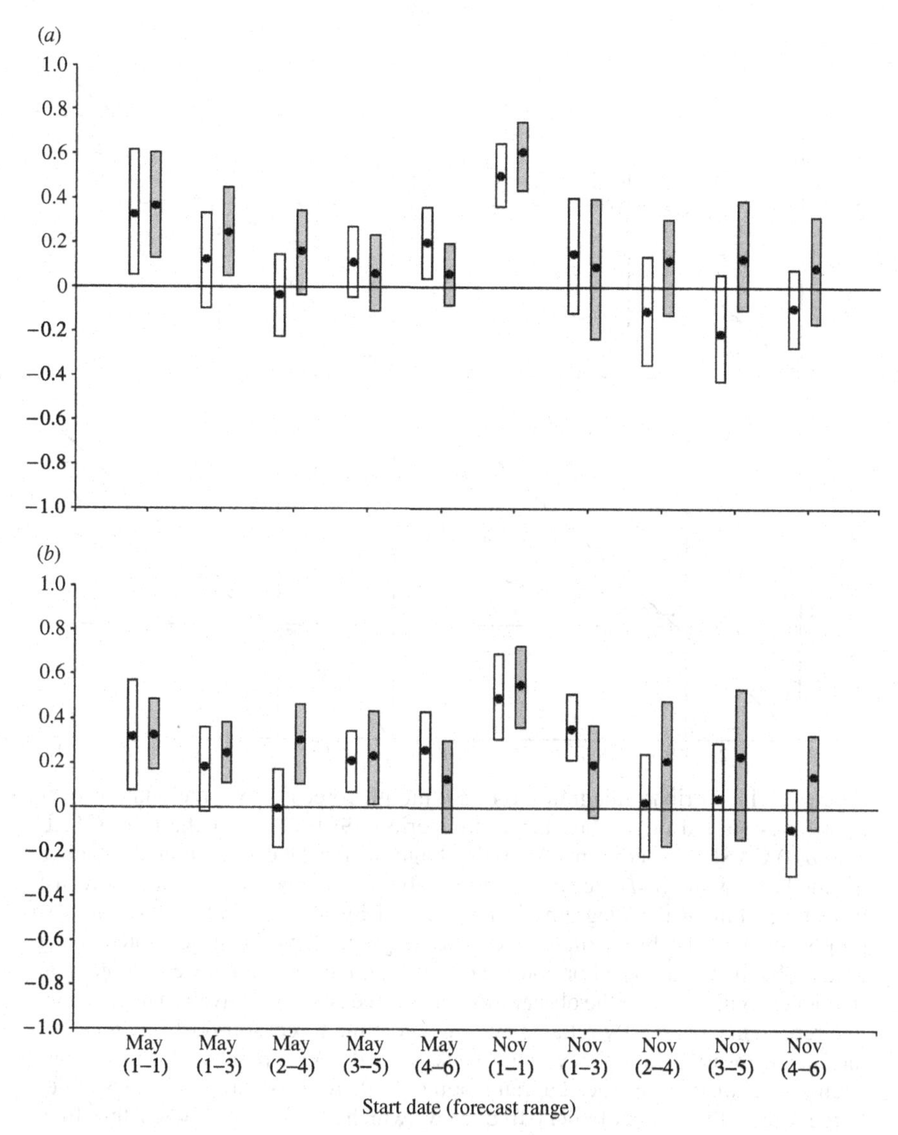

Figure 15.10 The ROCSS for the event 'anomalies above the upper tercile' of (*a*) 500 hPa geopotential height and (*b*) 850 hPa temperature for the CTRL (open bars) and CASBS (filled bars) experiments as a function of the start date and the forecast range (in brackets). The scores have been computed over Europe. The sample values are represented with a black circle, while the bars correspond to the 95% confidence, intervals obtained with a bootstrap method (see text for details).

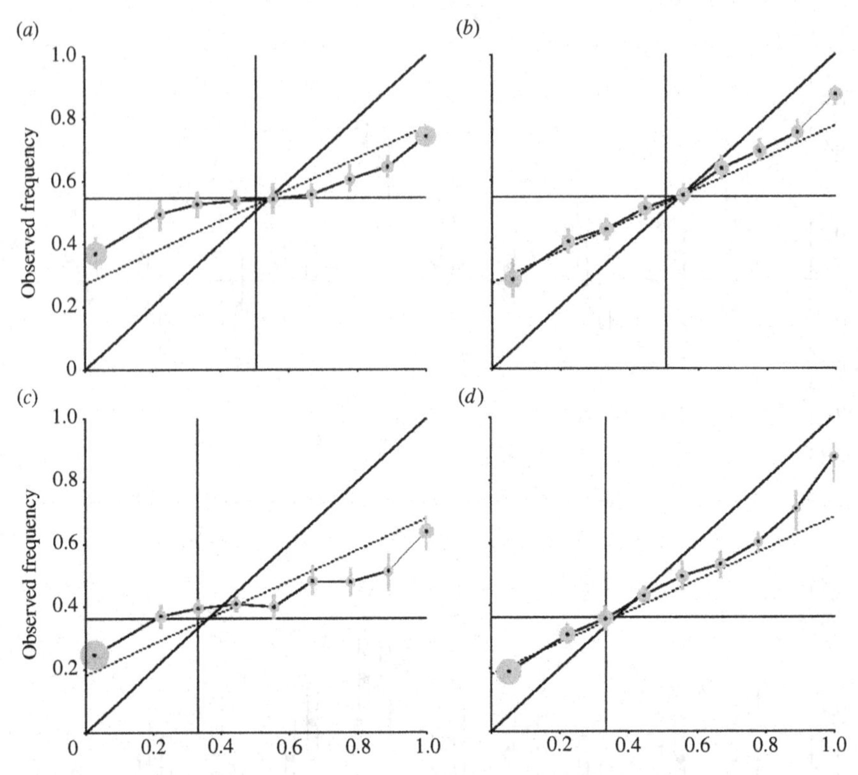

Figure 15.11 Attribute diagrams for first-month forecasts of precipitation over the tropics started on 1 May during the period 1991–2001 for the (*a,c*) CTRL and (*b,d*) CASBS experiments. (*a,b*) the diagrams for the event 'anomalies above the median', while (*c,d*) are for anomalies above the upper tercile. Each forecast probability bin in the diagrams is represented by a grey circle whose area is proportional to the bin sample size. The diagrams have been generated using nine probability bins. The horizontal and vertical lines indicate the climatological frequency of the event in the observations and forecasts, respectively. The diagonal line represents a perfectly reliable system where the forecast probability matches the mean observed frequency. The black dashed line separates skilful regions from unskilful ones in the diagram: points with forecast probabilities smaller (larger) than the climatological frequency, which fall below (above) this line, contribute to positive BSS; otherwise they contribute negatively to the BSS. Grey vertical bars over the dots indicate the 95% confidence intervals of the estimated observed frequency based on a 1000 bootstrap resampling procedure. The BSSs for infinite ensemble size and corresponding 95% confidence intervals are (*a*) −0.091 (−0.149, −0.034), (*b*) 0.153 (0.120, 0.187), (*c*) −0.085 (−0.146, −0.025) and (*d*) 0.160 (0.119, 0.203).

if a reliability curve were to be horizontal, the frequency of occurrence would not depend on the forecast probabilities and the system would have zero resolution. The black dashed line in the diagram separates skilful regions from unskilful ones in the diagram: points with forecast probabilities smaller (larger) than the climatological frequency, which fall below (above) this line, contribute to positive BSS; otherwise they contribute negatively to the BSS. Finally, the sharpness is the variance of the forecast probabilities and is shown in the diagram by the proportion of probability forecasts away from the average probability.

The diagrams in Fig. 15.11 show that the unreliable curves obtained for CTRL are much steeper for CASBS, bringing the curve into the skilful area determined by the black dashed line (note the significant improvement of the corresponding BSSI). This happens for both the events considered. The increased steepness improves both the reliability (by decreasing the distance to the diagonal) and the resolution (by increasing the distance to the horizontal line corresponding to the climatological frequency). This occurs at the expense of a reduction in the variance of the forecast probability, or sharpness, illustrated by a reduction in the size of the dots for the extreme forecast probabilities in CASBS with respect to CTRL. Although a sharp set of probabilities, i.e. forecast probabilities with larger variance, is a desirable feature of a forecast system, large sharpness may be harmful in the case of overconfident reliability diagrams, such as the ones for CTRL.

15.5 Summary and conclusions

In this chapter, we have studied the impact on the climate of the ECMWF coupled model of a CA scheme, as first proposed by Palmer (2001), with specific implementation by Shutts (2005), as a partial representation of stochastic subgrid processes. Two aspects of this impact have been studied: on the systematic error of the model and on seasonal forecast skill. For the former, some remarkable reductions in the error in tropical seasonal mean rainfall, extra-tropical blocking frequency, and temperature drift in the tropical Pacific were found. The skill of the model, measured both in terms of ensemble-mean accuracy in predicting tropical Pacific SST variability, and in terms of probabilistic seasonal predictions of temperature, precipitation and mean sea-level pressure, was notably improved by the addition of the CA scheme.

The reduction of systematic error can be understood by the fact that climate is a profoundly nonlinear dynamical system, and that the proposed CA acts as a source of subsynoptic-scale and mesoscale energy that can be backscattered to the planetary-scale components of the flow. A simple analogy is a double-well potential. By adding stochastic noise to the system, the subdominant well may be visited more frequently, thus altering the mean state of the system. The increase in forecast skill arises from a combination of two effects. The first is the reduction in systematic error as discussed above. The second arises from the fact that the model equations are a key source of forecast uncertainty for climate prediction and, through different random initialisations, the CA provides a representation of model uncertainty in ensemble predictions. Thus the CA increases the reliability of probabilistic seasonal forecasts, thereby impacting on standard skill scores.

In this paper, a relatively simple CA has been implemented in which the accurate representation of the dynamics of atmospheric subgrid motions is somewhat minimal. As such, one line of development in the CA model is the inclusion of more explicit atmospheric subgrid dynamics; for example, a 'Mexican wave' representation of equatorially trapped Kelvin wave dynamics can be coded very simply in a CA model. Hence, for example, one can envisage representing through a CA the sort of convectively driven wave modes in the tropical atmosphere which are not well resolved in a standard global climate model. This is work for the future.

Acknowledgements

This work was supported by the ENSEMBLES project (GOCE-CT-2003–505539). The authors would like to acknowledge Dr Thomas Jung for his contribution to the analysis of the experiments.

References

Adler, R. F. *et al.* 2003 The version 2 global precipitation climatology project (GPCP) monthly precipitation analysis (1979-present). *J. Hydrometeorol.*, **4**, 1147–1167 (doi:10.1175/1525–7541(2003) 004<1147:TVGPCP>2.0.CO;2).

Anderson, D. L. T., Stockdale, T., Balmaseda, M. A. *et al.* 2007 Development of the ECMWF seasonal forecast system 3. ECMWF Technical Memorandum 503. See http://www.ecmwf.int/publications/library/do/references/show?id=87744.

Balmaseda, M., Vidard, A. & Anderson, D. 2007 The ECMWF system 3 ocean analysis system. ECMWF Technical Memorandum 508. See http://www.ecmwf.int/publications/library/do/ references/show?id=87667.

Berner, J., Shutts, G. J., Leutbecher, M. & Palmer, T. N. 2009 A spectral stochastic kinetic energy backscatter scheme and its impact on flow-dependent predictability in the ECMWF ensemble prediction system. *J. Atmos. Sci.*, **66**, 603–626.

Buizza, R. & Palmer, T. N. 1995 The singular-vector structure of the atmospheric global circulation. *J. Atmos. Sci.*, **52**, 1434–1456 (doi:10.1175/1520–0469(1995)052<1434:TSVSOT>2.0.CO;2).

Ferro, C. A. T. 2007 Comparing probabilistic forecasting systems with the Brier score. *Weather and Forecasting*, **22**, 1076–1088 (doi:10.1175/WAF1034.1).

Frederiksen, J. S. & Davies, A. G. 1997 Eddy viscosity and stochastic backscatter parametrizations on the sphere for atmospheric circulation models. *J. Atmos. Sci.*, **54**, 2475–2492 (doi:10.1175/1520–0469(1997)054 <2475:EVASBP>2.0.CO;2).

Gardner, M. 1983 The game of life, parts I–III. In *Wheels, Life, and Other Mathematical Amusements*. W. H. Freeman.

Hsu, W.-R. & Murphy, A. H. 1986 The attributes diagram: a geometrical framework for assessing the quality of probability forecasts. *Int. J. Forecasting*, **2**, 285–293 (doi:10.1016/0169–2070(86)90048–8).

Jolliffe, I. T. & Stephenson, D. B. 2003 *Forecast Verification: a Practitioner's Guide in Atmospheric Science*. Wiley and Sons.

Jung, T., Palmer, T. N. & Shutts, G. J. 2005 Influence of stochastic parameterization on the frequency of occurrence of North Pacific weather regimes in the ECMWF model. *Geophys. Res. Lett.*, **32**, L23811 (doi:10.1029/2005GL024248).

Kraichnan, R. H. 1976 Eddy viscosity in two and three dimensions. *J. Atmos. Sci.*, **33**, 1521–1536. (doi:10.1175/1520–0469(1976)033<1521: EVITAT>2.0.CO;2).

Lanzante, J. R. 2005 A cautionary note on the use of error bars. *J. Climate*, **18**, 3699–3703 (doi:10.1175/JCLI3499.1).

Leith, C. E. 1990 Stochastic backscatter in a sub-grid-scale model: plane shear mixing layer. *Phys. Fluids A*, **2**, 297–299 (doi:10.1063/1.857779).

Mason, P. J. & Thomson, D. J. 1992 Stochastic backscatter in large-eddy simulations of boundary layers. *J. Fluid Mech.*, **242**, 51–78 (doi:10.1017/S0022112092002271).

Murphy, A. H. 1986 A new decomposition of the Brier score: formulation and interpretation. *Mon. Weather Rev.*, **114**, 2671–2673. (doi:10.1175/1520–0493(1986)114<2671: ANDOTB>2.0.CO;2).

Nicholls, N. 2001 The insignificance of significance testing. *Bull. Am. Meteorol. Soc.*, **82**, 981–986 (doi:10.1175/1520–0477(2001)082 <0981:CAATIO>2.3.CO;2).

Palmer, T. N. 2001 A nonlinear dynamical perspective on model error: a proposal for non-local stochastic-dynamic parameterization in weather and climate prediction. *Q. J. R. Meteorol. Soc.*, **127**, 279–304 (doi:10.1002/qj.49712757202).

Rodwell, M. & Doblas-Reyes, F. J. 2006 Predictability and prediction of European monthly to seasonal climate anomalies. *J. Climate*, **19**, 6025–6046 (doi:10.1175/JCLI3944.1).

Shutts, G. J. 2005 A kinetic energy backscatter algorithm for use in ensemble prediction systems. *Q. J. R. Meteorol. Soc.*, **131**, 3079–3102 (doi:10.1256/qj.04.106).

Shutts, G. J. & Palmer, T. N. 2007 Convective forcing fluctuations in a cloud-resolving model: relevance to the stochastic parameterization. *J. Climate*, **20**, 187–202 (doi:10.1175/JCLI3954.1).

Stephenson, D. B., Coelho, C. A. S. & Jolliffe, I. T. 2008. Two extra components of the Brier score decomposition. *Weather Forecasting*, **23**, 752–757.

Tibaldi, S. & Molteni, F. 1990 On the operational predictability of blocking. *Tellus,* **A42**, 343–365 (doi:10.1034/j.1600–0870.1990.t01-2–00003.x).

Uppala, S. M. *et al.* 2005 The ERA-40 reanalysis. *Q. J. R. Meteorol. Soc.*, **131**, 2961–3012 (doi:10.1256/qj.04.176).

16

Rethinking convective quasi-equilibrium: observational constraints for stochastic convective schemes in climate models

J. DAVID NEELIN, OLE PETERS, JOHNNY W.-B. LIN,
KATRINA HALES AND CHRISTOPHER E. HOLLOWAY

Convective quasi-equilibrium (QE) has for several decades stood as a key postulate for parameterisation of the impacts of moist convection at small scales upon the large-scale flow. Departures from QE have motivated stochastic convective parameterisation, which in its early stages may be viewed as a sensitivity study. Introducing plausible stochastic terms to modify the existing convective parameterisations can have substantial impact, but, as for so many aspects of convective parameterisation, the results are sensitive to details of the assumed processes. We present observational results aimed at helping to constrain convection schemes, with implications for each of conventional, stochastic or 'superparameterisation' schemes. The original vision of QE due to Arakawa fares well as a leading approximation, but with a number of updates. Some, like the imperfect connection between the boundary layer and the free troposphere, and the importance of free-tropospheric moisture to buoyancy, are quantitatively important but lie within the framework of ensemble-average convection slaved to the large scale. Observations of critical phenomena associated with a continuous phase transition for precipitation as a function of water vapour and temperature suggest a more substantial revision. While the system's attraction to the critical point is predicted by QE, several fundamental properties of the transition, including high precipitation variance in the critical region, need to be added to the theory. Long-range correlations imply that this variance does not reduce quickly under spatial averaging; scaling associated with this spatial averaging has potential implications for superparameterisation. Long tails of the distribution of water vapour create relatively frequent excursions above criticality with associated strong precipitation events.

16.1 Introduction

16.1.1 Overview

Moist convection is among the most important of the physical processes occurring at small scales not resolved at the grid size of climate models, currently on the order of 100 km in the horizontal. The bulk effects of these small scales must be parameterised as a function of the large-scale variables. Convective quasi-equilibrium (QE) postulates that fast removal of buoyancy by moist convective up/downdraughts at small scales establishes a response of the ensemble-mean convective heating, moisture sink and other properties that maintain statistical stationarity, balancing the large-scale flow, which is assumed to be slowly varying. Many variants of the associated closures lie at the heart of the parameterisations of moist convection in most current climate models, and similar considerations are relevant to numerical weather prediction models, despite their higher resolution.

Traditional parameterisations represent only the ensemble mean of the small scales as a deterministic function of a large-scale flow, disregarding fluctuations arising at the small scales. A sense that small-scale fluctuations are important motivated the development of stochastic convective parameterisations. The first generation of stochastic convective parameterisations may be viewed as a sensitivity study indicating that the impacts can indeed be substantial. However, stochastic parameterisation introduces new parameters and processes; the climate solution exhibits some of the same dismaying sensitivity to these as it does to the traditional convective parameterisation. Thus it becomes pressing to better characterise the statistics of convection in a manner that can inform stochastic convective parameterisation and our understanding of convection in general. Observational work with this aim is the main theme of this chapter.

After reviewing issues in QE stochastic convective parameterisation in Section 16.2, we present observational results from current projects in Sections 16.3 and 16.4. We argue that these affirm some aspects of QE, but that some of these results suggest a new interpretation of QE that may be highly compatible with stochastic convective parameterisation. In sections 16.5 and 16.6 we then discuss ways in which one might in the future bring these properties into representations of deep convection, discussing implications for both stochastic convection schemes and 'superparameterisation' approaches.

16.1.2 Convective QE and motivation for stochastic schemes

The original vision of convective QE is articulated in Arakawa & Schubert (1974): 'When the time scale of the large-scale forcing, is sufficiently larger than the [convective] adjustment time,... the cumulus ensemble follows a sequence of

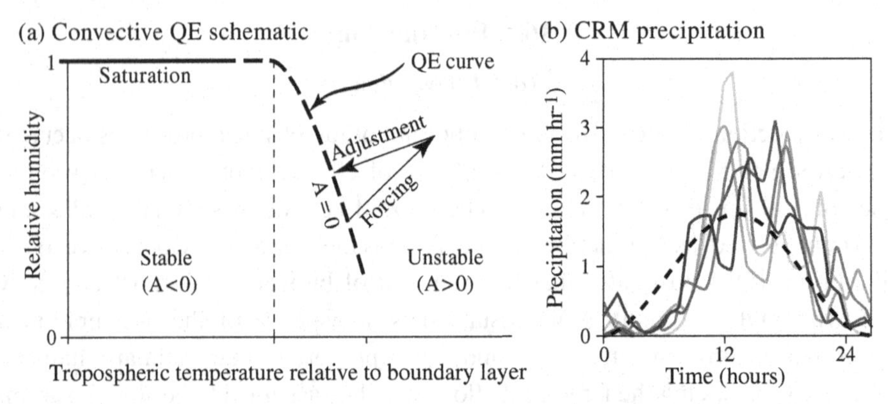

Figure 16.1 (*a*) Schematic of convective QE as envisioned by Arakawa & Schubert (1974). Adapted from Arakawa (2004). The line denoted 'QE curve' represents the statistical equilibrium towards which the system is postulated to be adjusted by deep convection, which is posited to act quickly relative to the large-scale forcing. The thick dashed part of this curve indicates zero buoyancy and the onset of convection through conditional instability (the solid line at relative humidity=1 corresponds to large-scale saturation). (*b*) Precipitation rates (solid curves) averaged over the 512 km domain from a two-dimensional cloud-resolving model (CRM) in response to an imposed large-scale forcing with cooling and moistening (dashed curve). A sequence of five different time intervals responding to repetitions of the same forcing are overlaid. The large variations from one instance to another, even for these domain averages, are departures from QE. Adapted from Xu *et al.* (1992).

quasi-equilibria with the current large-scale forcing. We call this... the quasi-equilibrium assumption.' 'The adjustment... will be towards an equilibrium state... characterized by... balance of the cloud and large-scale terms... ' Convection acts to reduce a measure of buoyancy, which in their parameterisation is the cloud work function A (for each of a spectrum of entraining plumes).

The system is slowly driven by large-scale forcing (moisture convergence, evaporation and radiative cooling) generating conditional instability, while small-scale convection provides a fast dissipation of buoyancy. Above the onset threshold, strong convection and precipitation greatly increase, pushing the system quickly back towards onset values. Thus, a statistical equilibrium tends to be established among buoyancy-related fields: temperature T and moisture q, including their vertical structures, are constrained. Arakawa (2004) provided a schematic, shown here in Fig. 16.1*a*, for the original QE postulate.

There are, however, several issues regarding QE that need to be addressed. First, using a finite adjustment time scale makes a difference in the behaviour, as does the value of the adjustment rate or plume closures affecting this (Betts & Miller 1986; Gregory & Rowntree 1990; Emanuel 1991, 1993; Moorthi & Suarez 1992; Emanuel *et al.* 1994; Yu & Neelin 1994; Zhang & McFarlane 1995; Pan & Randall

1998). Furthermore, the ensemble size of deep convective elements in a typical climate model (order of 100 km) grid box over a 10–30 min time increment is not large. One should expect variance about the ensemble-mean in such an average. This variance can drive large-scale variability (even more so in the presence of mesoscale organisation). Another issue relates to complex influences on vertical structures, such as convectively coupled wave modes (Straub & Kiladis 2003; Haertel & Kiladis 2004; Tian *et al.* 2006) discussed in Section 16.3. Raymond (1997) noted a tendency for the atmospheric boundary layer (ABL) to reach its own form of QE, mostly between surface fluxes and convective downdraughts, even as the free troposphere is being affected by remote wave dynamics.

Figure 16.1*b* shows an example from Xu *et al.* (1992) that motivated our own interest in stochastic convection schemes. Different instances of precipitation are shown from a cloud-resolving model (CRM) averaged over a domain that might roughly compare with a large-scale model grid (albeit from a two-dimensional CRM) for a case with shear, promoting mesoscale organisation. The following three features may be noted: (i) Xu *et al.* (1992) concluded that the similarity among realisations was sufficient for convection to be parameterisable, (ii) there is a substantial lag of the onset of rainfall relative to the large-scale forcing, suggesting that convective processes are not as quickly adjusting as originally hoped, and (iii) there is much variance among realisations for the same large-scale forcing – capturing these departures from QE is the motivation for stochastic parameterisations discussed in Section 16.2.

Superparameterisation (e.g. Grabowski 2001; Khairoutdinov & Randall 2001; Randall *et al.* 2003), in which a few small grid cells of a CRM are embedded within each large-scale grid point, retains (and potentially exaggerates) fluctuations associated with small scales. It also aims to address a host of issues such as the accurate representation of convective plumes and the level of detrainment for cloud water. We thus also discuss the implications of observations of precipitation variance for superparameterisation.

16.1.3 Simple considerations in the moisture and thermodynamic equations

For reference, the temperature and moisture equations are provided here, along with considerations used in subsequent sections:

$$\partial_t T + \nabla \cdot s\boldsymbol{v} + \partial_p(s\omega) - K_T \nabla^2 s + C_K = Q_c + \partial_p(F_R + F_T), \qquad (16.1.1)$$

$$\partial_t q + \nabla \cdot q\boldsymbol{v} + \partial_p(q\omega) - K_q \nabla^2 q = -Q_q + \partial_p F_q, \qquad (16.1.2)$$

where dry static energy $s = T + \phi$ and ϕ is geopotential, with heat capacity at constant pressure c_p absorbed into temperature T, latent heat of condensation absorbed into specific humidity q and acceleration due to gravity absorbed into pressure p; $\mathbf{v}$ is horizontal wind; Q_c is convective heating; $-Q_q$ is the moisture sink; F_R, F_T and F_q are vertical (diffusive for T and q) fluxes of radiation, sensible heat and moisture, respectively; and ω is vertical velocity in pressure coordinates. The terms K_T and K_q are horizontal diffusivities parameterising small-scale mixing; C_K is conversion to kinetic energy [sometimes neglected in (16.1.1)].

To provide a thumbnail sketch of how a QE scheme can be implemented, consider that one has some model of a convective plume that yields a temperature T_c of the convective element as a function of pressure, and of the large-scale T and q. In the simplest case of a plume rising adiabatically from the boundary layer, T_c is just the moist adiabat; inclusion of entrainment produces a dependence on environmental T and q in the free troposphere. The buoyancy depends on the difference $T_c - T$ (actually the virtual temperature, including direct effects of specific humidity on density, but for presentation purposes we omit the details of this). One such measure is convective available potential energy (CAPE), which can be written as $A = R_d \int (T_c - T)\, d\ln p$, where R_d is the gas constant of dry air (Emanuel 1994). Arakawa & Schubert (1974) and related schemes used a cloud work function for a spectrum of entraining plumes with differing T_c.

Quasi-equilibrium closures typically postulate that a weighted vertical integral measure of buoyancy is rapidly reduced. The simplest case is typified by the Betts & Miller (1986, BM) convective adjustment scheme, in which

$$Q_c = (T_c - T + \xi)/\tau_c, \quad Q_q = [q_c(T) - q + \xi]/\tau_c, \tag{16.1.3}$$

whenever the vertical integrals of the right-hand sides are positive. This scheme tends to reduce $(T_c - T)$ level by level when moisture exceeds a critical value $q_c(T)$. In (16.1.3), BM is shown with a stochastic term, ξ, to be described later ($\xi = 0$ for the standard BM scheme).

Parallels can be seen in the Zhang & McFarlane (1995, ZM) convective scheme in which the closure assumption $M_b F \equiv A/\tau_c$ creates a tendency for CAPE, A, to be reduced exponentially

$$\partial_t A + \cdots = -M_b F \equiv (A + \xi)/\tau_c, \tag{16.1.4}$$

where τ_c is an assumed convective time scale ($= 2$ h); M_b is the updraught cloud base mass flux; and F is the rate of CAPE removal per unit M_b from a cloud plume model. A stochastic component ξ will be defined later; $\xi = 0$ for the standard ZM scheme.

The energy constraint that the vertical mass-weighted integral $\hat{}$ of convective heating equals latent heat by moisture loss implies

$$\hat{Q}_c = \hat{Q}_q. \tag{16.1.5}$$

If most of the condensed water falls out of the system within the grid box, then the precipitation will be approximately given by $P = \hat{Q}_q$.

Several points may be noted regarding (16.1.5). First, in any stochastic term the convective heating should be matched by a corresponding moisture sink, respecting (16.1.5), or it will produce unphysical effects. Second, (16.1.5) can have substantial effects within convection schemes; in (16.1.3), for instance, it implies that T_c is determined by $[\hat{q}_c - \hat{q} + \hat{T}]$ (Neelin & Zeng 2000). Third, (16.1.5) implies that summing vertical integrals of (16.1.1) and (16.1.2) yields an equation (the moist static energy budget) in which the convective heating does not appear. Thus variations in the heating do not directly change the sum $(T + \hat{q})$. The column water vapour, $w = \hat{q}$, also known as precipitable water, will be important in observational analysis below, where we use the symbol w for consistency with other work.

If QE ties moisture and temperature closely together when convection is occurring, so that moisture is slaved to $\hat{T}$, an approximation used in some theoretical considerations (e.g. Emanuel *et al.* 1994), then the moist static energy equation governs the thermodynamics to a leading approximation. If so, stochastic variations in convective heating and precipitation would have little effect on the large-scale dynamics. This indicates the potential importance of departures from strict versions of QE. Variations in Q_c of sufficiently large amplitude relative to the mean will frequently shut down convection. This removes any QE constraints until the convection begins again. Smoothly posed QE schemes tend to operate rather more continuously than observed convection, so one principal effect of stochastic convection schemes is to intermittently break the QE constraints. We will argue below that the large ensemble average envisioned in the original posing of QE is a first approximation, rather than a constraint applying at short time scales. The intermittent departures from QE are inherent properties of convection, and it should be possible to set up the physics of stochastic schemes to include this in a manner that mimics convection.

16.2 Issues in stochastic deep convective parameterisation

16.2.1 Implementations of stochastic convective parameterisation

Given the deterministic convection schemes operating under QE constraints, a scheme that randomly perturbs the system from QE would be a reasonable first approach to stochastic parameterisation; this was indeed the method used in the earliest attempts. Buizza *et al.* (1999), for instance, perturbed all parameterised quantities using a uniform distribution in order to test the effects of such a representation of model error on ensemble spread. Yu & Neelin (1994) noted that noise could support tropical wave variability in a simplified primitive equation model.

Lin & Neelin (2000), seeking to represent fluctuations in convection arising at the small scales, such as those seen in Fig. 16.1*b*, introduced stochastic perturbations to large-scale CAPE in a BM convective parameterisation in an intermediate-complexity atmospheric model (Neelin & Zeng 2000). Noise ξ from a first-order autoregressive (Markov) process was added to the difference $T_c - T$ as in (16.1.3). In an alternate stochastic scheme, an empirically estimated precipitation distribution (log normal, with separately parameterised probability of zero precipitation), adjusted at each time step to match the mean of the deterministic convective scheme, provided stochastic convective heating (Lin & Neelin 2002).

Lin & Neelin (2003) described two stochastic parameterisations implemented in the National Center for Atmospheric Research Community Climate Model (CCM3): the CAPE-M_b and vertical structure of heating (VSH) schemes. In the CAPE-M_b scheme, random variations are introduced to the relationship between cloud-base mass flux (M_b) and CAPE. Physically, the random component may be thought of as grid-scale response to subgrid variability. Thus, a noise perturbation ξ is added to CAPE, as in (16.1.4). The VSH scheme is a first-cut look at effects of stochastic vertical structure variations, for instance, corresponding to variation in detrainment levels of convective elements. Noise ξ is introduced as a perturbation to temperature at each level, with $\xi = 0$ to ensure moist static energy conservation and to contrast with the CAPE-M_b scheme where ensemble-mean heating is directly perturbed.

Majda & Khouider (2002) define a stochastic model based on small-scale convective inhibition (CIN) and couple this model with a prototypical mass-flux convective parameterisation through area fraction for deep convection, boundary-layer equivalent potential temperature and stratiform mass flux. Inclusion of the scheme produces eastward-propagating convectively coupled waves that qualitatively resemble observations (Khouider *et al.* 2003).

Berner *et al.* (2005), following the suggestion of Palmer (2001), implemented an additional stochastic forcing on stream function based on a cellular automata model aiming to represent the mesoscale organisation of convective systems in the European Centre for Medium-Range Weather Forecasts (ECMWF) Integrated Forecasting System (IFS). Madden–Julian Oscillation (MJO)-like variability results, though without a spectral peak. Tompkins & Berner (2008) implement a scheme in the ECMWF IFS, applying a uniformly distributed stochastic input to humidity, which then influences triggering and updraught humidity in the deterministic convective parameterisation. This scheme results in increased ensemble spread, though less so than the Buizza *et al.* (1999) scheme. Combining the two schemes increases prediction skill in mid latitudes, though with mixed results in the tropics.

Teixeira & Reynolds (2008) apply a stochastic component to select tendencies related to the convective parameterisation in the Navy Operational Global

Atmospheric Prediction System weather prediction model. The stochastic component is drawn from a normal distribution and is applied to the tendency calculated by the deterministic parameterisation. This results in significant ensemble spread in dynamical variables such as 500 hPa geopotential height and winds. Plant & Craig (2008) draw random values in cloud-base mass-flux-related variables from distributions motivated by statistical mechanics considerations (Craig & Cohen 2006), and calculated convective tendencies using the deterministic Kain & Fritsch (1990) cloud model. Tests of the scheme in a single-column model yield results consistent with mean temperature and humidity profiles from CRM simulations.

Stochastic convective schemes so far have focused on the thermodynamic aspects of convection; it is worth noting parallel efforts to parameterise mesoscale circulations and momentum transport (e.g. Moncrieff & Klinker 1997; Moncrieff & Liu 2006) or to give subgrid-scale plumes more individual identity (Kuell *et al.* 2007), directions that may be highly useful to future stochastic parameterisation.

16.2.2 Sensitivity to stochastic parameterisation

Just as simulated climate and variability are sensitive to different deterministic convective schemes (Maloney & Hartmann 2001; Tost *et al.* 2006), one expects corresponding sensitivity to the introduction of the stochastic parameterisation schemes described in Section 16.2.1. Even the early simple tests of stochastic parameterisation produced substantial effects. Buizza *et al.* (1999) found improved forecast skill as a result of the inclusion of a stochastic representation for model error. Lin & Neelin (2000) found in their perturbations to model CAPE that, depending on the autocorrelation time, the stochastic scheme improved the simulated probability distribution of daily mean precipitation in tropical regions. For grid-scale noise at a one-day e-folding time scale, the scheme enhanced intraseasonal equatorial wavenumber-one 850 hPa zonal wind spectra power.

With substantial impacts comes substantial sensitivity, and new parameters requiring observational constraints. For instance, Lin & Neelin (2000) found their stochastic CAPE scheme to be sensitive to the autocorrelation time scale of the input noise, suggesting that time-scale parameters associated with prognostic convective or mesoscale elements will need to be quantified in future schemes. Tests with Lin & Neelin's (2002) empirical stochastic parameterisation showed such strong interaction of the heating with large-scale dynamics that they concluded that a stochastic parameterisation cannot be calibrated off-line. Rather, comparison to observations should be done with output from a full model simulation.

The importance of this model feedback can be seen in the intraseasonal response to stochastic convective schemes. If a deterministic climate model's interaction of convection with large-scale dynamics yields damped MJO-like intraseasonal

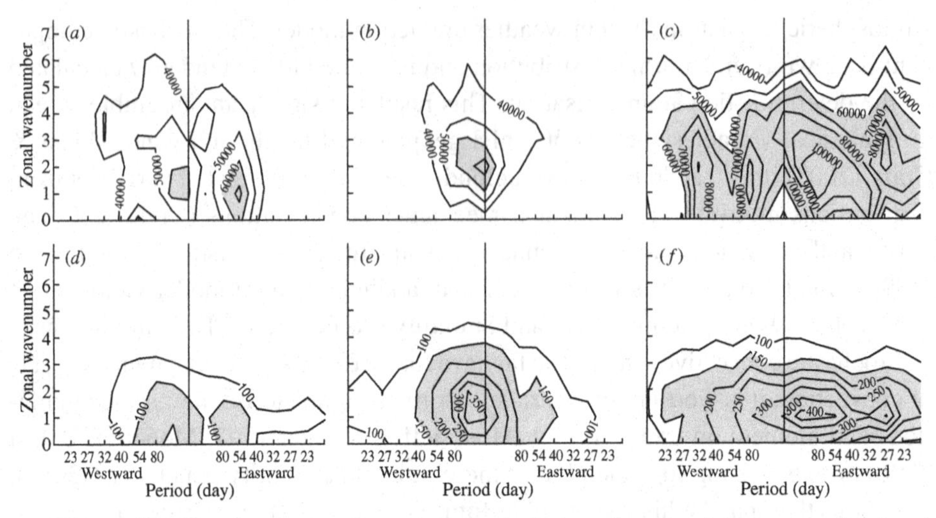

Figure 16.2 Zonal wavenumber versus period plot of equatorial spectra of (*a*) control run precipitation, *P*, (*b*) CAPE-M_b scheme precipitation, (*c*) VSH scheme precipitation, (*d*) control run 850 hPa zonal wind, u_{850}, (*e*) CAPE-M_b scheme u_{850} and (*f*) VSH scheme u_{850}. The spectra are shaded above 60 000 (W m^{-2})^2 for precipitation and above 150 (m s^{-1})^2 for u_{850}. The spectral estimator has a standard deviation of 10%. See Lin & Neelin (2003) for analysis details.

variability, a stochastic scheme will act to enhance that MJO-like variability to more closely resemble observations, as in the Lin & Neelin (2000) example. In contrast, if a deterministic climate model does not have a clear mode that yields MJO-like variability, the impact of a stochastic scheme will be less straightforward. As a detailed example, we consider the effects of the stochastic convective parameterisations of Lin & Neelin (2003), joined to the ZM convective parameterisation (as described in Section 16.1.3).

Figure 16.2 shows equatorial precipitation and 850 hPa zonal wind, u_{850}, spectral power for the CAPE-M_b and VSH schemes of Lin & Neelin (2003), compared to a CCM3 control run. The format follows Figs. 4 and 5 of Maloney & Hartmann (2001), which provide observations and deterministic convection scheme results. While the inclusion of stochastic variance can increase spectral power, it does not necessarily excite the portion of the wave spectrum corresponding to observed intraseasonal variability. For instance, in the CAPE-M_b scheme, spectral power in both u_{850} and precipitation is shifted to lower-frequency westward waves. Spectral power in u_{850} increases notably, but precipitation power remains nearly unchanged. The VSH scheme dramatically increases spectral power in u_{850} and precipitation, but it excites a broad range of wave frequencies, both eastward and westward propagating. Such results suggest that tuning stochastic convective parameterisations may be as difficult as tuning the deterministic parameterisations they are meant to improve.

Efforts have been made to constrain the formulation of stochastic parameterisations using arguments from first principles and statistics from cloud-resolving models. For instance, using principles from equilibrium statistical mechanics, for the case of negligible plume–plume interaction, Craig & Cohen (2006) derive an analytical expression for the probability distribution of total cloud mass flux over a region. Cohen & Craig (2006) confirmed these theoretical results using CRM simulations. Cloud-resolving modelling by Shutts & Palmer (2007) also confirmed the findings of Craig & Cohen (2006) that individual cloud mass flux follows a Boltzmann exponential distribution, though Shutts & Palmer (2007) found a different dependence between the mean and standard deviation of coarse-grained effective temperature tendencies than that is predicted by Craig & Cohen (2006).

While CRM studies are valuable, there is a clear need to derive additional constraints on the stochastic parameterisation problem from observations. In the next section, we discuss results aimed at better characterising the statistics of observed convection at short time scales and small space scales.

16.3 Vertical structure associated with QE

Several studies have indicated that tropical temperature profiles are constrained towards convective plume temperatures lifted from warm, moist ABL conditions, approximated by moist adiabats (Xu & Emanuel 1989; Brown & Bretherton 1997) or including entrainment (Kuang & Bretherton 2006), as predicted by QE. Figures 16.3*a*,*b* adapted from Holloway & Neelin (2007) show temperature perturbations from radiosondes at different pressure levels regressed on their free-tropospheric vertical averages along with correlations, as well as similar regressions made using an ensemble of reversible moist adiabats raised from typical tropical surface conditions. Both the monthly Comprehensive Aerological Reference Dataset (CARDS; Eskridge *et al.* 1995) averages and the daily Tropical Ocean Global Atmosphere Coupled Ocean–Atmosphere Response Experiment (TOGA-COARE) data (Ciesielski *et al.* 1997) show fairly good agreement with QE predictions (the moist-adiabatic curve) in the free troposphere. The negative deviations aloft are due to dynamical constraints as discussed in Holloway & Neelin (2007). The departures from QE expectations in the ABL [consistent with Brown & Bretherton (1997)] may be interpreted in terms of wave dynamics spreading the temperature signal horizontally from localised convection in the free troposphere, while competing effects from surface fluxes occur in the ABL. The lower correlations indicating less vertical coherence at higher time resolution fits with several recent studies showing that there are more complicated vertical temperature perturbation structures at these scales associated with tropical wave features (e.g. Straub & Kiladis 2003; Haertel & Kiladis 2004).

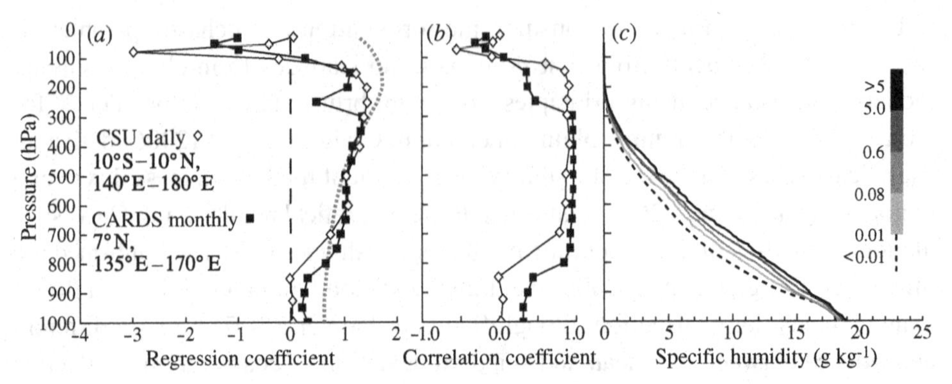

Figure 16.3 (*a*) Vertical temperature structures and (*b*) correlations, adapted from Holloway & Neelin (2007), associated with free tropospheric average temperature perturbations, using CARDS monthly anomalies and Colorado State University (CSU) TOGA-COARE daily averages. The dashed grey line in (*a*) shows similar regressions for the moist adiabat (CARDS case shown; CSU case nearly identical below 300 hPa). (*c*) Nauru ARM q profiles from twice-daily radiosondes conditionally averaged on 60 min average optical gauge precipitation (bins given in shaded bar in mm h^{-1}).

Overall, observations of tropical temperature perturbations such as those in Fig. 16.3*a,b* show fairly good agreement, especially at larger scales and in the free troposphere, with the idea of a dominant vertical structure that is not far from moist adiabatic as hypothesised from QE constraints. Figure 16.3*c* (adapted from Holloway & Neelin 2009) shows that moisture vertical structure, conditionally averaged on precipitation, also varies most in the free troposphere, whereas there are smaller changes in the ABL. These profiles are made using five years of radiosondes and optical gauge 60 min precipitation averages at the Nauru Island (0.5° S, 166.9° E) Atmospheric Radiation Measurement (ARM) Program observation site (Stokes & Schwartz 1994). Figure 16.3*c* supports growing evidence in the literature that traditional convective parameterisations, which tend to emphasise the ABL moist-static energy, are not giving enough weight to the role that lower-free-tropospheric dryness plays in suppressing deep convection (e.g. Parsons *et al.* 2000; Raymond 2000; Grabowski 2003; Derbyshire *et al.* 2004). While QE thinking has in principle always included entrainment (Arakawa & Schubert 1974), more emphasis must be placed on accurately taking free-tropospheric moisture into account. Recently there have been increasing efforts to include these processes in current schemes (e.g. Zhang & Mu 2005; Zhang & Wang 2006).

Analysis of the Nauru ARM sonde data suggests that most of the q variance associated with w variance occurs in the lower free troposphere above the ABL, although there is also a small but significant q variance in the ABL associated with w variance. Figure 16.4*a* shows precipitation and entraining CAPE conditionally averaged on w for the ARM data (Holloway & Neelin 2009). The sharp

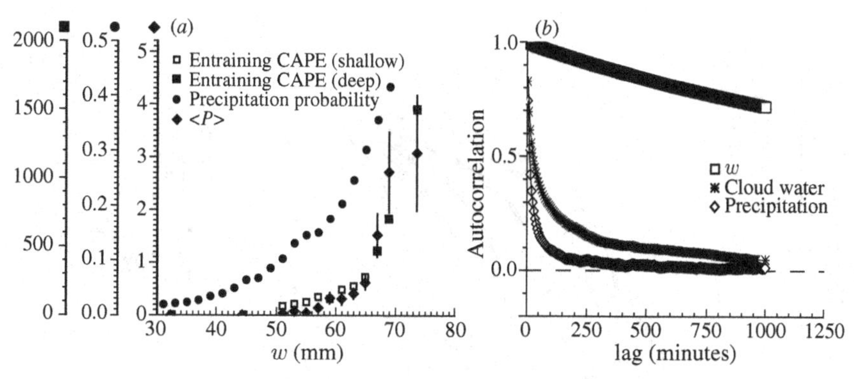

Figure 16.4 (*a*) Nauru ARM precipitation $\langle P \rangle$ (mm h^{-1}; w diamonds) as 60 min averages conditionally averaged on column water vapour w from radiosondes, with bars showing ± 1 standard error; entraining CAPE (J kg^{-1}) measured from lifting condensation level to level of neutral buoyancy (LNB) for 1000 hPa lifted parcels subjected to constant mixing, for the same w bins, with deep convective cases (LNB levels above 450 hPa) marked as filled squares (open squares otherwise); precipitation probability (dots) in next hour, given 5 min average w and 0 rain at time 0, from ARM MWR and optical gauge data. The y-axes are labelled by corresponding symbols. (*b*) Autocorrelations for 5 min average w, cloud water and precipitation from ARM MWR and optical gauge data. The w bin edge values for (*a*) (in mm) are: less than 35, 35–50, every 2 mm from 50 to 70, and greater than 70; for precipitation probability curve, 2 mm bins from 30 to 70.

pick-up of precipitation at a high enough value of w agrees with satellite analyses in Bretherton *et al.* (2004) and Peters & Neelin (2006). The entraining CAPE (Brown & Zhang 1997), taken for the positively buoyant part of the path of lifted 1000 hPa parcels entraining 0.1% of environmental air per hPa (so that the virtual T_c from Section 16.1.3 includes mixing with T), also shows a sharp pick-up, which appears to be due both to the moister free troposphere and the slightly higher ABL moist-static energy associated with larger w. The CIN is always below 6 J kg^{-1} for these composite profiles, and there is no obvious relationship between CIN and the precipitation pick-up. The three highest CAPE values have average levels of neutral buoyancy (which would theoretically correspond to cloud-top heights) approximately 150 hPa instead of below the 450 hPa level for the other bins (although these values are rather sensitive to choices of mixing profiles and adiabatic processes). Column water vapour w, which is readily available from satellites, therefore acts as more than a proxy for conditions conducive to deep convection; it is associated with both ABL and free-tropospheric contributions to buoyancy due to entrainment. As Fig. 16.4*b* illustrates, w also has significantly higher temporal autocorrelations than cloud water and precipitation [as seen in ARM microwave radiometer (MWR) and optical gauge data], suggesting that it can be thought of as a relatively slowly varying environmental control variable on which to compute statistics of sporadically occurring precipitation. Column water

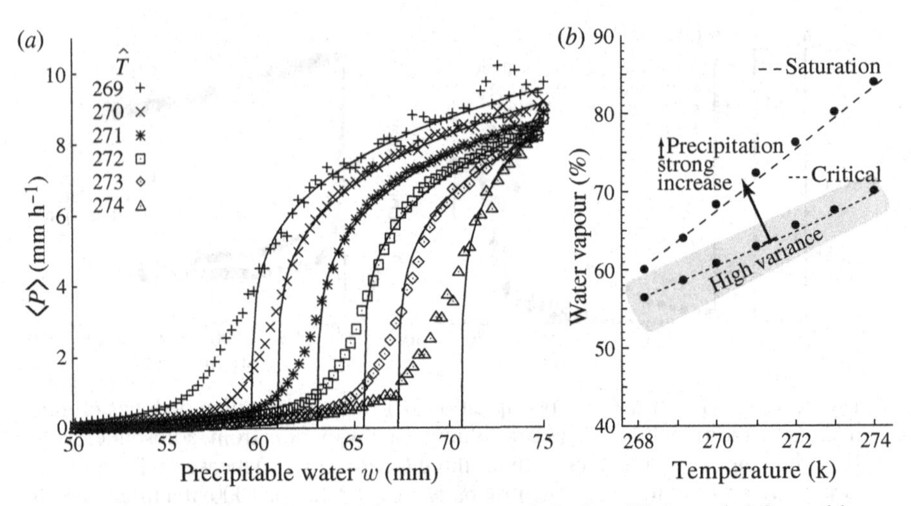

Figure 16.5 (*a*) Pickup of ensemble-average precipitation $\langle P \rangle$ in each 0.3 mm bin of column water vapour w for 1 K bins of the vertically averaged tropospheric temperature $\hat{T}$. Western Pacific $\langle P(w, \hat{T}) \rangle$ as a function of w for $\hat{T} = 269$ to 274 K; lines show power-law fits above criticality of the form (16.4.1). (*b*) The critical value of column water vapour w_c at which the onset of strong convection occurs, as a function of vertically averaged tropospheric temperature $\hat{T}$; dots give the values determined as in (*a*), here combined from the tropical West Pacific, East Pacific and Atlantic. The vertically integrated saturation value of water vapour is also shown. Schematic elements indicate the pick-up in precipitation and the band of high precipitation variance near criticality.

is also potentially useful for convective scheme transition probabilities (or trigger functions). The ARM MWR data in Fig. 16.4*a* show that probabilities of precipitation for the next hour, given no current precipitation, increase dramatically at high values of w.

16.4 Characteristics of the transition to strong convection

Examining the transition to strong convection at high time resolution, Peters & Neelin (2006) noted that the statistical properties of precipitating convection conform to those observed near continuous phase transitions, known as critical phenomena. The observed properties include a rapid pick-up of precipitation, P, above a critical value of precipitable water, w. The ensemble average, $\langle P \rangle$, conditioned on w, approaches a power law above criticality

$$\langle P \rangle(w) = a[(w - w_c)/w_c]^{\beta}, \quad if (w - w_c) > 0. \tag{16.4.1}$$

This is in contrast to a linear relation assumed in some convective parameterisations such as (16.1.3). Here this relationship is seen in Fig. 16.5*a* with column water

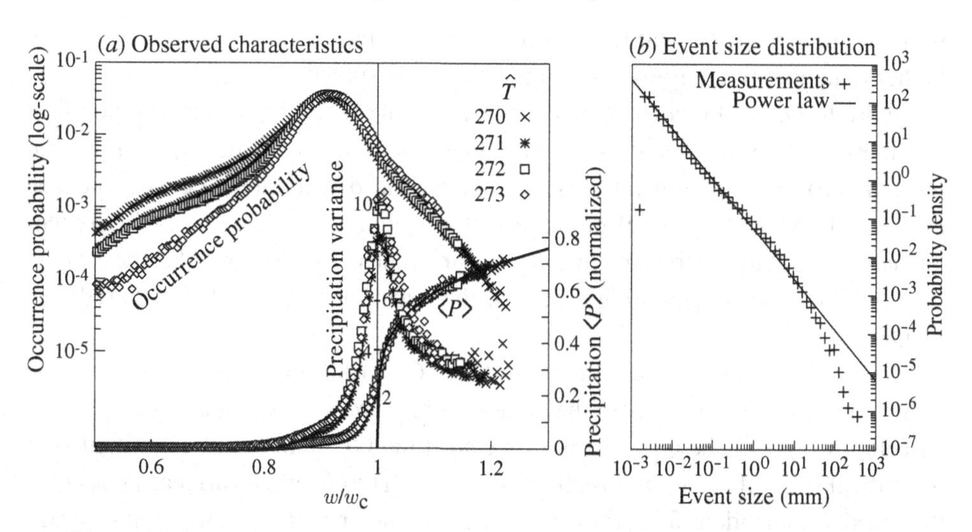

Figure 16.6 (*a*) Western Pacific observed characteristics from TMI data as a function of column water vapour normalised by the critical value w_c for each value of $\hat{T}$: probability density function of w for precipitating points (four upper curves), precipitation variance conditioned on w (four middle curves) and precipitation pick-up curve (non-dimensionalised by amplitude a from (16.4.1) for each $\hat{T}$). (*b*) The precipitation event-size distribution for the Nauru ARM site time series.

vapour and precipitation retrievals (Hilburn & Wentz 2007) from the Tropical Rainfall Measuring Mission Microwave Imager (TMI), with the added dimension that it is examined for various values of vertical average tropospheric temperature $\hat{T}$ from ERA-40 reanalysis (Uppala *et al.* 2005).

Another confirmed expectation from the theory of phase transitions is a peak in the precipitation variance, σ_P^2, near w_c (Fig. 16.6*a*). These properties are related to long-range spatial correlations in the critical region, observed as a non-trivial power-law dependence of the precipitation variance on the spatial averaging scale (Peters & Neelin 2006), implications of which will be discussed in Section 16.5.3. Power-law dependence of fluctuations on averaging scale has been previously observed in the time domain for CAPE (Yano *et al.* 2001).

The amplitude a and the critical value w_c, in (16.4.1), are expected to be sensitive to the details of the processes involved. The exponent β, on the other hand, is predicted to be robust. This was confirmed observationally: β is comparatively invariant under changes of ocean basin (Peters & Neelin 2006) or temperature. While critical values, $w_c(T)$, and amplitudes, $a(T)$, needed to be determined (see Neelin *et al.* 2009 for details) for each value of (vertically averaged) tropospheric temperature in Fig. 16.5*a*, no changes were made to β. Rescaling the water-vapour axis with $w_c(T)$ and the precipitation axis with $a(T)$, the curves collapse (Fig. 16.6*a*). The dependence of the critical value, w_c, on tropospheric temperature is shown in Fig. 16.5*b*. Rather than w_c being a constant fraction of saturation, at higher temperature the transition occurs at greater subsaturation.

Figure 16.6*a* is a direct observational assessment of the QE postulate. It shows the distribution of observed water-vapour values (the four upper curves), conditioned on non-zero precipitation rate. Most notably, the distribution is clearly peaked in the vicinity of the critical point, with the frequency of observed occurrences decreasing fast for increasing water-vapour values. In other words, the system spends most of its precipitating time near the transition to convection, as predicted by QE.

The precipitation variance, shown in Fig. 16.6*a* (the four middle curves), is an indicator of the susceptibility of the system. A high intrinsic variance (and therefore susceptibility) near the QE state is expected if the QE state can be identified with the critical point of a continuous phase transition. The observed variance peaks near the critical point identified from the order parameter pick-up (ensemble-average precipitation pick-up, also shown in Fig. 16.6*a*). In principle, this could be the result of unresolved variations in water vapour or other variables, coupled with the fact that precipitation changes most rapidly here. The theory of critical phenomena, however, suggests high variance to be an intrinsic property of the system which is not due to insufficient measurement resolution. Figure 16.6*b* shows the event-size distribution for the Nauru ARM site optical gauge precipitation. The event size is the rain integrated over an event, defined as consecutive measurements of non-zero rain rate (Peters *et al.* 2002). Event size is proportional to the energy released and has clear analogues in other self-organised critical systems. The scale-free distribution of event sizes is further evidence of the extreme sensitivity of the system.

16.5 Prototypes for these observed properties and implications

We use this section to discuss further some of the observed properties in the light of what is known from simpler prototype models.

16.5.1 A simple model exhibiting critical phenomena

System-wide self-organisation towards critical points of continuous phase transitions is a well-established field of research in statistical mechanics. In self-organised critical systems, of which moist convection appears to be an example, energy-dissipation rates (order parameters such as precipitation) pick up as a function of energy density (tuning parameters such as column water vapour). Perhaps the simplest model displaying such behaviour was introduced by Manna (1991). It is defined on a lattice to which particles are added. A hard-core repulsion is

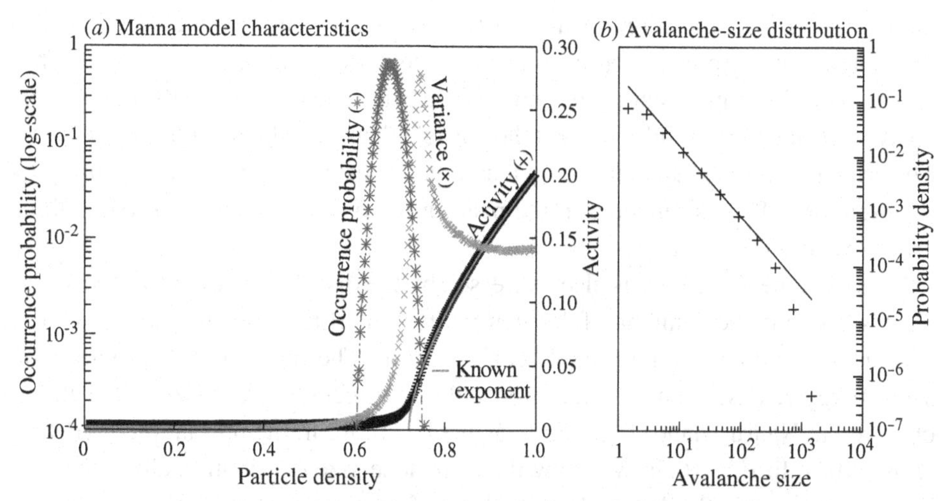

Figure 16.7 The characteristics of a two-dimensional Manna model of size 20 × 20. (*a*) As a function of the tuning parameter (the density of particles in the system) the order parameter pick-up (crosses) is shown along with its known exponent (grey line; Lübeck 2000), for the case of a closed system. This is similar to $\langle P \rangle$ in Fig. 16.6*a*. The probability for the system to be found at a given particle density under conditions of self-organisation (dark grey symbols, logarithmic scale) is Gaussian, like the core of the distribution in Fig. 16.6*a*. Also shown is the order parameter variance (closed system, light grey crosses, arbitrary units), which looks much like that in Fig. 16.6*a*. (*b*) Avalanche-size distribution. The straight line indicates the exponent known from numerical work, e.g. by Lübeck (2000).

implemented, such that each site can harbour no more than one particle. Wherever this local threshold is surpassed, particles topple to randomly chosen neighbouring sites, where they can induce further topplings, leading to system-wide avalanches that transport particles to the open boundaries where they are dissipated.

The overall effect of these dynamics is an attraction to the critical point of an underlying absorbing-state phase transition (Tang & Bak 1988; Dickman *et al.* 1998). These dynamics are best demonstrated using a small system where finite-size effects are similarly important as they seem to be in the atmosphere. Figure 16.7 shows data for a system of 20 × 20 sites. Although the model is clearly not a representation of moist convection, it exhibits properties that can be qualitatively compared with those observed for convection in Fig. 16.6.

A high energy density in the system leads to activity, defined as the density of unstable sites during avalanches (crosses in Fig. 16.7*a*). Activity in turn reduces the energy density through dissipation at the open boundaries. Since a new particle is added whenever the system has avalanched into a stable configuration, it hovers around criticality (dark grey symbols in Fig. 16.7*a*). In the case of the Manna model, the self-organisation towards criticality is so strong that we need to force the system away from criticality to explore the underlying phase transition. This is done here by implementing periodic boundaries such that no self-organisation takes place and we can fully control the particle density (which is now conserved). The critical point is characterised by a divergence of the system's susceptibility to perturbations. This is reflected in the peak in the order parameter variance (light grey cross in Fig. 16.7*a*).

This extreme sensitivity is also expressed by the distribution of avalanche sizes. Here we count the number of local topplings, starting from the addition of a particle and ending at a new stable configuration. The probability distribution of such energy-release events is scale-free and reminiscent of event-size distributions observed at Nauru Island, Fig. 16.6*b*, as well as those in the mid latitudes (Peters *et al.* 2002). In Fig. 16.7*b* we show the avalanche-size distribution along with its known exponent; in the thermodynamic limit of infinite system size, the distribution would follow a power law over an infinite range. The analogy with atmospheric event-size distributions suggests they can occur even for fixed, slow forcing – in other words, a scale-free range of precipitation events is associated with the organisation towards the critical point in QE.

16.5.2 Implications of the exponential tails

In Fig. 16.6*a* it was shown that the distribution of the atmospheric tuning parameter (the water vapour) has strongly non-Gaussian tails. There is a Gaussian-like core, but the tails are much better described by exponentials. One effect of these exponential tails is that we are able to observe the underlying phase transition. In the Manna model, the distribution is highly Gaussian, with the result that occurrences drop very rapidly above the critical point in the self-organising case. To observe the behaviour above criticality, we needed to introduce periodic boundaries. The question remains how it is possible that the atmosphere ever fluctuates as far from criticality, or QE, as it does.

A possible answer is provided by tracer dispersion in forced advection–diffusion problems, in which the tracer probability density distribution can have a Gaussian core with exponential tails (e.g. Gollub *et al.* 1991; Majda 1993; Shraiman & Siggia 1994). This can occur, for instance, in the two-dimensional case

$$\partial_t q + \mathbf{v} \cdot \nabla q - \kappa_0 \nabla^2 q = f, \tag{16.5.1}$$

with $\mathbf{v}$ a non-divergent random flow field, when a large-scale gradient of tracer q is maintained (Shraiman & Siggia 1994) or when including a forcing f, such as Gaussian random forcing or resetting q in strips (Pierrehumbert 2000).

There are parallels to this in the vertical integral of the moisture equation (16.1.2),

$$\partial_t \hat{q} + \widehat{\mathbf{v} \cdot \nabla q} + \widehat{q \nabla \cdot \mathbf{v}} - K_q \nabla^2 \hat{q} = -P + E, \qquad (16.5.2)$$

where P and E denote loss and gain terms due to precipitation and evaporation. Lower tropospheric advection is important (Parsons *et al.* 2000), with water vapour acting at first order like a passive tracer until convection begins. Evaporation and moisture divergence terms tend to set a large-scale gradient. The question of whether the complex loss term by precipitation and its feedbacks with moisture convergence act sufficiently like f in (16.5.1) to explain the exponential tails can presumably be answered by analysing models of various degrees of complexity between (16.5.1) and full CRM.

16.5.3 Variance about QE and the superparameterisation variance estimation problem

Superparameterisation approaches (Randall *et al.* 2003) aim to estimate the statistics for a large-scale model grid cell from an embedded cloud-resolving model. Computation time constraints imply that smaller grid cells of this model can cover only a small fraction of the area of the large-scale model cell – otherwise one would incur the computational cost of running the CRM globally. The variance scaling in the critical region has potential implications for this.

The following back-of-the-envelope argument suggests the likelihood that if the CRM were to produce the correct variance at the scale of its own grid cells, then the superparameterisation approach would tend to overestimate the variance appropriate to the larger grid size. Consider coarse-graining onto a larger $L \times L$ grid the results, say precipitation, of a CRM run at high resolution over a large domain. The averaging process reduces the variance. If the precipitation were spatially uncorrelated (from one high-resolution cell to the next), the precipitation variance would decrease by the trivial scaling factor L^{-2}. In superparameterisation, however, the averaging will be done over the far smaller number of high-resolution grid cells, M, that one can afford to run per coarse model grid box of size $L_G \times L_G$. The variance in the uncorrelated case will then decrease by a factor of M^{-1} instead of L_G^{-2}, where L_G is measured in units of the fine-resolution grid size.

This problem has the potential to be serious. If the CRM grid is 4 km, and the large-scale model grid is 200 km so that $L_G = 50$, for the uncorrelated case the variance at the 200 km scale should be a factor of $1/2500$ as large as at the 4 km scale. If only 16 CRM points are actually run in that region (assuming they continue to behave as they would in the fully resolved case), then the variance would decrease only by a factor 1/16, resulting in a 156-fold overestimate of the variance.

The long-range correlations found in the critical region change this dramatically. Peters & Neelin (2006) found that while subcritical microwave-estimated precipitation variance below criticality scales roughly by the trivial scaling above, in the critical region precipitation variance scales as $L^{-0.46}$. With the caveat that this has only been determined within the range from $L = 25$ to 200 km, we can examine the implications of this non-trivial scaling. For the example above, variance at the 200 km scale would then only be decreased by a factor of $50^{-0.46} = 0.17$. Using the same scaling for a three-dimensional CRM with 16 points in the horizontal yields a factor of $M^{-0.46/2} = 0.53$. The overestimate is thus only by a factor of three. Similar effects should apply for a two-dimensional CRM (e.g. Fig. 16.1*b*), although the scaling probably differs.

These examples provide only a very rough illustration of the issue. Many additional factors can enter: feedbacks between the large scale and the CRM in the superparameterisation may alter the statistics; numerical factors such as excess diffusivity may affect the CRM variance; and because superparameterisation (and convective parameterisation in general) interrupts turbulent cascades between the plume scale and large scale, more sophisticated treatment may be necessary. Nonetheless, knowledge of the empirical scaling may aid in assessing such impacts.

A converse of this quick calculation is to consider implications of the spatial scaling for the fluctuations about QE that one should aim to reproduce at different model grid sizes. Increasing L_G by a factor of 8 in the $L^{-0.46}$ scaling yields only a reduction of 0.4 in the variance near critical, and only 0.3 for a factor of 16 in L_G. Thus even on coarse model grids, variance about traditional QE remains an important effect to capture.

16.6 Discussion

16.6.1 Reinterpreting quasi-equilibrium

There are a number of important pragmatic issues regarding the implementation and consequences of convective QE, involving the boundary layer versus the free troposphere and the temperature and moisture structure. On short time scales, variations in temperature occur about the deep tropospheric structure characteristic of QE postulates, although at larger time and space scales the free troposphere tends to match QE expectations. Variations in the vertical structure of convective heating about that predicted by QE can have substantial impacts in stochastic schemes. The importance of free tropospheric water vapour places a premium on representation of entrainment; as a by-product, it makes column water vapour – for which we have

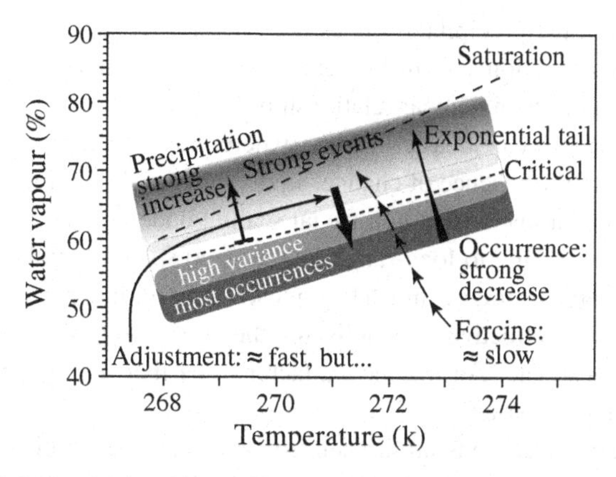

Figure 16.8 Schematic based on Fig. 16.5*b* illustrating the ingredients of an updated interpretation of convective QE. The critical value of column water vapour w_c at which the onset of strong convection occurs is shown as a function of vertically averaged tropospheric temperature $\hat{T}$. Under conventional QE, one expects the sharp increase in the expected value of precipitation above criticality and the system to spend much of its time near criticality. Additional properties are bundled with this: high precipitation variance near criticality, long-range correlations with associated clusters in precipitation near criticality and an event-size distribution with a scale-free range. While the probability of occurrence of a given $(w, \hat{T})$ state drops rapidly at criticality, long exponential tails of the distribution yield relatively frequent occurrence above criticality. These effects apply even when the adjustment process can be assumed to be fast.

abundant observations – a good indicator of conditions conducive to convection. While important for revising convective schemes, the issues of vertical structure are less fundamental to the principle of QE than are the following considerations, which suggest a reinterpretation of QE.

A number of these features are illustrated in Fig. 16.8, which overlays a schematic of the observed features on the empirically determined critical surface for column water vapour as a function of temperature from Fig. 16.5*b*. The sharp pick-up in precipitation at a particular set of values of temperature and moisture corresponds qualitatively to the original QE postulate. Interpreting precipitation as our observable for strong convection and dissipation of buoyancy and allowing for the fact that we have expressed our diagram with different temperature- and moisture-related variables, there is a qualitative correspondence to Arakawa's diagram (Fig.16.1*a*) for the original QE postulate.

However, as argued in Peters & Neelin (2006), the empirical properties of the onset of convection correspond closely to properties of a continuous phase transition. This implies that the transition associated with QE is inseparable from a number of important statistical properties.

- The ensemble-average precipitation does not increase linearly with the distance from criticality. With a functional form like equation (16.4.1), there would not be a single convective time scale as in a linear relationship.
- High precipitation variance near criticality (where the system spends much of its time) is expected even in the absence of changes in the large-scale forcing.
- This precipitation variance has a non-trivial scaling under spatial averaging in the critical region, corresponding to long-range (power-law) spatial correlations. The variance decreases with spatial averaging much less quickly than for independent points, so spatial averaging is not very effective at establishing the large-ensemble average envisioned in the original QE postulate. For instance, a factor of 64 increases in grid-box area yields only a factor of 0.4 in variance.
- These long-range correlations are associated with scale-free critical clusters reflected in the event sizes in Fig. 16.6*b* and Peters *et al.* (2002), as well as in the broad size distributions of mesoscale convective systems (Mapes & Houze 1993; Houze 2004; Nesbitt *et al.* 2006; Peters *et al.* 2009).

The transition further exhibits dynamics characteristic of self-organised criticality as described in Sections 16.4 and 16.5. While the above properties are shared by all (including equilibrium) critical systems, below we present dynamical aspects specific to self-organised non-equilibrium critical systems.

- The probability density function for column water vapour exhibits a sharp drop near critical. This is associated with the sharply increasing sink for water vapour/buoyancy above criticality.
- For precipitating points, the probability density function (PDF) has a peak below criticality, i.e. the system spends much of its time close to the critical point, as envisioned under QE. However, the width of the distribution is substantial. For the approximately Gaussian core, plus or minus 1 standard deviation covers the region from roughly $0.8w_\mathrm{c}$ to w_c, for points measured at a scale of $(25\ \mathrm{km})^2$.
- Above criticality, the PDF for water vapour has an approximately exponential tail. Relative to a Gaussian, this slow decay above criticality implies relatively frequent occurrences of high precipitation. This behaviour is commonly observed in forced advection–diffusion processes.

The high variance near criticality is one of the clearest differences from convective QE as originally conceived. A large fraction of the precipitation occurs in the critical region, with another substantial contribution from the excursions above criticality (approximately 70 and 25%, respectively, for the observations in Section 16.4). It is thus not surprising that QE schemes have tended to be deficient in precipitation variance, with both temperature and moisture pinned too close to the postulated equilibrium. These are aspects that stochastic convective schemes and superparameterisation aim to address, and both have the potential to incorporate information from the observations presented here.

16.6.2 Conjectures for stochastic convection schemes

Given that conventional aspects of convection schemes are in need of revision to include such information as the dependence on tropospheric water vapour, and given the unfortunate sensitivity to the details of implementation in both types of parameterisation, it makes sense to rethink both conventional and stochastic aspects together. Approaches seeking to constrain the stochastic processes from observations and cloud-resolving models appear promising, along with basic principles such as care to conserve moist-static energy.

The manner in which observations of precipitation, water vapour and temperature collapse to simple dependences expected from critical phenomena summarised above is highly suggestive. The fact that some of these properties occur commonly in simple systems, including lattice models as seen in the parallels between Figs. 16.6 and 16.7, suggests that we ought to be able to capture them in our convective schemes. That some properties are universal implies that they do not provide a test of whether a system contains the correct physics for convection, but systems that aim to represent the statistics of convection should include them. It also suggests that systems simpler than fully resolving convection can be devised that will include some of these properties that have now been so clearly observed for convection.

One promising approach, adapting lattice models for use at the subgrid scale, has been initiated (Majda & Khouider 2002; Khouider *et al.* 2003). While the specifics of the moist physics based on convective inhibition do not match the observations here, the approach is adaptable to different physics, whether from meteorology or other systems (Katsoulakis *et al.* 2003). The key points of neighbour interactions at the subgrid scale are included at low computational cost – the challenge is to accurately implement the complex physics of convection onset into a binary on/off representation at the convective scale. Empirical evaluation of transition probabilities should aid in this endeavour.

On the more computationally intensive side of the proposed solutions, these properties may be approximated in models that attempt to coarsely resolve convection (Kuang *et al.* 2005; Pauluis & Garner 2006; Knutson *et al.* 2007), and this should be investigated. Checking these properties in cloud-resolving models will also help to isolate the physics, in particular whether or not the plume scale is sufficient or if cloud microphysics, such as aggregation and fallout processes, are involved. There is reason to believe that the network of superparameterisation CRM points envisioned by Randall *et al.* (2003) should be able to capture some aspects of the long-range correlation seen in the observations, and it is posited that the long-range correlations should help with the issue of correctly estimating grid-scale variance from a small sample of points. Examination of superparameterisation strategies to determine if they can yield the same exponents as a full grid will be of interest.

Between these two approaches in terms of computational intensity, there is room for schemes similar to recent stochastic approaches, adapting convective plume models akin to those from conventional parameterisations, but including stochastic representations of the subgrid-scale fluctuations. Conjectures based on the observational results here suggest that carrying water vapour at slightly higher resolution than the wind field to interact with parameterised convective plumes might aid in capturing the exponential tail of the water-vapour distribution above criticality.

An appealing conjecture for such next-generation schemes comes from analogy with certain other systems exhibiting critical phenomena. Conventional parameterisations are the equivalent of mean field theory in that individual elements interact only with the large-scale mean. While mean field theory can approximate the pick-up of an order parameter in a phase transition, albeit with incorrect exponent, its usefulness is often limited (e.g. Yeomans 1992) by non-negligible effects of fluctuations. The mesoscale literature already provides a list of positive near-neighbour feedbacks [e.g. Mapes (1993), Redelsperger *et al.* (2000), Tompkins (2001), Houze (2004), Moncrieff & Liu (2006) and references therein], including: uplift by cold pools; mesoscale surface-flux enhancement; seeding from mesoscale anvils; and water-vapour increases in the vicinity of previous plumes by detrainment, rain re-evaporation or baroclinic wave convergence. Inclusion of interaction among subgrid-scale plumes may thus be a key to capturing the phenomena noted here. Fortunately, the long-range correlations imply that the exact details of the physics should not be necessary to obtain the properties of Section 16.6.1, opening the door to parameterisations that approximate these interactions cheaply.

Acknowledgements

This work was supported under National Science Foundation ATM-0082529 and OPP-0129800 (JWL), National Oceanic and Atmospheric Administration NA05-OAR4311134 and NA05-OAR4310007 and the Guggenheim Foundation (JDN). The TMI data are from Remote Sensing Systems. We thank A. Arakawa, C. Bretherton, K. Emanuel, M. Moncrieff and D. Randall for discussions and J. Meyerson for graphical assistance.

References

Arakawa, A. 2004 The cumulus parameterization problem: past present and future. *J. Climate*, **17**, 2493–2525 (doi:10.1175/1520–0442(2004)017<2493:RATCPP>2.0.CO;2).

Arakawa, A. & Schubert, W. H. 1974 Interaction of a cumulus cloud ensemble with the large-scale environment Part I. *J. Atmos. Sci.*, **31**, 674–701 (doi:10.1175/1520–0469(1974)031<0674:IOACCE>2.0.CO;2).

Berner, J., Shutts, G. & Palmer, T. 2005 Parameterising the multiscale structure of organised convection using a cellular automaton. In Proceedings of the ECMWF Workshop on Representation of Sub-grid Processes Using Stochastic-Dynamic Models, Reading, UK, June 2005.

Betts, A. K. & Miller, M. J. 1986 A new convective adjustment scheme. Part II: single column tests using GATE wave, BOMEX, ATEX and Arctic air-mass data sets. *Q. J. R. Meteorol. Soc.*, **112**, 693–709 (doi:10.1256/smsqj.47307).

Bretherton, C. S., Peters, M. E. & Back, L. 2004 Relationships between water vapor path and precipitation over the tropical oceans. *J. Climate*, **17**, 1517–1528 (doi:10.1175/1520–0442(2004)017<1517:RBWVPA>2.0.CO;2).

Brown, R. G. & Bretherton, C. S. 1997 A test of the strict quasi-equilibrium theory on long space and time scales. *J. Atmos. Sci.*, **5**, 624–638 (doi:10.1175/1520–0469(1997)054<0624:ATOTSQ>2.0.CO;2).

Brown, R. G. & Zhang, C. 1997 Variability of midtropospheric moisture and its effect on cloud-top height distribution during TOGA COARE. *J. Atmos. Sci.*, **54**, 2760–2774 (doi:10.1175/1520–0469(1997)054<2760:VOMMAI>2.0.CO;2).

Buizza, R., Miller, M. & Palmer, T. N. 1999 Stochastic representation of model uncertainties in the ECMWF ensemble prediction system. *Q. J. R. Meteorol. Soc.*, **125**, 2887–2908 (doi:10.1256/smsqj.56005).

Ciesielski, P. E., Hartten, L. & Johnson, R. H. 1997 Impacts of merging profiler and rawinsonde winds on impacts of merging profiler and rawinsonde winds on TOGA COARE analyses. *J. Atmos. Oceanic Technol.*, **14**, 1264–1279 (doi:10.1175/1520–0426(1997)014<1264:IOMPAR>2.0.CO;2).

Cohen, B. G. & Craig, G. C. 2006 Fluctuations in an equilibrium convective ensemble. Part II: numerical experiments. *J. Atmos. Sci.*, **63**, 2005–2015 (doi:10.1175/JAS3710.1).

Craig, G. C. & Cohen, B. G. 2006 Fluctuations in an equilibrium convective ensemble. Part I: theoretical formulation. *J. Atmos. Sci.*, **63**, 1996–2004 (doi:10.1175/JAS3709.1).

Derbyshire, S. H., Beau, I., Bechtold, P. *et al.* 2004 Sensitivity of moist convection to environmental humidity. *Q. J. R. Meteorol. Soc.*, **130**, 3055–3079 (doi:10.1256/qj.03.130).

Dickman, R., Vespignani, A. & Zapperi, S. 1998 Self-organized criticality as an absorbing-state phase transition. *Phys. Rev. E*, **57**, 5095–5105(doi:10.1103/PhysRevE.57.5095).

Emanuel, K. A. 1991 A scheme for representing cumulus convection in large-scale models. *J. Atmos. Sci.*, **48**, 2313–2335 (doi:10.1175/1520–0469(1991)048<2313:ASFRCC>2.0.CO;2).

Emanuel, K. A. 1993 The effect of convective response time on WISHE modes. *J. Atmos. Sci.*, **50**, 1763–1775 (doi:10.1175/1520–0469(1993)050<1763:TEOCRT>2.0.CO;2).

Emanuel, K. A. 1994 *Atmospheric Convection.* Oxford University Press.

Emanuel, K. A., Neelin, J. D. & Bretherton, C. S. 1994 On large-scale circulations in convecting atmospheres. *Q. J. R. Meteorol. Soc.*, **120**, 1111–1143 (doi:10.1002/qj.49712051902).

Eskridge, R. E., Alduchov, O. A., Chernykh, I. V., Panmao, Z., Polansky, A. C. & Doty, S. R. 1995 A comprehensive aerological reference data set (CARDS): rough and systematic errors. *Bull. Am. Meteorol. Soc.*, **76**, 1759–1775 (doi:10.1175/1520–0477(1995)076<1759:ACARDS>2.0.CO;2).

Gollub, J. P., Clarke, J., Gharib, M., Lane, B. & Mesquita, O. N. 1991 Fluctuations and transport in a stirred fluid with a mean gradient. *Phys. Rev. Lett.*, **67**, 3507–3510 (doi:10.1103/PhysRevLett.67.3507).

Grabowski, W. W. 2001 Coupling cloud processes with the large-scale dynamics using the cloud-resolving convection parameterization (CRCP). *J. Atmos. Sci.*, **58**, 978–997 (doi:10.1175/1520–0469(2001)058<0978:CCPWTL>2.0.CO;2).

Grabowski, W. W. 2003 MJO-like coherent structures: sensitivity simulations using the cloud-resolving convection parameterization (CRCP). *J. Atmos. Sci.*, **60**, 847–864 (doi:10.1175/1520–0469(2003)060<0847:MLCSSS>2.0.CO;2).

Gregory, D. & Rowntree, P. R. 1990 A mass flux convection scheme with representation of cloud ensemble characteristics and stability-dependent closure. *Mon. Weather Rev.*, **118**, 1483–1506 (doi:10.1175/1520–0493(1990)118<1483:AMFCSW>2.0.CO;2).

Haertel, P. T. & Kiladis, G. N. 2004 Dynamics of 2-day equatorial waves. *J. Atmos. Sci.*, **61**, 2707–2721 (doi:10.1175/JAS3352.1).

Hilburn, K. A. & Wentz, F. J. 2007 Intercalibrated passive microwave rain products from the unified microwave ocean retrieval algorithm (UMORA). *J. Appl. Meteorol. Clim.*, **47**, 778–794 (doi:10.1175/2007JAMC1635.1).

Holloway, C. E. & Neelin, J. D. 2007 The convective cold top and quasi equilibrium. *J. Atmos. Sci.*, **64**, 1467–1487 (doi:10.1175/JAS3907.1).

Holloway, C. E. & Neelin, J. D. 2009 Moisture vertical structure, column water vapor, and tropical deep convection. *J. Atmos. Sci.*, **66**, 1665–1683.

Houze, R. A. J. 2004 Mesoscale convective systems. *Rev. Geophys.*, **42**, RG4003. (doi:10.1029/2004RG000150).

Kain, J. S. & Fritsch, J. M. 1990 A one-dimensional entraining/detraining plume model and its application in convective parameterization. *J. Atmos. Sci.*, **47**, 2784–2802 (doi:10.1175/1520–0469(1990)047<2784:AODEPM>2.0.CO;2).

Katsoulakis, M. A., Majda, A. J. & Vlachos, D. G. 2003 Coarse-grained stochastic processes for microscopic lattice systems. *Proc. Natl. Acad. Sci. USA*, **100**, 782–787 (doi:10.1073/pnas.242741499).

Khairoutdinov, M. F. & Randall, D. A. 2001 A cloud resolving model as a cloud parameterization in the NCAR community climate system model: preliminary results. *Geophys. Res. Lett.*, **28**, 3617–3720 (doi:10.1029/2001GL013552).

Khouider, B., Majda, A. J. & Katsoulakis, M. A. 2003 Coarse-grained stochastic models for tropical convection and climate. *Proc. Natl Acad. Sci. USA*, **100**, 11941–11946 (doi:10.1073/pnas1634951100).

Knutson, T. R., Sirutis, J. J., Garner, S. T., Held, I. M. & Tuleya, R. E. 2007 Simulation of the recent multidecadal increase of Atlantic hurricane activity using an 18-km-grid regional model. *Bull. Am. Meteorol. Soc.*, **88**, 1549–1565 (doi:10.1175/BAMS-88–10–1549).

Kuang, Z. & Bretherton, C. S. 2006 A mass-flux scheme view of a high-resolution simulation of a transition from shallow to deep cumulus convection. *J. Atmos. Sci.*, **63**, 1895–1909 (doi:10.1175/JAS3723.1).

Kuang, Z., Blossey, P. N. & Bretherton, C. S. 2005 A new approach for 3D cloud-resolving simulations of large-scale atmospheric circulation. *Geophys. Res. Lett.*, **32**, L02809. (doi:10.1029/2004GL021024).

Kuell, V., Gassmann, A. & Bott, A. 2007 Towards a new hybrid cumulus parametrization scheme for use in non-hydrostatic weather prediction models. *Q. J. R. Meteorol. Soc.*, **133**, 479–490 (doi:10.1002/qj.28).

Lin, J. W.-B. & Neelin, J. D. 2000 Influence of a stochastic moist convective parameterization on tropical climate variability. *Geophys. Res. Lett.*, **27**, 3691–3694 (doi:10.1029/2000GL011964).

Lin, J. W.-B. & Neelin, J. D. 2002 Considerations for stochastic convective parameterization. *J. Atmos. Sci.*, **59**, 959–975 (doi:10.1175/1520–0469(2002)059<0959:CFSCP>2.0.CO;2).

Lin, J. W.-B. & Neelin, J. D. 2003 Toward stochastic moist convective parameterization in general circulation models. *Geophys. Res. Lett.*, **30**, 1162. (doi:10.1029/2002GL016203).

Lübeck, S. 2000 Moment analysis of the probability distribution of different sandpile models. *Phys. Rev. E.*, **61**, 204–209 (doi:10.1103/PhysRevE.61.204).

Majda, A. J. 1993 The random uniform shear layer: an explicit example of turbulent diffusion with broad tail probability distributions. *Phys. Fluids A: Fluid Dyn.*, **5**, 1963–1970 (doi:10.1063/1.858823).

Majda, A. J. & Khouider, B. 2002 Stochastic and mesoscale models for tropical convection. *Proc. Natl Acad. Sci. USA*, **99**, 1123–1128 (doi:10.1073/pnas.032663199).

Maloney, E. D. & Hartmann, D. L. 2001 The sensitivity of intraseasonal variability in the NCAR CCM3 to changes in convective paramterisation. *J. Climate*, **14**, 2015–2034 (doi:10.1175/1520–0442(2001)014<2015:TSOIVI>2.0.CO;2).

Manna, S. S. 1991 Two-state model of self-organized criticality. *J. Phys. A Math. Gen.*, **24**, L363-L369 (doi:10.1088/0305–4470/24/7/009).

Mapes, B. E. 1993 Gregarious tropical convection. *J. Atmos. Sci.*, **50**, 2026–2037 (doi:10.1175/1520–0469(1993)050<2026:GTC>2.0.CO;2).

Mapes, B. E. & Houze, R. A. 1993 Cloud clusters and superclusters over the oceanic warm pool. *Mon. Weather Rev.*, **121**, 1398–1415 (doi:10.1175/1520–0493(1993)121<1398:CCASOT>2.0.CO;2).

Moncrieff, M. W. & Klinker, E. 1997 Organized convective systems in the tropical western pacific as a process in general circulation models: a TOGA-COARE case-study. *Q. J. R. Meteorol. Soc.*, **123**, 805–827 (doi:10.1002/qj.49712354002).

Moncrieff, M. W. & Liu, C. 2006 Representing convective organization in prediction models by a hybrid strategy. *J. Atmos. Sci.*, **63**, 3404–3420 (doi:10.1175/JAS3812.1).

Moorthi, S. & Suarez, M. J. 1992 Relaxed Arakawa–Schubert: a parameterization of moist convection for general circulation models. *Mon. Weather Rev.*, **120**, 978–1002 (doi:10.1175/1520–0493(1992)120<0978:RASAPO>2.0.CO;2).

Neelin, J. D. & Zeng, N. 2000 A quasi-equilibrium tropical circulation model – formulation. *J. Atmos. Sci.*, **57**, 1741–1766 (doi:10.1175/1520–0469(2000)057<1741:AQETCM>2.0.CO;2).

Neelin, J. D., Peters, O. & Hales, K. 2009. The transition to strong convection. *J. Atmos. Sci.*, **66**, 2367–2384.

Nesbitt, S., Cifelli, W. R. & Rutledge, S. A. 2006 Storm morphology and rainfall characteristics of TRMM precipitation features. *Mon. Weather Rev.*, **134**, 2702–2721 (doi:10.1175/MWR3200.1).

Palmer, T. N. 2001 A nonlinear dynamical perspective on model error: a proposal for non-local stochastic-dynamic parameterisation in weather and climate prediction models. *Q. J. R. Meteorol. Soc.*, **127**, 279–304 (doi:10.1002/qj.49712757202).

Pan, D.-M. & Randall, D. A. 1998 A cumulus parameterization with a prognostic closure. *Q. J. R. Meteorol. Soc.*, **124**, 949–981 (doi:10.1256/smsqj.54713).

Parsons, D. B., Yoneyama, K. & Redelsperger, J.-L. 2000 The evolution of the tropical western Pacific atmosphere–ocean system following the arrival of a dry intrusion. *Q. J. R. Meteorol. Soc.*, **126**, 517–548 (doi:10.1002/qj.49712656307).

Pauluis, O. & Garner, S. 2006 Sensitivity of radiative–convective equilibrium simulations to horizontal resolution. *J. Atmos. Sci.*, **63**, 1910–1923 (doi:10.1175/JAS3705.1).

Peters, O. & Neelin, J. D. 2006 Critical phenomena in atmospheric precipitation. *Nat. Phys.*, **2**, 393–396 (doi:10.1038/nphys314).

Peters, H., Hertlein, C. & Christensen, K. 2002 A complexity view of rainfall. *Phys. Rev. Lett.*, **88**, 018701 (doi:10.1103/PhysRevLett.88.018701).

Peters, O., Neelin, J. D. & Nesbitt, S. W. 2009 Mesoscale convective systems and critical clusters. *J. Atmos. Sci.*, in press.

Pierrehumbert, R. T. 2000 Lattice models of advection–diffusion. *Chaos Interdiscip. J. Nonlin. Sci.*, **10**, 61–74 (doi:10.1063/1.166476).

Plant, R. S. & Craig, G. C. 2008 A stochastic parameterization for deep convection based on equilibrium statistics. *J. Atmos. Sci.*, **65**, 87–105 (doi:10.1175/2007JAS2263.1).

Randall, D., Khairoutdinov, M., Arakawa, A. & Grabowski, W. 2003 Breaking the cloud parameterization deadlock. *Bull. Am. Meteorol. Soc.*, **84**, 1547–1564 (doi:10.1175/BAMS-84–11–1547).

Raymond, D. J. 1997 Boundary layer quasi-equilibrium (BLQ). *The Physics and Paramterisation of Moist Atmospheric Convection*, ed. R. K. Smith, pp. 387–397. Kluwer Academic Publishers.

Raymond, D. J. 2000 Thermodynamic control of tropical rainfall. *Q. J. R. Meteorol. Soc.*, **126**, 889–898 (doi:10.1256/smsqj.56405).

Redelsperger, J., Guichard, F. & Mondon, S. 2000 A parameterization of mesoscale enhancement of surface fluxes for large-scale models. *J. Climate*, **13**, 402–421 (doi:10.1175/1520–0442(2000)013<0402:APOMEO>2.0.CO;2).

Shraiman, B. I. & Siggia, E. D. 1994 Lagrangian path integrals and fluctuations in random flow. *Phys. Rev.*, **49**, 2912–2927 (doi:10.1103/PhysRevE.49.2912).

Shutts, G. J. & Palmer, T. N. 2007 Convective forcing fluctuations in a cloud-resolving model: relevance to the stochastic parameterization problem. *J. Climate*, **20**, 187–202 (doi:10.1175/JCLI3954.1).

Stokes, G. M. & Schwartz, S. E. 1994 The Atmospheric Radiation Measurement (ARM) program: programmatic background and design of the cloud and radiation testbed. *Bull. Am. Meteorol. Soc.*, **75**, 1201–1221 (doi:10.1175/1520–0477(1994)075<1201:TARMPP>2.0.CO;2).

Straub, K. H. & Kiladis, G. N. 2003 The observed structure of convectively coupled Kelvin waves: comparison with simple models of coupled wave instability. *J. Atmos. Sci.*, **60**, 1655–1668 (doi:10.1175/1520–0469(2003)060<1655:TOSOCC>2.0.CO;2).

Tang, C. & Bak, P. 1988 Critical exponents and scaling relations for self-organized critical phenomena. *Phys. Rev. Lett.*, **60**, 2347–2350 (doi:10.1103/PhysRevLett.60.2347).

Teixeira, J. & Reynolds, C. 2008 Stochastic nature of physical parameterizations in ensemble prediction: a stochastic convection approach. *Mon. Weather Rev.*, **136**, 483–496 (doi:10.1175/2007MWR1870.1).

Tian, B., Waliser, D., Fetzer, E. *et al.* 2006 Vertical moist thermodynamic structure and spatial-temporal evolution of the MJO in AIRS observations. *J. Atmos. Sci.*, **63**, 2462–2485 (doi:10.1175/JAS3782.1).

Tompkins, A. M. 2001 Organization of tropical convection in low vertical wind shears: the role of water vapor. *J. Atmos. Sci.*, **58**, 529–545 (doi:10.1175/1520–0469(2001)058<0529:OOTCIL>2.0.CO;2).

Tompkins, A. M. & Berner, J. 2008 A stochastic convective approach to account for model uncertainty due to unresolved humidity variability. *J. Geophys. Res.*, **113**, D18101 (doi:10.1029/2007JD009284).

Tost, H., Jöckel, P. & Lelieveld, J. 2006 Influence of different convection parameterisations in a GCM. *Atmos. Chem. Phys.*, **6**, 5475–5493.

Uppala, S. M. *et al.* 2005 The ERA-40 re-analysis. *Q. J. R. Meteorol. Soc.*, **131**, 2961–3012 (doi:10.1256/qj.04.176).

Xu, K.-M. & Emanuel, K. A. 1989 Is the tropical atmosphere conditionally unstable?. *Mon. Weather Rev.*, **117**, 1471–1479 (doi:10.1175/1520–0493(1989)117<1471:ITTACU>2.0.CO;2).

Xu, K.-M., Arakawa, A. & Krueger, S. K. 1992 The macroscopic behavior of cumulus ensembles simulated by a cumulus ensemble model. *J. Atmos. Sci.*, **49**, 2402–2420 (doi:10.1175/1520–0469(1992)049<2402:TMBOCE>2.0.CO;2).

Yano, J.-I., Fraedrich, K. & Blender, R. 2001 Tropical convective variability as 1/f noise. *J. Climate*, **14**, 3608–3616 (doi:10.1175/1520–0442(2001)014<3608:TCVAFN>2.0.CO;2).

Yeomans, J. 1992 *Statistical Mechanics of Phase Transitions*. Oxford University Press.

Yu, J.-Y. & Neelin, J. D. 1994 Modes of tropical variability under convective adjustment and the Madden–Julian oscillation. Part II: numerical results. *J. Atmos. Sci.*, **51**, 1895–1914 (doi:10.1175/1520–0469(1994)051<1895:MOTVUC>2.0.CO;2).

Zhang, G. J. & McFarlane, N. A. 1995 Sensitivity of climate simulations to the parameterization of cumulus convection in the Canadian Climate Centre general circulation model. *Atmos. Oceanogr.*, **33**, 407–446.

Zhang, G. J. & Mu, M. 2005 Simulation of the Madden–Julian oscillation in the NCAR CCM3 using a revised Zhang–McFarlane convection parameterization scheme. *J. Climate*, **18**, 4046–4064 (doi:10.1175/JCLI3508.1).

Zhang, G. J. & Wang, H. 2006 Toward mitigating the double ITCZ problem in NCAR CCSM3. *Geophys. Res. Lett.*, **33**, L06709 (doi:10.1029/2005GL025229).

17

Comparison of stochastic parameterisation approaches in a single-column model

MICHAEL A. BALL AND ROBERT S. PLANT

We discuss and test the potential usefulness of single-column models (SCMs) for the testing of stochastic physics schemes that have been proposed for use in general circulation models (GCMs). We argue that although single-column tests cannot be definitive in exposing the full behaviour of a stochastic method in the full GCM, and although there are differences between SCM testing of deterministic and stochastic methods, SCM testing remains a useful tool. It is necessary to consider an ensemble of SCM runs produced by the stochastic method. These can usefully be compared with deterministic ensembles describing initial condition uncertainty and also with combinations of these (with structural model changes) into poor man's ensembles. The proposed methodology is demonstrated using an SCM experiment recently developed by the GCSS (GEWEX Cloud System Study) community, simulating transitions between active and suppressed periods of tropical convection.

17.1 Introduction

In recent years, increasing attention has been given to the potential usefulness (Palmer 2001; Wilks 2005) of introducing some stochastic component(s) to the physical parameterisations used in general circulation models (GCMs). For example, many GCMs are known to have insufficient high-frequency, small-scale variability of convective heating rates and precipitation in the tropics, which may damage their ability to represent low-frequency, large-scale climate variability (Ricciardulli & Garcia 2000; Horinouchi *et al.* 2003). A wide variety of plausible stochastic methods continue to be suggested and actively investigated, including perturbing the inputs to a parameterisation (e.g. Tompkins & Berner 2008), perturbing the parameters used within it (e.g. Byun & Hong 2007), perturbing its outputs (e.g. Teixeira & Reynolds, 2008) and even constructing new parameterisations designed explicitly to be stochastic from the outset (e.g. Plant & Craig 2008). There is a growing acceptance that the use of stochastic elements in GCMs may be desirable for both theoretical and practical reasons (e.g. Penland 2003; Williams 2005). Thus, the time may soon be approaching when the key question changes from *why a stochastic method* to *which stochastic method*. Here, we explore whether single-column modelling might be able to provide some insights that could inform such decision making.

The aim of a stochastic scheme is to introduce variability into the numerical representation of the climate system. In order to determine the variability of some climate phenomenon in the GCM, either multiple or long integrations are likely to be required. Further lengthy explorations would also be required if one wished to assess the impact of that variability on other aspects of the model climate. How then, in practice, should one choose the stochastic method(s) to be used in a GCM? The difficulty is not simply the range of possible schemes available in the literature, but also (at least) two other important considerations.

First, we do not know how well various methods might combine. The motivations behind various schemes, and the uncertainties they attempt to address, may often appear to be very different. At first sight then it may be attractive to use several methods. However, there may be a danger in this of some 'double counting', particularly if attempting to combine some of the more generic methods to address parameterisation uncertainty. For instance, taking a single parameterisation and perturbing its inputs, parameters and outputs simultaneously might not be totally unreasonable, but it would be extremely naive to expect good performance by implementing three such methods directly 'off the shelf'.

Second, one's difficulties are compounded by the fact that many (if not all) of the stochastic schemes in the literature themselves contain free parameters and structural uncertainties. We can use the random parameters approach to offer a simple example. Suppose that one wished to choose random values for the entrainment rate and the convective available potential energy (CAPE) closure time scale in the parameterisation of deep convection. Should those choices be correlated; and if so, then how?

It would surely be impractical to conduct full GCM testing of all plausible stochastic physics schemes and all possible variations on their basic themes. However, it should be possible to do better than testing some best guesses. As a first step, we describe in this paper essentially a test of concept for the idea that single-column model (SCM) experiments might have some useful value for comparing stochastic schemes. We are not at this stage attempting an assessment of the relative performance of various stochastic schemes. Rather, our objective is to demonstrate that simple methods to improve one's understanding of the behaviour of various stochastic schemes are both possible and worth pursuing. It does not seem unreasonable to hope that current best guesses could evolve into educated guesses.

This chapter is organised as follows. In Section 17.2, we introduce some issues in single-column modelling and their implications for testing stochastic methods. Section 17.3 describes the modelling framework used in this study, including the stochastic methods (17.3.1) and the ensemble approach (17.3.2). Results are shown for the sensitivity to initial condition (IC) perturbations (Section 17.4.1) and for the mean states (Section 17.4.2) and variabilities (Sections 17.4.3 and 17.4.4) of various stochastic methods. Conclusions are drawn in Section 17.5.

17.2 An SCM approach for stochastic schemes

Single-column modelling has a long history as a useful guide towards understanding and testing the behaviour of deterministic parameterisations within a GCM. In the full GCM, a parameterisation interacts with model dynamics and with the other parameterisations. Essentially, the SCM is a means to understand the latter, which may or may not dominate in the full GCM. Usually, the dynamical forcing of the SCM is determined beforehand, perhaps based on an observational campaign. The forcing is independent of the current model state, which constrains the possible responses. Thus the SCM may behave differently from the corresponding full GCM if dynamical feedback is an important aspect of the situation modelled. One consequence is that parameterisation errors in a GCM which adversely affect model dynamics may not be apparent in the SCM, which is kept on track by the prescribed dynamics.

It is not immediately apparent how an SCM might be used to make meaningful comparisons of stochastic methods. In some cases, the use of a single column may simply not be viable because the stochastic terms cannot easily be applied (e.g. Shutts 2005). Indeed, it has been suggested that an ideal stochastic method would probably be non-local (Palmer 2001; Craig *et al.* 2005; Ghil *et al.* 2005). For the present though, the majority of stochastic methods can be formulated for a single column. Nonetheless, an obvious objection to SCM comparisons remains: feedback from the introduced variability to the dynamics may be a key feature of the behaviour in a full GCM (e.g. Lin & Neelin 2002). This will be missing in a traditional SCM experiment with specified dynamics. Single-column models can include appropriate dynamical feedbacks by using a parameterised dynamics formulation, such as a weak temperature gradient approximation for the tropical atmosphere (e.g. Sobel *et al.* 2007), or by coupling vertical advection to the parameterised diabatic heating via a gravity wave model (Bergman & Sardeshmukh 2004). Ultimately, we believe that these and similar frameworks would be particularly well suited to studying stochastic physics schemes, but do not pursue them further here.

It may nonetheless be possible to gain some useful insights into the behaviour of stochastic methods through an SCM comparison. The results for each method must be considered in the form of an ensemble of SCM runs, each run having a different set of random numbers. Our proposal is to compare such ensembles with the SCM results obtained from multiple deterministic parameterisations, the suite of deterministic configurations being treated as a so-called poor man's ensemble (e.g. Mylne *et al.* 2002).

17.3 Experimental set-up

Experiments have been carried out using the single-column form of the UK Met Office Unified Model (UM, Cullen 1993). The model runs are based on GCSS PCCS case 5, the design of which is described by Petch *et al.* (2007). Specifically, we study here the consecutive time periods B and C. Model intercomparison cases have been a major part of the Global Energy and Water cycle EXperiment (GEWEX) Cloud System Study (GCSS), which aims to support the development of physically based parameterisations for cloud processes. An overview of the precipitating convective cloud systems (PCCS) working group can be found in Moncrieff *et al.* (1997).

Case 5 simulates a column of the atmosphere in the tropical west Pacific warm pool region, at 2 °S and 156° E, and the model runs presented here span the period 9–28 January 1993. The forcing data set is derived from observations taken in the TOGA–COARE campaign (Webster & Lukas 1992). It contains temperature and moisture increments due to large-scale vertical and horizontal advection.[1] Also prescribed are sea-surface temperatures and time series of observed winds, towards which the SCM is strongly relaxed, with a time scale of one hour. Any changes to the winds in these runs are therefore limited. The forcings and ICs are derived directly from surface and radiosonde measurements averaged over the TOGA–COARE IFA (Intensive Flux Array, see figure 14 of Webster & Lukas 1992).

The focus of case 5 is the transition of tropical convection from suppressed to active phases, and two such transitions occur during these SCM runs. See Fig. 17.1 in which the periods are defined as in Petch *et al.* (2007). Here we label ActB, a very active period with heavy rain, SupC, in which convection is suppressed by the large-scale forcing, and the following active phase, ActC. Rain rates are effectively constrained by the large-scale forcing and are similar to those found in other SCMs (Woolnough *et al.*, unpublished work).

The SCM runs use a time step of 30 min and there are 38 levels in the vertical. The performance of the default UM SCM for this case in comparison with other

[1] These data sets are available for the whole of the TOGA-COARE observing period, along with further information about their derivation, at http://tornado.atmos.colostate.edu/togadata/data/ifa_data.html.

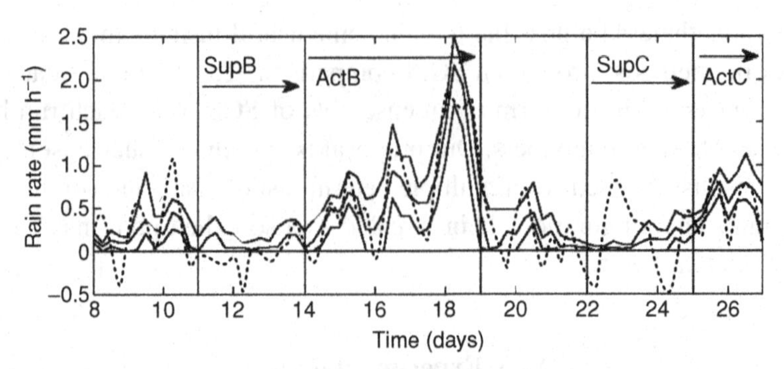

Figure 17.1 Six-hourly mean rainfall rates: 5th, 50th and 95th percentiles from an ensemble of SCM runs using the default UM configuration (solid lines), along with a budget-derived estimate for the TOGA–COARE IFA (dotted line). This estimate has some negative values, since the observations were insufficient to derive an accurate moisture budget (Petch *et al.* 2007). The time axis is labelled in whole days since the start of the month of January 1993, as in later figures. Annotations are explained in the text.

models is discussed by Petch *et al.* (2007) and Woolnough *et al.* (unpublished work). It is more consistent with CRM simulations than some of the SCMs, which were somewhat dry.

17.3.1 Model variants

Taking the UM SCM as a basis, several model configurations have been tested. These differ through either the convection parameterisation or the stochastic method used, and are described below. Most of the stochastic methods have been implemented by introducing a stochastic element to the pre-existing UM parameterisations. We quantify the variability associated with a stochastic method using the spread of an ensemble, with a different set of random numbers drawn for the stochastic component of each ensemble member. Small IC perturbations are also included in the ensembles; these are discussed in Section 17.3.2.

Default UM. The default UM configuration contains parameterisations for layer cloud microphysics, radiation, boundary-layer processes and convection. Martin *et al.* (2006) provide an overview of the current set of schemes. Convection is represented by a deterministic bulk mass-flux scheme based on Gregory & Rowntree (1990), but which has since been modified (Martin *et al.* 2006). There are prognostic moisture variables for specific humidity, cloud liquid water content and cloud ice water content.

Kain–Fritsch convection scheme. An alternative deterministic mass-flux scheme for convection is that of Kain & Fritsch (1990, KF). The version described by Kain (2004) has been implemented here. For a discussion of the differences between the schemes of Gregory & Rowntree (1990) and KF in the UM in a forecasting context, see Done (2002).

Multiplicative noise scheme. This scheme follows the method of Buizza *et al.* (1999) and is designed to represent model uncertainty. At each time step, the total parameterised tendencies for each model variable are multiplied by a random number ε_1 chosen from a uniform distribution between $1-k$ and $1+k$. The random number is the same for each model variable and at each vertical level. Temporal correlation is enforced by keeping the same random number for multiple time steps. Buizza *et al.* (1999) found that the greatest improvements to the performance of the ECMWF ensemble prediction system occurred for $k = 0.5$ and a new random number every six hours. The same choices are made here. Total tendencies from the default UM parameterisations are multiplied by ε_1 at the end of each time step, with a check to restore moisture to zero if the stochastic perturbation implies a negative value.

Random parameters scheme. General circulation model parameterisations include parameters for which the appropriate value is not well determined. The random parameters scheme attempts to account for parameterisation uncertainty by allowing such parameters to vary within a plausible range. This scheme follows the system used (Arribas 2004) in the Met Office Global and Regional Ensemble Prediction System (MOGREPS, Mylne *et al.* 2005). The relevant parameters and ranges can be found in Arribas (2004), but include the entrainment rate and CAPE closure time scale from the UM convection scheme. Temporal correlations are described by a first-order auto-regression model,

$$P_{n+1} = \mu_P + r(P_n - \mu_P) + k_P \varepsilon_2, \tag{17.3.1}$$

in which the parameter is labelled P and the update number n; μ_P is the default value of P, r is an auto-correlation coefficient and $k_P \varepsilon_2$ is a stochastic shock term (see Section 3.2 of Mylne *et al.* 2005) in which ε_2 is a random number uniformly distributed between -1 and 1, and k_P is a parameter-dependent normalisation. Each parameter is subject to maximum and minimum acceptable bounds, and the same random number ε_2 is used for all parameters at each update, every three hours.

Random but constant parameters. Another approach to sampling parameter uncertainty is to run an ensemble in which each run has a fixed, but different, parameter set. This approach has been used to make probabilistic predictions of future climate (e.g. Murphy *et al.* 2004). For this study, we simply adapt the random parameters scheme above by choosing initial parameter values randomly within the acceptable range and holding these fixed. Our method does not explore parameter space in an unbiased way, as it is constrained by the correlations between parameters assumed in the random parameters scheme above. Nonetheless, it allows for an interesting test of the temporal correlations in that scheme.

Plant & Craig stochastic convection scheme. In the Plant & Craig (2008) parameterisation, a finite number of distinct plumes are present in a grid-box area at any instant, resulting in a random sampling of the full spectrum for an ensemble of cumulus clouds. The spectrum used is based on an equilibrium exponential distribution of cloud base mass flux (Craig & Cohen 2006) and plumes are produced at random, with the properties of each based on an adaptation of the KF plume model. The smaller the grid-box size, the more limited the sampling and the larger the fluctuations from statistical equilibrium. A sounding averaged over nearby grid points and recent time steps provides a smoothed input for the CAPE closure calculation. Of course, spatial averaging is not possible in an SCM. Preliminary tests showed that the scheme behaved sensibly in the SCM when averaging over 20 time steps. This choice represents a compromise between providing a smooth input profile and the need to capture variations in the dynamical forcings.

Deterministic limit of the Plant & Craig scheme. The Plant & Craig (2008) scheme can operate as a spectral convective parameterisation by running the plume model for every category of cloud and weighting the tendencies according to the probability of that cloud occurring. This corresponds to the deterministic limit of a very large grid box in which the cumulus ensemble is well sampled.

17.3.2 IC ensembles and ensemble size

It should be noted that a stochastic scheme is not required in order for a parameterisation embedded in an SCM to exhibit variability: even if the prescribed forcings remain constant, a purely deterministic SCM will vary from one time step to the next. This is particularly associated with switches in parameterisations, the most important of which is the trigger function in the convection scheme. This can often exhibit exaggerated on–off behaviour (e.g. Willett & Milton 2006), and the exact set of time steps on which the convection scheme triggers can be very sensitive to small changes in the model state (this was found to be true in deterministic SCM ensembles used in this study, not shown). Such unsteady behaviour inherent to convective and other parameterisations provides a source of variability in deterministic and stochastic SCMs alike.

Part of the variability in a stochastic physics SCM ensemble[2] may arise simply because the stochastic (ST) perturbations force each ensemble member to follow a different realisation, with the convection being triggered on a different set of time steps. Such realisations can also be explored in a deterministic model by running an ensemble with IC perturbations. We suggest that such IC ensembles should be run in order to make meaningful comparisons of stochastic schemes with their deterministic counterparts. Hack & Pedretti (2000) suggest that an ensemble approach is appropriate for SCM studies as an SCM can be sensitive to small differences in the ICs.

The perturbations for an IC SCM ensemble should be small enough not to introduce significant bias to any of the ensemble members, but large enough to force the ensemble members to diverge into an unbiased sample of probable realisations

[2] That is, an ensemble in which each member has the same stochastic parameterisation scheme but draws a different set of random numbers for it.

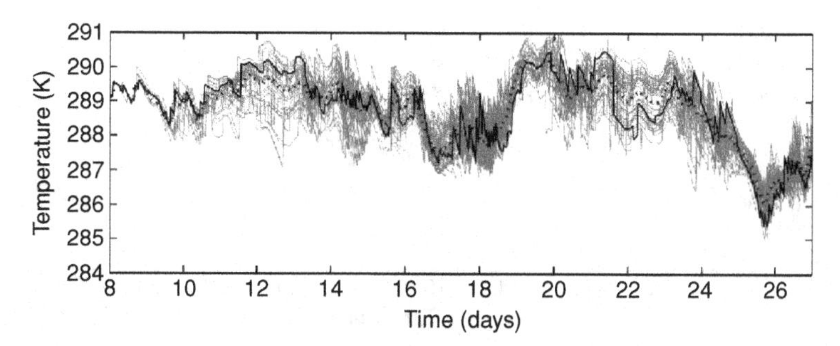

Figure 17.2 Ensemble plume plot of temperature on model level 10 (800 hPa) for the default UM with the set 1 IC perturbations. The solid line denotes the control run (with no IC perturbations) and the dotted line denotes the ensemble mean.

early in the model run. It should be emphasised that the perturbations are only used in this study to provide a sample of realisations and are not necessarily intended to represent realistic IC uncertainty. Thus, there is no requirement for the perturbations to match instrumental and sampling errors in the observations that provide the ICs. The results of such ensemble tests are discussed in Section 17.4.1.

Another aspect to consider is the ensemble size required. We use 39 member ensembles, which appears to be sufficient to produce useable results. The robustness of results derived from ensembles of this size can be estimated from a brief statistical consideration.

Assuming some model variable to be approximately normally distributed, its ensemble mean has a standard error of $\sigma/\sqrt{K}$, where σ is the standard deviation and K is the number of ensemble members. For example, in the IC ensemble for the default UM, the ensemble standard deviation for temperature is of the order 0.5 K. This gives an error of approximately 0.08 K, and a 95% confidence interval (CI) of approximately 0.16 K, which is less than 10% of the amplitude of typical temperature variations during the model runs (Fig. 17.2).

It can be shown that a reasonable approximation to the sampling distribution of the ensemble standard deviation is a normal distribution with standard deviation $\sigma/\sqrt{2K}$, for ensemble sizes $K \gtrsim 25$. We have checked explicitly that the normal distribution holds and in our case it leads to 95% CI of approximately 22% of σ. An interval this broad suggests that ensemble spreads calculated for single variables should be interpreted with care.

More accurate ensemble spreads occur for an error norm which sums the ensemble spread over the model column. We present below results for the total column root-mean-square ensemble spread (TCES), given by

$$\mathrm{TCES} = \left[\frac{\int_{p_{\mathrm{top}}}^{p_{\mathrm{surf}}} \sum_{k=1}^{K} (F_k - \overline{F})^2 \mathrm{d}p}{(p_{\mathrm{surf}} - p_{\mathrm{top}})(K-1)} \right]^{1/2}, \qquad (17.3.2)$$

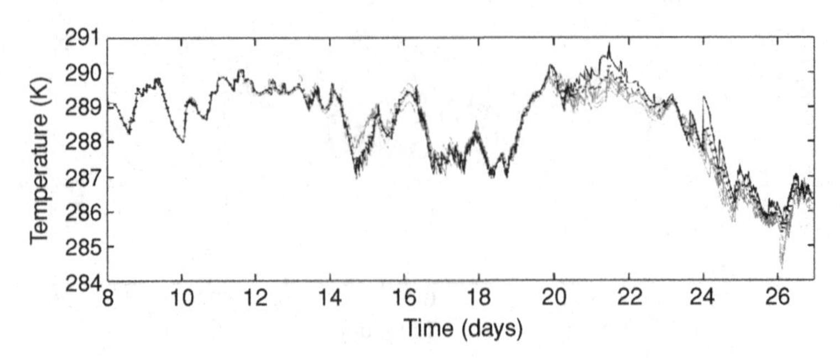

Figure 17.3 Similar plot to Fig. 17.2 but with the Kain–Fritsch convection scheme.

where F is a model variable; $\overline{F}$ its ensemble mean; and k labels ensemble members. Assuming hydrostatic balance, the TCES is the square root of the mass-weighted vertical integral of estimated model field variance. We have performed some simple tests to estimate a sampling error for the TCES for temperature of approximately 10%, roughly half that for a single level.

17.4 Results

17.4.1 Ensemble sensitivity to IC perturbations

We have constructed IC ensembles from two sets of IC perturbations. In set 1, random temperature perturbations are applied to the lowest model level, chosen from a uniform distribution between ± 0.25 K. Temperature perturbations for set 2 are larger and cover a greater vertical extent. A uniform distribution is again used with amplitude 0.5 K at the surface and decreasing exponentially above with a height scale of 1 km. For these larger perturbations, it is desirable to ensure that no spurious supersaturation occurs, and so corresponding perturbations are applied to the specific humidity field in order to maintain the relative humidity.

Figure 17.2 shows spaghetti plots of temperature for a single model level in the lower troposphere. Set 1 IC perturbations have been added to the default UM. Figure 17.3 is equivalent, but for the Kain–Fritsch convection scheme. For the default UM, the perturbations appear to produce a good spread of realisations, but using the KF scheme the ensemble members fail to diverge. Even at the end of the 19-day runs, they remain clustered in six distinct groups, the members of each group triggering convection on the same set of time steps (not shown). Figure 17.4 shows the temperature plume for the Kain–Fritsch scheme using the set 2 IC perturbations. It is clear that the ensemble members are less tightly clustered than with the set 1 perturbations, producing a more representative sample of realisations. But the spread is still smaller than in the default UM using set 1. It is

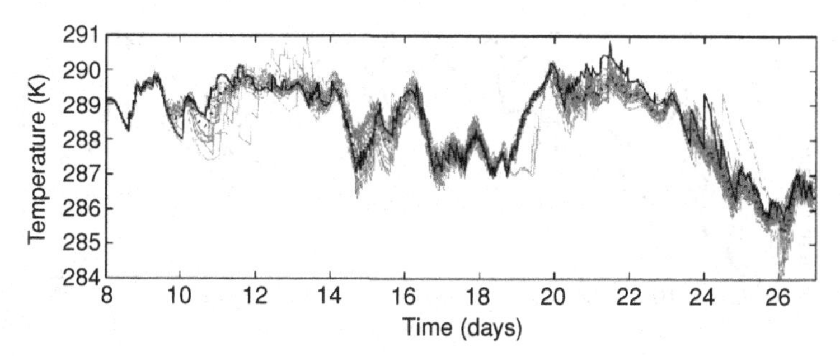

Figure 17.4 Similar plot to Fig. 17.2 but with the Kain–Fritsch convection scheme and the set 2 IC perturbations.

perhaps slightly surprising that the SCM responds so differently to IC perturbations when different convection schemes are used. This is in contrast to Hume & Jakob (2005).

An important point to note from the simulations of Hack & Pedretti (2000) is their observation of bifurcations in SCM solutions (e.g. their figure 4), with ensemble members dividing into two or more preferred modes. Clearly, this raises issues with the representativeness of statistics such as the ensemble mean, since the mean state may lie between modes and never actually occur. Little evidence for multimodal behaviour was found in the present study. Clearly separated modes do occasionally occur, as seen for example, using the Kain–Fritsch convection scheme around the 19th (Fig. 17.4). However, these persist for no longer than a day or so. The presence or absence of bifurcations is presumably related to the character of either (or both) the SCM or the large-scale forcing. We do not speculate further here, but rather note that the ensemble mean and standard deviation appear to be genuinely useful diagnostics for the present study.

Figure 17.5 shows a time series of the TCES of temperature, for the default UM ensemble with set 1 and set 2 IC perturbations, and also for an ensemble that includes a multiplicative-noise scheme, both with and without set 2 IC perturbations added. The corresponding plots for relative humidity are shown in Fig. 17.6.

Looking first at the two default UM TCES ensembles, there is more spread over the first six days using the larger set 2 IC perturbations. However, the set 1 and 2 ensembles look very similar beyond 6 days, suggesting that the ensemble spread has saturated in both. This is reassuring as it suggests that the saturated level of ensemble spread in temperature is independent of the size and nature of the IC perturbations, but rather provides a measure of the inherent variability of the SCM. For the Kain–Fritsch scheme, the larger IC perturbations produce larger ensemble spreads throughout (Figs. 17.3 and 17.4), but this is because the spread did not saturate when the set 1 IC perturbations were used.

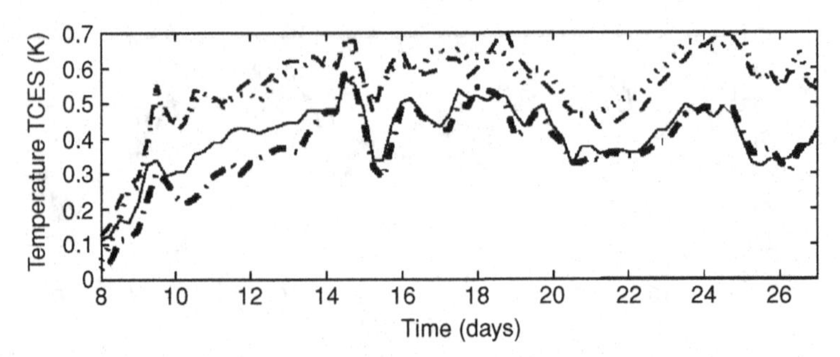

Figure 17.5 The TCES of temperature in the default UM with set 1 (dash–dotted line) and set 2 IC perturbations (solid line), with multiplicative-noise perturbations (dotted line) and with both set 2 IC perturbations and multiplicative noise (dashed line).

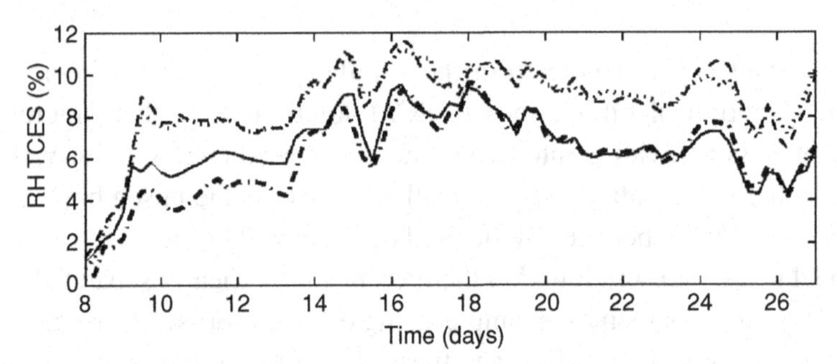

Figure 17.6 As in Fig. 17.5, but for relative humidity (RH) rather than temperature.

In a stochastic physics (ST) SCM ensemble, the stochastic method provides some physically motivated source of variability. One might anticipate that the physics perturbations would allow the ST ensemble to explore at least those realisations accessible to its deterministic analogue.[3] If this is true, then IC perturbations should have little effect on ensemble spread when implemented in an ST ensemble. It is clear from Figs. 17.5 and 17.6 that beyond the first 36 hours or so, the ensembles including multiplicative-noise ST perturbations have spreads that are consistently larger than those occurring in the IC-only ensembles, typically by a factor of about a third. The inclusion of IC perturbations in addition to multiplicative noise slightly increases the spread during the first day, but has no significant effect thereafter. This is consistent with the idea that the IC perturbations allow one to sample different realisations, but do not affect the underlying distribution of probable realisations,

[3] By which we mean the equivalent configuration with the stochastic component disabled, providing of course that such an equivalent is well defined. For example, for a stochastic method in which model parameters are selected randomly, the deterministic analogue is simply a simulation with the default parameter set.

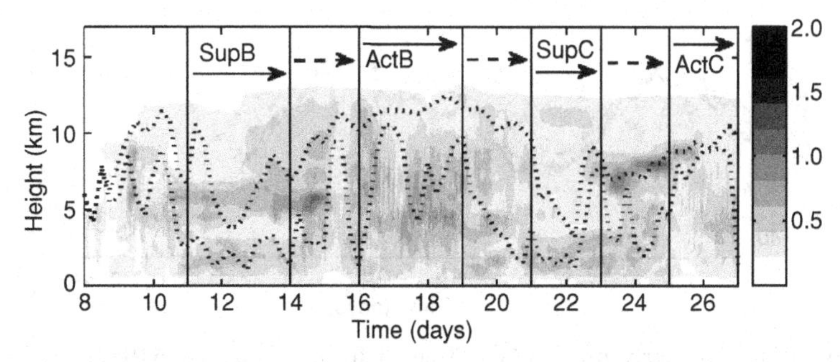

Figure 17.7 Ensemble RMS spread in temperature (K) for the default UM ensemble using set 2 IC perturbations (shaded). Overlaid are the 25th and 75th percentiles of the convective cloud-top height (dotted lines). These values are determined at each time step from the subset of ensemble members in which the convection scheme triggers and are smoothed for presentation here using six-hourly averaging. Annotations are explained in the text.

which emerges once the spread of the ensemble saturates. Similar conclusions apply for the other stochastic methods used (not shown).

Comparisons of the effects of IC and ST perturbations have been made before in the context of global GCM ensemble prediction systems. Buizza *et al.* (1999) found that IC-only ensembles produced consistently larger spread than ST-only ensembles, and that ensembles with IC and ST perturbations produced greater spread still.[4] Teixeira & Reynolds (2008) found similar results over the tropics using a multiplicative-noise scheme applied only to the moist convective tendencies (their figure 7*a*). Although these results differ from ours in placing far greater emphasis on IC perturbations, this is not surprising given the context. In particular, we focus on the saturated level of ensemble spread due to IC perturbations whereas in the cited studies, the runs do not reach saturation.[5] Also those studies used much larger IC perturbations designed to sample IC uncertainty.

It is interesting to note in Fig. 17.5 and 17.6 the time variability of the ensemble spreads. The spread clearly has some dependence on the large-scale forcing, with a peak followed by a sudden drop in spread occurring at the start of each convectively active phase. To study this in more detail, we show in Fig. 17.7 a time-height plot of the ensemble spread in temperature in the set 2 default UM ensemble. The spread appears to follow different characteristic regimes during suppressed and active phases, while behaving in a more unsteady manner during transition periods between the two. Note that in Fig. 17.7, the active and suppressed phases labelled in Fig. 17.1 have been redefined in order to separate out the transition periods. This

[4] This is shown for forecast days 3, 5 and 7 in their table 1a.
[5] See figure 7a of Teixeira & Reynolds (2008), for example.

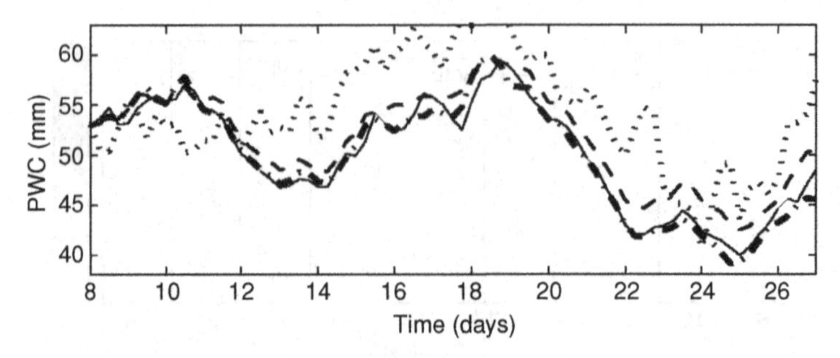

Figure 17.8 Six-hourly mean PWC derived from TOGA–COARE observations (dotted line), and ensemble means for the default UM (dashed line), Kain–Fritsch (solid line) and Plant & Craig deterministic mode (dash–dotted line) configurations.

will allow the ensemble variability characteristic of suppressed and active phases to be analysed in Section 17.4.3.

Most notably during transitions from suppressed to active phases, the pattern of ensemble spread appears to be related to the convective cloud-top height. For example, peaks in TCES on the 15th (Figs. 17.5 and 17.6) correspond to large spreads in the mid troposphere where the ensemble produces a broad range of convective cloud tops. During the following day, this range suddenly narrows and the ensemble spread drops throughout the troposphere. Another interesting feature is the sloping layer of high ensemble spread around the 24th and 25th, which increases in height from roughly 7 to 9 km during the transition from SupC to ActC. This layer closely follows the 75th percentile of convective cloud-top height, indicating an ascending lid on the convection. The ensemble spread is large here because the ensemble members produce a range of different heights for this lid, which has a sharp temperature gradient across it (not shown).

17.4.2 Intercomparison of ensemble-mean states

An ensemble was produced for each of the SCM configurations described in Section 17.3.1. In the case of the stochastic Plant & Craig scheme, two separate ensembles were produced for columns with horizontal scales Δx of 50 and 100 km. (As explained in Section 17.3.1, the stochastic fluctuations in that scheme depend on the column size.) To ensure a consistent comparison between stochastic physics SCM ensembles and their deterministic analogues, we included the set 2 IC perturbations in all ensembles, although beyond the first day they make very little difference to the stochastic ones.

We show results here for the precipitable water content (PWC), the mass-weighted integral of specific humidity through the column. Figure 17.8 shows the time series of observation-derived PWC and ensemble means for three deterministic SCM configurations. The SCMs exhibit large systematic biases from the

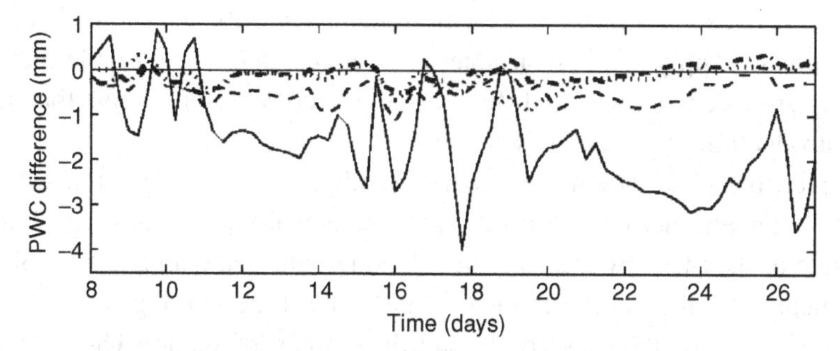

Figure 17.9 Six-hourly mean difference in ensemble mean PWC relative to the default UM, for the multiplicative noise (dotted line), time-varying random parameters (dashed line), constant random parameters (dash–dotted line) and Kain–Fritsch (solid line) configurations.

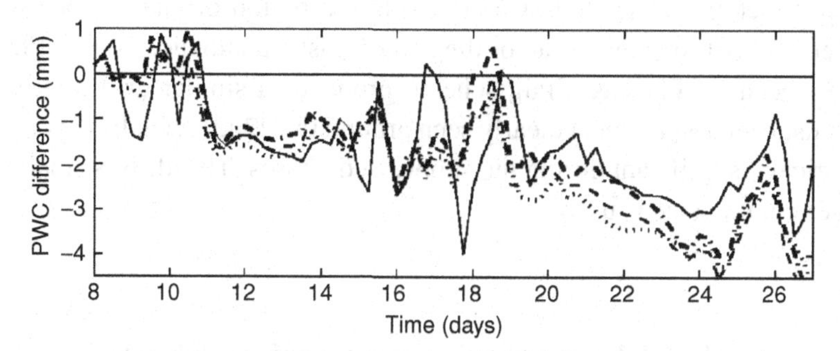

Figure 17.10 Similar plot to Fig. 17.9, but for the Plant & Craig scheme with grid lengths of 50 km (dotted line) and 100 km (dashed line) and in deterministic mode (dash–dotted line), and for the Kain–Fritsch scheme (solid line).

observed PWC: this is probably due to discrepancies in the large-scale forcings used to drive the SCMs, as found in other SCM and CRM studies which use advective forcings derived from observations (e.g. Krueger & Lazarus 1999). The Kain–Fritsch and Plant & Craig schemes (which both use the same plume model) produce a drier state than the default UM, although well within the range of values seen when comparing various SCMs (Woolnough *et al.* unpublished work). There is also tropospheric cooling in these schemes relative to the default UM (not shown).

Several of the SCM configurations in this study include ST perturbations but are based on the UM convection scheme. In Fig. 17.9, we show the difference in ensemble-mean PWC between these configurations and their deterministic analogue, the default UM. Also shown is the difference between the Kain–Fritsch scheme and the default UM. Figure 17.10 shows similar plots for the Plant & Craig scheme. In terms of ensemble-mean PWC, the difference between the two convection parameterisations (default UM and Kain–Fritsch) is several times the difference between any of the stochastic schemes and its deterministic analogue. Similar remarks apply to other variables and suggest that the ensemble-mean fields are more sensitive to structural differences in the convection scheme than they are to the introduction of stochastic schemes.

However, the observation that stochastic physics schemes designed to represent model uncertainty or departures from statistical equilibrium can change the mean state of the SCM by even a relatively small amount is interesting. Statistical tests indicate that the ensemble mean state of the time-varying random parameters ensemble is significantly cooler and dryer than the default UM for much of the model run, especially during periods of suppressed convection. However, the constant random parameters scheme did not produce this deviation despite sampling the same range of values for model parameters (Fig. 17.9). This suggests noise-induced drift; i.e. the random noise introduced by the time variation of the model parameters, causes the SCM to explore a region of phase space which is asymmetric about the mean state of the deterministic analogue. Note in Fig. 17.10 that the stochastic Plant & Craig scheme produces a similar drift relative to its deterministic analogue (most clearly seen around the 22nd), which is also found to be statistically significant during the suppressed phases. This drift is smaller when the larger column size is used.

17.4.3 Intercomparison of ensemble variability

Figure 17.11 shows time-mean vertical profiles of ensemble spread in temperature for the active and suppressed periods ActB, SupC and ActC, as labelled in Fig. 17.7. There are marked differences between active and suppressed phases. This is most apparent in the mid troposphere where the spread tends to be higher during active phases, whereas in the lower troposphere most of the SCM configurations exhibit greater spread during the suppressed phase (the Plant & Craig scheme is an exception during ActC). These observations are consistent with the notion that convective variability is a key ingredient in producing spread.

There are distinct differences between the profiles in Fig. 17.11*a–c*, which are for configurations using the default UM convection scheme, and those in Fig. 17.11*d–f*, which are for configurations based on the Kain–Fritsch plume model. The latter grouping exhibits large peaks in ensemble spread in the upper troposphere and lower stratosphere regions, presumably associated with convective overshoots. Such peaks are absent for the first grouping, which tend to have greater spread in the mid troposphere. The vertical structure of the ensemble spread profile appears to be primarily dependent on the convective plume model used, with the ST perturbations primarily affecting its amplitude.

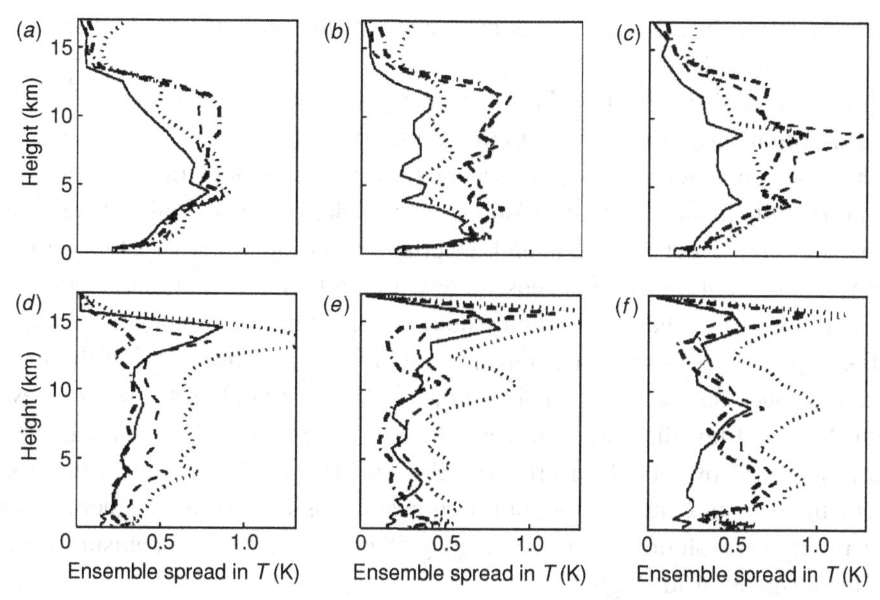

Figure 17.11 Vertical profiles of ensemble RMS spread in temperature, for eight different SCM ensembles, averaged over the periods (*a*,*d*) ActB, (*b*,*e*) SupC and (*c*,*f*) ActC. The periods are as defined in Fig. 17.7. The profiles in (*a*–*c*) are for the default UM configuration (solid line), the multiplicative-noise scheme (dotted line), the time-varying random parameters scheme (dashed line) and the constant random parameters scheme (dot–dashed line). Those in (*d*–*f*) are for the Kain–Fritsch configuration (solid line), and the Plant & Craig scheme with a grid length of 50 km (dotted line), 100 km (dashed line), and in deterministic mode (dot–dashed line).

In the troposphere, the default UM ensemble produces more spread than the Kain–Fritsch ensemble. These profiles confirm that the convection parameterisation is an important source of variability and also that different deterministic convection parameterisations produce rather different variabilities in the host model. Thus, if the high-frequency variability of a model does have important effects on climate, one should introduce some (stochastic) method to control the high-frequency variability, or at least should investigate the on–off characteristics of the GCM convection parameterisation.

The schemes that represent model uncertainty (multiplicative noise, random parameters and constant random parameters) tend to scale up the profile of ensemble spread produced by their deterministic analogue in the mid and upper troposphere, but have relatively little effect on the lower troposphere. The multiplicative-noise scheme also affects the stratosphere, as it directly perturbs the radiative tendencies that dominate there. The stochastic Plant & Craig scheme also tends to scale up the profile of spread produced by its deterministic mode, but differs from the other methods in that during ActB and SupC it creates substantial increases in spread in the lower troposphere.

The deterministic Plant & Craig scheme generally produces small ensemble spreads, often smaller than those in the Kain–Fritsch ensemble. This is consistent with its design, since it uses time-averaged profiles to reduce time-step-to-time-step variability in its closure calculations. The stochastic form of this scheme is not designed to represent generic model uncertainty, but rather the variability arising from subsampling the cumulus ensemble within a finite area. For an area of side $\Delta x = 100$ km, the scheme is certainly more spread than in deterministic mode, but still comparable with the deterministic Kain–Fritsch ensemble and in the troposphere is much less spread than any of the model uncertainty schemes. However, with $\Delta x = 50$ km the ensemble spread has tropospheric values comparable to those produced by model uncertainty schemes. These results suggest that local fluctuations about convective equilibrium become as important as generic model uncertainty at resolutions of approximately 50 km, and a key mechanism for variability at smaller grid lengths.

17.4.4 Comparison of stochastic physics SCM ensemble spread to model uncertainty

Although the stochastic physics schemes used in this study do produce significant ensemble spread, it remains to determine whether or not the levels of spread are appropriate. To examine this point, it is useful to compare the ensembles that are designed to represent model uncertainty with the full range of model states produced by different deterministic structural configurations. A poor man's ensemble is produced by combining the 39-member IC ensembles produced by the default UM, the Kain–Fritsch scheme and the deterministic Plant & Craig scheme, each with equal weighting. The spread of this combined 117-member ensemble can be used as a simple measure of the spread of model states associated with model uncertainty. Certainly, the representativeness of such an ensemble is questionable, but we would suggest that a stochastic scheme that aims to represent model uncertainty should produce at least comparable levels of spread. Figure 17.12 shows the time series of several ensemble percentiles of PWC for each of the stochastic schemes and for the constant random parameters scheme, compared with the same percentiles of the combined deterministic ensemble. (The ensemble-mean PWC was shown in Fig. 17.8.)

It is encouraging to find that the three schemes designed to represent model uncertainty do indeed produce spread comparable to the combined deterministic ensemble (Fig. 17.12*a–c*). However, these schemes tend simply to broaden the ensemble about the ensemble mean state of their deterministic analogue. Thus, they

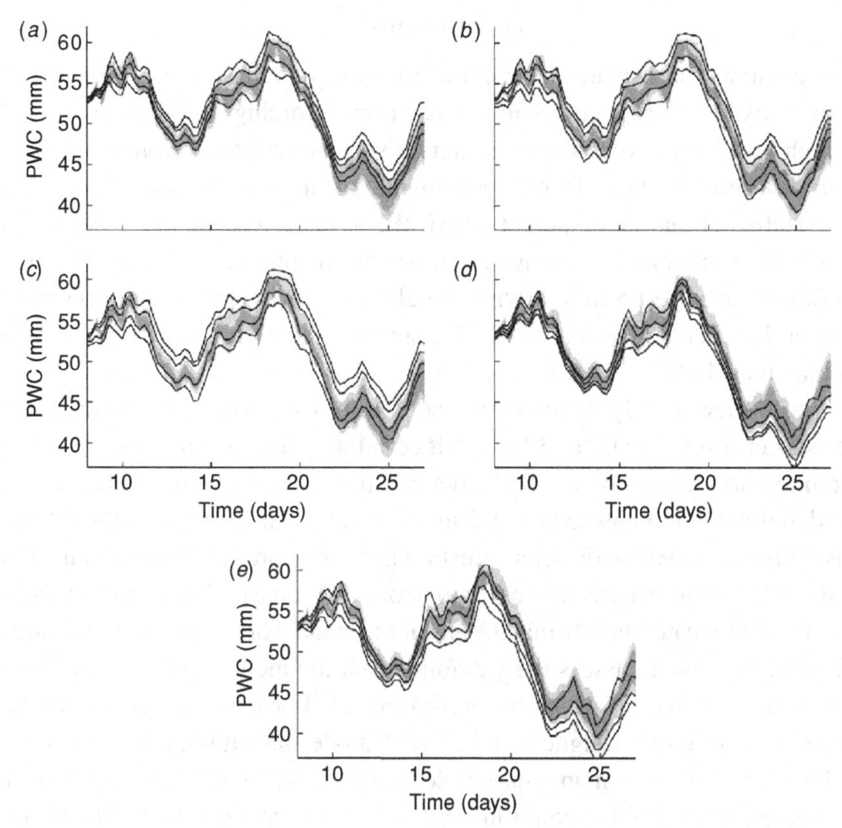

Figure 17.12 The 5th, 25th, 75th and 95th percentiles of PWC (solid lines) for ensembles using (*a*) the multiplicative-noise scheme, (*b*) the time-varying random parameters scheme, (*c*) the constant random parameters scheme, and (*d,e*) the Plant & Craig scheme with a grid length of (*d*) 100 km and (*e*) 50 km. Each panel also shows (shaded) the same ensemble percentiles for the combined deterministic ensemble.

fail to explore regions of phase space which are accessible to the other deterministic schemes. The model uncertainty scheme that looks most promising in this study is the random parameters scheme. As discussed in Section 17.4.2, it produces some noise-induced drift, and this appears to be favourable, in the sense that the distribution of PWC is nudged towards that of the combined deterministic ensemble. It only fails to encompass the full range of model uncertainty during the last day or two of the runs.

Figure 17.12*d,e* show results for the stochastic Plant & Craig scheme. These confirm the points made in Section 17.4.3 and show that fluctuations about convective equilibrium result in ensemble spread similar to key aspects of model uncertainty on scales of approximately 50 km.

17.5 Conclusions

Single-column tests isolate the parameterisation schemes of a GCM and allow one to study their interactions under prescribed forcings. The strengths of the approach are also its weaknesses: it can be very helpful to explore the behaviour of parameterisations in a clean arrangement, but the behaviour is not necessarily representative of that in the parent GCM. We have performed single-column tests of tropical convection, comparing various stochastic physics methods. The interactions of stochastic perturbations with model dynamics are likely to be an important aspect of their behaviour in a full GCM, but we wished to consider whether SCM tests may nonetheless have value.

It is necessary to study stochastic methods with ensembles of SCM runs. Here we have used an ensemble size of 39 which is certainly practical, and sufficient to make some inferences about the methods, but additional members might have allowed more definitive statements to be made in some cases. The stochastic ensembles can be usefully compared with deterministic ensembles produced by IC uncertainty and also with combinations of these into poor man's ensembles. Such comparisons allow one to judge not merely that the ensemble mean from some stochastic method is sensible, but also to assess the variability that the method produces through its interactions with the GCM parameterisation set. For example, if the spread of a stochastic ensemble designed to represent model uncertainty was much larger (smaller) than the poor man's ensemble, then the SCM would imply that for a good performance of the method in a full GCM the interactions of the stochastic perturbations with GCM dynamics should be such as to strongly dampen (amplify) the variability introduced.

In agreement with Hack & Pedretti (2000), deterministic SCM runs were found to be sensitive to small IC perturbations. The variability of IC ensembles depends on the convection parameterisation used, according to the timings and frequency of triggering. The perturbations chosen allowed the runs to diverge into a set of independent realisations within a few days. The same IC perturbations had very little effect on the ensembles produced by stochastic methods, beyond the first day or two.

Three methods designed to represent model uncertainty appeared to perform well, the ensemble spreads being broadly similar to that of the deterministic poor man's ensemble. The ensemble-mean states were close to the ensemble means of the deterministic analogue (differences in the convective parameterisation produced substantially larger changes to the mean). For the random parameters scheme, however, there was a statistically significant noise-induced drift of ensemble-mean PWC. The Plant & Craig scheme produced levels of spread similar to the model uncertainty approaches for $\Delta x = 50$ km, suggesting that fluctuations about convective equilibrium form an important component of variability at and below this scale.

Although there are some changes in the methodology and philosophy for SCM testing of stochastic methods, we are inclined to view our results as encouraging and to speculate that SCM testing may have a useful role to play in studying stochastic parameterisations. For example, it seems clear from our results that a 2.5° GCM integration would not be a good test of the impact of the convective fluctuations parameterised in the Plant & Craig scheme. Given that comprehensive testing of all details of all plausible stochastic methods will remain impractical, we contend that the indications of potential impact that may be gleaned from SCM tests are far preferable to no indications at all.

Acknowledgements

We are grateful to A. Arribas, N. Bowler and K. Mylne for their discussions of stochastic methods, to S. Woolnough for discussions of GCSS case 5 and for providing the observational data, to R. Wong for providing us the default UM SCM configuration for that case, and to NCAS Computer Modelling Support. M.A.B. is funded by the NERC award NER/S/A/2006/14189 with CASE support from the Met Office.

References

Arribas, A. 2004 Results of an initial stochastic physics scheme for the Met Office Unified Model. Forecasting research technical report no. 452. Met Office, UK.

Bergman, J. W. & Sardeshmukh, P. D. 2004 Dynamic stabilisation of atmospheric single column models. *J. Climate*, **17**, 1004–1021 (doi:10.1175/1520-0442(2004)017<1004:DSOASC>2.0.CO;2).

Buizza, R., Miller, M. & Palmer, T. N. 1999 Stochastic representation of model uncertainties in the ECMWF ensemble prediction system. *Q. J. R. Meteorol. Soc.*, **125**, 2887–2908 (doi:10.1256/smsqj.56005).

Byun, Y. -H. & Hong, S. -Y. 2007 Improvements in the subgrid-scale representation of moist convection in a cumulus parameterisation scheme: the single-column test and its impact on seasonal prediction. *Mon. Weather Rev.*, **135**, 2135–2154 (doi:10.1175/MWR3397.1).

Craig, G. C. & Cohen, B. G. 2006 Fluctuations in an equilibrium convective ensemble. Part I: theoretical formulation. *J. Atmos. Sci.*, **63**, 1996–2004 (doi:10.1175/JAS3709.1)

Craig, G., Fraedrich, K., Lin, J. *et al.* 2005 Working group 1 report: issues in convection. In Proceedings of the ECMWF Workshop on Representation of Sub-grid Processes using Stochastic-Dynamic Models, Reading, 6–8 June, pp. ix–xii.

Cullen, M. J. P. 1993 The unified forecast/climate model. *Meteorol. Mag.*, **122**, 89–94

Done, J. M. 2002 Predictability and representation of convection in a mesoscale model, p. 180. Ph.D. thesis, University of Reading, UK.

Ghil, M., Berloff, P., Buizza, R. *et al.* 2005 Working group 3 report: the parameterisation problem in weather and climate models. In Proceedings of the ECMWF Workshop on Representation of Sub-grid Processes using Stochastic-Dynamic Models, Reading, 6–8 June, pp. xvii–xxii.

Gregory, D. & Rowntree, P. R. R. 1990 A mass flux convection scheme with representation of cloud ensemble characteristics and stability dependent closure. *Mon. Weather Rev.*, **118**, 1483–1506 (doi:10.1175/1520-0493(1990)118<1483:AMFCSW>2.0.CO;2).

Hack, J. J. & Pedretti, J. A. 2000 Assessment of solution uncertainties in single-column modeling frameworks. *J. Climate*, **13**, 352–365 (doi:10.1175/1520-0442(2000)013<0352:AOSUIS>2.0.CO;2).

Horinouchi, T., Pawson, S., Shibata, K. *et al.* 2003 Tropical cumulus convection and upward-propagating waves in middle-atmospheric GCMs. *J. Atmos. Sci.*, **60**, 2765–2782 (doi:10.1175/1520-0469(2003)060<2765:TCCAUW>2.0.CO;2).

Hume, T. & Jakob, C. 2005 Ensemble single column modeling (ESCM) in the tropical western Pacific: forcing data sets and uncertainty analysis. *J. Geophys. Res.*, **110**, D13109 (doi:10.1029/2004JD005704).

Kain, J. S. 2004 The Kain–Fritsch convective parameterization: an update. *J. Appl. Meteorol.*, **43**, 170–181 (doi:10.1175/1520-0450(2004)043<0170:TKCPAU>2.0.CO;2).

Kain, J. S. & Fritsch, J.M. 1990 A one-dimensional entraining/detraining plume model and its application in convective parameterization. *J. Atmos. Sci.*, **47**, 2784–2802 (doi:10.1175/1520-0469(1990)047<2784:AODEPM>2.0.CO;2).

Krueger, S. K. & Lazarus, S. M. 1999 Intercomparison of multi-day simulations of convection during TOGA COARE with several cloud-resolving and single-column models. In *Twenty-third Conference Hurricanes and Tropical Meteorology, Dallas, TX*. Preprint, pp. 643–647. American Meteorological Society.

Lin, J. W.-B. & Neelin, J. D. 2002 Considerations for stochastic convective parameterization. *J. Atmos. Sci.*, **59**, 959–975 (doi:10.1175/1520–0469(2002)059<0959:CFSCP>2.0.CO;2).

Martin, G. M., Ringer, M. A., Pope, V. D. *et al.* 2006 The physical properties of the atmosphere in the new Hadley Centre Global Environmental Model (HadGEM1). Part I: model description and global climatology. *J. Climate*, **19**, 1274–1301 (doi:10.1175/JCLI3636.1).

Moncrieff, M. W., Krueger, S. K., Gregory, D. *et al.* 1997 GEWEX cloud system study (GCSS) working group 4: precipitating convective cloud systems. *Bull. Amer. Meteorol. Soc.*, **78**, 831–845 (doi:10.1175/1520-0477(1997)078<0831:GCSSGW>2.0.CO;2).

Murphy, J., Sexton, D. M. H., Barnett, D. N. *et al.* 2004 Quantification of modelling uncertainties in a large ensemble of climate change simulations. *Nature*, **430**, 768–772 (doi:10.1038/nature02771).

Mylne, K. R., Evans, R. E. & Clark, R. T. 2002 Multi-model multi-analysis ensembles in quasi-operational medium-range forecasting. *Q. J. R. Meteorol. Soc.*, **128**, 361–384 (doi:10.1256/00359000260498923).

Mylne, K., Bowler, N., Arribas, A. & John, S. 2005 MOGREPS (Met Office Global and Regional Ensemble Prediction System). Met Office Scientific Advisory Committee, 10 to 11 November 2005, Paper No. 10.5.

Palmer, T. N. 2001 A nonlinear dynamical perspective on model error: a proposal for non-local stochastic–dynamic parameterisation in weather and climate prediction models. *Q. J. R. Meteorol. Soc.*, **127**, 279–304 (doi:10.1002/qj.49712757202).

Penland, C. 2003 A stochastic approach to nonlinear dynamics: a review. *Bull. Amer. Meteorol. Soc.*, **84**, ES43–ES52 (doi:10.1175/BAMS-84-7-Penland).

Petch, J.C., Willett, M., Wong, R. Y. & Woolnough, S. J. 2007 Modelling suppressed and active convection. Comparing a numerical weather prediction, cloud-resolving and single-column model. *Q. J. R. Meteorol. Soc.* **133**, 1087–1100 (doi:10.1002/qj.109).

Plant, R. S. & Craig, G. C. 2008 A stochastic parameterization for deep convection based on equilibrium statistics. *J. Atmos. Sci.*, **65**, 87–105 (doi:10.1175/2007JAS2263.1).

Ricciardulli, L. & Garcia, R.R. 2000 The excitation of equatorial waves by deep convection in the NCAR Community Climate Model (CCM3). *J. Atmos. Sci.*, **57**, 3461–3487 (doi:10.1175/1520-0469(2000)057<3461:TEOEWB>2.0.CO;2).

Shutts, G. 2005 A kinetic energy backscatter algorithm for use in ensemble prediction systems. *Q. J. R. Meteorol. Soc.*, **131**, 3079–3102 (doi:10.1256/qj.04.106).

Sobel, A. H., Bellon, G. & Bacmeister, J. 2007 Multiple equilibria in a single-column model of the tropical atmosphere. *Geophys. Res. Lett.*, **34**, L22804 (doi:10.1029/2007GL031320).

Teixeira, J. & Reynolds, C.A. 2008 The stochastic nature of physical parameterisations in ensemble prediction: a stochastic convection approach. *Mon. Weather Rev.*, **136**, 483–496.

Tompkins, A. M. & Berner, J. 2008. A stochastic convective approach to account for model uncertainty due to unresolved humidity variability. *J. Geophys. Res.*, **113**, D18101 (doi:10.1029/2007JD009284).

Webster, P. J. & Lukas, R. 1992 TOGA – COARE: the coupled ocean–atmosphere response experiment. *Bull. Amer. Meteorol. Soc.*, **73**, 1377–1416 (doi:10.1175/1520-0477(1992)073<1377:TCTCOR>2.0.CO;2).

Wilks, D. S. 2005 Effects of stochastic parameterisations in the Lorenz '96 system. *Q. J. R. Meteorol. Soc.*, **131**, 389–407 (doi:10.1256/qj.04.03).

Willett, M. R. & Milton, S. F. 2006 The tropical behaviour of the convective parameterisation in aquaplanet simulations and the sensitivity to timestep. Forecasting research technical report no. 482. Met Office, UK.

Williams, P. D. 2005 Modelling climate change: the role of unresolved processes. *Phil. Trans. R. Soc. A*, **363**, 2931–2946 (doi:10.1098/rsta.2005.1676).

18

Stochastic parameterisation of multiscale processes using techniques from computer game physics

THOMAS ALLEN, GLENN J. SHUTTS AND
CHRISTOPHER J. SMITH

Inspired by techniques used by animators, this chapter argues that parameterisations (such as those representing deep convection and mountain drag) could be computed on a finer grid than the general circulation model, and use the same dynamical equations but with reduced-complexity physics and reduced-accuracy numerical techniques. The fine-scale parameterisation grid would correctly capture the spatial and temporal correlation scales of the modelled processes, as well as non-equilibrium aspects that are usually absent from conventional parameterisation schemes (e.g. the diurnal cycle in convection). Reducing the accuracy requirement at the fine scale, and reducing the algorithmic complexity of multiple interacting physical processes, could bring forward in time some of the benefits of global convective-scale resolution that are beyond current computational resources.

18.1 Introduction

The design of numerical weather prediction (NWP) and climate models has always been constrained by the need for speed so that a numerical forecast can be delivered in a matter of hours. Numerical techniques have been employed that permit the use of large time steps at the expense of some tolerable loss of accuracy. Animators in the computer games industry are confronted with a similar need for speed so that natural physical phenomena can be modelled convincingly from a visual perspective. This demand for realism in computer simulation of clouds and fluid motion has increasingly led programmers to use the same physical equations as those used to forecast weather and climate, as well as in other areas of classical mechanics (see, for example, the book by Erleben *et al.* 2005). However, in the case of a game, accuracy is not the primary issue: visual plausibility and computational speed are the goals. It is interesting to note that this drive for realism led Sony, along with IBM and Toshiba, to develop the Cell Broadband Engine for the Playstation 3 console. This processor is currently being used for the Los Alamos supercomputer (Roadrunner), which was the first machine to exceed a Petaflop.

Current subgrid-scale parameterisation techniques used in numerical weather prediction and climate modelling assume local statistical equilibrium or steady-state conditions. For instance in convection parameterisation, the supposed existence of a unique relationship between the state of the atmosphere in a grid column and the fluxes carried by the clouds requires there to be a large number of convective clouds in that column. For deep convection this assumption is unrealistic at current NWP resolutions; i.e. horizontal grid lengths ~40 km (Shutts & Palmer 2007). In the tropics, deep convective cloud systems span hundreds of kilometres and their dynamical influence extends beyond that, making the column-based convection parameterisation difficult to justify. The mutual independence of neighbouring grid columns in so far as parameterisation is concerned makes for poor numerics and a failure to capture the long-range influence and organisation of cloud systems. These deficiencies are ameliorated somewhat by the tendency of topographic and mesoscale dynamical forcing to define the pattern of parameterised convection.

Tomita *et al.* (2005) have carried out global atmospheric simulation on the Earth Simulator using a 3.5 km grid without parameterised convection. Its realism in terms of tropical circulation is very encouraging but the computational expense is too prohibitive for climate simulation. Grabowski (2000) showed the potential benefits of the 'superparameterisation' approach which embeds a two-dimensional cloud-resolving model (CRM) within each grid column of a coarse-resolution global climate model, dispensing with the need for conventional convective parameterisation. The need to end the 'parameterisation deadlock' using approaches like Grabowski's superparameterisation has been forcefully argued by Randall *et al.* (2003) although the computational burden is still a barrier.

In the very different world of computer games technology there is a inexorable drive towards greater graphical realism and interactivity with the simulated natural environment. The Hollywood film industry is increasingly using computer-generated imagery for special effects involving fluid motion and cloud/smoke rendering. For instance, flight simulators need to render smoke and cloudscapes at usable interactive frame rates for an immersive user experience (e.g. 30 frames per second). This has led to the development of fluid and particle motion simulators that can simulate the growth of clouds and describe the interaction of a moving aircraft with the cloud itself. This is possible through a combination of powerful graphics processor technology, highly simplified physics and efficient coding techniques (e.g. Harris 2003). From the perspective of computer visualisation, predictive accuracy (in the sense of forecasting where a cloud will be at a particular time and what form it might take) is not required, only visual plausibility. However code stability and speed are paramount to the games programmer. Interestingly, the semi-Lagrangian advection algorithm used in numerical weather prediction (e.g. Staniforth & Coté 1991) has been adopted in real-time fluid simulation for visualisation because of its stability with respect to choice of time step (Stam 1999). Much of the computational overhead in fluid simulation comes from the elliptic pressure solver which enforces the continuity of mass. Considerable savings can be obtained by reducing the accuracy requirement, e.g. by fixing the number of iterations and not checking for convergence.

The use of cellular automata (e.g. John Conway's 'Game of Life'; see Gardner 1970) was suggested by Palmer (1997) as a means of describing near-grid-scale variability associated with unresolved physical processes like deep convection. He envisaged a probability-based cellular automaton (CA) where the state of a cell (alive or dead) is governed by a probability distribution function (PDF) that depends on the state of the cell's nearest neighbours and is also a function of an associated specification of topographic data, e.g. land/sea, orographic height and land type. Organised convection could be modelled on this fine grid of cells by making convection (represented by a living cell) more likely if neighbouring cells were already convecting. The need for stochastic physical parameterisations in NWP and climate models was suggested by Palmer (2001) and the CA approach is one of many that could be employed.

The idea of using a CA as a pattern generator was taken up by Shutts (2005) in the development of a kinetic backscatter scheme that replaces energy lost by model dissipation. A time-evolving CA pattern (see Fig. 18.1) was used in place of a scheme based on smoothed random number fields (more commonly used in turbulent backscatter formulations, e.g. Mason & Thomson 1992). Although not used in that work, the CA approach allowed the possibility of a mutual coupling between the CA grid and the forecast model grid. For instance the PDF could be made a function of some model state parameters such as the convective available potential energy when modelling convection. In general one would imagine only a weak coupling between the CA grid and the NWP grid so that the forcing coming from the CA gently nudges the NWP model flow. For instance, the simple 'Mexican wave' rule (stand up if the person to your right is already standing up) could form the basis of an equatorial Kelvin wave forcing function that would modulate convective parameterisation and drive an eastward-moving convectively coupled wave response.

The outline of this chapter is as follows. The computer visualization theme is developed in Section 18.2 and is suggested to be following an ever-converging path with techniques and approximations used in NWP. In Section 18.3 a highly simplified and efficient cloud-resolving model is described that could form the basis of a more realistic stochastic forcing algorithm for convection parameterisation. This use of more realistic simulation models for stochastic parameterisation is

Figure 18.1 Pattern generated by a cellular automaton where each cell can have 50 'lives' with a life being lost at every step. The grey-scale intensity is a measure of the number of lives remaining with 50 being white and 0 being black. The resulting animated pattern is reminiscent of convective cloud organisation in a cold air outbreak over the sea.

discussed in Section 18.4 and linked with the idea of dual-grid model configurations. A general discussion follows in Section 18.5.

18.2 Computer games animation techniques

Early computer landscape modelling or scene generator software used fractal methods to define natural phenomena like mountains, forests and clouds. Related to this is the work of Barnsley (2000) on iterated function systems that attempt to find a set of rules (functions) which, when randomly combined, produce a rendering of the desired object. Determining these rules is a difficult and somewhat controversial task, however. Recent emphasis is on real-time simulation and rendering of clouds and smoke for flight simulators using fast algorithms operating on graphics processor hardware. Rendering of realistic cloudscapes from a moving viewpoint requires efficient algorithms that describe the multiple forward scattering and attenuation of sunlight (Harris 2003).

When it comes to simulating the motion of fluid flow, however, animators have turned to using the Navier–Stokes equations and have copied methods used by the computational fluid dynamics community with one important exception; they are not concerned about accuracy, only numerical stability and visual likeness to the real

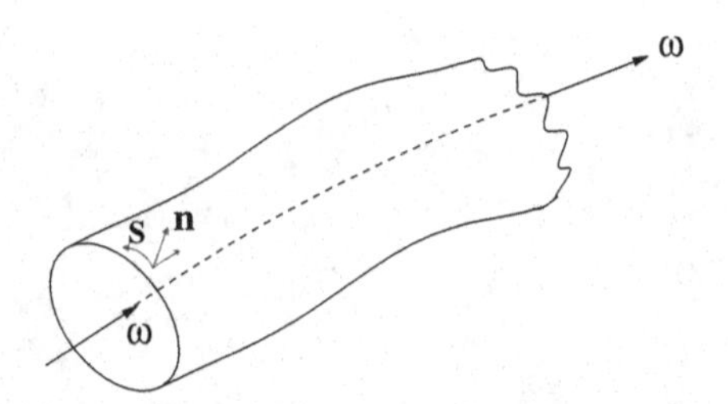

Figure 18.2 A sketch of a vortex tube with vortex line passing along its axis. The vector triad comprises a normal to the vortex tube **n**, a vector equal to the local vorticity lying in the vortex tube surface and the resulting vector cross product **S**.

phenomenon (Stam 1999). Following current NWP practice, Stam uses the semi-Lagrangian advection algorithm with very large time steps to simulate neutrally buoyant fluid flow carrying tracers that might represent smoke or clouds. Since low-order interpolation of flow variables to the departure point acts to unnaturally smooth the fields, Fedkiw *et al.* (2001) have adopted the 'vorticity confinement' technique of Steinhoff & Underhill (1994) to oppose spurious dissipation and maintain vorticity features close to the model's grid scale. Steinhoff's technique has even been used by the Hollywood film industry in the creation of realistic computer-generated smoke and fluid motion scenes.

Vorticity confinement can be achieved in different ways but the simplest way of viewing it is the addition of an apparent force to the momentum equation that is perpendicular to the local vorticity. The circulation around a vortex tube is increased by a magnitude proportional to the vorticity (see Fig. 18.2).

Specifically, the momentum equation can be written in the form:

$$\frac{\partial \mathbf{u}}{\partial t} + (\mathbf{u} \cdot \nabla)\,\mathbf{u} + \frac{1}{\rho}\nabla p = \nu \nabla^2 \mathbf{u} + \epsilon \mathbf{S}, \tag{18.1}$$

where $\mathbf{u} = (u, v, w)$ is the velocity, p is the pressure, ρ is the density and ν is the kinematic viscosity. The vorticity confinement term $\epsilon\mathbf{S}$ is defined such that ϵ is a constant and

$$\mathbf{S} = \mathbf{n} \times \omega, \tag{18.2}$$

where $\omega = \nabla \times \mathbf{u}$ is the vorticity and the unit normal $\mathbf{n}$ is given by

$$\mathbf{n} = \frac{\nabla\,|\omega|}{|\nabla\,|\omega||}. \tag{18.3}$$

Here the constant ϵ, which is usually chosen by trial and error, is proportional to the horizontal grid spacing and inversely proportionally to the time step. Energetically the vorticity confinement term tends to counterbalance numerical diffusion and ϵ must be small enough to prevent net energy input.

An alternative to vorticity confinement is to estimate more precisely the diffusion of the first-order, semi-Lagrangian scheme. By subtracting this 'diffusive' contribution from the equation of motion the desired effect of reducing the numerical dissipation can be achieved. Although such a step is not entirely equivalent to using a higher-order interpolant, it has a similar effect but at a much reduced computational cost.

The basic idea behind this method is to perform a 'modified equation' analysis (see, for example, Morton & Mayers 2005) for the first-order semi-Lagrangian advection scheme. When applied to the one-dimensional advection equation for ϕ:

$$\phi_t + u\phi_x = 0 \tag{18.4}$$

one obtains

$$\hat{\phi}_t + u\hat{\phi}_x = \kappa\hat{\phi}_{xx} + R \tag{18.5}$$

where $\hat{\phi}$ is the equivalent numerical solution, R represents residual terms in the analysis and the numerical diffusivity κ is given by:

$$\kappa = \frac{r(1-r)\cdot(\delta x)^2}{2\delta t} \tag{18.6}$$

(see also McCalpin 1988). Here δx and δt are the horizontal grid spacing and time step respectively while the parameter $r \in [0, 1]$ is the fractional part of the Courant number. The fact that $\kappa \geq 0$ is independent of the size of the Courant number is why the method is stable. Note that for non-constant advection there is also a dispersive correction to (18.5) which is ignored since we are only interested in the dissipation. The fact that r is calculated before the semi-Lagrangian advection is performed suggests that the numerical dissipation terms could be used in place of the vorticity confinement term. The 'ill-posed' nature of this anti-diffusion means that some care must be taken with the implementation of such a term in order to maintain numerical stability (e.g. by slightly reducing the anti-diffusivity so that the net diffusivity is positive on average). However the potential advantage of the method is that it can be applied to all fields. The beneficial effects of this method can be seen in Fig. 18.3 which shows the vorticity field for flow over a series of boxes/buildings. The grey area indicates the dispersion of a passive tracer. The four simulations in this figure comprise:

- the standard first-order scheme,
- the first-order scheme applied to momentum but with monotone cubic interpolation of the tracer,
- cubic interpolation of all variables,
- the first-order scheme with the proposed anti-diffusion.

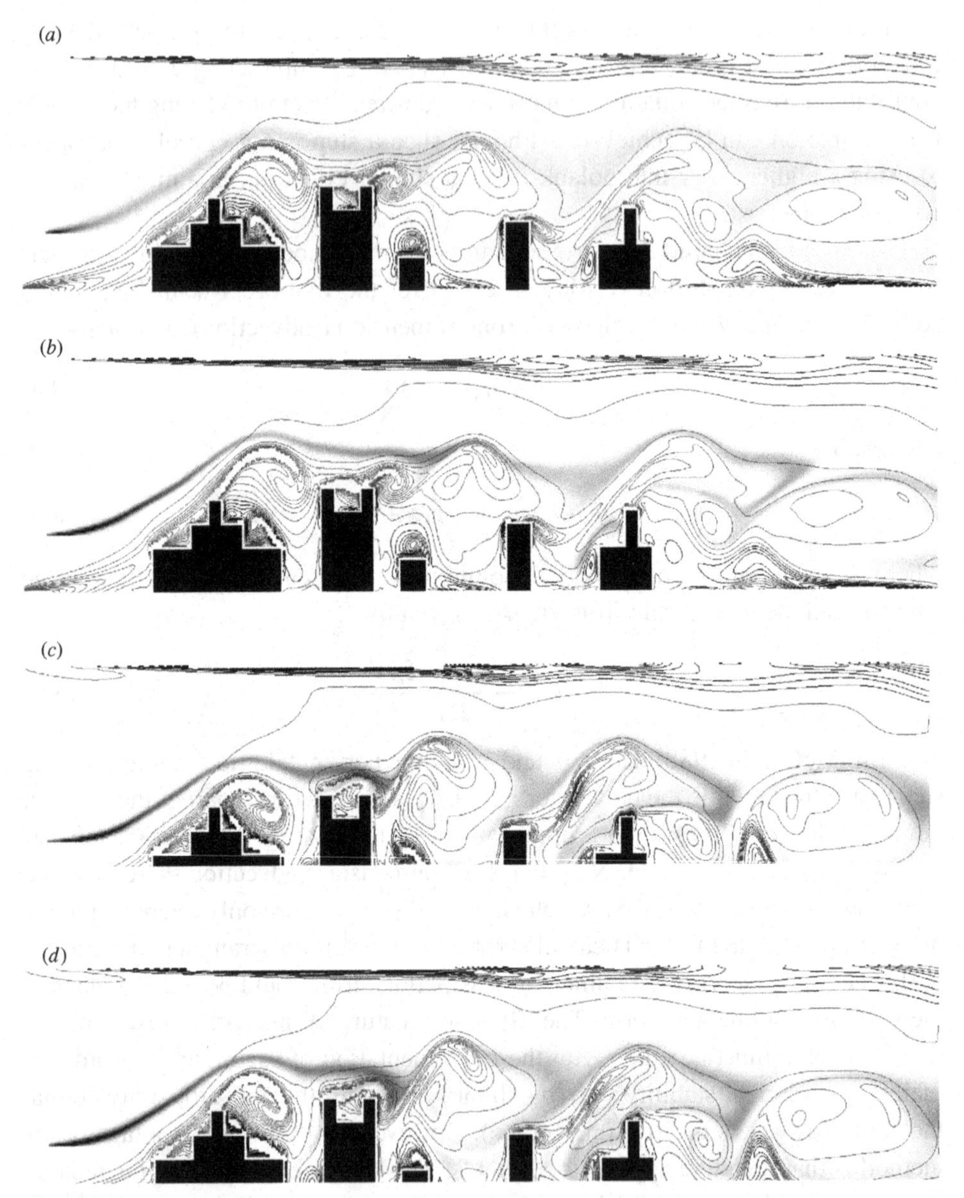

Figure 18.3 A snapshot of the vorticity field from an interactive movie of flow past a series of buildings; 500 × 125 points were used and the location of buildings is enforced by masking the flow to zero. The grey streak is a tracer field introduced at its leading edge on the left of the domain. On a 2.4 GHz Pentium 4 personal computer using OpenGL graphics the animation was able to run at 15 frames a second. (*a*) Standard linear scheme; (*b*) using monotone cubic interpolation on tracer; (*c*) monotone cubic on all variables; (*d*) linear interpolation with anti-diffusion terms.

The simulation that used the anti-diffusion scheme was computationally as cheap as the standard first-order scheme and typically about four or five times faster than those using cubic interpolation.

As with vorticity confinement, the emphasis is on creating the effect of higher resolution without the attendant pointwise accuracy. The additional terms are like deterministic backscatter and probably do much to improve the statistical properties of the flow without attempting to have extra accuracy in the Lagrangian advection step.

18.3 A simple cloud-resolving model

As a test of the ideas from the previous section, a simple cloud model was developed. This model consisted of the Navier–Stokes equation (18.1) with the addition of a buoyancy force $\mathbf{b}$ along with equations for the (perturbation) potential temperature θ, and three states for the water vapour: vapour (q_{v}), condensate (q_{c}) and rain (q_{r}). From the reference potential temperature profile, background reference profiles of pressure and density can be determined and used within the continuity equation (anelastic approximation) and the calculation of the saturation vapour (q_{vs}).

When $q_{\mathrm{v}} > q_{\mathrm{vs}}$ the excess water vapour is converted to condensate and θ corrected to account for the latent heat release. Similarly if $q_{\mathrm{c}} > 0$ and $q_{\mathrm{v}} < q_{\mathrm{vs}}$ condensate is removed. This process can be concisely written as

$$\begin{aligned} C &= \frac{q_{\mathrm{vs}} - q_{\mathrm{v}}}{1 + L\partial q_{\mathrm{vs}}/\partial\theta}, \\ Q &= q_{\mathrm{c}} - \max(0, q_{\mathrm{c}} - C), \\ q_{\mathrm{v}} &\to q_{\mathrm{v}} + Q, \\ q_{\mathrm{c}} &\to q_{\mathrm{c}} - Q, \\ \theta &\to \theta - LQ, \end{aligned} \tag{18.7}$$

where $L = L_t/(C_p\Pi)$, Π is the Exner function, C_p is the specific heat capacity and L_t is the latent heat of vaporisation. Note that the denominator in the definition of C arises from a single Newton iteration and is there to ensure that $q_{\mathrm{v}} \leq q_{\mathrm{vs}}$ at the updated temperature. The conversion of condensate to rain is also relatively simple and can be written as

$$\begin{aligned} C_{\mathrm{r}} &= \max(0, A_{\mathrm{r}}(q_{\mathrm{c}} - Q_{\mathrm{precip}})), \\ q_{\mathrm{r}} &\to q_{\mathrm{r}} + C_{\mathrm{r}}, \\ q_{\mathrm{c}} &\to q_{\mathrm{c}} - C_{\mathrm{r}}. \end{aligned} \tag{18.8}$$

Here A_{r} is a constant fixed here to be 0.001 times the time step while Q_{precip} was chosen to be a constant (0.001) divided by the reference density normalised by its surface value. The rain was also allowed to evaporate with a time scale of

5 minutes. Finally the fall speed of the rain was set to be inversely proportional to the reference density and chosen to be 3 m s^{-1} at the ground.

A major problem that occurs in the present situation but not in that of Harris (2003) arises because of the background stratification, which not only requires stable integration of the internal gravity waves but must also respect hydrostatic balance that is jeopardised by the incomplete convergence of the pressure solver. The semi-implicit treatment of the gravity waves is accomplished using time splitting and is achieved by solving the equations

$$\begin{aligned} \dot{w} &= \beta\tilde{\theta} + (b - \overline{b} - \beta\tilde{\theta}), \\ \dot{\theta} &= -\theta_0' w, \end{aligned} \tag{18.9}$$

where $b = g(\theta_v/\theta_0 - q_h)$ is the buoyancy, $\theta_v \sim (1 + \alpha q_v)(\theta_0 + \theta)$ is the virtual potential temperature, $q_h = q_c + q_r$ is the water loading, $\alpha = 0.61$ is a constant and g is the acceleration due to gravity. The parameter $\overline{b}$ is the horizontally averaged values of b which accounts for the hydrostatic pressure contribution. The parameter $\tilde{\theta}$ is the difference between θ and some reference value, which is chosen to be the previous time-step value. The term involving $\beta = g(1 + \alpha q_v)/\theta_0$ allows for the implicit treatment of the gravity waves. Basically the terms in brackets for the w equation are all calculated at the previous time step while the remaining terms are calculated using an off-centred finite difference method, which results in a 2×2 matrix problem at each grid point.

Because of the long (relative to a conventional CRM) time steps used in the semi-Lagrangian scheme (30 seconds in this example) it was decided to run the thermodynamics using two shorter time steps of 15 seconds. The basic algorithm can be summarised as follows.

- Calculate departure points and interpolation weights.
- Calculate field increments due to Coriolis and damping layers.
- Calculate latent heat increments and phase transitions for a half time step.
- Using interpolation weights from the departure point calculation apply the 'anti-diffusion' increment (18.5) to relevant fields.
- Advect fields using bilinear interpolation.
- Perform the second-half time step of the thermodynamics.
- Diffuse fields if necessary (not used in this model).
- Approximately solve the Poisson equation for pressure and update velocity field.
- Go to the next time step.

As a test of the model a simulation was performed in a domain of width 100 km and depth 10 km with a uniform grid whose spacing was 200 m in the horizontal and 100 m in the vertical. The reference profile was fixed to be as shown in Fig. 18.4. The background geostrophic wind was set to −4 m s^{-1} (westward flow) with the value of the Coriolis parameter set to a representative mid-latitude value of

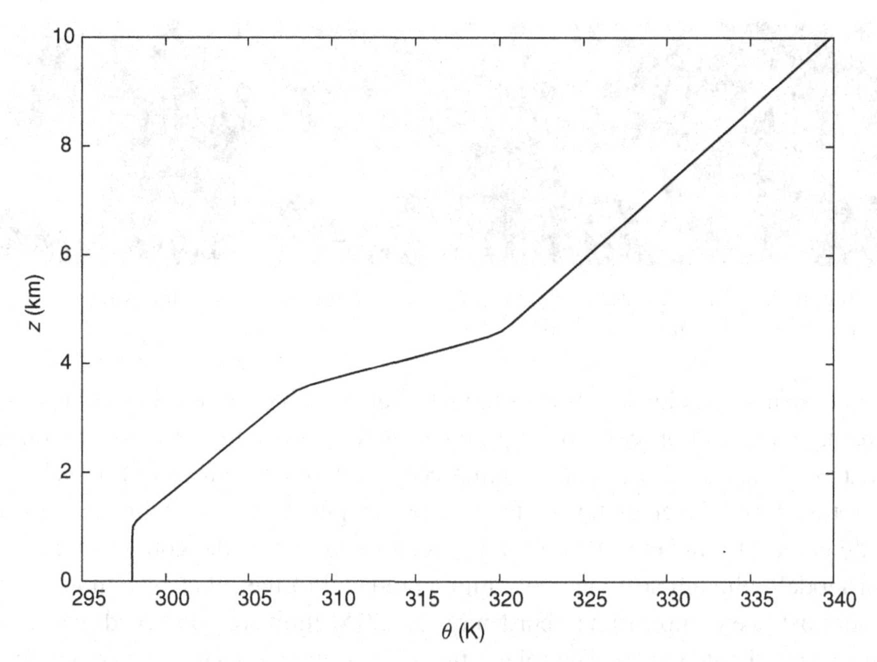

Figure 18.4 Reference potential temperature profile used in the simple cloud model.

10^{-4} s^{-1}. For the initial conditions q_c was set to zero while the potential temperature perturbation was zero everywhere except in a small bubble where it had a maximum value of 0.01 K. The water vapour, q_v, was set to 95% of the saturation value at the surface, 90% of q_{vs} up to 2500 m and then linearly tailed (as a percentage) to zero at a height of 7500 m. The initial horizontal components of velocity were set equal to the geostrophic wind while the vertical component was set to zero.

An illustration of the cloud cover, as measured by q_c, is shown in Fig. 18.5 after five hours of simulation. Not only is this a plausible looking simulation for a CRM but other features, such as relative humidity (not shown), also produce quite realistic features. These results are strongly suggestive of real flows and imply that there is scope for the computer games approach to be exploited in atmospheric parameterisations.

18.4 Proposal for a 'dual-grid' class of stochastic parameterisation

The combined stochastic pattern generator (or emulator) and the forecast model constitute a dual-grid system in which the former acts to force statistical fluctuations due to subfilter scale processes. The potential level of complexity of the emulator could range from a simple two-dimensional cellular automaton to a

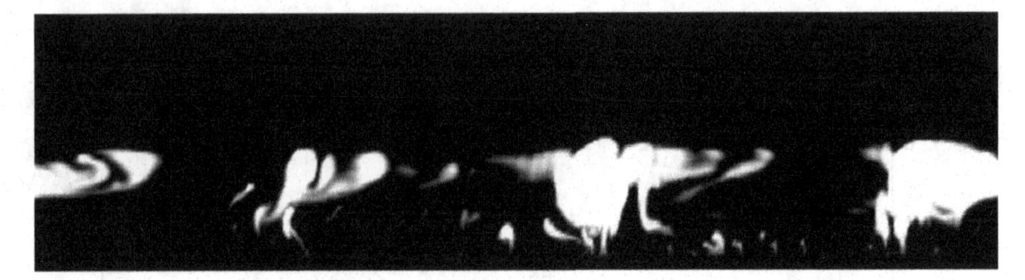

Figure 18.5 Liquid water content from the fast cloud model. Here the white areas correspond to values of $q_c \geq 0.001$.

three-dimensional cloud-resolving model with simplified physics (similar to the superparameterisation concept of Randall *et al.* 2003). In the computationally cheapest scenario, all the conventional column-based parameterisation schemes are retained and the emulator modulates their output. The computational demands of this kind of emulator are very small relative to that of the conventional forecast model although still relatively high compared to the 'pure' dynamical core. In contrast, the computational burden of the CRM limit would exceed that of the forecast or climate model. The role of the forecast model would then be merely to orchestrate sub-mesoscale processes in the fine-scale model and remove the need for many of the subgrid scale parameterisations (e.g. deep convection and mountain drag). Wherever the computational demand lies, it is important to remember that the fine-scale emulator has substantial compromises in its pointwise accuracy relative to the coarser NWP or climate model. In effect, the combined system does its dynamics and physics at different resolutions and on different grids with the physics being computed at higher resolution but degraded accuracy for speed.

The coupling issue is now explored by considering the evolution of some scalar variable q, defined to be q^{F} on the fine grid and q^{C} on the coarse grid. If the scalar is advected by an incompressible three-dimensional vector velocity field $\mathbf{u}$ for which

$$\nabla \cdot \mathbf{u} = 0 \tag{18.10}$$

then q^{F} satisfies

$$q_t^{\mathrm{F}} + \nabla \cdot (\mathbf{u} q^{\mathrm{F}}) = F_q \tag{18.11}$$

where F_q is some unspecified source term and the subscript t denotes differentiation. Let $\mathbf{u}$ and q^{F} be expanded as

$$\mathbf{u} = \overline{\mathbf{U}} + \tilde{\mathbf{U}}, \tag{18.12}$$

with

$$q^{\mathrm{F}} = \overline{q}^{\mathrm{F}} + \tilde{q}^{\mathrm{F}}, \tag{18.13}$$

where $\overline{\tilde{q}^{\mathrm{F}}} = 0$.

Now taking the coarse-grain average of (18.11) leads to

$$\overline{q}_t^{\mathrm{F}} + \nabla \cdot (\overline{\mathbf{U}}\overline{q}^{\mathrm{F}}) = \overline{F_q} - \nabla \cdot (\overline{\tilde{\mathbf{U}}\tilde{q}^{\mathrm{F}}}) \tag{18.14}$$

where the overlines refer to coarse-grain spatial averages and the tildes are the deviation therefrom. The difference (Q) between the coarse model q field and its coarse-grained equivalent in the fine model is

$$Q = q^{\mathrm{C}} - \overline{q}^{\mathrm{F}}. \tag{18.15}$$

We assume that Q obeys a conservation equation of the form

$$Q_t + \nabla \cdot (\mathbf{U}^{\mathrm{C}} Q) = F_Q, \tag{18.16}$$

where F_Q is an unspecified forcing function. Substituting for Q from (18.15) yields

$$q_t^{\mathrm{C}} + \nabla \cdot (\mathbf{U}^{\mathrm{C}} q^{\mathrm{C}}) - \overline{q}_t^{\mathrm{F}} - \nabla \cdot (\mathbf{U}^{\mathrm{C}} \overline{q}^{\mathrm{F}}) = F_Q, \tag{18.17}$$

which can be written as

$$q_t^{\mathrm{C}} + \nabla \cdot (\mathbf{U}^{\mathrm{C}} q^{\mathrm{C}}) - \overline{q}_t^{\mathrm{F}} - \nabla \cdot (\overline{\mathbf{U}}^{\mathrm{F}} \overline{q}^{\mathrm{F}}) - \nabla \cdot \left[(\mathbf{U}^{\mathrm{C}} - \overline{\mathbf{U}}^{\mathrm{F}}) \overline{q}^{\mathrm{F}} \right] = F_Q. \tag{18.18}$$

Finally, using (18.14), the above equation can be written as,

$$q_t^{\mathrm{C}} + \nabla \cdot (\mathbf{U}^{\mathrm{C}} q^{\mathrm{C}}) = F_Q + \overline{F_q} - \nabla \cdot (\overline{\tilde{\mathbf{U}}\tilde{q}^{\mathrm{F}}}) + \nabla \cdot \left[(\mathbf{U}^{\mathrm{C}} - \overline{\mathbf{U}}^{\mathrm{F}}) \overline{q}^{\mathrm{F}} \right]. \tag{18.19}$$

Equation (18.19) can be regarded as the coarse model equivalent of (18.11) but with the right-hand side containing (from left to right): the source term for Q; the coarse-grained source of q; an eddy flux divergence of q on the fine grid; and an advective correction term. Thus far (18.19) is no more than a mathematical manipulation motivated by the idea that the evolution equation for q obeys conservation-style laws on both coarse and fine grids. A potentially important aspect of the coupling lies in the specification of the term F_Q which represents a form of model error source. One approach would be to represent F_Q as a first-order autoregressive process characterised by a judiciously chosen mean and variance. Note that it is not desirable to force q^{C} to exactly equal $\overline{q}^{\mathrm{F}}$ (i.e. $Q = 0$) because the fine-grid calculation is inaccurate. In contrast the coarse model uses accurate numerical techniques and in an NWP context ingests 'observational truth' – particularly at the largest scales. Ideally the fine-scale source F_q should therefore include scale-selective relaxation terms that relax the largest scales of the fine model towards the coarse model's accurate q^{C} field.

As a proof-of-concept exercise, a simple fluid dynamical problem has been modelled on a fine grid and coupled to a coarse grid. The fine grid has ten times more resolution in both directions than the coarse model but uses the same time step, therefore implying a Courant number ten times larger.

The simulated flow consists of three convective plumes ascending through an isentropic atmosphere at rest. The plumes are forced by three heat sources situated

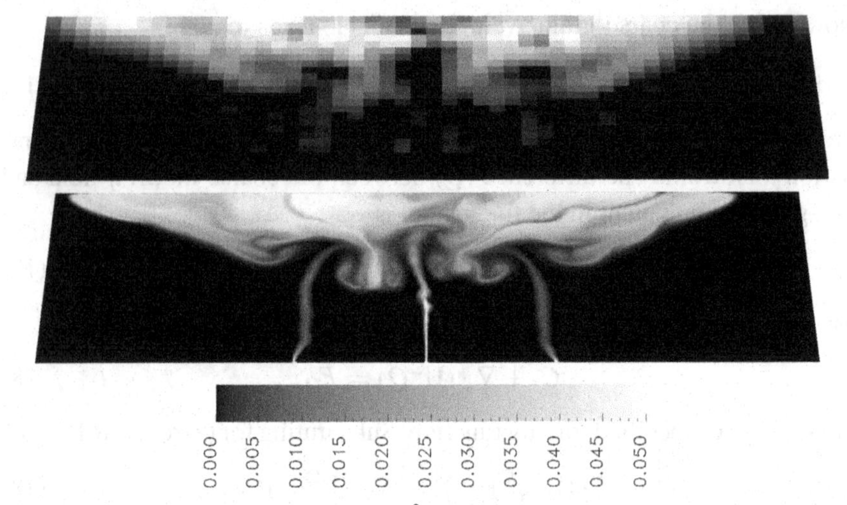

Figure 18.6 The buoyancy field (m s^{-2}) on the coarse (upper) and fine (lower) model grids after the plume has ascended to the top of the model domain (upper edge) from three surface heat sources. The model has 25 $\times$ 50 points and uses a 25 second time step.

at the surface, each of which spans three grid points on the fine mesh. The heat sources are entirely subgrid-scale for the coarse model and therefore have no direct influence on the coarse-model fields in the absence of diffusion and a boundary-layer scheme. Coupling in the momentum and buoyancy equations is achieved using the equivalent terms to those on the right-hand side of (18.19) with $F_Q = 0$ and F_q corresponding to very weak relaxation towards the coarse model.

Figure 18.6 shows the buoyancy on both the coarse- and fine-mesh model at some time after the plume has ascended up to the model's lid. From this figure it can be seen that, even though the fine-mesh solution would be considered highly inaccurate, many of the actual features of an accurate converged solution are present. It can also be seen that the coarse model, which only sees the plumes through the coupling to the fine grid, reproduces most of the features of the rising plume. It is important to note that, apart from the relaxation term (which could be set to zero in this instance), there are no tunable parameters in this form of parameterisation. It is unlikely that a column-based parameterisation could generate the upscale influence found in the coupled-grid simulation and the additional computational expense, provided not too prohibitive, would be worthwhile.

The feasibility of the dual-grid approach will depend primarily on the speed achievable for the fine-grid model and on the nature of the coupling between the two grids. In the context of ensemble prediction systems the fine-grid model would necessarily have to be simple because the computational burden would be too high for each forecast member to have a three-dimensional cloud-resolving model associated with it. More realistically, one might consider using a fast two-dimensional fluid emulator (with stochastic forcing) as a fine-grid model associated with each member. One could relax the large scales of the fine model to the forecast model whilst relaxing the small scales of the forecast model towards the emulator.

18.5 Discussion

This chapter supports the growing desire in the climate and NWP modelling communities to move away from column-based physical parameterisation and use some intermediate approach where higher spatial resolution can be used at an affordable computational cost.

Grabowski (2000) and Randall *et al.* (2003) discussed the the technique of embedding conventional cloud-resolving models in the grid boxes of a climate model and found encouraging improvements, though at a very high cost. Jung & Arakawa (2005) discussed the limitation caused by the use of cyclic lateral boundary conditions in superparameterisation and proposed a modified coupling method between the climate model and the cloud-resolving model. Computational-cost savings are proposed in this paper through the use of cheap advection and physics algorithms in the manner adopted by the computer animation industry. Devices like vorticity confinement help to maintain statistical accuracy in spite of compromises made to the accuracy of the Lagrangian conservation and pressure solver steps.

Even if a fast, simplified, CRM is developed the memory requirements for running it in on a global grid would be very high (e.g. about 80 gigabytes per model field for a 2 km grid with 50 vertical levels). Perhaps a less-ambitious objective initially would be to couple a 10 km fine grid to a climate model with 100 or 200 km resolution. Other computational savings in the fine model could be made using adaptive mesh refinement as shown in Fig. 18.7 where the grid is allowed to refine or coarsen depending on the nature of the flow. This would significantly reduce the cost of the fine-grid simulation, since the grid would only need refining in regions of convection with a few fixed high-resolution regions near regions of significant unresolved orography. In the polar regions longitudinal resolution could also be significantly reduced (e.g. Hubbard & Nikiforakis 2003). Grid refinement could be permanently enabled for mountainous terrain and perhaps an equatorial strip (e.g. from 20 degrees north to 20 degrees south). By limiting oneself to first-order (at worst) accuracy in space and time, many of the complex programming issues related to adaptive mesh refinement are greatly simplified.

This class of coupled model is an example of the heterogeneous, multi-scale method (HMM; Engquist *et al.* 2003; Ren & E 2005) which is a general

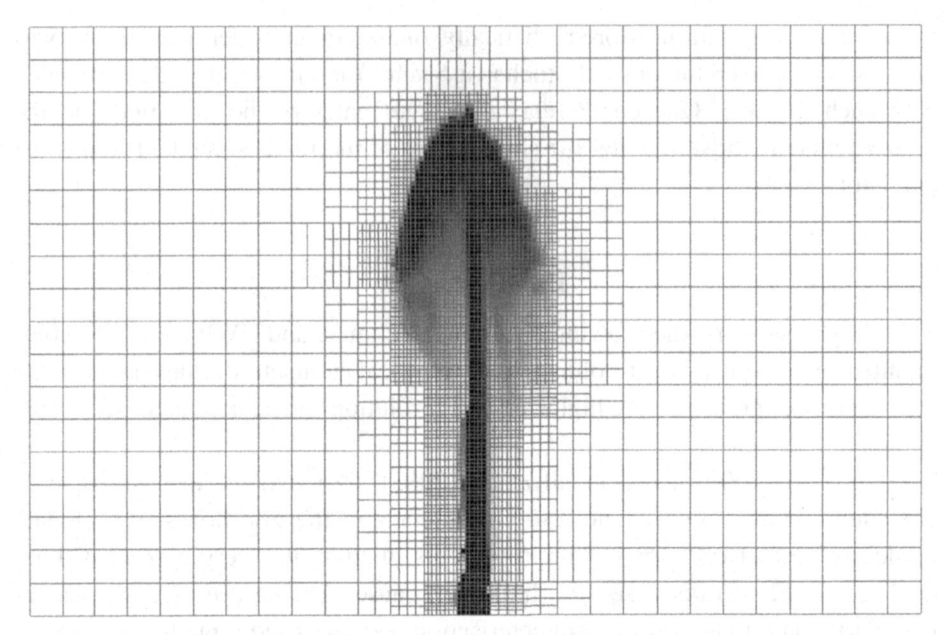

Figure 18.7 A simple adaptive grid model of a plume rising from a surface heat source. As the plume rises from the fixed-surface heat source the grid is adapted using a modified quadtree where the horizontal and vertical grids are allowed to change independently.

methodology for efficient computation in coupled macroscopic and microscopic models. Another area in physics where HMM is relevant is the simulation of polymeric fluid flow (polymers in a solvent). Here the internal stress is made up of contributions from the solvent viscosity and that due to the bending and stretching of polymers. The dynamics of the immersed polymers are modelled as a microscale process (e.g. using a springy dumbbell idealisation) that couples to a macroscale fluid model. This method contrasts with the classical generalised Newtonian model for the viscosity of the polymeric fluid.

The schematic diagram in Fig. 18.8 summarises two potential strategies for future parameterisation. The left-hand diagram represents the proposed dual-grid approach with a post-processor lying between the coarse and fine models; the right-hand diagram represents the conventional column-based parameterisation approach with the addition of a stochastic modulator. An example of the latter is the European Centre for Medium-Range Weather Forecasts' (ECMWF's) stochastic physics scheme (Buizza *et al.* 1999) which multiplies parameterisation tendency fields by a random number drawn from the range 0.5 to 1.5 and with uniform PDFs.

Apart from having a crude description of moist thermodynamic and cloud microphysical processes, the fine model of the dual-grid method could include a crude

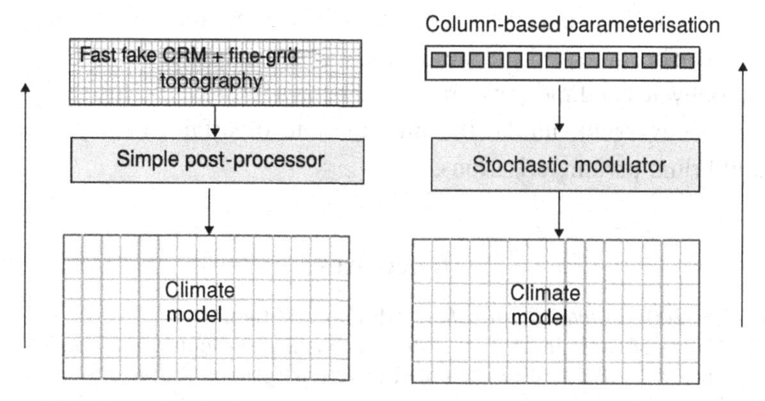

Figure 18.8 Parameterisation strategies: (left) dual-grid approach; (right) conventional column-based approach with additional stochastic terms.

specification of orography, as in the buildings model discussed earlier, together with land-type or sea-state information. Global NWP model grid lengths of ~40 km are insufficient to resolve mesoscale mountain ranges like the Swiss Alps and the Pyrenees and it is well known that they exert a major influence on the flow through very high drag forces. The fine model could include the specification of these mountains quite accurately and perform a more detailed surface drag calculation than that achievable on the coarse grid. Indeed initial simulations, like those described here (see Shutts & Allen 2007), indicate that the upscale influence of subgrid orographic features, namely the formation of large-scale von Karman streets, is possible within the present dual-grid formalism. However, extra terms are required within the coarse model to correct for the presence of the immersed obstacle; i.e. to correct for the fact that the volume of fluid on the coarse and fine meshes is different. Similarly, regions of islands and lakes could be better defined on the fine grid and provide more accurate coarse-grained surface energy fluxes to the coarse NWP model.

At this stage it is difficult to say how practical the dual-grid method will be and how much tuning will be required in both the fine model's physics scheme and in the subsequent post-processing step. More careful consideration of the coupling method will be required to ensure an optimal transmission of information between the two grids.

In the context of climate modelling, the dual-grid technique would be likely to provide a better representation of the upscale influence of tropical cloud systems and coupling to equatorially trapped waves. The ability of climate models to represent low-frequency tropical variability like the Madden–Julian Oscillation is linked to convective parameterisation since different schemes have varying impacts on the phenomenon. Nevertheless the task of coupling and tuning the dual-grid approach is likely to be a formidable one. As computational resources increase, the resolution of the fine grid can be increased and the accuracy of advection, moist thermodynamics, cloud microphysics and radiative transfer schemes improved. This strategy would be properly convergent, unlike the present state of affairs with column-based, quasi-equilibrium parameterisations.

References

Barnsley, M. F. 2001 *Fractals Everywhere*. Morgan Kaufmann.

Buizza, R., Miller, M. & Palmer, T. N. 1999 Stochastic representation of model uncertainty in the ECMWF Ensemble Prediction System, *Q. J. R. Meteorol. Soc.*, **125**, 2887–2908.

Engquist, E. W., Engquist, B. & Huang, Z. 2003 Heterogeneous multiscale method: A general methodology for multiscale modeling. *Physical Review B*, **67**(9): 092101.

Erleben, K., Sporring, J., Henriksen, K. & Dohlmann, H. 2005. *Physics Based Animation*. Charles River Media.

Fedkiw, R., Stam, J. & Jensen, H. W. 2001 Visual simulation of smoke. *SIGGRAPH Conf. Proc.*, 23–30.

Gardner, M. 1970 Mathematical games. The fantastic combinations of John Conway's new solitaire game "life", *Sci. Am.*, **223**, 120–123.

Grabowski, W. W. 2000 Coupling cloud processes with the large-scale dynamics using Cloud-Resolving Convection Parameterisation. *J. Atmos. Sci.*, **58**, 978–997.

Harris, M. J. 2003 *Real-Time Cloud Simulation and Rendering*. University of North Carolina USA.

Hubbard, M. E. & Nikiforakis, N. 2003 A three-dimensional, adaptive, Godunov-type model for global atmospheric flows. *Mon. Weather Rev.*, **131**, 1848–1864.

Jung, J-H. & Arakawa, A. 2005 Preliminary tests of multiscale modelling with a two-dimensional framework: sensitivity to coupling methods. *Mon. Weather Rev.*, **133**, 649–662.

McCalpin, J. D. 1988 A quantitative analysis of the dissipation inherent in semi-Lagrangian advection. *Mon. Weather Rev.*, **116**, 2330–2336.

Mason, P. J. & Thomson D. J. 1992 Stochastic backscatter in large-eddy simulations of boundary layers. *J. Fluid Mech.*, **242**, 51–78.

Morton, K. W. & Mayers, D. F. 2005 *Numerical Solution of Partial Differential Equations: An Introduction.* Cambridge University Press.

Palmer, T. N. 1997 On parametrizing scales that are only somewhat smaller than the smallest scales, with application to convection and orography. Proceedings of the ECMWF Workshop on New Insights and Approaches to Convective Parameterization, 4–7 November, Reading.

Palmer, T. N. 2001 A nonlinear dynamical perspective on model error: A proposal for non-local stochastic-dynamic parameterization in weather and climate prediction models. *Q. J. R. Meteorol. Soc.*, **127**, 279–304.

Randall, D., Khairoutdinov, M., Arakawa A. & Grabowski, W. W. 2003 Breaking the cloud parameterization deadlock. *Bull. Am. Meteorl. Soc.*, **84**, 1547–1564.

Ren, W. & E., W. 2005 Heterogeneous multiscale method for the modelling of complex fluids and micro-fluidics. *J. Compar. Physiol.*, **204**, 1–26.

Shutts, G. J. 2005 A kinetic energy backscatter algorithm for use in ensemble prediction systems, *Q. J. R. Meteorol. Soc.*, **131**, 3079–3102.

Shutts, G. J. & Palmer, T. N. 2007 Convective forcing fluctuations in a cloud-resolving model: relevance to the stochastic parameterization problem. *J. Climate*, **20**, 187–202.

Shutts, G. J. & Allen, T. 2007 Sub-gridscale parameterization from the perspective of a computer games animator. *Atmos. Sci. Lett.*, **8**, 85–92.

Stam, J. 1999 Stable fluids. *SIGGRAPH Conf. Proc.*, 121–128.

Staniforth, A. & Coté, J. 1991 Semi-Lagrangian integration schemes for atmospheric models – a review. *Mon. Weather Rev.*, **119**, 2206–2223.

Steinhoff, J. & Underhill, D. 1994 Modification of the Euler equations for "vorticity confinement:" application to the computation of interacting vortex rings, *Phys. Fluids*, **6**, 2738–2744.

Tomita, H., Miura, H., Iga, S., Nasumo, T. & Satoh, M. 2005 A global cloud-resolving simulation: preliminary results from an aqua planet experiment. *Geophys. Res. Lett.*, **32**, L08805, doi: 10.1029/2005GL022459.

Index

Printed in the United States
By Bookmasters